The Oxford Dictionary of
Abbreviations

Editors

Market House Books Ltd

Fran Alexander
Peter Blair
John Daintith
Alice Grandison
Valerie Illingworth
Elizabeth Martin
Anne Stibbs

Oxford University Press

Judy Pearsall
Sara Tulloch

Oxford Paperback Reference

The most authoritative and up-to-date reference books for both students and the general reader.

The Oxford Dictionary of

Abbreviations

SECOND EDITION

Oxford   New York

OXFORD UNIVERSITY PRESS

1998

Oxford University Press, Clarendon Street, Oxford OX2 6DP

Oxford   New York
Athens   Auckland   Bangkok   Bogota   Bombay   Buenos Aires
Calcutta   Cape Town   Dar es Salaam   Delhi
Florence   Hong Kong   Istanbul   Karachi
Kuala Lumpur   Madras   Madrid   Melbourne
Mexico City   Nairobi   Paris   Singapore
Taipei   Tokyo   Toronto

and associated companies in
Berlin   Ibadan

Oxford is a trade mark of Oxford University Press

© *Market House Books Ltd, 1992, 1998*

First published 1992
First issued as an Oxford University Press paperback 1993
Reissued in new covers 1996
Second edition 1998

British Library Cataloguing in Publication Data

Data available

ISBN 0-19-280073-6

10 9 8 7 6 5 4 3 2 1

Printed in Great Britain by
Cox & Wyman Ltd, Reading, Berkshire

Preface

During the last few decades abbreviations have proliferated to such an extent that they now form a major – and still increasing – part of the language. Most English dictionaries, however, do not fully reflect this important addition to our vocabulary, typically including only a selection of the better known abbreviations. The *Oxford Dictionary of Abbreviations* is designed to fill this gap, providing explanations for the bewildering variety of abbreviations in use today.

This dictionary caters for all levels of usage, from the colloquial, as in **yuppie** and **POETS day**, to the highly specialized abbreviations used in scientific and technical fields. The advances in all aspects of science and technology, notably computing, are reflected in the extensive coverage of abbreviations in these fields in the dictionary. Similar coverage is given to the body of abbreviations generated by the business and financial communities, the medical profession (especially in the fields of immunology and genetics), and the world of sport. For this new edition over 1700 new abbreviations and senses have been added in these and other fields. The second edition also takes account of the recent proliferation of abbreviations used in personal and other advertisements. Field labels or brief explanations are provided for these and other abbreviations when necessary. As well as abbreviations used in English-speaking countries, the dictionary also includes a selection of abbreviations of foreign terms and organizations, especially those likely to be encountered by businessmen and others with contacts abroad; translations or brief explanations are given for these entries. When the country of origin of an abbreviation is not evident from the definition, a geographical label precedes the definition.

The entries cover all types of shortened forms of words and phrases, including initialisms, such as **BBC** and **FBI**; acronyms (i.e. initialisms that are pronounced as words, rather than as a series of letters), such as **Aids** and **NATO**; shortenings, for example **Gen.** (for General, Geneva, etc.); and symbols for units of measurement, chemical elements, airlines, postal districts, vehicle registration offices, and so on. A selection of symbols that are not formed from letters of the Roman alphabet is given in an Appendix to the dictionary. A pronunciation guide (in brackets preceding the definition), using the International Phonetic Alphabet, is provided for acronyms; an explanation of the guide is given on page ix.

Often there are several possible forms for an abbreviation, with variations in upper and lower case and in the use of points. In deciding the forms of the entry words in this dictionary, the editors have used a number of conventions. Points have been omitted for acronyms (e.g. **Aids**), for all-capital abbreviations (e.g. **NUJ**), for degrees and other qualifications consisting of upper and lower-case letters (e.g. **BSc**, **FIBiol**), and in other cases when this can be done without flouting convention or causing confusion with a real word (e.g. **plc** rather than p.l.c.). Shortenings are usually followed by a point unless the last letter of the shortened form is the last letter of the word (e.g. **gdn** = garden, guardian). When two or more forms are equally common, variants are included in brackets before the definition.

1998

Contents

Pronunciation Guide

The symbols used in indicating pronunciations of acronyms and certain other abbreviations are those of the International Phonetic Alphabet. The following consonants have their usual English sounds: b, d, f, h, k, l, m, n, p, r, s, t, v, w, z. Additional symbols are as follows:

ɑː *as in* bar (bɑː)
æ *as in* cat (kæt)
aɪ *as in* buy (baɪ)
aʊ *as in* gout (gaʊt)
tʃ *as in* cheese (tʃiːz)
ɛ *as in* get (gɛt)
ə *as in* paper ('peɪpə)
ᵊ *as in* able (eɪbᵊl)
ɜː *as in* turn (tɜːn)
eɪ *as in* pay (peɪ)
ɛə *as in* fair (fɛə)
ɪ *as in* pin (pɪn)
iː *as in* be (biː)
ɪə *as in* hear (hɪə)

j *as in* yet (jɛt)
dʒ *as in* job (dʒɒb)
ŋ *as in* bring (brɪŋ)
ɒ *as in* got (gɒt)
əʊ *as in* go (gəʊ)
ɔː *as in* law (lɔː)
ɔɪ *as in* coin (kɔɪn)
ʃ *as in* shin (ʃin)
ʒ *as in* measure ('mɛʒə)
θ *as in* thaw (θɔː)
ð *as in* then (ðɛn)
ʊ *as in* wool (wʊl)
uː *as in* new (njuː)
ʌ *as in* cup (kʌp)

For acronyms of two or more syllables the stressed syllables are preceded by the marks ' or ˌ in the pronunciation. The most strongly stressed syllable is preceded by ', as in 'peɪpə (paper); weakly stressed syllables are preceded by ˌ as in 'kɒmɪˌkɒn (Comecon). Unstressed syllables are not preceded by stress marks, as in 'mɛmərɪ (memory).

A

a (ital.) *Physics, symbol for* acceleration • *symbol for* are(s) (metric measure of land) • *symbol for* atto- (prefix indicating 10^{-18}, as in **am**, attometre) • *symbol for* year (Latin *annus* or French *année*) • *indicating* the first vertical row of squares from the left on a chessboard

a. about • accepté (French: accepted; written on a bill of exchange) • accepted • acre(s) • acreage • acting • actual • address • adjective • advance(d) • afternoon • age • akzeptiert (German: accepted; written on a bill of exchange) • *Music* alto • amateur • anno (Latin: in the year) • anonymous • answer • ante (Latin: before) • aqua (Latin: water) • arrive(s) • arriving • *Baseball* assist(s)

a 2 *Music* a due (Italian: for two (voices or instruments))

A *British fishing port registration for* Aberdeen • *Physics, symbol for* absolute (temperature) • (ital.) *Physics, symbol for* absorbance • *Cards, symbol for* ace • *Chem., symbol for* acid-catalysed reaction • (ital.) *Chem., symbol for* activity • *Biochem., symbol for* adenine • *Biochem., symbol for* adenosine • *Films* adult (former certification) • *Education* Advanced (in **A level**) • (ital.) *Chem., symbol for* affinity • *symbol for* ammeter (in circuit diagrams) • *Physics, symbol for* ampere(s) • *Physics, symbol for* ampere-turn • *Botany, symbol for* androecium (in a floral formula) • anode • (ital.) *Maths., symbol for* area • *symbol for* an arterial road (as in **A1**, **A40**, etc.) • assault (in **A-day**) • atom(ic) (as in **A-bomb**) • austral (former monetary unit of Argentina) • *international vehicle registration for* Austria • avancer (French: go faster; on a clock or watch regulator) • (ital.) *Chem., symbol for* Helmholtz function • (bold ital.) *Physics, symbol for* magnetic vector potential • *Physics, symbol for* mass number • (ital.) *Physics, symbol for* nucleon number • (ital.) *Electronics,*

symbol for Richardson constant • *symbol for* ten (in hexadecimal notation) • *Med., indicating* a blood group or its associated antigen (*see also under* ABO) • *Education, indicating* the highest mark or grade for work • *indicating* managerial, professional (occupational grade) • *indicating* a musical note or key • *Astronomy, indicating* a spectral type • *indicating* a standard paper size (*see under* A0–A10) • *Logic, indicating* a universal affirmative categorical proposition

Å *Physics, symbol for* angstrom unit

A. (*or* **A**) academician • academy • acreage • acre(s) • adjutant • admiral • air • Altesse (French: Highness) • *Music* alto • America(n) • *Bible* Amos • *Horticulture* annual • answer • anterior • April • armoured • article • artillery • art(s) • assistant • associate • athletic • August • Australia(n)

A0 *indicating* a standard paper size, 841×1189 mm

A1 *indicating* a standard paper size, 594×841 mm • *indicating* (in Lloyd's Register) ships maintained in good and efficient condition

A2 *international civil aircraft marking for* Botswana • *indicating* a standard paper size, 420×594 mm

A3 *international civil aircraft marking for* Tonga • *indicating* a standard paper size, 297×420 mm

A4 *indicating* a standard paper size, 210×297 mm

A5 *international civil aircraft marking for* Bhutan • *indicating* a standard paper size, 148×210 mm

A6 *international civil aircraft marking for* United Arab Emirates • *indicating* a standard paper size, 105×148 mm

A7 *international civil aircraft marking for* Qatar • *indicating* a standard paper size, 74×105 mm

A8 *airline flight code for* Aerolineas Paraguayas • *indicating* a standard paper size, 52×74 mm

A9 *indicating* a standard paper size, 37×52 mm

A9C *international civil aircraft marking for* Bahrain

A10 *indicating* a standard paper size, 26×37 mm

A40 *international civil aircraft marking for* Oman

3A *international civil aircraft marking for* Monaco

5A *international civil aircraft marking for* Libya

9A *international civil aircraft marking for* Croatia

āa (*or* **āā**) *Med.* ana (Greek: of each; specifying quantities of ingredients in prescriptions)

a.a. absolute alcohol • acting appointment • after arrival • *Shipping* always afloat • approximate absolute • area administrator • arithmetic average • attendance allowance • author's alteration • *Photog., indicating* that one print should be made of each negative

AA administrative assistant • advertising agency • Advertising Association • age allowance (for taxation) • Air Attaché • Alcoholics Anonymous • *British fishing port registration for* Alloa • *Chem.* allyl alcohol • *airline flight code for* American Airlines • American Aviation (aircraft) • *Biochem.* amino acid • *Chem.* andesine anorthosite • Anglers' Association • antiaircraft • *Physics* antiproton accumulator • *Biochem.* arachidonic acid • Architectural Association • *Chem.* aristolochic acid • Army Act • *Biochem.* ascorbic acid • Associate in Accounting • Associate in Agriculture • Associate in (*or of*) Arts • Association of Agriculture • Astronautica Acta (journal of the International Astronautical Federation) • *Physics* atomic absorption • *Physics* atomic adsorption • Augustinians of the Assumption • author's alteration • Automobile Association • *Chem.* azelaic acid • *British vehicle registration for* Bournemouth • *indicating* a certificate issued for a ship stating that all crew matters are in order • (formerly) *indicating* a film not be shown to a child under 14 unless accompanied by an adult • *Electricity, indicating* a size of battery or cell • *Angling, indicating* a size of fine split weights

AA1 (*or* **AA-one**) *Finance, indicating* a very high quality rating for credit

AAA *Med.* abdominal aortic aneurysm • (USA) Agricultural Adjustment Act • (USA) Agricultural Adjustment Administration • Allied Artists of America • Amateur Athletic Association • American Accounting Association • American Automobile Association • *Biochem.* amino acid analogue(s) • (*or* **triple A**) antiaircraft artillery • (USA) Army Audit Agency • Association of Average Adjusters • Australian Association of Accountants • Australian Automobile Association • Automobile Association of America • *Finance, indicating* the highest quality rating for credit

AAAA Amateur Athletic Association of America • American Association of Advertising Agencies • Associated Actors and Artistes of America (trade union) • Australian Association of Advertising Agencies

AAAC all-aluminium alloy conductor

AAAI American Association for Artificial Intelligence • Associate of the Institute of Administrative Accountants

AAAL American Academy of Arts and Letters

AAAM American Association of Aircraft Manufacturers

AA&QMG Assistant Adjutant and Quartermaster-General

AAAS American Academy of Arts and Sciences • American Academy of Asian Studies • American Association for the Advancement of Science • Associate of the American Antiquarian Society

AAB Aircraft Accident Board • Association of Applied Biologists

AABC American Amateur Baseball Congress

AABL Associated Australian Banks in London

AABM Australian Association of British Manufacturers

AABW Antarctic Bottom Water

AAC *US Air Force* Alaskan Air Command • Aeronautical Advisory Council • Agricultural Advisory Council • Amateur Athletic Club • anno ante Christum (Latin: in the year before Christ) • Army Air Corps • *Electronics* automatic amplitude control

AACB (USA) Aeronautics and Astronautics Coordinating Board • Association of African Central Banks

AACC American Association of Clinical Chemists • *Aeronautics* area-approach control centre

AACE American Association of Cost Engineers • Association for Adult and Continuing Education

AACP American Association for Child Psychiatry • Anglo-American Council on Productivity

AACR *Bibliog.* Anglo-American Cataloguing Rules

AACS Aberdeen-Angus Cattle Society • (USA) Airways and Air Communications Service • *Computing* automated-access control system

AAD *Med.* atlantoaxial dislocation

AADC air aide-de-camp • (USA) Army–Air Defense Command

AADFI Association of African Development Finance Institutions

AADS (USA) Army–Air Defense System

AAE American Association of Engineers

AAEC Australian Atomic Energy Commission

AAeE Associate in Aeronautical Engineering

AAEE (*or* A & AEE) Aircraft and Armament Experimental Establishment • American Association of Electrical Engineers

AAES Association of Agricultural Education Staffs • Australian Army Education Service

AAEW Atlantic airborne early warning

AAF Allied Air Forces • (USA) Army Air Forces

AAFC anti-aircraft fire control

AAFCE Allied Air Forces in Central Europe

AAFIS Army and Air Force Intelligence Staff

AAFNE Allied Air Forces in Northern Europe

AAFSE Allied Air Forces in Southern Europe

AAgr Associate in Agriculture

AAG Air Adjutant-General • Assistant Adjutant-General • Association of American Geographers

AAGS Association of African Geological Surveys

AAHPER American Association for Health, Physical Education, and Recreation

AAIA Associate of the Association of International Accountants • Associate of the Australian Institute of Advertising • Association of American Indian Affairs

AAIB Associate of the Australian Institute of Bankers

AAII Associate of the Australian Insurance Institute

AAIL American Academy and Institute of Arts and Letters

AAL *Meteorol.* above aerodrome level • Academy of Art and Literature • (USA) Arctic Aeromedical Laboratory • Association of Assistant Librarians

AALA American Association for Laboratory Accreditation

AALD Australian Army Legal Department

AALL American Association of Law Libraries

AAM air-to-air missile • American Association of Microbiology • American Association of Museums • (formerly) Anti-Apartheid Movement • Association of Assistant Mistresses in Secondary Schools • Australian Air Mission

AAMI *Med.* age-associated memory impairment

AAMW (USA) Association of Advertising Men and Women

AAMWS Australian Army Medical Women's Service

AANA Australian Association of National Advertisers

A & A additions and amendments

A&AEE Aeroplane and Armament Experimental Establishment

a&b assault and battery

A & C addenda and corrigenda • (ital.) Antony and Cleopatra (Shakespeare)

A&E accident and emergency (medical specialty *or* hospital department)

a&h *Insurance* accident and health

a&i *Insurance* accident and indemnity

A & I (USA) Agricultural and Industrial (college)

A & M (USA) Agricultural and Mechanical (college) • Ancient and Modern (hymn book)

A & N Army and Navy (Club *or* Stores)

A & NI Andaman and Nicobar Islands

A & P advertising and promotion • (New Zealand) Agricultural and Pastoral (Association, Show, etc.)

a&r assault and robbery

A & R artists and recording • artists and repertoire *or* repertory (as in **A & R man**)

a&s *Insurance* accident and sickness

A & SH Argyll and Sutherland Highlanders

AANS Australian Army Nursing Service

AAO American Association of Orthodontists • Anglo-Australian Observatory

AAOC Australian Army Ordnance Corps

AAOM American Academy of Occupational Medicine

AAOO American Academy of Ophthalmology and Otolaryngology

AAP affirmative action programme • *Chem.* alumina–aluminum phosphate • Association of American Publishers • Australian Associated Press • *Astronautics* (USA) Apollo Applications Program

AAPA Advertising Agency Production Association • American Association of Port Authorities

AAPB American Association of Pathologists and Bacteriologists

AAPC All African People's Conference

AAPE American Academy of Physical Education

AAPG American Association of Petroleum Geologists

AAPHI Associate of the Association of Public Health Inspectors

AAPM American Association of Physicists in Medicine

AAPS American Association for the Promotion of Science

AAPSO Afro-Asian People's Solidarity Organization

AAPSS American Academy of Political and Social Science

AAPSW Associate of the Association of Psychiatric Social Workers

AAPT American Association of Physics Teachers • Associate of the Association of Photographic Technicians

a.a.r. after action report • *Insurance* against all risks • aircraft accident record (*or* report) • average annual rainfall

AAR Association of American Railroads

AARP American Association of Retired Persons

AAS *RC Church* Acta Apostolicae Sedis (Latin: Acts of the Apostolic See; publication) • American Academy of Sciences • American Antiquarian Society • American Astronautical Society • American Astronomical Society • Andersen air sampler • Army Air Service • Associate in Applied Science • Association of Asian Studies • atomic absorption spectrometry (*or* spectroscope *or* spectroscopy *or* spectrophotometry) • Australian Academy of Science • Auxiliary Ambulance Service • Fellow of the American Academy of Arts and Sciences (Latin *Academiae Americanae Socius*)

AASA Associate of the Australian Society of Accountants

AASB *Aeronautics* aerodynamically air-staged burner • American Association of Small Businesses

AASC Allied Air Support Command • Australian Army Service Corps

AASF Advanced Air Striking Force

AASG Association of American State Geologists

A'asia Australasia

AASM Associated African States and Madagascar (treaty)

AASR Airport and Airways Surveillance Radar (Raytheon)

AASS Associate of the American Antiquarian Society (Latin *Americanae Antiquarianae Societatis Socius*)

AAT *Psychol.* achievement anxiety test • *Med.* alpha-1-antitrypsin (treatment for cystic fibrosis) • Anglo-Australian Telescope (Siding Spring, NSW) • Association of Accounting Technicians • *Med.* asymptomatic autoimmune thyroiditis

AATA (eɪˈɑːtə) Anglo-American Tourist Association

AATNU Administration de l'assistance technique des Nations Unies (French: United Nations Technical Assistance Administration)

AATT American Association of Textile Technology

AATTA Arab Association of Tourism and Travel Agents

AATUF All-African Trade Union Federation (replaced by OATUU)

AAU (USA) Amateur Athletic Union • Association of American Universities

AAUN American Association for United Nations • Australian Association for United Nations

AAUP American Association of University Presses • American Association of University Professors (trade union)

AAUW American Association of University Women

AAV *Microbiol.* adeno-associated virus • assault amphibious vehicle

AAVC Australian Army Veterinary Corps

AAVS American Anti-Vivisection Society

AAWC Australian Advisory War Council

AAZPA American Association of Zoological Parks and Aquariums

ab. about • abridgment

a.b. anchor bolt • *Baseball* (times) at bat

a/b airborne

a.B. *Commerce* auf Bestellung (German: on order)

Ab antibody • (formerly) *Chem.*, symbol for astatine (originally called alabamine; *see under* At)

AB *postcode for* Aberdeen • *British fishing port registration for* Aberystwyth • able(-bodied) seaman *or* rating • advisory board • air board • airborne • *US Air Force* Airman Basic • Alberta (Canada) • Alliance balkanique (French: Balkan Alliance) • ammonia for bees (applied for stings) • antiballistic (missile) • asthmatic bronchitis • *Baseball* (times) at bat • *Physics* atomic bremsstrahlung • automated bibliography • (USA) Bachelor of Arts (Latin *Artium Baccalaureus*) • *British vehicle registration for* Worcester • *indicating* a human blood group containing A and B antigens (*see also under* ABO)

A/B (Sweden) Aktiebolaget (joint-stock company)

ABA *Botany* abscisic acid • Amateur Boxing Association • American Badminton Association • American Bankers' Association • American Bar Association • American Basketball Association (now merged with the National Basketball Association) • American Book Award • American Booksellers Association • Antiquarian Booksellers' Association • Associate in Business Administration • Association of British Archaeologists • Australian Bankers' Association

ABAA Antiquarian Booksellers' Association of America

ABAC Association of British Aero Clubs and Centres

abb. abbey • *Commerce* abbuono (Italian: allowance *or* discount)

Abb. Abbey • Abbot (*or* Abbess)

ABB *Accounting* activity-based budgeting

ABBA ('æbə) American Board of Bio-Analysis

abbrev. (*or* **abbr.**) abbreviation

ABC *Accounting* activity-based costing (as in **ABC method**) • Advance Booking Charter (airline ticket) • advanced bio-medical capsule • air bridge to Canada • *Med.* airway, breathing, circulation (in first aid) • America–Britain–Canada • American book-prices current • American Broadcasting Companies • analysis of benefits and costs • Argentina–Brazil–Chile • Aruba, Bonaire, Curaçao (in **ABC Islands**) • Associated British Cinemas • Atanasoff–Berry computer • atomic, biological, and chemical (weapons or warfare) • Audit Bureau of Circulation • Australian Broadcasting Corporation • automatic binary computer • automatic brake control

ABCA American Business Communication Association • Army Bureau of Current Affairs

ABCB Association of Birmingham Clearing Banks

ABCC Association of British Chambers of Commerce • Association of British Correspondence Colleges

ABCD American, British, Chinese, Dutch (powers in the Pacific in World War II) • atomic, biological, and chemical protection and damage control

ABCFM American Board of Commissioners for Foreign Missions

ABCM Associate of Bandsmen's College of Music

abd. abdicate(d) • abdomen • abdominal

ABD *US Navy* advanced base depot • all but dissertation (of a candidate reading for a higher degree) • (*or* **abd**) average body dose (of radiation)

ABDA ('æbdə) American, British, Dutch, Australian (Command in the Pacific; World War II)

abdom. abdomen • abdominal

ABDP Association of British Directory Publishers

ABE *Chem.* acetone–butanol–ethanol (solvent)

ABEd (USA) Bachelor of Arts in Education

Aber. Aberdeen • Aberdonian

ABERCOR ('æbə,kɔː) Associated Banks of Europe Corporation

abf *Colloquial* absolute bloody final (drink)

Abf. Abfahrt (German: departure)

ABF Actors' Benevolent Fund • Associated British Foods

ABFD Association of British Factors and Discounters

ABFM American Board of Foreign Missions

ABGB Allgemeines Bürgerliches Gesetzbuch (German; the Austrian Civil Code)

abgk. abgekürzt (German: abbreviated)

ABGWIU (USA) Aluminum Brick and Glass Workers International Union

Abh. Abhandlungen (German: transactions *or* treatises)

ABH actual bodily harm

ABI *Computing* application binary interface • Associate of the Institute of Book-keepers • Association of British Insurers

ABIA Association of British Introduction Agencies

ab init. ab initio (Latin: from the beginning)

ABINZ Associate of the Bankers' Institute of New Zealand

ABIOS *Computing* advanced basic input/output system

abl. ablative

ab' abril (Spanish: April)

ABL atmospheric boundary layer

ABLA ('æblə) American Business Law Association

ABLJ adjustable buoyancy lifejacket

ABLS Association of British Library Schools • (USA) Bachelor of Arts in Library Science

ABM antiballistic missile • Associate in Business Management • Australian Board of Missions • automatic batch mix

ABMA Army Ballistic Missile Agency

ABMAC ('æbmæk) Association of British Manufacturers of Agricultural Chemicals

ABMEWS ('æbmjuːz) antiballistic missile early warning system

ABMPM Association of British Manufacturers of Printers' Machinery

ABMT *Med.* allogeneic bone-marrow transplantation

abn airborne

Abn Aberdeen

ABN *Chem.* acid–base–neutral • Anti-Bolshevik Bloc of Nations

ABNE Association for the Benefit of Non-contract Employees

ABO American Board of Ophthalmology • American Board of Orthodontics • American Board of Otolaryngology • *Med., indicating* a blood group system based on the presence or absence of antigens A and B (*see also under* A; AB; B; O)

ABOF Association of British Organic Fertilizers

ABOG American Board of Obstetrics and Gynecology

A-bomb atom bomb

ABOS American Board of Oral Surgery • American Board of Orthopedic Surgery

Abp Archbishop

ABP American Board of Pathology • American Board of Pediatrics • American Board of Peridontology • American Board of Prosthodontics • *Biochem.* androgen-binding protein • *Med.* arterial blood pressure • Associated Book Publishers • Associated British Ports

ABPA Australian Book Publishers' Association

ABPC American Book Publishers' Council • Associated British Picture Corporation

ABPI Association of the British Pharmaceutical Industry

ABPN American Board of Psychiatry and Neurology • Association of British Paediatric Nurses

ABPO Advanced Base Personnel Officer

ABPS American Board of Plastic Surgery

ABPVM Association of British Plywood and Veneer Manufacturers

abr. abridge(d) • abridgment

ABR American Board of Radiology • *Med.* auditory brainstem recording

ABRACADABRA (ˌæbrəkəˈdæbrə) Abbreviations and Related Acronyms Associated with Defense, Astronautics, Business, and Radio-electronics (publication of Raytheon Company, Lexington, USA)

ABRC Advisory Board for the Research Councils

ABRES advanced ballistic re-entry system

ABRO ('æbrəʊ) Animal Breeding Research Organization • Army in Burma Reserve of Officers

ABRS Association of British Riding Schools

ABRSM Associated Board of the Royal Schools of Music

ABRV advanced ballistic re-entry vehicle

abs. absence (or absent) • absolute • absorbent • abstract

a.b.s. aux bons soins (de) (French: care of, c/o; on postal addresses)

ABS *Chem.* acrylonitrile–butadiene–styrene (type of plastic) • *Chem.* alkyl benzene sulphonate (usually of sodium) • American Bible Society • American Board of Surgery • American Bureau of Shipping • Antiblockiersystem (in **ABS brake**; German: antilocking system) • antilock braking system • Architects' Benevolent Society • Australian Bureau of Statistics

ABSA Association of Business Sponsorship of the Arts

abse. re. *Law* absente reo (Latin: the defendant being absent)

abs. feb. *Med.* absente febre (Latin: when there is no fever)

ABSIE ('æbsɪ) American Broadcasting Station in Europe (in World War II)

ABSM Associate of the Birmingham School of Music

ABSM(TTD) Associate of the Birmingham School of Music (Teacher's Training Diploma)

absol. absolute

abs. re. *Law* absente reo (*see* abse. re.)

abstr. abstract(ed)

abs. visc. absolute viscosity

ABSW Association of British Science Writers

abt about

Abt. Abteilung (German: division or part)

ABT advanced backplane technology • advanced bi-CMOS technology • Association of Building Technicians

ABTA ('æbtə) Allied Brewery Traders' Association • Association of British Travel Agents • Australian British Trade Association

ABTAC ('æbtæk) Australian Book Trade Advisory Committee

ABTUC All-Burma Trade Union Congress

ABU Asian Broadcasting Union • Assembly of the Baptist Union

A-BU Anglo-Belgian Union

abv. above

ABWR *Nuclear physics* advanced boiling water reactor

ac. acre • activity

a.c. a capo (Italian: new line) • (or **à.c.**) à compte (French: on account) • advisory committee • *Physics* alternating current • année courante (French: this year) • anno corrente (Latin: this year) • *Med.* ante cibum (Latin: before meals; in prescriptions) • assegno circolare (Italian: banker's draft or cashier's check) • author's correction

a/c account • account current • ao cuidado de (Portuguese: care of, c/o; on postal addresses)

aC antes de Cristo (Portuguese: before Christ; BC) • avanti Cristo (Italian: before Christ; BC)

Ac *Chem.*, *symbol for* acetyl (ethanoyl) group • *Chem.*, *symbol for* actinium • (ital.) *Meteorol.* altocumulus

AC *airline flight code for* Air Canada • *Chem.* activated carbon • *Chem.* activated charcoal • *Chem.* active centre • *Med.* acute cholecystitis • *Chem.* acyl-oxygen cleavage • *Biochem.* adenylate cyclase • advanced CMOS • Aero Club • Air Command • Air Commodore • air conditioning • air control • Air Corps • Air Council • aircraftman • Alcohol Concern (a charity) • Alpine Club • *Physics* alternating current • *Meteorol.* altocumulus • Ambulance Corps • analogue computer • analytical chemist • angular correlation • annual conference • ante Christum (Latin: before Christ) • *Law* Appeal Case(s) • Appeal Court • appellation contrôlée (*see under* AOC) • Army Corps • Army Council • artillery college • Arts Council • Assistant Commissioner • Athletic Club • Atlantic Charter • Auditor Camerae (Latin: Auditor of the Papal Treasury) • Azione Cattolica (Italian: Catholic Action; social organization) • Companion of the Order of Australia • *British vehicle registration for* Coventry

A/C account • account current • air conditioning • aircraft • aircraftman

ACA *Telecom.* adjacent-channel attenuation • advanced combat aircraft • Agricultural Cooperative Association • American Camping Association • American Canoe Association • *Chem.* ammoniacal copper arsenate • Anglers' Cooperative Association • Arts Council of America • Associate of the Institute of Chartered Accountants • Association of Consulting Actuaries • Australian Consumers' Association • Australian Council for Aeronautics

7-ACA *Chem.* 7-aminocephalosporanic acid

ACAA Agricultural Conservation and Adjustment Administration • American Coal Ash Association • Associate of the Australasian Institute of Cost Accountants

ACAB ('eɪkæb) (USA) Army Contract Adjustment Board

acac ('æk,æk) *Chem.* acetylacetonate ion (used in formulae)

acad. academic(al) • (*or* **Acad.**) academy

Acad. fran. Académie française (French literary academy)

ACADI (France) Association des cadres dirigeants de l'industrie (Association of Industrial Executives)

Acad. med. Academy of medicine

Acad. mus. Academy of music

Acad. sci. Academy of science

ACAE Advisory Council for Adult and Continuing Education

ACAL Advanced Computer Architecture Laboratory

AC & U (USA) Association of Colleges and Universities

ACAO Assistant County Advisory Officer

Acap. Acapulco

ACAR *Physics* angular correlation of annihilation radiation

ACAS (*or* **Acas**; 'eɪkæs) Advisory Conciliation and Arbitration Service • Assistant Chief of Air Staff

ACB Association Canadienne des Bibliothèques (French: Canadian Library Association) • Australian Cricket Board

ACBG *Med.* aorto-coronary bypass graft

ACBi Associate of the Institute of Bookkeepers

ACBS Accrediting Commission for Business Schools

ACBSI Associate of the Chartered Building Societies Institute

ACBWS automatic chemical–biological warning system

acc. accelerate • acceleration • accent • *Commerce* acceptance • *Commerce* accepted • *Music* accompanied (by) • *Music* accompaniment • according (to) • *Book-keeping* account • accountant • *Grammar* accusative

ACC (New Zealand) Accident Compensation Corporation • *Computing* accumulator • *Chem.* acid copper chromate • *Med.* acute cardiovascular collapse • Administrative Coordination Committee • advanced communications course • Agricultural Credit Corporation Ltd • Anglican Consultative Council • annual capital charge • *Med.* antibody-dependent complement-mediated cytotoxicity • army cadet college • Army Catering Corps • *Scouting* Assistant County Commissioner • Associated Chemical Companies • Association of County Councils

ACCA Aeronautical Chamber of Commerce of America • Agricultural Central Cooperative Association • Associate of the Chartered Association of Certified Accountants • Association of Certified and Corporate Accountants

ACCC Association of Canadian Community Colleges

ACCD American Coalition of Citizens with Disabilities

acce. *Commerce* acceptance

accel. *Music* accelerando (Italian; with increasing speed)

access. accessory

ACCHAN Allied Command Channel (in NATO)

ACCM Advisory Council for the Church's Ministry

ACCO Association of Child Care Officers

accom. accommodation

accomp. accompanied • accompaniment • accompany

ACCP American College of Chest Physicians

accred. accredited

ACCS Associate of the Corporation of Secretaries (formerly Corporation of Certified Secretaries)

acct. accountant • account(ing)

ACCT Association of Cinematograph, Television, and Allied Technicians

accum. accumulative

accus. *Grammar* accusative

accy accountancy

acd. accord

ACD *Med.* acid citrate dextrose (as in **ACD solution**) • *Med.* acquired cystic disease of the kidney • *Chem.* anode–cathode distance

ACDA Advisory Committee on Distinction Awards • (USA) Arms Control and Disarmament Agency • Aviation Combat Development Agency

ACDAS Automatic Control and Data Acquisition System

AC/DC (*or* **AC-DC**) *Physics* alternating current/direct current • *Colloquial, indicating* bisexual

ACDCM Archbishop of Canterbury's Diploma in Church Music

ACdre Air Commodore

ACDS Assistant Chief of Defence Staff

acdt accident

ACE *Astronomy* Advanced Composition Explorer • advanced computing environment • *Engineering* advanced cooled engine • (ɛɪs) Advisory Centre for Education • *Med.* alcohol–chloroform–ether (anaesthetic) • (ɛɪs) Allied Command Europe • American Council on Education • (ɛɪs) *Biochem., Med.* angiotensin-converting enzyme (as in **ACE inhibitor**) • (USA) Army Corps of Engineers • Association for the Conservation of Energy • (Member of the) Association of Conference Executives • Association of Consulting Engineers • Association of Cultural Exchange • Australian College of Education • (ɛɪs) *Computing* Automatic Computing Engine

ACEA Action Committee for European Aerospace

ACEd. Associate in Commercial Education

ACEEE American Council for an Energy-Efficient Economy

ACEF Australian Council of Employers' Federations

ACER Australian Council for Educational Research

ACERT Advisory Council for Education of Romanies and other Travellers

ACES *Chem.* N-(2-acetamido)-2-aminoethanesulphonic acid

acet. acetone

ACET Advisory Committee on Electronics and Telecommunications • *Med.* Aids care, education, and treatment

ACEWEX Allied Command Europe Weather Exchange

ACF (Canada) Académie Canadienne Française (French Canadian Academy) • (South Africa) Active Citizen Force • *Physics* angular correlation function • Army Cadet Force • *Chemical engineering* autogenous-carrier flotation process • Automobile Club de France

ACFA Army Cadet Force Association

ACFAS (Canada) Association Canadienne-Française pour l'avancement des sciences (French-Canadian Association for the Advancement of Science)

ACFHE Association of Colleges of Further and Higher Education

ACFI Associate of the Clothing and Footwear Institute

ACFM *Physics* alternating-current field measurement

acft aircraft

ACG An Comunn Gaidhealach (Gaelic: The Gaelic Society; the Highland Association) • Assistant Chaplain-General • automatic control gear

ACGB Arts Council of Great Britain

ACGBI Automobile Club of Great Britain and Ireland

ACGI Associate of the City and Guilds Institute

ACGS Assistant Chief of the General Staff

ACh *Biochem.* acetylcholine

ACH *Building trades* air changes per hour • *Banking* automated clearing house

AChE *Biochem.* acetylcholinesterase

ACHR American Society of Human Rights

AChS Associate of the Society of Chiropodists

ACI Alliance coopérative internationale (French: International Cooperative Alliance) • Alloy Castings Institute • American Concrete Institute • army council instruction • Associate of the Institute of Commerce • Automobile Club d'Italia

ACIA Associate of the Corporation of Insurance Agents • *Computing* asynchronous communications interface adapter

ACIAA Australian Commercial and Industrial Artists' Association

ACIArb Associate of the Chartered Institute of Arbitrators

ACIB Associate of the Chartered Institute of Bankers • Associate of the Corporation of Insurance Brokers

ACIC (USA) Aeronautical Charting and Information Center

acid. *Med.* acidulated drop

ACIGS Assistant Chief of the Imperial General Staff

ACII Associate of the Chartered Insurance Institute

ACILA Associate of the Chartered Institute of Loss Adjusters

ACINF (USA) Advisory Committee on Irradiated and Novel Foods

ACIOPJF Association catholique internationale des œuvres de protection de la jeune fille (French; International Catholic Girls' Society)

ACIS American Committee for Irish Studies • Associate of the Institute of Chartered Secretaries and Administrators (formerly Chartered Institute of Secretaries)

ACIT Associate of the Chartered Institute of Transport

ACJP airways corporations joint pensions

ack. acknowledge • acknowledgment

ACK (or **ack**) *Computing, Telecom.* acknowledgment

ack-ack ('æk,æk) anti-aircraft (World War I phonetic alphabet for AA)

ackgt (or **ackt**) acknowledgment

ACL *Computing* access control list • action-centred leadership (devised by the Industrial Society) • Analytical Chemistry Laboratory • Atlas Computer Laboratory

ACLA Anti-Communist League of America

ACLANT (æk'lænt) Allied Command Atlantic

ACLC Air Cadet League of Canada

ACLP Association of Contact Lens Practitioners

ACLS American Council of Learned Societies • automatic carrier landing system

ACLU American Civil Liberties Union • American College of Life Underwriters

ACM Air Chief Marshal • air combat manoeuvring • American Campaign Medal • asbestos-containing material • (USA) Assistant Chief of Mission • Association for Computing Machinery • authorized controlled material

ACMA Agricultural Cooperative Managers' Association • Associate of the Chartered Institute of Management Accountants • Associate of the Institute of Cost and Management Accountants

ACMC Association of Canadian Medical Colleges

ACME advanced computer for medical research • (USA) Advisory Council on Medical Education • (USA) Association of Consulting Management Engineers

ACMET Advisory Council on Middle East Trade

ACMF (USA) Air Corps Medical Forces • Allied Central Mediterranean Forces • Australian Commonwealth Military Forces

ACMM Associate of the Conservatorium of Music, Melbourne

acmp. accompany

ACMP Advisory Committee on Marine Pollution • Assistant Commissioner of the Metropolitan Police

ACMRR Advisory Committee on Marine Resources Research (in the FAO)

ACMT American College of Medical Technologists

a.c.n. all concerned notified

ACN *Chem.* alkane carbon number • American College of Neuropsychiatrists • (small caps.) ante Christum natum (Latin: before the birth of Christ)

ACNA (USA) Advisory Council on Naval Affairs • Arctic Institute of North America

ACNS Assistant Chief of Naval Staff • Associated Correspondents News Service

ACNY Advertising Club of New York

ACO Admiralty Compass Observatory • Association of Children's Officers

AC of S Assistant Chief of Staff

ac.o.g. aircraft on ground

ACom (or **AComm**) Associate in Commerce

Acops ('ækɒps) Advisory Commission on Pollution of the Sea

ACOR American Center for Oriental Research

ACORD (or **Acord**) (ə'kɔːd) Advisory Council on Research and Development

ACORN ('eɪkɔːn) (ital.) A Classification of Residential Neighbourhoods (directory) • *Computing* associative content retrieval network • *Computing* automatic checkout and recording network

ACOS Advisory Committee on Safety • American College of Osteopathic Surgeons • Assistant Chief of Staff

ACOST (or **Acost**) (ə'kɒst) Advisory Council on Science and Technology

ACP *Biochem.* acyl carrier protein • advanced computer program • advanced computer project • African, Caribbean, and Pacific (countries) • American College of Pharmacists • American College of Physicians • Associate of the College of Preceptors • Association of Circus Proprietors of Great Britain •

Association of Clinical Pathologists • Automóvel Clube de Portugal (Automobile Club of Portugal)

ACPA (USA) Associate of the Institute of Certified Public Accountants

ACPAR ('æk,pɑ:) *Physics* angular correlation of positron-annihilation radiation

ACPD alternating-current plasma detector • alternating-current potential difference

ACPM American Congress of Physical Medicine and Rehabilitation

ACPO (*or* **Acpo**) ('ækpəʊ) Association of Chief Police Officers

acpt. *Commerce* acceptance

acq. acquire • acquittal

ACR Admiral Commanding Reserves • advanced capabilities radar • aircraft control room • American College in Rome • *Aeronautics* approach control radar • Association of College Registrars • audio cassette recorder

ACRB Aero-Club Royal de Belgique (French: Royal Belgian Aero Club)

acrd accrued

ACRE ('eɪkə) Action with Rural Communities in England • automatic climatological recording equipment

ACRF Advanced Computing Research Facility

acrg. acreage

ACRI (USA) Air-Conditioning and Refrigeration Institute

ACRL (USA) Association of College and Research Libraries

acron. acronym

ACRR American Council on Race Relations

ACRS *Taxation* Accelerated Cost Recovery System • Advisory Committee on Reactor Safeguards

a/cs aircraft security vessel

ACS *Aeronautics* active control system • Additional Curates Society • Admiral Commanding Submarines • Admiralty computing service • advanced communications system • airport catering services • air-conditioning system • *Aeronautics* altitude control system • American Cancer Society • American Ceramic Society • American Chemical Society • American College of Surgeons • assembly control system • Associate in Commercial Science • Association of Caribbean States • Australian Computer Society •

Finance automated confirmation service • *Computing* automated control system

ACSA Allied Communications Security Agency

ACSE (Australia) Association of Consulting Structural Engineers

ACSEA Allied Command South East Asia

ACSET Advisory Committee on the Supply and Training of Teachers

ACSL *Computing* Advanced Continuous Simulation Language • assistant cub-scout leader

ACSM American Congress on Surveying and Mapping • Associate of the Camborne School of Mines

ACSN Association of Collegiate Schools of Nursing

ACSOS Acoustical Society of America

ACSPA Australian Council of Salaried and Professional Associations

a/cs pay. accounts payable

a/cs rec. accounts receivable

ACSS *Printing* automated colour separation system

act. acting • *Grammar* active • activities • actor • actual • actuary

ACT activated-complex theory • *Meteorol.* active (*or* activated) • *Aeronautics* active control technology • advance corporation tax • *Computing* advanced CMOS with TTL inputs • advanced coal technology • Advisory Council on Technology • Agricultural Central Trading Ltd • Air Council for Training • American College Test • *Computing* analogical circuit technique • Associated Container Transportation • Associate of the College of Technology • Association of Corporate Treasurers • Australian Capital Territory • Australian College of Theology

a cta. a cuenta (Spanish: on account)

ACTC Art Class Teacher's Certificate

actg acting

ACTH *Biochem.* adrenocorticotrophic hormone

ACTO Advisory Council on the Treatment of Offenders

ACTP American College Testing Program

ACTS (USA) Acoustic Control and Telemetry System • Associate of the Society of Certified Teachers of Shorthand • Australian Catholic Truth Society

ACTT Association of Cinematograph, Television, and Allied Technicians (*see* BECTU)

ACTU Australian Council of Trade Unions

ACTWU (USA) Amalgamated Clothing and Textile Workers Union

ACU Actors' Church Union • American Congregational Union • Asian Clearing Union • Association of Commonwealth Universities • Auto-Cycle Union • *Computing* automatic calling unit

ACUA Association of Cambridge University Assistants

ACUE American Committee of United Europe

ACUS Atlantic Council of the United States

ACV (*or* **acv**) actual cash value • air-cushion vehicle • Associated Commercial Vehicles • Associate of the College of Violinists

ACW agricultural crop waste • Air Command (*or* Control) and Warning • aircraftwoman • (*or* **a.c.w.**) *Physics* alternating continuous wave • automatic car wash

ACWS aircraft control and warning system

ACWW Associated Country Women of the World

ACY average crop yield

ad (æd) advertisement • *Tennis* (North America) advantage

ad. adapt(ation) • adapter • *Med.* addatur *or* adde (Latin: let there be added *or* add; in prescriptions) • adverb

a.d. (*or* **A/d**) after date • ante diem (Latin: before the day) • autograph document

a.D. ausser Dienst (German; retired, on half pay)

AD *Insurance* accidental damage • (Venezuela) Acción Democrática (Spanish: Democratic Action Party) • accumulated dose (of radiation) • *Military* active duty • administrative department • *Military* air defence • air dried • Algerian dinar (monetary unit; *see also under* DA) • Alzheimer's disease • (small caps.) anno Domini (Latin: in the year of the Lord) • *British fishing port registration for* Ardrossan • armament depot • Art Director • assembly district • Assistant Director • *Politics* Australian Democrat(s) • autograph document • average deviation • Dame of the Order of Australia • *British vehicle registration for* Gloucester

A-D Albrecht Dürer (1471–1528, German painter and engraver)

A/D aerodrome • after date • *Computing* analog–digital (conversion *or* converter)

ADA *Chem.* N-(2-acetamido)iminodiacetic acid • *Chem.* acetone dicarboxylic acid • *Biochem., Med.* adenosine deaminase (as in **ADA deficiency**) • Agricultural Development Association • (USA) Air Defense Agency • Aluminium Development Association • American Dental Association • American Diabetes Association • Americans for Democratic Action • *Chem.* ammonium dihydrogen arsenate • Association of Drainage Authorities • Atomic Development Authority • Australian Dental Association • (*or* **Ada**; ˈeɪdə) *Computing, indicating* a programming language (after Ada Lovelace (1815–52), British mathematician)

ADAA Art Dealers Association of America

ADAD (ˈeɪdæd) (USA) automatic telephone dialing-announcing device

adag. (*or* **adago, adgo**) *Music* adagio

adap. adapted

ADAPSO (əˈdæpsəu) (USA) Association of Data Processing Organizations

Adapt (əˈdæpt) Access for Disabled People to Arts Today

ADAPTS (əˈdæpts) (USA) Air-Deliverable Antipollution Transfer System (in the Coast Guard)

ADAS (ˈeɪdæs) Agricultural Development and Advisory Service • *Computing* automatic data acquisition system • *Computing* auxiliary data annotation set

ADAWS *Military* action data automation and weapons systems • Assistant Director of Army Welfare Services

A-day assault day

ADB accidental death benefit • African Development Bank • Asian Development Bank • Associate of the Drama Board (Education) • (Canada) Atlantic Development Board • (USA) Bachelor of Domestic Arts

ADC advanced developing country (*or* countries) • *Telecom.* advice of duration and charge • aerodrome control • (Jamaica) Agricultural Development Corporation • aide-de-camp • Aid to Dependent Children • (USA) Air Defense Command • *Chem.* allyl diglycol carbonate • Amateur Dramatic Club • *Computing* analog–digital converter • apparent diffusion coefficient • Art Directors' Club • *Scouting* Assistant District Commissioner • Association of District

Councils • *Computing* automated distribution control • automatic digital calculator

ADCC (USA) Air Defense Control Center • *Med.* antibody-dependent cellular (*or* cell-mediated) cytotoxicity

ADCCP *Computing* advanced data communication control procedure

ADCI American Die Casting Institute

ADCM Archbishop of Canterbury's Diploma in Church Music

ADCONSEN ('ædkɒn,sen) (USA) (with) advice and consent of the Senate

ADC(P) personal aide-de-camp to HM the Queen

ADCT Art Directors' Club, Toronto

add. *Med.* add *or* let there be added (in prescriptions; Latin *adde* or *addantur*) • addendum (*or* addenda) • addition(al) • address(ed)

ADD *Aeronautics* airstream direction detector • *Med.* attention deficit disorder

ADDC (USA) Air Defense Direction Center

ADDF Abu Dhabi Defence Force

addl (*or* **addnl**) additional

addn addition

addsd addressed

ADDS Association of Directors of Social Services

ADE *Immunol.* antibody-dependent enhancement

ADEF (France) Agence d'évaluation financière (rating agency)

ad effect. *Med.* ad effectum (Latin: until effective; in prescriptions)

ADEME Assistant Director, Electrical and Mechanical Engineering

ad eund. ad eundem gradum (Latin: to the same degree)

a.d.f. after deducting freight

ADF *Finance* (Australia) approved deposit fund • Asian Development Fund • Australian Defence Force • *Navigation* automatic direction finder

ad fin. ad finem (Latin: at *or* near the end)

ADFManc Art and Design Fellow, Manchester

ADFW Assistant Director of Fortifications and Works

ADG Assistant Director-General

ADGB Air Defence of Great Britain

ADGMS Assistant Director-General of Medical Services

ad gr. acid. *Med.* ad gratam aciditatem

(Latin: to an agreeable acidity; in prescriptions)

ad gr. gust. *Med.* ad gratum gustum (Latin: to an agreeable taste; in prescriptions)

ADH *Biochem.* antidiuretic hormone • Assistant Director of Hygiene • Association of Dental Hospitals

ADHD *Med.* attention deficit hyperactivity disorder

ad h.l. ad hunc locum (Latin: at this place)

ADI *Med.* acceptable daily intake (of a toxic substance in food) • approved driving instructor

ad inf. ad infinitum (Latin: to infinity)

ad init. ad initium (Latin: at the beginning)

ad int. ad interim (Latin: in the meantime)

ADIZ (USA) air defense identification zone

adj. adjacent • adjective • adjoining • *Maths.* adjoint • adjourned • adjudged • adjunct • *Insurance, Banking, etc.* adjustment • adjutant

ADJ *Meteorol.* adjacent

Adj.A (USA) Adjunct in Arts

ADJAG Assistant Deputy Judge Advocate-General

adjt (*or* **Adjt**) adjutant

Adjt-Gen Adjutant-General

ADK (Malaysia) Order of Ahli Darjah Kinabalu

ADL *Electronics* acoustic delay line • *Med.* activity (*or* activities) of daily living • *Computing* Ada design language • (USA) Anti-Defamation League (of B'nai B'rith) • assistant director of labour

ad lib. ad libitum (Latin: according to pleasure, i.e. freely)

ad loc. ad locum (Latin: at the place)

ADLP Australian Democratic Labor Party

adm. administration • administrative • administrator • admission • admitted

Adm. Admiral • Admiralty

ADM Advanced Diploma in Midwifery • annual delegate meeting • Association of Domestic Management • atomic demolition munitions • average daily membership

ADMA American Drug Manufacturers' Association • (USA) Aviation Distributors' and Manufacturers' Association

Ad. Man. Advertisement Manager

admin. administration • (*or* **Admin.**) administrator

Admin. Apost. Administrator Apostolic

Adml Admiral

ADMO(CA) Assistant Director of the Meteorological Office (Civil Aviation)

ADMOS ('ædmɒs) *Computing* automatic device for mechanical order selection

admov. *Med.* admoveatur (Latin: let it be applied; in prescriptions)

Adm. Rev. Admodum Reverendus (Latin: Very Reverend)

ADMS Assistant Director of Medical Services

ADMT Association of Dental Manufacturers and Traders of the United Kingdom

admx administratrix

ADN Allgemeiner Deutscher Nachrichtendienst (German news agency)

ADNA Assistant Director of Naval Accounts

ADNC Assistant Director of Naval Construction

ad neut. *Med.* ad neutralizandum (Latin: until neutral)

ADNI Assistant Director of Naval Intelligence

ADNOC Abu Dhabi National Oil Company

ADNS Assistant Director of Nursing Services

ADO (USA) advanced development objective • air defence officer • assistant district officer • Association of Dispensing Opticians • automotive diesel oil

ADOF Assistant Director of Ordnance Factories

ADOS Assistant Director of Ordnance Services

ADP *Biochem.* adenosine diphosphate • air defence position • *Physics* area density of particles • Association of Dental Prosthesis • automatic data processing

ADPA Associate Diploma of Public Administration

ADPCM *Computing* adaptive differential pulse code modulation

ADPLAN ('æd,plæn) advanced planning

ADPR Assistant Director of Public Relations

ADPSO Association of Data Processing Service Organizations

ADR accident data recorder • *Law* alternative dispute resolution • *Stock exchange* American Depository Receipt

ADRA Animal Diseases Research Association (Edinburgh)

ad s. *Law* ad sectam (Latin: at the suit of)

a.d.s. autograph document signed

A.d.S. Académie des Sciences (French: Academy of Sciences)

ADS *Computing* accurately defined systems • advanced dressing station • (USA) air defense sector • American Dialect Society • *Engineering* articulated drill string • Astrophysics Data System • *Engineering* automatic depressurization system

ad saec. ad saeculum (Latin: to the century)

ADS&T Assistant Director of Supplies and Transport

ad sat. *Med.* ad saturandum (Latin: to saturation)

ADSOC ('ædsɒk) (USA) Administrative Support Operations Center

ADSS Australian Defence Scientific Service

ADST *Finance* approved deferred share trust

adst. feb. *Med.* adstante febre (Latin: when fever is present)

ADT American District Telegraph • Assistant Director of Transport • Atlantic Daylight Time • *Electronics* automatic double tracking • average daily traffic

ADTECH ('æd,tɛk) advanced decoy technology

ADTS Automated Data and Telecommunications Service

ad us. ad usum (Latin: according to custom)

ad us. ext. *Med.* ad usum externum (Latin: for external use)

adv. advance • adverb(ial) • adversus (Latin: against) • advertisement • advertising • advice • advise • adviser • advisory • advocate

Adv. Advent • Advocate

ad val. ad valorem (Latin: according to value)

advb adverb

adven. adventure

adv. pmt. advance payment

ADVS Advanced Diploma in Voice Studies • Assistant Director of Veterinary Services

advt advertisement

advv. adverbs

ADW air defence warning • Assistant Director of Works

ADWE&M Assistant Director of Works, Electrical and Mechanical

ADX *Computing* automatic data exchange • *Telecom.* automatic digital exchange

ADXPS *Chem., Physics* angular dependent X-ray photoelectron spectroscopy

ae. *Numismatics* aeneus (Latin: made of copper) • aetatis (Latin: at the age of; aged)

AE account executive • *Chem.* acid ester • *Physics* acoustic emission • adult education • aeronautical engineer(ing) • *Taxation* age exemption • agricultural engineer(ing) • Air Efficiency Award • (Sweden) Aktiebolaget Atomenergi (Atomic Energy Corporation) • All England • American English • army education • Associate in Education • Associate in Engineering • atomic energy • *Photog.* autoexposure • *British vehicle registration for* Bristol • *pseudonym of* George Russell, 1867–1935, Irish poet • *airline flight code for* Mandarin Airlines • *Insurance, indicating* (in Lloyd's Register) third-class ships

AEA *Physics* acoustic emission analysis • (USA) Actors' Equity Association • *Engineering* air entraining agent • American Economic Association • (USA) Atomic Energy Act • Atomic Energy Authority

AEAA Asociación de Escritores y Artistas Americanos (Spanish: Association of Spanish-American Writers and Artists)

AEAF Allied Expeditionary Air Force

AE & MP Ambassador Extraordinary and Minister Plenipotentiary

AE & P Ambassador Extraordinary and Plenipotentiary

AEAUSA Adult Education Association of the United States of America

AEB Area Electricity Board • Associated Examining Board

AEBP atmospheric equivalent boiling point

AEC *Insurance* additional extended coverage • Agricultural Executive Council • American Express Company • *Chem.* anion-exchange capacity • Army Electronics Command • Associated Equipment Company • Association of Education Committees • (USA) Atomic Energy Commission • (South Africa) Atomic Energy Corporation • (USA)

Atomic Energy Council • *Photog.* automatic exposure control

AECB (USA) Atomic Energy Control Board

AECCG African Elephant Conservation Coordinating Group

AECI African Explosives and Chemical Industries • Associate of the Institute of Employment Consultants

AECL Atomic Energy of Canada Limited

AEd Associate in Education

AED advanced electronics design • *Computing* Algol Extended for Design (programming language) • Association of Engineering Distributors • automated engineering design • Doctor of Fine Arts (Latin *Artium Elegantium Doctor*)

AEDE Association européenne des enseignants (French: European Teachers' Association)

AEDOT Advanced Energy Design and Operation Technologies

AEDP (USA) Alternative Energy Development Program

AEDS (USA) Association for Educational Data Systems • Atomic Energy Detection Systems

AEDU Admiralty Experimental Diving Unit

AeE Aeronautical Engineer

AEE airborne evaluation equipment • *Chem.* alkaline-earth element • Atomic Energy Establishment

AEEC (USA) Airlines Electronic Engineering Committee

AEEE (Canada) Army Equipment Engineering Establishment

AEEN Agence européenne pour l'energie nucléaire (French: European Agency for Atomic Energy)

AEEU Amalgamated Engineering and Electrical Union (formed from merger of AEU and EETPU)

AEF Allied Expeditionary Force • American Expeditionary Force(s) • Australian Expeditionary Force

A-effect *Theatre* alienation effect

AEFM Association européenne des festivals de musique (French: European Association of Music Festivals)

a.e.g. (*or* **aeg**) ad eundem gradum (Latin: to the same degree) • *Bookbinding* all edges gilt

AEG (Germany) Allgemeine Elektrizitäts-Gesellschaft (General Electric Company)

AEGIS ('iːdʒɪs) Aid for the Elderly in Government Institutions • *Computing* an existing general information system

AEGM Anglican Evangelical Group Movement

AEH A(lfred) E(dward) Housman (1859–1936, British poet)

AEI (USA) Alternative Energy Institute • American Express International • Associated Electrical Industries • Association des écoles internationales (French: International Schools Association)

AEIOU Austriae Est Imperare Orbi Universi (Latin: it is given to Austria to rule the whole world) • Austria Erit In Orbe Ultima (Latin: Austria will be the world's last survivor)

AEJ Association for Education in Journalism

AEL Admiralty Engineering Laboratory • aeronautical (*or* aircraft) engine laboratory • Associated Engineering, Limited • automation engineering laboratory

AELE (France) Association européenne de libre-échange (European Free Trade Association; EFTA)

AELTC All England Lawn Tennis Club

AEM Air Efficiency Medal • analytical electron microscopy

AEMT Association of Electrical Machinery Trades

AEMU *Astronautics* Advanced Extra-vehicular Mobility Unit

aen. *Numismatics* aeneus (Latin: made of copper)

Aen. (ital.) The Aeneid (Virgil's epic)

AEn Associate in English

AENA All England Netball Association

AEng Associate in Engineering

Aen. Nas. Aenei Nasi (Brasenose College, Oxford; Latin: of the Brazen Nose)

AEO *Chem.* alkaline-earth oxide • Assistant Education Officer • Assistant Experimental Officer • Association of Exhibition Organizers

AEP Agence européenne de productivité (French: European Production Agency) • American Electric Power

AEPI American Educational Publishers Institute

aeq. aequalis (Latin: equal)

aer. aeronautics • aeroplane

AER Army Emergency Reserve

aera. aeration

AERA American Educational Research Association • Associate Engraver, Royal Academy • Associate of the British Electrical and Allied Industries Research Association

AerE Aeronautical Engineer

AERE Atomic Energy Research Establishment (Harwell)

AERI Agricultural Economics Research Institute (Oxford University)

AERNO aeronautical equipment reference number

aero. aeronautic(al)

AERO Air Education and Recreation Organization

aeron. aeronautical • aeronautics

aerosp. aerospace

Aes. Aesop (6th century BC, Greek author of fables)

AES advanced energy system • (USA) Aerospace Electrical Society • Agricultural Economics Society (Reading University) • (USA) Airways Engineering Society • *Physics* atomic emission spectrum (spectroscopy, spectrometry, *or* spectrography) • Audio Engineering Society • *Physics* Auger electron spectrum (*or* spectroscopy)

AESE (USA) Association of Earth Science Editors

AESL (USA) Aerospace Energy Systems Laboratory

aesth. aesthetics

aet. (*or* **aetat.**) aetatis (Latin: at the age of; aged)

AET Associate in Electrical (*or* Electronic) Technology

AETR advanced engineering test reactor

AEU Amalgamated Engineering Union (formerly AUEW; *see* AEEU) • American Ethical Union

AEU(TASS) Amalgamated Engineering Union (Technical, Administrative, and Supervisory Section)

AEV aerothermodynamic elastic vehicle • air evacuation • *Microbiol.* avian erythroblastosis virus

AEW Admiralty Experimental Works • airborne early warning (aircraft)

AEW & C airborne early warning and control

AEWHA All England Women's Hockey Association

AEWLA All England Women's Lacrosse Association

AEWS advanced earth satellite weapons system • aircraft early warning system

a.f. advanced freight • (*or* **a/f**) *Commerce* a favor (Spanish: in favour) • anno futuro (Latin: next year) • audiofrequency

Af *symbol for* afgháni (monetary unit of Afghanistan)

Af. Africa(n) • Afrikaans • *symbol for* Aruban florin (monetary unit of Aruba)

AF Académie Française (French literary academy) • (*or* **A/F**) *Engineering* across flats (used in denoting sizes of nuts, as in ½**AF**) • (France) Action Française (former journal and political party) • Admiral of the Fleet • air force • (USA) Air Foundation • *airline flight code for* Air France • Anglo-French • Armée Française (French Army) • Associated Fisheries • Associate Fellow • audiofrequency • *Photog.* autofocus • automatic (smoke) filter • *British vehicle registration for* Truro

A/F antiflooding • as found (in auction catalogues)

AFA *Nuclear engineering* advanced fuel assembly • African Football Association • Air Force Act • (USA) Air Force Association • Amateur Fencing Association • Amateur Football Alliance • American Foundrymen's Association • Associate in Fine Arts • (Scotland) Associate of the Faculty of Actuaries

AFAC American Fisheries Advisory Committee

AFAEP Association of Fashion Advertising and Editorial Photographers

AFAFC (USA) Air Force Accounting and Finance Center

AFAIAA Associate Fellow of the American Institute of Aeronautics and Astronautics

AFAIM Associate Fellow of the Australian Institute of Management

AFAITC (USA) Armed Forces Air Intelligence Training Center

AFAL (USA) Air Force Astronautical Laboratory

AFAM (*or* **AF & AM**) Ancient Free and Accepted Masons

AFAS (France) Association française pour l'avancement des sciences (French Association for the Advancement of the Sciences)

AFASIC Association for all Speech-Impaired Children

AFB *Microbiol.* acid-fast bacilli • (USA) Air Force Base • American Federation for the Blind

AFBC *Chemical engineering* atmospheric fluidized-bed combustor (*or* combustion)

AFBD Association of Futures Brokers and Dealers

AFBF American Farm Bureau Federation

AFBMD (USA) Air Force Ballistic Missile Division

AFBPsS Associate Fellow of the British Psychological Society

AFBS American and Foreign Bible Society

AFC Air Force Cross • Amateur Football Club • *Immunol.* antibody-forming cell • Association Football Club • Australian Flying Corps • (*or* **a.f.c.**) automatic flight control • (*or* **a.f.c.**) automatic frequency control

AFCAI Associate Fellow of the Canadian Aeronautical Institute

AFCE Associate in Fuel Technology and Chemical Engineering • automatic flight control equipment

AFCEA (USA) Armed Forces Communications and Electronics Association

AFCENT (*or* **AFCent**) ('æf‚sεnt) Allied Forces in Central Europe (of NATO)

afco ('æf‚kəʊ) *Engineering* automatic fuel cut-off

AFCO Admiralty Fleet Confidential Order

AFCRL (USA) Air Force Cambridge Research Laboratories

AFCS Air Force Communications Service • automatic flight control system

AFCU American and Foreign Christian Union

AFCW Association of Family Case Workers

AFD accelerated freeze dried (*or* drying) • air force depot • (USA) Doctor of Fine Arts

AFDC (USA) Aid to Families with Dependent Children

AFDCS Association of First Division Civil Servants

AFDM *Engineering* advanced fluid-dynamics model

AFDS Air Fighting Development Squadron

AFDW *Chem.* ash-free dry weight

AFEA American Farm Economics Association • American Film Export Association

AFEE Airborne Forces Experimental Establishment

AFEIS Advanced Further Education Information Service

AFERO Asia and the Far East Regional Office (of the FAO)

AFES Admiralty Fuel Experimental Station

AFESC (USA) Air Force Engineering and Services Center

AFESD Arab Fund for Economic and Social Development

AFEX (USA) Air Forces Europe Exchange • (USA) Armed Forces Exchange

aff. affairs • affectionate • (or **affil.**) affiliate(d) • affirmative • affix

AFF *Photog.* aberration-free focus

Aff.IP Affiliate of the Institute of Plumbing

AFFL Agricultural Finance Federation, Ltd

afft affidavit

AFFTC (USA) Air Force Flight Test Center

Afg. Afghanistan

AFG *international vehicle registration for* Afghanistan

AFGE American Federation of Government Employees

Afgh. (or **Afghan.**) Afghanistan

AFGL (USA) Air Force Geophysics Laboratory

AFGM American Federation of Grain Millers

AFH American Foundation for Homeopathy

AFHC Air Force Headquarters Command

AFHQ Allied Force(s) Headquarters • Armed Forces Headquarters

AFI American Film Institute

AFIA American Foreign Insurance Association • (Australia) Associate of the Federal Institute of Accountants

AFIAP Artiste de la Fédération internationale de l'art photographique (French: Artist of the International Federation of Photographic Art)

AFICD Associate Fellow of the Institute of Civil Defence

AFII American Federation of International Institutes

AFIIM Associate Fellow of the Institution of Industrial Managers

AFIMA Associate Fellow of the Institute of Mathematics and its Applications

AFIPS American Federation of Information Processing Societies, Inc.

AFL *Computing* abstract family of languages • Air Force List • American Federation of Labor • American Football League • (Portugal) Associação de Futebol de Lisboa (Lisbon Football Association)

AFLA Amateur Fencers' League of America • Asian Federation of Library Associations

AFLC (USA) Air Force Logistics Command

AFL-CIO American Federation of Labor and Congress of Industrial Organizations

aflt afloat

AFM *Engineering* abrasive flow machining • Air Force Medal • American Federation of Musicians of the United States and Canada • assistant field manager • Associated Feed Manufacturers • *Chem., Physics* atomic force microscope (or microscopy) • *Telecom.* audiofrequency modulation

AFMA Armed Forces Management Association • Artificial Flower Manufacturers' Association of Great Britain

AFMDC (USA) Air Force Missile Development Center

AFMEC African Methodist Episcopal Church

AFMED (or **AFMed**) Allied Forces Mediterranean

afmo. afectísimo (Spanish: very affectionately; in correspondence)

AFMTC (USA) Air Force Missile Test Center

AFN American Forces Network • Armed Forces Network • Association of Free Newspapers

AFNE Allied Forces Northern Europe

AFNIL ('æfnɪl) (France) Agence francophone pour la numérotation internationale du livre (French Agency for International Standard Book Numbers; ISBN)

AFNOR ('æf,nɔː) Association française de normalisation (the standards organization of France)

AFNORTH (or **AFNorth**) ('æf,nɔːθ) Allied Forces Northern Europe

AFO Admiralty Fleet Order • army forwarding officer

AFOAR (USA) Air Force Office of Aerospace Research

AFOAS (USA) Air Force Office of Aerospace Sciences

AF of M American Federation of Musicians of the United States and Canada

AFOM Associate of the Faculty of Occupational Medicine

AFOSR (USA) Air Force Office of Scientific Research

AFP Agence France Press (French independent news agency) • (*or* **afp**) *Med.* alpha-fetoprotein

AFPRB Armed Forces Pay Review Board

Afr. Africa(n)

AFr. (*or* **A-Fr.**) Anglo-French

AFR accident frequency rate (in industry) • air–fuel ratio • automatic fingerprint recognition

AFRA American Farm Research Association • American Federation of Television and Radio Artists

AFRASEC ('æfrə,sɛk) Afro-Asian Organization for Economic Cooperation

AFRC Agricultural and Food Research Council

Afrik. Afrikaans

AFRO Africa Regional Office (of the FAO)

AFRTS (USA) Armed Forces Radio and Television Service

AFS Advanced Flying School • air force station • Alaska Ferry Service • American Field Service • Army Fire Service • Atlantic Ferry Service • *Chem., Physics* atomic fluorescence spectrometry • auxiliary fire service • *Physics* axial field spectrometer

AFSA American Foreign Service Association • (USA) Armed Forces Security Agency

AFSBO American Federation of Small Business Organizations

AFSC Air Force Systems Command • American Friends Service Committee • Armed Forces Staff College

AFSCME American Federation of State, County, and Municipal Employees (trade union)

afsd aforesaid

AFSIL ('æfsɪl) Accommodation for Students in London

AFSLAET Associate Fellow of the Society of Licensed Aircraft Engineers and Technologists

AFSOUTH (*or* **AFSouth**) ('æf,sauθ) Allied Forces in Southern Europe

aft. after • afternoon

AFT American Federation of Teachers • axial-flow turbine

AFTE American Federation of Technical Engineers

AFTM American Foundation of Tropical Medicine

AFTP *Computing* anonymous file transfer protocol

aftn afternoon

AFTN Aeronautical Fixed Telecommunications Network

AFTR American Federal Tax Reports

AFTRA American Federation of Television and Radio Artists (trade union)

AFU advanced flying unit

AFULE Australian Federated Union of Locomotive Enginemen

AFV armoured fighting vehicle • armoured force vehicle

AFVG *Aeronautics* Anglo-French variable geometry

AFVPA Advertising Film and Videotape Producers' Association

AFW army field workshop

AFWAL (USA) Air Force Wright Aeronautical Laboratories

AFWL (USA) Air Force Weapons Laboratory

ag. agent • agreement • agriculture

a.g. *Chem., Military* anti-gas

Ag antigen • *Chem., symbol for* silver (Latin *argentum*)

Ag. August

AG Accountant-General • (Nigeria) Action Group Party • Adjutant-General • Agent-General (of colonies) • air gunner • (Germany) Aktiengesellschaft (public limited company; plc) • (Italy) Alberghi per la Gioventù (youth hostels) • art gallery • Attorney-General • *British vehicle registration for* Hull

Aga ('ɑːɡə) Aktiebolaget Gasackumulator (cooker)

AGA *Military* air-to-ground-to-air • Amateur Gymnastics Association • American Gas Association • American Genetic Association • *Obstetrics* appropriate for gestational age • Australian Garrison Artillery

AGAC American Guild of Authors and Composers

AGACS automatic ground-to-air communications system

AGARD Advisory Group for Aerospace Research and Development (in NATO)

a.g.b. a (*or* any) good brand

AGBI Artists' General Benevolent Institution

AGC (USA) advanced graduate certificate • American Grassland Council • (*or* **a.g.c.**) *Electronics* automatic gain control

AGCA *Aeronautics* automatic ground-controlled approach

AGCAS Association of Graduate Careers Advisory Service

AGCC Arab Gulf Cooperation Council

AGCL *Aeronautics* automatic ground-controlled landing

AGCM *Meteorol.* atmospheric general circulation model

agcy agency

agd agreed

a.g.d. *Engineering* axial gear differential

AGDC *Freemasonry* Assistant Grand Director of Ceremonies

AGDL Attorney-General of the Duchy of Lancaster

AgE Agricultural Engineer

AGE Admiralty Gunnery Establishment • aerospace ground equipment • Associate in General Education • automatic guidance electronics

AGES Association of Agricultural Education Staffs

AGF Adjutant-General to the Forces • army ground forces

ag. feb. *Med.* aggrediente febre (Latin: when the fever increases)

agg. (*or* **aggr.**) aggregate

AGH Australian General Hospital

AGI (USA) adjusted gross income (in income-tax returns) • American Geographical Institute • American Geological Institute • Année Géophysique Internationale (French: International Geophysical Year, 1 July 1957–31 Dec 1958) • annual general inspection • Artistes graphiques internationaux (French: International Graphic Artists) • Associate of the Institute of Certificated Grocers

AGIP Agenzia Generale Italiana Petroli (Italian national oil company)

agit. *Med.* agitatum (Latin: shaken)

agit. ante sum. *Med.* agita ante sumendum (Latin: shake before taking)

Agitprop ('ædʒɪt,prɒp) Agitpropbyuro (formerly, Soviet bureau in charge of agitation and propaganda)

AGK Astronomische Gesellschaft Katalog (German: Astronomical Society Catalogue)

AGL above ground level

aglm. agglomerate

AGM advisory group meeting • *Med.* African green monkey • air-to-ground missile • annual general meeting • Award of Garden Merit (awarded by the Royal Horticultural Society)

AGMA American Guild of Musical Artists • Athletic Goods Manufacturers' Association

agn again • agnomen

AGN *Astronomy* active galactic nucleus

agnos. agnostic

ago. agosto (Italian: August)

AGOR Auxiliary-General Oceanographic Research

AGP Academy of General Practice • *Computing* accelerated (*or* advanced) graphics port • aviation general policy

AGPA American Group Psychotherapy Association

AGPL Administração-Geral do Pôrto de Lisboa (Portuguese: Port of Lisbon Authority)

agr. agreement • agricultural • agriculture • agriculturist

AGR *Nuclear physics* advanced gas-cooled reactor • Association of Graduate Recruiters

AGRA Army Group Royal Artillery • Association of Genealogists and Record Agents

AGREE (ə'griː) Advisory Group on Reliability of Electronic Equipment

AGRF American Geriatric Research Foundation

agric. agricultural • agriculture • agriculturist

AGRM Adjutant-General, Royal Marines

agron. agronomy

AGRS *Chemical engineering* acid-gas removal system

AGS *Astronautics* abort guidance system • aircraft general standard • Air Gunnery School • Alpine Garden Society • American Geographical Society • Associate in General Studies

AGSM Associate of the Guildhall School of Music and Drama • Australian Graduate of the School of Management

AGSRO Association of Government Supervisors and Radio Officers

AGSS American Geographical and Statistical Society

agst against

agt agent • agreement

AGT advanced gas turbine • Art Gallery of Toronto • (USA) Association of Geology Teachers • *Physics* average generation time

ag^{to} agosto (Italian, Portuguese, and Spanish: August)

AGU American Geophysical Union

AGV *Pharmacol.* aniline gentian violet • automated guided vehicle(s) • *Electronics* autonomous guided vehicle

AGVA American Guild of Variety Artists • *Aeronautics* Anglo-German variable geometry

a.g.w. (*or* **AGW**) actual gross weight

AGWAC Australian Guided Weapons and Analogue Computer

AGWI *Marine insurance* Atlantic, Gulf, West Indies

agy agency

a.h. *Shipping* aft (*or* after) hatch • ampere-hour • armed helicopter

AH *airline flight code for* Air Algerie • (small caps.) anno Hebraico (Latin: in the Jewish year) • (small caps.) anno Hegirae (Latin: in the year of the Hegira; in the Muslim calendar) • (*or* **A/H**) *Marine insurance* Antwerp–Hamburg coastal ports • *British fishing port registration for* Arbroath • *Chem.* aromatic hydrocarbon • *Med.* artificial hyperglycaemia • *British vehicle registration for* Norwich

AHA American Heart Association • American Historical Association • American Hospitals Association • (formerly, in the NHS) Area Health Authority • Australian Hotels Association

AHAM (USA) Association of Home Appliance Manufacturers

AH & FITB Agricultural, Horticultural, and Forestry Industry Training Board

AHAUS Amateur Hockey Association of the United States

AHC Accepting Houses Committee • American Horticultural Council • Army Hospital Corps

ahd ahead

AHD (ital.) American Heritage Dictionary

AHDL *Med.* alveolar hydatid disease of the liver

AHE (USA) Associate in Home Economics • (USA) Association for Higher Education

AHEM Association of Hydraulic Equipment Manufacturers

AHF *Chem.* anhydrous hydrogen fluoride • *Med.* antihaemophilic factor

AHG *Med.* antihaemophilic globulin

AHGMR Ad Hoc Group on Missile Reliability

AHH *Med., Biochem.* aryl hydrocarbon hydroxylase (used in tobacco research)

AHI American Health (*or* Hospital) Institute

AHIL (USA) Association of Hospital and Institutional Libraries

a.h.l. ad hunc locum (Latin: on this passage)

AHL American Hockey League

AHMC Association of Hospital Management Committees

AHMPS Association of Headmistresses of Preparatory Schools

AHMS American Home Mission Society

AHNCV Area of High Nature Conservation Value

AHP *Engineering* absorption heat pump • air horsepower • assistant house physician

AHPR Association of Health and Pleasure Resorts

AHQ Air Headquarters • Allied Headquarters • Army Headquarters

a.h.r. acceptable hazard rate (in risk analysis)

AHRC Australian Humanities Research Council

AHRHS Associate of Honour of the Royal Horticultural Society

AHRIH(NZ) Associate of Honour of the Royal Institute of Horticulture (New Zealand)

AHS adult health study • American Helicopter Society • (small caps.) anno humanae salutis (Latin: in the year of human salvation) • annual housing survey • assistant house surgeon

AHSA American Horse Shows Association

AHSB Authority, Health, and Safety Branch (of the Atomic Energy Authority)

AHSM Associate of the Institute of Health Services Management

AHT absorption heat transformer • acoustic homing torpedo • Animal Health Trust • Association of Highway Technicians

a.h.v. ad hanc vocem (Latin: at this word)

AHWA Association of Hospital and Welfare Administrators

AH-WC Associate of the Heriot-Watt College

a.i. ad interim (Latin: for the meantime)

AI Admiralty instruction • Admiralty Islands • *Computing* Adobe Illustrator • *airline flight code for* Air India • air interdiction • Altesse Impériale (French: Imperial Highness) • American Institute • Amnesty International • (small caps.) anno inventionis (Latin: in the year of the discovery) • Anthropological Institute • Army Intelligence • artificial insemination • artificial intelligence • *Physics, Chem.* associative ionization • *Physics* avalanche injection • *Irish vehicle registration for* Meath

AIA Abrasive Industries Association • (USA) Aerospace Industries Association • American Institute of Aeronautics • American Institute of Architects • American Insurance Association • *Chem.* aminoimidazoazaarene • Anglo-Indian Association • Archeological Institute of America • Associate of the Institute of Actuaries • Association of International Artists • automated image analysis • (New Zealand) Aviation Industry Association

AIAA Aircraft Industries Association of America • American Institute of Aeronautics and Astronautics • Association of International Advertising Agencies

AIAB Associate of the International Association of Book-keepers

AIAC Air Industries Association of Canada • Association internationale d'archéologie classique (French: International Association for Classical Archaeology)

AIAE Associate of the Institution of Automobile Engineers

AIAgrE Associate of the Institution of Agricultural Engineers

AIAL Associate Member of the International Institute of Arts and Letters

AIAS Associate Surveyor Member of the Incorporated Association of Architects and Surveyors • Australian Institute of Agriculture and Science

AIB accidents investigation branch (of insurance companies) • American Institute of Banking • Association of Independent Businesses • (Italy) Associazione Italiana Biblioteche (Italian Library Association)

AIBA Association internationale de boxe amateur (French: International Amateur Boxing Association)

AIBC Architectural Institute of British Columbia

AIBD Associate of the Institute of British Decorators • Association of International Bond Dealers

AIBE Associate of the Institute of Building Estimators

AIBM Associate of the Institute of Baths Management • Association internationale des bibliothèques musicales (French: International Association of Music Libraries)

AIBP Associate of the Institute of British Photographers

AIBS American Institute of Biological Sciences • (South Africa) Associate of the Institute of Building Societies

AIBScot Associate of the Institute of Bankers in Scotland

AIC Agricultural Improvement Council • Agricultural Institute of Canada • *Physics* Alfvén ion cyclotron • American Institute of Chemists • artificial insemination centre • Art Institute of Chicago

AICA Associate Member of the Commonwealth Institute of Accountants • Association internationale des critiques d'art (French: International Association of Art Critics)

AICB Association internationale contre le bruit (French: International Association Against Noise)

AICBM anti-intercontinental ballistic missile

AICC All India Congress Committee

AICE American Institute of Chemical Engineers • American Institute of Consulting Engineers • Associate of the Institution of Civil Engineers

AICeram Associate of the Institute of Ceramics

AIChE American Institute of Chemical Engineers

AICMA Association internationale des constructeurs de matériel aérospatial (French: International Association of Aerospace Equipment Manufacturers)

AICPA American Institute of Certified Public Accountants

AICRO Association of Independent Contract Research Organizations

AICS Associate of the Institute of

Chartered Shipbrokers · Association internationale du cinéma scientifique (French: International Scientific Film Association)

AICTA (Trinidad) Associate of the Imperial College of Tropical Agriculture

AICV armoured infantry combat vehicle

AID acute infectious disease · Aeronautical Inspection Directorate · (USA) Agency for International Development · agricultural industrial development · Aircraft Intelligence Department · American Institute of Decorators · American Institute of Interior Designers · Army Intelligence Department · *Med.* artificial insemination (by) donor · Association internationale pour le développement (French: International Development Association)

AIDA Association internationale de droit africain (French: International African Law Association) · Association internationale de la distribution des produits alimentaires (French: International Association of Food Distribution) · *Advertising* attention, interest, desire, action (of the customer)

AIDAS Agricultural Industry Development Advisory Service

AIDB Association of International Bond Dealers

AIDD American Institute for Design and Drafting

AIDF African Industrial Development Fund

AIDIA Associate of the Industrial Design Institute of Australia

AIDL (New Zealand) Auckland Industrial Development Laboratory

AIDP Associate of the Institute of Data Processing · Association internationale de droit pénal (French: International Association of Penal Law)

Aids (*or* **AIDS**) (eɪdz) acquired immune deficiency syndrome

AIDS accident information display system · aircraft integrated data system · air force intelligence data-handling system

AIE Associate of the Institute of Education

AIEA Agence internationale de l'énergie atomique (French: International Atomic Energy Agency)

AIED Association internationale des étudiants dentaires (French: International Association of Dental Students)

AIEE Associate of the Institution of Electrical Engineers · Association des instituts d'études européennes (French: Association of Institutes for European Studies)

AIEP (USA) Association of Independent Electricity Producers

AIF Agenzia Internazionale Fides (Vatican City State news agency) · Alliance internationale des femmes (French: International Alliance of Women) · (USA) Atomic Industrial Forum · (formerly) Australian Imperial Forces

AIFA Associate of the International Faculty of Arts

AIFireE Associate of the Institution of Fire Engineers

AIFM Associate of the Institute of Factory Managers · Association internationale des femmes médecins (French: International Association of Women Doctors)

AIG Adjutant (*or* Assistant) Inspector-General · *Chem.* aerosol-ionization gas analyser · Assistant Instructor of Gunnery

AIGA American Institute of Graphic Arts

AIGCM Associate of the Incorporated Guild of Church Musicians

AIGS Agricultural Investment Grant Schemes

AIH all in hand · *Med.* artificial insemination (by) husband · Association internationale de l'hôtellerie (French: International Hotel Association)

AIHA American Industrial Hygiene Association

AIHsg Associate of the Institute of Housing

AIIA Associate of the Indian Institute of Architects · Associate of the Insurance Institute of America · Australian Institute of International Affairs

AIIAL Associate of the International Institute of Arts and Letters

AIIE American Institute of Industrial Engineers

AIInfSc Associate of the Institute of Information Scientists

AIITech Associate Member of the Institute of Industrial Technicians

AIJD Association internationale des juristes démocrates (French: International Association of Democratic Lawyers)

AIJPF Association internationale des journalistes de la presse féminine et familiale

(French: International Association of Women's Press Journalists)

AIL air intelligence liaison • Associate of the Institute of Linguists

AILAS *Aeronautics* automatic instrument-landing approach system

AILO air intelligence liaison officer

AILocoE Associate of the Institution of Locomotive Engineers

AIM Africa Inland Mission • Alternative Investment Market • (eɪm) American Indian Movement (civil-rights organization) • American Institute of Management • Amsterdam International Market • analytical ion microscopy • Association of Industrial Machinery Merchants • *Presbyterianism* Australian Inland Mission • Australian Institute of Management

a.i.m.a. as interest may appear

AIMarE Associate of the Institute of Marine Engineers

AIMCO (USA) Association of Internal Management Consultants

AIME American Institute of Mechanical Engineers • American Institute of Mining Engineers

AIMechE Associate of the Institution of Mechanical Engineers

AIMI Associate of the Institute of the Motor Industry

AIMinE Associate of the Institution of Mining Engineers

AIMM Australasian Institute of Mining and Metallurgy

AIMO Association of Industrial Medical Officers

AIMPA Association internationale de météorologie et de physique de l'atmosphère (French: International Association of Meteorology and Atmospheric Physics)

AIMPE Australian Institute of Marine and Power Engineers

AIMS (eɪmz) Association for Improvement in Maternity Services

AIMSW Associate of the Institute of Medical Social Workers

AIMU American Institute of Marine Underwriters

AIN American Institute of Nutrition

AINEC All-India Newspaper Editors' Conference

AInstBCA Associate of the Institute of Burial and Cremation Administration

AInstCE Associate of the Institution of Civil Engineers

AInstExE Associate of the Institute of Executive Engineers and Officers

AInstFF Associate of the Institute of Freight Forwarders Ltd

AInstMO Associate of the Institute of Market Officers

AInstP Associate of the Institute of Physics

AInstPI Associate of the Institute of Patentees and Inventors

AINucE Associate of the Institution of Nuclear Engineers

AIP American Institute of Physics • Associate of the Institute of Plumbing • Association of Independent Producers

AIPA Associate Member of the Institute of Practitioners in Advertising • Association internationale de psychologie appliquée (French: International Association of Applied Psychology)

AIPO American Institute of Public Opinion

AIPR Associate of the Institute of Public Relations

AIProdE Associate of the Institution of Production Engineers

AIPS Association internationale de la presse sportive (French: International Sports Press Association) • astronomical image-processing system

AIQS Associate Member of the Institute of Quantity Surveyors

a-i-r artist-in-residence

AIR All India Radio • American Institute of Refrigeration

AIRA (USA) air attaché

AIRC Association of Independent Radio Companies

AIRCENT ('ɛəsɛnt) Allied Air Forces, Central Europe

AIRCOM ('ɛəkɒm) airways communications system

AIREP ('ɛərɛp) *Meteorol.* coded air to ground report

AIRG Agency for Intellectual Relief in Germany

AIRH Association internationale de recherches hydrauliques (French: International Association of Hydraulic Research)

AIRMET ('ɛəmɛt) *Meteorol.* regional forecast for civil aviation

airmiss ('ɛəmɪs) aircraft miss

AIRPASS ('εəpɑ:s) aircraft interception radar and pilots attack sight system

AIRS *Physics, Electronics* acoustic-imaging recognition system

AIRTE Associate of the Institute of Road Transport Engineers

AIRTO Association of Independent Research and Technology Organizations

AIS *Commerce* agreed industry standard • androgen insensitivity syndrome • Anglo-Italian Society • Association internationale de sociologie (French: International Sociological Association) • Australian Illawarra Shorthorn (cattle breed)

AISA Associate of the Incorporated Secretaries Association

AISE Association internationale des sciences économiques (French: International Economics Association)

AISI American Iron and Steel Institute • Associate of the Iron and Steel Institute

AISJ Association internationale des sciences juridiques (French: International Association of Legal Science)

AISS Association internationale de la sécurité sociale (French: International Social Security Association)

AIST (USA) Agency of Industrial Science and Technology • Associate of the Institute of Science and Technology

AIStructE Associate of the Institution of Structural Engineers

AIT Alliance internationale de tourisme (French: International Tourist Alliance) • *Meteorol.* artificial ionospheric turbulence • Asian Institute of Technology • Association of HM Inspectors of Taxes • Association of Investment Trusts

AITA Association internationale du théâtre d'amateurs (French: International Amateur Theatre Association)

AITC Association internationale des traducteurs de conférence (French: International Association of Conference Translators) • Association of Investment Trust Companies

AITI Associate of the Institute of Translators and Interpreters

AITO Association of Independent Tour Operators

AIU Association internationale des universités (French: International Association of Universities)

AIV Association internationale de volcanologie (French: International Association of Volcanology) • *Agriculture, indicating* a method of preserving fodder (initials of its inventor, Artturi Ilmari Virtanen, 1895–1973, Finnish chemist)

AIW (International Union of) Allied Industrial Workers of America • Atlantic-Intercoastal Waterway (Cape Cod to Florida Bay)

AIWC All-India Women's Conference

AIWEM Associate of the Institution of Water and Environmental Management

AIWM American Institute of Weights and Measures

AIX *Computing* advanced interactive executive

AJ *Telecom.* anti-jam • (ital.) Archaeological Journal • Associate in Journalism • *British vehicle registration for* Middlesbrough

AJA American of Japanese ancestry • Anglo-Jewish Association • Australian Journalists' Association

AJAG Assistant Judge Advocate-General

AJC Australian Jockey Club

AJCC American Joint Committee on Cancer

AJE *Accounting* adjusting journal entry

AJEX Association of Jewish Ex-Service Men and Women

AJPM ad Jesum per Mariam (Latin: to Jesus through Mary)

AJR Association of Jewish Refugees • Australian Jurist Reports (Victoria)

AJY Association for Jewish Youth

AK *US postcode for* Alaska • automatic Kalashnikov (rifle) • Knight of the Order of Australia • *British vehicle registration for* Sheffield

a.k.a. (*or* **AKA**) also known as

Akad. Akademie (German: academy)

AKC American Kennel Club • Associate of King's College, London

AKEL (Cyprus) Anorthotikon Komma Ergazomanou Laou (Greek: Progressive Party of the Working People)

AKFM (Madagascar) Ankoton'ny Kongresy ny-Fahaleovantena Malagasy (Congress Party for Malagasy Independence)

AKOE (Cyprus) Anti-Killer Organization of Expatriates

AKR auroral kilometric radiation (from the earth)

Aktb (*or* **Akt, Aktieb**) (Sweden) Aktiebolaget (joint-stock company)

al. alcohol(ic) • alia (Latin: other things)

a.l. *Finance* allotment letter • *Commerce* après livraison (French: after delivery) • autograph letter

Al (ital.) *Physics, symbol for* Alfvén number • *Chem., symbol for* aluminium

AL Abraham Lincoln (1809–65, US president 1861–65) • *Med.* activity of living • Admiralty letter • *US postcode for* Alabama • *international vehicle registration for* Albania • *Chem.* alkyl-oxygen cleavage • América Latina (Spanish: Latin America) • *Baseball* American League • American Legion • Anglo-Latin • (small caps.) anno lucis (Latin: in the year of light) • *Physics* antistokes luminescence • army list • *British vehicle registration for* Nottingham • *postcode for* St Albans

Ala (*or* **ala**) *Biochem.* alanine

Ala. Alabama

ALA Air Licensing Authority • (*or* **ala**) all letters answered • American Library Association • Associate in Liberal Arts • Associate of the Library Association • Association of London Authorities • Authors' League of America

ALAA Associate of Library Association of Australia • Associate of the London Association of Certified Accountants

ALAC Artificial Limb and Appliance Centre

ALACP American League to Abolish Capital Punishment

ALADI Asociación Latino-Americana de Integración (Spanish: Latin-American Integration Association)

ALAI Association littéraire et artistique internationale (French: International Literary and Artistic Association)

ALA-ISAD American Library Association—Information Science and Automation Division

ALAM Associate of the London Academy of Music and Dramatic Art

Alap (*or* **ALAP**) ('ælæp) as low as practicable (describing radiation doses or levels)

Alara (*or* **ALARA**) ('ælərə) as low as reasonably achievable (describing radiation doses or levels)

Alas. Alaska

alb. *Med.* albumin

Alb. Albania(n) • (*or* **Alba.**) Alberta • Albion

Alban. Albanensis (Latin: (Bishop) of St Albans)

ALBM air-launched ballistic missile

Albq. Albuquerque

ALBSU Adult Literacy and Basic Skills Unit

alc. alcohol

a.l.c. à la carte

ALC Agricultural Land Commission • Associate of Loughborough College of Advanced Technology

ALCAN Aluminium Company of Canada

ALCD Associate of the London College of Divinity

alch. alchemy

ALCM air-launched cruise missile • Associate of London College of Music

ALCS Authors' Lending and Copyright Society

Ald. Alderman

ALD *Computing* automatic line drawing

ALDEV African Land Development

Aldm. Alderman

ALE *Insurance* additional living expense • Association for Liberal Education • *Electronics* atomic-layer epitaxy

A level *Education* Advanced level

ALF Animal Liberation Front • Arab Liberation Front (Iraq) • (*or* **Alf**) automatic letter facer (Post Office sorting machine)

ALFSEA Allied Land Forces South-East Asia

alg. algebra(ic)

a.l.g. advanced landing ground

Alg. Algeria(n) • Algiers

ALG *Med.* antilymphocyte globulin

ALGES Association of Local Government Engineers and Surveyors

ALGFO Association of Local Government Financial Officers

Algol (*or* **ALGOL**) ('ælgɒl) *Computing* algorithmic language

ALH Australian Light Horse

ALI American Library Institute • Argyll Light Infantry • Associate of the Landscape Institute

ALICE ('ælɪs) Autistic and Language-Impaired Children's Education

align. alignment

ALJ Administration Law Judge • Australian Law Journal

ALJR Australian Law Journal Reports

alk. alkali

ALL acute lymphatic (lymphoblastic, lymphocytic, *or* lymphoid) leukaemia

ALLC Association for Literary and Linguistic Computing

alleg. allegory

All H. All Hallows

all'ingr. all'ingrosso (Italian: wholesale)

allo. *Music* allegro

all'ott. *Music* all'ottava (Italian; play an octave higher)

All S. All Souls

All SS All Saints

ALM *Med.* alveolar lining material · Association of Lloyd's Members · audiolingual method (in teaching a foreign language) · Master of the Liberal Arts (Latin *Artium Liberalium Magister*)

ALN *Med.* axillary lymph nodes

ALNA Armée de libération nationale d'Angola (French: Angolan National Liberation Army)

ALO air liaison officer · allied liaison officer

ALOE A Lady of England (pseudonym of Charlotte M. Tucker, 1821–93, British novelist)

alp. alpine

ALP American Labor Party · Australian Labor Party · automated learning process · (USA) automated library program

ALPA Air Line Pilots' Association

ALPAL *Computing* A Livermore Physics Applications Language

alph. alphabetical

ALPO Association of Land and Property Owners · Association of Lunar and Planetary Observers

ALPSP Association of Learned and Professional Society Publishers

ALPURCOMS all-purpose communications system

alr aliter (Latin: otherwise)

ALR American Law Reports

ALRC Anti-Locust Research Centre

ALRI airborne long-range input

ALS accident localization system · Agricultural Land Service · *Med.* amyotrophic lateral sclerosis · *Immunol.* antilymphocytic serum · *Aeronautics* approach lighting systems · Associate of the Linnean Society · (*or* **a.l.s.**) autograph letter signed · automated library system

Alsat. Alsatian

al seg. *Music* al segno (Italian: to (*or* at) the sign)

ALSEP Apollo Lunar Surface Experiment Package

ALSTTL advanced low-power Schottky transistor transistor logic

alt. alteration · alternate · alternative · alternator · altimeter · altitude · alto

Alt. Altesse (French: Highness)

ALT Agricultural Land Tribunal · *Biochem., Med.* alanine aminotransferase

Alta. Alberta

alt. dieb. *Med.* alternis diebus (Latin: every other day)

alter. alteration

alt. hor. *Med.* alternis horis (Latin: every other hour)

alt. noct. *Med.* alternis noctibus (Latin: every other night)

ALTPR Association of London Theatre Press Representatives

ALTU Association of Liberal Trade Unionists

ALU *Computing* arithmetic and logic unit

alum. aluminium · alumna(e) · alumnus (*or* alumni)

ALV *Microbiol.* avian leucosis virus

ALWR *Nuclear physics* advanced light-water reactor

am *Physics, symbol for* attometre(s)

am. ammeter · ammunition

a.m. amplitude modulation · ante meridiem (Latin: before noon)

a/m above mentioned

aM (*or* **a/M**) am Main (German: on the River Main; after place names)

Am *Chem., symbol for* americium

Am. America(n) · *Bible* Amos

AM Academy of Management · administrative memorandum · *airline flight code for* Aeromexico · air mail · *Astronautics* airlock module · Air Marshal · Air Ministry · Albert Medal · (*or* **A-M**) Alpes-Maritimes (department of France) · *Radio* amplitude modulation · *Physics* angular momentum · (small caps.) anno mundi (Latin: in the year of the world) · annus mirabilis (Latin: year of wonders; 1666) · ante meridiem (*see under* a.m.) · area manager · army manual · assistant manager · Associate Member · *Insurance* assurance mutuelle (French: mutual insurance) · Ave Maria (Latin: Hail Mary) · Award of Merit (from the Royal Horticultural Society) ·

(USA) Master of Arts (Latin *Artium Magister*) • Member of the Order of Australia • *British vehicle registration for Swindon*

AMA against medical advice • American Management Association • American Marketing Association • American Medical Association • American Missionary Association • American Motorcycle Association • Assistant Masters Association • Associate of the Museums Association • Association of Metropolitan Authorities • Australian Medical Association

AMAB Army Medical Advisory Board

AMAE American Museum of Atomic Energy

amal. amalgamated

AMAmIEE Associate Member of the American Institute of Electrical Engineers

AMARC Associated Marine and Related Charities

AMASCE Associate Member of the American Society of Civil Engineers

amat. amateur

AMAusIMM Associate Member of the Australasian Institute of Mining and Metallurgy

Amb. (*or* **Ambas.**) Ambassador • ambulance

AMB Air Ministry bulletin • (USA) Airways Modernization Board

AMBAC Associate Member of the British Association of Chemists

ambig. ambiguous

AMBIM Associate Member of the British Institute of Management

AMC (USA) Aerospace Manufacturers' Council • Agricultural Mortgage Corporation Ltd • American Motors Corporation • (USA) Army Missile Command • (USA) Army Mobile Command • (USA) Army Munitions Command • Art Master's Certificate • Association of Management Consultants • Association of Municipal Corporations

AMCA Architectural Metal Craftsmen's Association

AMCIB Associate Member of the Corporation of Insurance Brokers

AMCIOB Associate Member of the Chartered Institute of Building

AMCL Association of Metropolitan Chief Librarians

AMCOS ('æmkɒs) Aldermaston Mechanized Cataloguing and Ordering System (in the Atomic Energy Authority)

AMCS airborne missile control system

AMCT Associate of the Manchester College of Technology

am. cur. *Law* amicus curiae (Latin: a friend of the court)

amd amend

AMD acid-mine drainage • Admiralty machinery depot • aerospace medical division • air movement data • *Meteorol.* amend • Army Medical Department

AMDB Agricultural Machinery Development Board

AMDEA Association of Manufacturers of Domestic Electrical Appliances

AMDEC Agricultural Marketing Development Executive Committee

AMDG ad majorem Dei gloriam (Latin: to the greater glory of God; the Jesuit motto)

amdt amendment

AME (USA) Advanced Master of Education • African Methodist Episcopal • Association of Municipal Engineers

AMEC Australian Minerals and Energy Council

AMEDS (*or* **AMedS**) (USA) Army Medical Service

AMEE Admiralty Marine Engineering Establishment • Association of Managerial Electrical Executives

AMEIC Associate Member of the Engineering Institute of Canada

AMEM African Methodist Episcopal Mission

Am. Emb. American Embassy

AMEME Association of Mining Electrical and Mechanical Engineers

amend. (*or* **amendt**) amendment

Amer. Ind. American Indian

Amer. Std. American Standard

AMES Air Ministry Experimental Station • Association of Marine Engineering Schools

AMet Associate of Metallurgy

AMEWA Associated Manufacturers of Electric Wiring Accessories

Amex ('æmɛks) American Express • American Stock Exchange

AMEZC African Methodist Episcopal Zion Church

AMF Australian Marine Force • Australian Military Forces

amg among

a.m.g. *Aeronautics* automatic magnetic guidance

AMG Allied Military Government

AMGO Assistant Master-General of Ordnance

AMGOT Allied Military Government of Occupied Territory (in World War II)

a.m.i. *Military* advanced manned interceptor • *Aeronautics* air mileage indicator

AMI *Med.* acute myocardial infarction • American Meat Institute • American Military Institute • Ancient Monuments Inspectorate • Association Montessori internationale (French: International Montessori Association)

AMIAE Associate Member of the Institution of Automobile Engineers

AMIAP Associate Member of the Institution of Analysts and Programmers

AMIBF Associate Member of the Institute of British Foundrymen

AMICE Associate Member of the Institution of Civil Engineers

AMICEI Associate Member of the Institution of Civil Engineers of Ireland

AMIChemE Associate Member of the Institution of Chemical Engineers

AMICW Associate Member of the Institute of Clerks of Works of Great Britain

AMIE Associate Member of the Institute of Engineers and Technicians

AMIE(Aust) Associate Member of the Institution of Engineers, Australia

AMIED Associate Member of the Institute of Engineering Designers

AMIEI Associate Member of the Institution of Engineering Inspection

AMIE(Ind) Associate Member of the Institution of Engineers, India

AMIElecIE Associate Member of the Institution of Electrical and Electronic Incorporated Engineers

AMIERE Associate Member of the Institute of Electronic and Radio Engineers

AMIEx Associate Member of the Institute of Export

AMIFireE Associate Member of the Institution of Fire Engineers

AMIGasE Associate Member of the Institution of Gas Engineers

AMIH Associate Member of the Institute of Housing

AMIHT Associate Member of the Institution of Highways and Transportation

AMII Association of Musical Instrument Industries

AMIIM Associate Member of the Institution of Industrial Managers

AMILocoE Associate Member of the Institution of Locomotive Engineers

AMIMarE Associate Member of the Institute of Marine Engineers

AMIMechE Associate Member of the Institution of Mechanical Engineers

AMIMGTechE Associate Member of the Institution of Mechanical and General Technician Engineers

AMIMI Associate Member of the Institute of the Motor Industry

AMIMinE Associate Member of the Institution of Mining Engineers

AMIMM Associate Member of the Institution of Mining and Metallurgy

Am. Ind. American Indian

AMInIsTech Associate in Minerals Technology

AMInstBE Associate Member of the Institution of British Engineers

AmInstEE American Institute of Electrical Engineers

AMInstR Associate Member of the Institute of Refrigeration

AMInstTA Associate Member of the Institute of Traffic Administration

AMINucE Associate Member of the Institution of Nuclear Engineers

AMIOP Associate Member of the Institute of Printing

AMIPA Associate Member of the Institute of Practitioners in Advertising

AMIPE Associate Member of the Institution of Production Engineers

AMIPM Associate Member of the Institute of Personnel Management

AMIPRE Associate Member of the Incorporated Practitioners in Radio and Electronics

AMIProdE Associate Member of the Institution of Production Engineers

AMIQM Associate Member of the Institute of Quality Assurance

AMIRA Australian Mineral Industries Research Association

AMIRSE Associate Member of the Institute of Railway Signalling Engineers

AMIStructE Associate Member of the Institution of Structural Engineers

AMITA Associate Member of the Industrial Transport Association

AMITE (USA) Associate Member of the Institute of Traffic Engineers

AMIWEM Associate Member of the Institution of Water and Environmental Management

AMIWM Associate Member of the Institution of Works Managers

AMJ Assemblée mondiale de la jeunesse (French: World Assembly of Youth)

AML abandoned mine land • acute myelocytic (myeloid, *or* myelogenous) leukaemia • Admiralty materials laboratory • (New Zealand) applied mathematics laboratory

AMLS (USA) Master of Arts in Library Science

amm. ammunition

AMM *Physics* anomalous magnetic moment • antimissile missile • assistant marketing manager • Association médicale mondiale (French: World Medical Association)

AMMA (*or* **Amma**) ('æmə) Assistant Masters' and Mistresses' Association

AMMI American Merchant Marine Institute

amn ammunition

AMNECInst Associate Member of the North East Coast Institution of Engineers and Shipbuilders

AMNILP Associate Member of the National Institute of Licensing Practitioners

AMNZIE Associate Member of the New Zealand Institution of Engineers

AMO Air Ministry order • *Biochem.* anaerobic methane oxidation • area medical officer • assistant medical officer • Association of Magisterial Officers

AMOB automatic meteorological oceanographic buoy

AMORC Ancient Mystical Order Rosae Crucis (Rosicrucians)

amort. amortization

AMOS automatic meteorological observing station

amp. amperage • ampere • amplified • amplitude

AMP *Biochem.* adenosine monophosphate • Air Member for Personnel (in the RAF) • Associated Master Plumbers and Domestic Engineers • Australian Mutual Provident Society

AMPAS (USA) Academy of Motion Picture Arts and Sciences

AMPC (USA) automatic message processing center • auxiliary military pioneer corps

amph. amphibian • amphibious

AMPHIBEX ('æmfɪbɛks) amphibious exercise

ampl. amplifier

AMPS *Finance* auction market preferred stock • automatic message processing system

AMPSS advanced manned precision strike system

AMPTE *Astronomy* active magnetospheric particle tracer explorer

AMQ American medical qualification

AMR Atlantic missile range • automated meter reading • automatic message routing

Amraam ('æm,rɑːm) advanced medium-range air-to-air missile

AMRINA Associate Member of the Royal Institution of Naval Architects

Amrit. Amritsar

AMRO Association of Medical Record Officers

AMRS Air Ministry radio station

AMS *Physics* accelerator mass spectrometry (*or* spectrometer) • accident-monitoring system • (USA) Agricultural Marketing Service • American Mathematical Society • American Meteorological Society • American Microscopical Society • American Musicological Society • Ancient Monuments Society • *Physics* anisotropy of magnetic susceptibility • *Astronautics* Apollo mission simulator • army map service • army medical services (*or* staff) • Assistant Military Secretary • Australian medical services • automatic music search

AMSA advanced manned strategic aircraft

AMSAIEE Associate Member of the South African Institution of Electrical Engineering

Am. Sam. American Samoa

AMSAM antimissile surface-to-air missile

AMSE Associate Member of the Society of Engineers

AMSEF antiminesweeping explosive float

AMSERT Associate Member of the Society of Electronic and Radio Technicians

AMSGA Association of Manufacturers and Suppliers for the Graphic Arts

AMSL above mean sea level

AMSO Air Member for Supply and Organization (in the RAF) • Association of Market Survey Organizations

AMSST Associate of the Society of Surveying Technicians

Amst. Amsterdam

Amstrad ('æmstræd) Alan Michael Sugar Trading

AMSW (USA) Master of Arts in Social Work

amt amount

AMT Academy of Medicine, Toronto • (*or* **a.m.t.**) airmail transfer • Air Member for Training (in the RAF) • (USA) alternative minimum tax • area management team • Associate in Mechanical Technology • Associate in Medical Technology • Association of Marine Traders • (USA) Master of Arts in Teaching

AMTA Association of Multiple Travel Agents

AMTC Academic Member of the Trinity College of Music • Art Master's Teaching Certificate

AMTDA Agricultural Machinery Tractor Dealers' Association

AMTE Admiralty Marine Technology Establishment

AMTI airborne moving target indicator

AMTRI Advanced Manufacturing Technology Research Institute

AMTS Associate Member of the Television Society

amu (*or* **a.m.u.**) *Physics* atomic mass unit

AMU Associated Metalworkers' Union • (USA) Associated Midwestern Universities • Association of Master Upholsterers

AMUA (Australia) Associate of Music, University of Adelaide

AMus Associate in Music

AMusD Doctor of Musical Arts

AMusLCM Associate in Music, London College of Music

AMusTCL Associate in Music, Trinity College of Music, London

AMV Association mondiale vétérinaire (French: World Veterinary Association) • *Microbiol.* avian myeloblastosis virus

AMVAP Associated Manufacturers of Veterinary and Agricultural Products

AMVERS automated merchant vessel report system

AMVETS ('æmvɛts) American Veterans (of World War II and subsequent wars)

AMW average molecular weight

an. anno (Latin: in the year) • anonymous • answer

a.n. above named

An *Chem., symbol for* actinon

An. Annam

AN *Chem.* acid number • *Chem.* amino-1-naphthol • *Chem.* ammonium nitrate • Anglo-Norman • *airline flight code for* Ansett Australia • *Med.* antenatal • Associate in Nursing • *Physics* audible noise • *Med.* avascular necrosis • *British vehicle registration for* Reading

A/N advice note • alphanumeric

ANA All Nippon Airways • American Nature Association • American Neurological Association • American Newspaper Association • American Numismatic Association • American Nurses' Association • *Commerce* Article Number Association • (USA) Associate National Academician • Association of Nurse Administrators • Australian Natives' Association

anac. (*or* **anacr.**) anacreon

anaes. anaesthesia • anaesthetic

anag. anagram

anal. analogous • analogy • analyse • analysis • analytic

ANAPO (Colombia) Alianza Nacional Popular (Spanish: National Popular Alliance)

ANARE Australian National Antarctic Research Expedition

anat. anatomical • anatomy

anc. ancient

ANC *Chem.* acid-neutralizing capacity • advanced nuclear computer • African National Congress • Army Nurse Corps • Australian Newspapers Council

anch. anchored

ANCOM ('ænkɒm) Andean Common Market

anct. ancient

ANCUN Australian National Committee for the United Nations

and. *Music* andante (Italian; at a moderate speed)

And *Astronomy* Andromeda

AND *international vehicle registration for* Andorra • (ænd) *Computing, Electronics* and (from its sense in logic, as in **AND gate**, **AND operation**)

ANDB (USA) Air Navigation Development Board

ANDI ('ændɪ) *Med.* abnormal development and involution

ANEC American Nuclear Energy Council

ANECInst Associate of the North-East Coast Institution of Engineers and Shipbuilders

ANERI (USA) Advanced Nuclear Equipment Research Institute

anes. (USA) anesthetic

ANF (USA) Advanced Nuclear Fuels (Corporation) • antinuclear factor • Atlantic Nuclear Force • *Med.* atrial natriuretic factor • Australian National Flag Association

ang. angle • angular

Ang. Anglesey

ANG (USA) Air National Guard • Australian Newspaper Guild

ANGB (USA) Air National Guard Base

Angl. Angleterre (French: England) • Anglican • Anglicized

Anglo-Fr. Anglo-French

Anglo-Ind. Anglo-Indian

Anglo-Ir. Anglo-Irish

Anglo-L. Anglo-Latin

Anglo-Sax. Anglo-Saxon

ANGUS Air National Guard of the United States

Anh. Anhang (German: appendix; of a book, etc.)

anhyd. (*or* **anhydr.**) *Chem.* anhydrous

ANI (Portugal) Agência de Notícias de Informações (news agency) • (Uruguay) Agencia Nacional de Informaciones (national press agency) • (Portugal) Agência Nacional de Informações (national information agency) • American Nuclear Insurers

anim. *Music* animato (Italian: in a lively manner)

Ank. Ankunft (German: arrival)

ANL (ital.) Archaeological News Letter • (USA) Argonne National Laboratory • *Engineering* automatic noise limiting • National Library of Australia

Anm. Anmerkung (German: note)

ANM Admiralty Notices to Mariners

ann. annals • anni (Latin: years) • anno (Latin: in the year) • annual • annuity

ANN *Telecom.* all-figures numbers now • *Computing* artificial neural network

ANNA (USA) Army–Navy–NASA–Air Force satellite

anniv. anniversary

annot. annotate • annotation • annotator

annuit. annuitant

annul. annulment

Annunc. Annunciation

ANO (USA) Association of Nuclear Operators

anon. anonymous(ly)

ANOVA ('ænəʊvə) *Maths.* analysis of variance

ANP advanced nursing practice • aircraft nuclear propulsion • (Netherlands) Algemeen Nederlands Persbureau (independent news agency) • ammonium nitrate–phosphate (fertilizer) • Australian National Party

ANPA American Newspaper Publishers' Association • Australian National Publicity Association

anr another

anrac ('ænræk) aids navigation radio control

ANRC American National Red Cross • Australian National Research Council

ANRE advanced nuclear rocket engine

ANRPC Association of Natural Rubber Producing Countries

ans. answer

a.n.s. autograph note signed

ANS *Physics* advanced neutron source • American Nuclear Society • Army News Service • Army Nursing Service • *Computing* artificial neural system • Astronomical Netherlands Satellite • *Anatomy, Zoology* autonomic nervous system

ANSA (Italy) Agenzia Nazionale Stampa Associata (national press agency)

ANSI ('ænsɪ) American National Standards Institute

ANSI–SPARC American National Standards Institute/Systems Planning and Requirements Committee

ANSL Australian National Standards Laboratory

ANSP Academy of Natural Sciences of Philadelphia • Australian National Socialist Party

ANSR advanced neutron source reactor

ANSTI African Network of Scientific and Technological Institutions

ANSTO Australian Nuclear Science and Technology Organization

ant. antenna • anterior • antilog • antiquarian • antique • antiquity • antonym

Ant *Astronomy* Antlia

Ant. Antarctica • Antigua • Antrim

ANTA ('æntə) American National Theatre and Academy • Australian National Travel Association

Antarc. Antarctic

anthol. anthology

anthrop. (or **anthropol.**) anthropological • anthropology

Antig. Antigua

antilog ('æntɪˌlɒg) antilogarithm

antiq. antiquarian • antiquity

Ant. Lat. antique Latin

ant. ld. *Printing* antique laid (paper)

anton. antonym

ANTOR Association of National Tourist Office Representatives

Antr. Antrim

Ants. J. (ital.) Antiquaries Journal

ant. wo. *Printing* antique wove (paper)

ANU Australian National University

ANWR Arctic National Wildlife Refuge

anx annex

ANZ Australia and New Zealand Banking Group

ANZAAS ('ænzəs, -zæs) Australian and New Zealand Association for the Advancement of Science

Anzac ('ænzæk) Australian and New Zealand Army Corps (in World War I)

ANZAM ('ænzæm) Australia, New Zealand, and Malaysia (defence strategy for SE Asia)

ANZAMRS Australian and New Zealand Association for Medieval and Renaissance Studies

ANZCAN Australia, New Zealand, and Canada

ANZIA Associate of the New Zealand Institute of Architects

ANZIC Associate of the New Zealand Institute of Chemistry

ANZLA Associate of the New Zealand Library Association

ANZUK ('ænzək) Australian, New Zealand, and United Kingdom (defence force in Singapore and Malaysia)

ANZUS ('ænzəs) Australia, New Zealand, and the United States (referring to the security alliance between them)

aO (or **a/O**) an der Oder (German: on the River Oder; after place names)

AO Accountant Officer • *Chem.* acridine orange (dye) • Air Officer • air ordnance • (small caps.) anno ordinis (Latin: in the year of the order) • *Chem.* anthracene oil • area office • army order • *Physics* atomic orbital • Australian Opera • *airline flight code for* Aviaco • *British vehicle registration for* Carlisle • Officer of the Order of Australia

A/O (or **a/o**) *Accounting* account of • and others

AOA activity-on-arrow • (USA) Administration on Aging • Aerodrome Owners' Association • Air Officer in charge of Administration • American Ordnance Association • American Orthopedic Association • American Osteopathic Association • American Overseas Association

a.o.b. any other business • at or below

AOB advanced operational base • Antediluvian Order of Buffaloes • any other business

AOC Air Officer Commanding • (small caps.) anno orbis Conditi (Latin: in the year of the Creation) • *Chem.* anodic oxide coating • (or **AC**) appellation (d'origine) contrôlée (French: controlled place of origin; wine classification) • Army Ordnance Corps • Artists of Chelsea

AOCB any other competent business

AOC-in-C Air Officer Commanding-in-Chief

AOCM *US Air Force* aircraft out of commission for maintenance

AOD acousto-optic device • advanced ordnance depot • Ancient Order of Druids • Army Ordnance Department

AOER Army Officers' Emergency Reserve

AOF Afrique Occidentale Française (French: French West Africa) • Ancient Order of Foresters • *Chem.* anode oxide film • Australian Olympic Federation

A of F Admiral of the Fleet

AOFP absolute open-flow potential

A of S Academy of Science

AOG aircraft on ground

AOH Ancient Order of Hibernians

aoi angle of incidence

AoI aims of industry

a.o.i.v. *Engineering* automatically operated inlet valve

A-OK (USA) all (systems) OK

AOL (USA) absent over leave • Admiralty Oil Laboratory • *Computing* America Online • (USA) Atlantic Oceanographic Laboratories (Environmental Science Services Administration) • (Canada)

Atlantic Oceanographic Laboratory (Bedford Institute of Oceanography)

AOM amorphous organic matter

AON activity-on-node

AONB area of outstanding natural beauty

AOP Association of Optical Practitioners

AOPU Asian Oceanic Postal Union

AOQ average outgoing quality

AOQL average outgoing quality limit

aor angle of reflection

aor. aorist

a/or and/or

AOR acetate-oxidizing rod-shaped (bacterium) • *Music* adult-oriented rock • *Law* (USA) advice of rights • *Music* album-oriented rock • *Music* (USA) album-oriented radio

AOS *Astronautics* acquisition of signal • American Opera Society • American Ophthalmological Society • Ancient Order of Shepherds • automated office system

AOSIS Association of Small Island States

AOSM (New Zealand) Associate of the Otago School of Mines

AOSO advanced orbiting solar observatory

AOSS Fellow of the American Oriental Society (Latin *Americanae Orientalis Societatis Socius*)

AOSTRA Alberta Oil Sands Technology and Research Authority

AOSW Association of Official Shorthand Writers

AOT *Engineering* allowed outage (*or* out-of-service) time

AOU American Ornithologists' Union

ap. apothecary • apparent • apud (Latin: in the works of *or* according to)

a.p. above proof • additional premium • advanced post • *Med.* ante prandium (Latin: before a meal; in prescriptions) • *Maths.* arithmetical progression • author's proof

Ap. apostle • April

AP *Physics* acoustic plasmon • *Physiol.* action potential • (*or* **A/P**) *Insurance* additional premium • *Grammar* adjective phrase • aerosol particle • (USA) airplane • (USA) Air Police • air pollution • air publication (Ministry of Defence) • (Spain) Alianza Popular (Popular Alliance; political party) • *Chem.* alkaline permanganate • *Biochem.* alkaline phosphatase • (USA) American plan (of paying hotel bills) • *Chem.* ammonium

perchlorate • *Surveying* Amsterdamsch Peil (German: Amsterdam level; mean level used in the Netherlands, Belgium, and N Germany) • Andhra Pradesh • *Med.* angina pectoris • *Med.* anteroposterior • *Med.* anterior pituitary • antipersonnel • *Commerce* à protester (French: to be protested; on bills of exchange) • armour-piercing • *Computing* array processor • *Med.* arterial (blood) pressure • Associated Presbyterian • Associated Press • atomic power • (*or* **A/p**) authority to pay (*or* purchase) • automotive products • *Insurance* average payable • *British vehicle registration for* Brighton • *international civil aircraft marking for* Pakistan

APA *Taxation* additional personal allowance • Alaska Power Authority • (Australia) All Parties Administration • *Computing* all points addressable (as in **APA mode**) • American Philological Association • American Physicists Association • American Pilots Association • American Press Association • American Protestant Association • American Psychiatric Association • American Psychological Association • *Telecom.* annular phased array • Associate in Public Administration • Association for the Prevention of Addiction • Association of Public Analysts • Australian Physiotherapy Association • Austria Presse Agentur (Austrian Press Agency)

6-APA *Biochem.* 6-aminopenicillanic acid

APACL Asian People's Anti-Communist League

APACS Association for Payment Clearing Services

APAE Association of Public Address Engineers

APANZ Associate of the Public Accountants of New Zealand

apart. apartment

a-part. *Physics* alpha particle

APB Accounting Principles Board • all-points bulletin (police alert) • Auditing Practices Board

APBA American Power Boat Association

APBF Accredited Poultry Breeders' Federation

APC *Engineering* advanced process control • *Bacteriol.* aerobic plate count • air-pollution control • (Sierra Leone) All People's Congress • American Philatelic Congress • *Immunol.* antigen-presenting cell • Appalachian Power Company • armoured personnel carrier • Assistant Principal Chaplain • Associated Portland Cement • Auditing Practices Committee • automatic phase control • *Electronics* automatic power control • automatic public convenience

APCA (USA) Air Pollution Control Association • Anglo-Polish Catholic Association

APCIMS Association of Private Client Investment Managers and Stockbrokers

APCK Association for Promoting Christian Knowledge (in the Church of Ireland)

APCN (small caps.) anno post Christum natum (Latin: in the year after the birth of Christ)

APCO Association of Pleasure Craft Operators

APCOL All-Pakistan Confederation of Labour

apd approved

APD Administrative Planning Division • (USA) Air Pollution Division • Army Pay Department

APDC Apple and Pear Development Council

Ap. Deleg. Apostolic Delegate

AP/DOS advanced pick/disk operating system

APE Amalgamated Power Engineering • automatic photomapping equipment • *Physics* available potential energy

APEC ('eɪpɛk) Asia-Pacific Economic Cooperation Conference

APEX ('eɪpɛks) Advance-Purchase Excursion (reduced airline or rail fare) • Association of Professional, Executive, Clerical, and Computer Staff (now amalgamated with GMB)

APF Association for the Propagation of the Faith

APFC Asia-Pacific Forestry Commission

APFIM *Chem., Physics* atom-probe field-ion microscopy

APG *US Air Force* air proving ground

APGA American Public Gas Association

aph. aphorism

APH *Optics* adjusting, principal axis, hyperspherical • A(lan) P(atrick) Herbert (1890–1971, British writer and politician) • *Obstetrics* antepartum haemorrhage • *Biochem.* anterior pituitary hormone

APHA American Public Health Association

aphet. aphetic

APHIS ('eɪfɪs) (USA) Animal and Plant Health Inspection Service

API air-pollution index • air-position indicator • American Petroleum Institute (*see also* API scale) • *Computing* application programmer interface • Association phonétique internationale (French: International Phonetic Association) • atmospheric pressure ionization

APIS Army Photographic Intelligence Service

API scale American Petroleum Institute scale (for measuring the specific gravity of petroleum products)

ap. J-C (France) après Jésus-Christ (after Jesus Christ; AD)

Apl April

APL Alberta Power Limited • alternating polarization laser • (USA) Applied Physics Laboratory (Johns Hopkins University) • *Computing* A Programming Language

APLA ('æplə) Azanian People's Liberation Army

APLE Association of Public Lighting Engineers

APM (USA) Academy of Physical Medicine • airborne particulate matter • Assistant Paymaster • Assistant Provost-Marshal • *Astronomy* automatic plate-measuring machine

APMC Allied Political and Military Commission

APMG Assistant Postmaster-General

APMI Associate of the Pensions Management Institute

apmt appointment

APN (ex-USSR) Agentstvo Pechati Novosti (news press agency)

APNEC All-Pakistan Newspaper Employees' Confederation

apo. apogee

APO acting pilot officer • African People's Organization • Armed Forces (*or* Army) Post Office • Asian Productivity Organization

Apoc. Apocalypse • (*or* **Apocr.**) Apocrypha(l)

apog. apogee

apos. apostrophe

APOTA automatic positioning telemetering antenna

apoth. apothecary

app. apparatus • apparent(ly) • appeal • appended • appendix (of a book) • applied • appointed • apprentice • approved • approximate(ly)

App. Apostles

APP African People's Party (of Kenya; now part of KANU) • *Computing* application portability profile

APPA African Petroleum Producers Association

appar. apparatus • apparent(ly)

appd approved

APPES American Institute of Chemical Engineers Physical Properties Estimation System

APPITA (*or* **Appita**) ('æpɪtə) Australian Pulp and Paper Industries Technical Association

appl. appeal • appellant • applicable • applied

appos. appositive

appr. apprentice

APPR (USA) Army Package Power Reactor

appro ('æprəʊ) approbation • approval

approx. approximate(ly)

apps appendices

appt appoint • appointment

apptd appointed

APPU Australian Primary Producers' Union

appurts appurtenances

appx appendix (of a book)

Apr. April

APR Accredited Public Relations Practitioner • annualized percentage rate (of interest) • annual progress report • annual purchase rate (in hire-purchase schemes)

APRA Air Public Relations Association

APRC (small caps.) anno post Romam conditam (Latin: in the year after the foundation of Rome)

APRI Associate of the Plastics and Rubber Institute

A/Prin. Assistant Principal

apr. J-C (France) après Jésus-Christ (*see* ap. J-C)

APRS *Navigation* acoustic position reference system • Association for the Protection of Rural Scotland • Association of Professional Recording Studios

a.p.s. autograph poem signed

Aps *Astronomy* Apus

APS (Australia) Aborigines' Protection

Society • *Physics* advanced photon source • Advanced Photo System • American Peace Society • American Philatelic Society • American Philosophical Society • American Physics Society • American Physiological Society • American Protestant Society • *Chem., Physics* appearance-potential spectroscopy • Arizona Public Service (Company) • army postal service • Assistant Private Secretary • Associate of the Philosophical Society

APSA American Political Science Association • Associate of the Photographic Society of America • Australian Political Studies Association

APSE (æps) *Computing* Ada programming support environment

APSL Acting Paymaster Sublieutenant

APsSI Associate of the Psychological Society of Ireland

APST Association of Professional Scientists and Technologists

APSW Association of Psychiatric Social Workers

apt. apartment

APT advanced passenger train • advanced process technology • *Nautical* after peak tank • *Med.* alum-precipitated (diphtheria) toxoid • Association of Polytechnic Teachers • Association of Printing Technologists • Association of Private Traders • automatic picture transmission (from satellites)

APTC Army Physical Training Corps

APTI Association of Principals of Technical Institutions

APTIS all-purpose ticket-issuing systems

APTS *Astronautics* Automatic Picture Transmission Subsystem (in NASA)

APTU African Postal and Telecommunications Union

APU acute psychiatric unit (of a prison, etc.) • Arab Postal Union • *Education* Assessment of Performance Unit (body monitoring pupil performance) • *Aeronautics* auxiliary power unit

APUC Association for Promoting Unity of Christendom

APUD amine-precursor uptake and decarboxylation (in **APUD cell**)

APW *Optics* approximation of plane waves

APWA All Pakistan Women's Association • American Public Welfare (*or* Works) Association

APWR advanced pressurized water reactor

APWU American Postal Workers Union

aq. aqua (Latin: water) • aqueous

Aq. *Horticulture* aquatic

AQ *Psychol.* accomplishment (*or* achievement) quotient • *Military* Administration and Quartering • *airline flight code for* Aloha Airlines

aq. bull. *Pharmacol.* aqua bulliens (Latin: boiling water)

AQC Associate of Queen's College, London

aq. cal. *Pharmacol.* aqua calida (Latin: warm water)

aq. com. *Pharmacol.* aqua communis (Latin: tap water)

aq. dest. *Pharmacol.* aqua destillata (Latin: distilled water)

aq. ferv. *Pharmacol.* aqua fervens (Latin: hot water)

aq. frig. *Pharmacol.* aqua frigida (Latin: cold water)

AQI air-quality index

Aql *Astronomy* Aquila

AQL acceptable quality level

AQM *Physics* additive quark model

AQMG Assistant Quartermaster-General

aq. m.pip. *Pharmacol.* aqua menthae piperitae (Latin: peppermint water)

aq. pur. *Pharmacol.* aqua pura (Latin: pure water)

Aqr *Astronomy* Aquarius

aq. tep. *Pharmacol.* aqua tepida (Latin: tepid water)

aque. aqueduct

ar. arrival • arrive(s) *or* arrived

a.r. (*or* **a/r**, **A/r**) *Insurance* all risks • anno regni (Latin: in the year of the reign)

Ar *Chem., symbol for* argon • *Chem., symbol for* aryl group

Ar. Arabia(n) • Arabic • Aramaic

AR *Psychol.* accomplishment (*or* achievement) ratio • (*or* **A/R**) account receivable • acid resisting • acrylic rubber • *Med.* acute rejection • *Physiol.* adrenergic receptor (*or* adrenoceptor) • advice of receipt • *airline flight code for* Aerolineas Argentinas • (Poland) Agencja Robotnicza (Workers' Press Agency) • (USA) airman recruit • Altesse Royale (French: Royal Highness) • *Chem.* analytical reagent • *Physiol.* androgen receptor • Anna Regina (Latin: Queen Anne) • (ital.) Annual Register of World Events • annual report • *Taxation* annual return •

Med. aortic regurgitation • *Numismatics* argentum (Latin: silver) • *US postcode for* Arkansas • Army Regulations • *Image technol.* aspect ratio • Assistant Resident • Associated Rediffusion • Autonomous Region • Autonomous Republic • *British fishing port registration for* Ayr • *British vehicle registration for* Chelmsford

ARA Aircraft Research Association • American Railway Association • Army Rifle Association • Associate of the Royal Academy • (New Zealand) Auckland Regional Authority

Arab. Arabia(n) • Arabic

ARAC Associate of the Royal Agricultural College

arach. arachnology

ARACI Associate of the Royal Australian Chemical Institute

ARAD Associate of the Royal Academy of Dancing

ARAeS Associate of the Royal Aeronautical Society

ARAIA Associate of the Royal Australian Institute of Architects

Aram. Aramaic

ARAM Associate of the Royal Academy of Music

ARAMCO Arabian-American Oil Company

ARAS Associate of the Royal Astronomical Society • atomic resonance absorption spectroscopy

arb. arbiter • arbitrageur • arbitrary • arbitration • arbitrator

ARB *Civil aviation* Air Registration Board • Air Research Bureau

ARBA Associate of the Royal Society of British Artists

ARBE (Belgium) Académie Royale des Beaux-Arts, École Supérieure des Arts Décoratifs et École Supérieure d'Architecture de Bruxelles (Brussels Royal Academy of Fine Arts)

arbor. arboriculture

ARBS Associate of the Royal Society of British Sculptors

arc. *Music* arcato *or* coll'arco (Italian: with the bow)

ARC Aeronautical Research Council • Agricultural Research Council (now the AFRC) • *Med.* Aids-related complex • American Red Cross • (USA) Ames Research Center • *Optics* antireflection coating • Archaeological Resource Centre

(York) • Architects' Registration Council • Arthritis and Rheumatism Council • Astrophysical Research Consortium • Atlantic Research Corporation • *Physics* average resonance capture

ARCA (*or* **ARCamA**) Associate of the Royal Cambrian Academy • Associate of the Royal Canadian Academy (of Arts) • Associate of the Royal College of Art

arccos ('ɑːˌkɒs) *Maths.* arc (inverse) cosine

arccosec ('ɑːkəʊˌsɛk) *Maths.* arc (inverse) cosecant

arccot ('ɑːˌkɒt) *Maths.* arc (inverse) cotangent

arch (ɑːtʃ) *Maths.* arc (inverse) hyperbolic cosine

arch. archaic • archaism • archery • archipelago • architect • architectural • architecture

Arch. Archbishop • Archdeacon • Archduke

archaeol. archaeology

Archbp Archbishop

Archd. Archdeacon • Archduke

archit. architecture

archt. architect

ARCIC Anglican/Roman Catholic International Commission

ARCL *Chem., Physics* allowable residual contamination level

ARCM Associate of the Royal College of Music

Arcnet ('ɑːkˌnɛt) attached resource computing architecture network

ARCO Associate of the Royal College of Organists

ARCO(CHM) Associate of the Royal College of Organists with Diploma in Choir Training

arcos ('ɑːˌkɒs) *Maths.* arc (inverse) cosine

ARCOS Anglo-Russian Cooperative Society

arcosech ('ɑːkəʊˌsɛtʃ) arc (inverse) hyperbolic secant

arcosh ('ɑːˌkɒʃ) *Maths.* arc (inverse) hyperbolic cosine

arcoth ('ɑːˌkɒθ) *Maths.* arc (inverse) hyperbolic cotangent

ARCPsych Associate of the Royal College of Psychiatrists

ARCS Associate of the Royal College of Science • Associate of the Royal College of Surgeons (of England) • Australian Red Cross Society

arcsec ('ɑːkˌsɛk) *Maths.* arc (inverse) secant • arc second

arcsin ('ɑːkˌsaɪn) *Maths.* arc (inverse) sine

ARCST Associate of the Royal College of Science and Technology (Glasgow)

arctan ('ɑːkˌtæn) *Maths.* arc (inverse) tangent

ARCUK Architects' Registration Council of the United Kingdom

ARCVS Associate of the Royal College of Veterinary Surgeons

ARD acute radiation disease

ARDC (USA) Air Research and Development Command

ARDEC (USA) Army Research Development and Engineering Center

ARDMS automated route design and management system

ARDS *Med.* adult respiratory distress syndrome

ARE activated reactive evaporation • Admiralty Research Establishment • Arab Republic of Egypt • Associate of the Royal Society of Painter-Etchers and Engravers

AREI Associate of the Real Estate and Stock Institute of Australia

ARELS Association of Recognized English Language Schools

ARENA (əˈriːnə) (Brazil) Aliança Renovadora Nacional (Spanish: National Renewal Alliance)

ARF *Med.* acute renal failure • *Med.* acute respiratory failure • Advertising Research Foundation

Arg (*or* **arg**) *Biochem.* arginine

Arg. *Heraldry* argent • Argentina • Argentine (*or* Argentinian) • Argyll(shire)

a. Rh. (*or* **a/Rh, a.R.**) am Rhein (German: on the River Rhine; after place names)

ARHA Associate of the Royal Hibernian Academy of Painting, Sculpture, and Architecture

ARHS Associate of the Royal Horticultural Society

Ari *Astronomy* Aries

ARI acute respiratory infection

ARIAS Associate of the Royal Incorporation of Architects in Scotland

ARIBA Associate of the Royal Institute of British Architects (now Member; *see* RIBA)

ARICS Professional Associate of the Royal Institution of Chartered Surveyors

ARIEL ('ɛərɪəl) Automated Real-time

Investments Exchange Limited (former computerized share-dealing system)

ARIMA *Computing* autoregressive integrated moving average

ARINA Associate of the Royal Institution of Naval Architects

ARIPHH Associate of the Royal Institute of Public Health and Hygiene

ARIS advanced range instrumentation ship

Arist. Aristotle (384–322 BC, Greek philosopher)

Aristoph. Aristophanes (*c.* 445–*c.* 380 BC, Greek dramatist)

arith. arithmetic(al) • arithmetician

Ariz. Arizona

Ark. Arkansas

ARL Admiralty Research Laboratory • Aeronautical Research Laboratory • Arctic Research Laboratory • (USA) Association of Research Libraries • Australian Rugby

ARLL *Computing* advanced run length limited

Arm (ɑːm) *Computing* Acorn RISC machine

Arm. Armagh • Armenia(n) • Armoric(an)

ArM Master of Architecture (Latin *Architecturae Magister*)

ARM (ɑːm) (USA) adjustable-rate mortgage • Alliance réformée mondiale (French: Worldwide Presbyterian Alliance) • antiradar missile • antiradiation missile • *Obstetrics* artificial rupture of membranes • atomic resolution microscope • Australian Republican Movement

ARMA Association of Residential Managing Agents • *Computing* autoregressive moving average

ARMCM Associate of the Royal Manchester College of Music

armd armoured

ARMET ('ɑːmɛt) *Meteorol.* area forecast for upper wind and temperature

ARMIT Associate of the Royal Melbourne Institute of Technology

ARMS (ɑːmz) Action for Research into Multiple Sclerosis • Associate of the Royal Society of Miniature Painters

ARNA Arab Revolution News Agency

ARO army routine order • Asian Regional Organization • Associate Member of the Register of Osteopaths

AROD airborne remote operated device

AROS African Regional Organization for Standardization

arp. *Music* arpeggio

ARP *Finance* (USA) adjustable rate preferred (stock) • air-raid precautions • Associated Reformed Presbyterian

ARPA ('ɑːpə) Advanced Research Projects Agency (*see* DARPA)

ARPANET (*or* **Arpanet**) ('ɑːpəˌnɛt) *Computing* Advanced Research Projects Agency Network

ARPO Association of Resort Publicity Officers

ARPS Associate of the Royal Photographic Society • Association of Railway Preservation Societies

arr. *Music* arrangement • *Music* arranged (by) • arrival • arrive(s) • (*or* **arrd**) arrived

ARR accounting rate of return • (small caps.) anno regni Regis *or* Reginae (Latin: in the year of the King's (*or* Queen's) reign) • Association of Radiation Research

ARRC Associate of the Royal Red Cross

arron. (France) arrondissement (administrative district)

ARRS American Roentgen Ray Society

ARS *Med.* acute radiation syndrome • (USA) Agricultural Research Service • American Records Society • American Recreation Society • American Rocket Society • (small caps.) anno reparatae salutis (Latin: in the year of our redemption) • *Med.* aortic regurgitation and stenosis • Army Radio School

ARSA Associate of the Royal Scottish Academy • Associate of the Royal Society of Arts

ARSAP ('ɑːsæp) (USA) Advanced Reactor Severe Accident Program

ARSCM Associate of the Royal School of Church Music

arsech ('ɑːˌsɛtʃ) *Maths.* arc (inverse) hyperbolic secant

arsh (ɑːʃ) *Maths.* arc (inverse) hyperbolic sine

ARSH Associate of the Royal Society for the Promotion of Health

arsinh ('ɑːˌsaɪn) *Maths.* arc (inverse) hyperbolic sine

ARSL Associate of the Royal Society of Literature

ARSM Associate of the Royal School of Mines

ARSR air route surveillance radar

ARSW Associate of the Royal Scottish Society of Painting in Watercolours

art. article • artificer • artificial • artillery • artist

Art. Artemis (Greek goddess)

ART *Computing* algebraic reconstruction technique • *Computing* automated reasoning tool

artanh ('ɑː,tæn) *Maths.* arc (inverse) hyperbolic tangent

ARTC air route traffic control

arth (ɑːθ) *Maths.* arc (inverse) hyperbolic tangent

artif. artificer

art. pf. artist's proof

arty artillery

ARU American Railway Union • *Computing* audio response unit

ARV *Bible* American (Standard) Revised Version

ARVA Associate of the Rating and Valuation Association

ARVIA Associate of the Royal Victoria Institute of Architects

ARWA Associate of the Royal West of England Academy

ARWS Associate of the Royal Society of Painters in Water-Colours

ARXPS *Chem., Physics* angle-resolved X-ray photoelectron spectroscopy

As (ital.) *Meteorol.* altostratus • *Chem., symbol for* arsenic • *Chem.* asphaltene

As. Asia(n) • Asiatic

AS Academy of Science • Admiral Superintendent • admittance spectroscopy • *Education* Advanced Supplementary (in **AS level**) • air speed • air staff • *airline flight code for* Alaska Airlines • *Insurance* all sections • *Music* al segno (Italian: to (or àt) the sign) • *Meteorol.* altostratus • *Geology* aluminium silicate • American Samoa • *Computing* analogue states • (or **A-S**) Anglo-Saxon • (small caps.) anno salutis (Latin: in the year of salvation) • (small caps.) anno Salvatoris (Latin: in the year of the Saviour) • *Chem.* anodic stripping • antisubmarine • Assistant Secretary • assistant surgeon • Associate in Science • *British vehicle registration for* Inverness • *Taxation, symbol for* personal allowance

A/S account sales • *Education* Advanced Supplementary (in **A/S level**) • *Banking* after sight • (Denmark) Aktieselskab (joint-stock company) • (Norway) Aktjeselskap (limited company; Ltd) • alongside

ASA Acoustic Society of America • (USA) Addiction Services Agency • Advertising Standards Authority • Amateur Swimming Association • American Standards Association • American Statistical Association • Army Sailing Association • Associate Member of the Society of Actuaries • (USA) Associate of the Society of Actuaries • Australian Society of Accountants

ASAA Associate of the Society of Incorporated Accountants and Auditors

ASAB Association for the Study of Animal Behaviour

ASAI Associate of the Society of Architectural Illustrators

ASAM Associate of the Society of Art Masters

AS&TS of SA Associated Scientific and Technical Societies of South Africa

a.s.a.p. as soon as possible

ASAP automated shipboard aerological programme

ASAT (or **Asat**) ('eɪsæt) *Military* antisatellite (interceptor)

asb. asbestos

a.s.b. aircraft safety beacon

ASB Accounting Standards Board • (South Africa) Afrikaans Studentebond (students' union) • *Church of England* Alternative Service Book • American Society of Bacteriologists

ASBAH Association for Spina Bifida and Hydrocephalus

ASBM air-to-surface ballistic missile

ASBSBSW Amalgamated Society of Boilermakers, Shipwrights, Blacksmiths, and Structural Workers

asc. ascend (or ascent)

Asc. *Astrology* Ascendant

ASc Associate in Science

ASC Accounting Standards Committee • Administrative Staff College, Henley • (USA) Air Service Command • altered state of consciousness • American Society of Cinematographers • Anglo-Soviet Committee • Asian Socialist Conference

ASCA Associate of the Society of Company and Commercial Accountants

ASCAB Armed Services Consultant Approval Board

ASCAP ('æskæp) American Society of Composers, Authors, and Publishers

ASCC Accounting Standards Steering Committee • Association of Scottish

Climbing Clubs • *Computing* Automatic Sequence Controlled Calculator

ASCE American Society of Civil Engineers

ASCII ('æskɪ) *Computing* American Standard Code for Information Interchange

ASCM Australian Student Christian Movement

ASCS *Engineering* advanced Stirling conversion system

ASCU (USA) Association of State Colleges and Universities

ASD *Engineering* adjustable speed drive • Admiralty Salvage Department • *Engineering* antislip drive • Armament Supply Department • *Med.* atrial septal defect

ASDAR ('æzdɑː) aircraft-to-satellite data relay

ASDC Associate of the Society of Dyers and Colourists

a/s de aux soins de (French: care of, c/o; in postal addresses)

ASDE Airport Surface Detection Equipment

Asdic (*or* **ASDIC**) ('æzdɪk) Allied Submarine Detection Investigation Committee • (USA) Armed Services Documents Intelligence Center

ASE Admiralty Signal Department • American Stock Exchange • *Physics* amplified spontaneous emission • Army School of Education • Associate of the Society of Engineers • Association for Science Education • automotive Stirling engine

ASEA Association of South East Asia

ASEAN ('æsɪˌæn) Association of South East Asian Nations

ASEC (USA) Applied Solar Energy Corporation

ASEE American Society for Engineering Education • Association of Supervisory and Executive Engineers

ASF *Chem.* acid-soluble fraction • *Computing* aspect source flag • Associate of the Institute of Shipping and Forwarding Agents

ASG Acting (*or* Assistant) Secretary-General

ASGB Aeronautical Society of Great Britain • Anthroposophical Society in Great Britain

ASGBI Anatomical Society of Great Britain and Ireland

asgd assigned

asgmt assignment

ASH (æʃ) Action on Smoking and Health

ashp airship

ASHRAE American Society of Heating, Refrigeration, and Air-Conditioning Engineers

ASI airspeed indicator • Association soroptimiste internationale (French: Soroptimist International Association)

ASIA Airlines Staff International Association

ASIAD Associate of the Society of Industrial Artists and Designers

ASIA(Ed) Associate of the Society of Industrial Artists (Education)

ASIC ('æsɪk) *Electronics* application-specific integrated circuit

ASIF Amateur Swimming International Federation

ASIO Australian Security Intelligence Organization

ASIP *Astronomy* all-sky imaging photometer

ASIRC (USA) Aquatic Sciences Information Retrieval Center (Rhode Island)

ASIS *Astronautics* abort sensing and implementation system • American Society for Information Science

a.s.l. above sea level

ASL Acting Sublieutenant • Advanced Student in Law • American Sign Language • American Soccer League • assistant scout leader

ASLA American Society of Landscape Architects

ASLB (USA) Atomic Safety and Licensing Board

ASLE American Society of Lubrication Engineers

ASLEF (*or* **Aslef**) ('æzlɛf) Associated Society of Locomotive Engineers and Firemen

ASLEP *Astronautics* Apollo surface lunar experiments package

A/S level (*or* **AS level**) *Education* Advanced Supplementary level

ASLIB (*or* **Aslib**) ('æzlɪb) Association for Information Management (formerly Association of Special Libraries and Information Bureaux)

ASLO American Society of Limnology and Oceanography • Australian Scientific Liaison Office

ASLP Amalgamated Society of Lithographic Printers

ASLW Amalgamated Society of Leather Workers

ASM Acting Sergeant-Major • air-to-surface missile • *Computing* algorithmic state machine • American Society for Metals • assistant sales manager • assistant scoutmaster • assistant stage manager • assistant station master • Association of Senior Members

ASME American Society of Mechanical Engineers • Association for the Study of Medical Education

As. Mem. Associate Member

ASMO Arab Organization for Standardization and Metrology

ASMP American Society of Magazine Photographers

Asn (*or* **asn**) *Biochem.* asparagine

ASN army service number • average sample number

ASN. 1 *Computing* abstract syntax notation 1

ASNE American Society of Newspaper Editors

ASO Air Staff Officer • American Symphony Orchestra • area supplies officer

ASOS automatic storm observation service

Asp (*or* **asp**) *Biochem.* aspartic acid

ASP *Commerce* accepté sous protêt (French: accepted under protest) • *US Air Force* aerospace plane • African Special Project (of the IUCN) • (Tanzania) Afro-Shirazi Party • American selling price • (æsp) (USA) Anglo-Saxon Protestant • Astronomical Society of the Pacific

ASPA Australian Sugar Producers' Association

ASPAC ('æz,pæk) Asian and Pacific Council

ASPC *Commerce* accepté sous protêt, pour compté (French: accepted under protest for account) • Association of Swimming Pool Contractors

ASPCA American Society for Prevention of Cruelty to Animals

ASPEP Association of Scientists and Professional Engineering Personnel

ASPF Association of Superannuation and Pension Funds

ASPI *Computing* advanced SCSI programming interface

ASR airport surveillance radar • air-sea rescue • *Meteorol.* altimeter setting region • *Computing* automatic send and receive

ASRE Admiralty Signal and Radar Establishment

A/SRS air-sea rescue service

ass. assembly • assistant • association • assurance

ASS automatic space station

ASSC Accounting Standards Steering Committee

Ass-Com-Gen Assistant-Commissary-General

ASSET ('æset) *US Air Force* aerothermodynamic-elastic structural systems environmental tests

ASSGB Association of Ski Schools in Great Britain

AssIE Associate of the Institute of Engineers and Technicians

assigt assignment

assim. assimilate

assmt assessment

assn association

assoc. associate(d) • association

AssocEng Associate of Engineering

AssocIMinE Associate of the Institution of Mining Engineers

AssocISI Associate of the Iron and Steel Institute

AssocMCT Associateship of Manchester College of Technology

AssocMIAeE Associate Member of the Institution of Aeronautical Engineers

assocn association

AssocSc Associate in Science

ASSR Autonomous Soviet Socialist Republic

asst (*or* **ass/t**) assistant

asstd assorted

ASSU American Sunday School Union

assy assembly

Assyr. Assyrian

AST above-ground storage tank • air service training • *Biochem., Med.* aspartate aminotransferase • assured shorthold tenancy • Atlantic Standard Time • automated screen trading

ASTA American Society of Travel Agents • (New Zealand) Auckland Science Teachers' Association

ASTC Administrative Service Training Course • Associate of the Sydney Technical College

ASTIA (USA) Armed Services Technical Information Agency

ASTM American Society for Testing Materials

ASTMS ('æztɛmz *or* 'eɪ 'ɛs 'tiː 'ɛm 'ɛs) Association of Scientific, Technical, and Managerial Staffs (now part of the MSF)

ASTOR antisubmarine torpedo ordnance rocket

ASTP Apollo-Soyuz test project

astr. astronomer • astronomical • astronomy

astro. astronautics • astronomer • astronomy

ASTRO (USA) Air Space Travel Research Organization

astrol. astrologer • astrological • astrology

astron. astronomer • astronomical • astronomy

astrophys. astrophysical

Ast. T. astronomical time

ASTTL *Computing* advanced Schottky transistor transistor logic

ASU American Students Union • Arab Socialist Union

ASUA Amateur Swimming Union of the Americas

ASV aircraft-to-surface vessel • *Bible* American Standard Version • *Microbiol.* avian sarcoma virus

ASVA Associate of the Incorporated Society of Valuers and Auctioneers

ASVU Army Security Vetting Unit

ASW Amalgamated Society of Wood Workers • antisubmarine warfare • antisubmarine work • Association of Scientific Workers • Association of Social Workers • *Physics* augmented spherical wave

ASWDU Air Sea Warfare Development Unit

ASWE Admiralty Surface Weapons Establishment

at. atmosphere • atomic • *Navigation* attitude • attorney

At *symbol for* ampere-turn (unit) • *Chem., symbol for* astatine • (æt) *Colloquial* (a member of the) Auxiliary Territorial Service

AT *Physics* accelerator technology • achievement test • Advanced Technologies (as in **IBM-AT**) • *Immunol.* agglutination test • *Biochem.* alkyl transferase • alternative technology • *Chem.* amino-1,2,4-triazole • (ital.) Angling Times • antitank • apparent time • *Physics* appearance time • (USA) appropriate

technology • arrival time • *Med.* ataxia telangiectasia • Atlantic Time • *Education* attainment target • *Computing* attention Australia Telescope • *British vehicle registration for* Hull • *airline flight code for* Royal Air Maroc

A/T American terms

ATA *Nuclear physics* advanced test accelerator • Air Transport Association • Air Transport Auxiliary • American Translators' Association • Amusement Trades Association • Animal Technicians' Association • Associate Technical Aide • *Computing* Advanced Technologies attachment • Atlantic Treaty Association

ATAC Air Transport Advisory Council

ATAE Association of Tutors in Adult Education

ATAF Allied Tactical Air Force

ATAM Association for Teaching Aids in Mathematics

AT&T American Telephone and Telegraph Company

Atapi Advanced Technologies attachment packet interface

ATAS Air Transport Auxiliary Service

ATB advanced technology bomber • all-terrain bike • at the time of the bomb (*or* bombing)

ATBC acetyltributyl citrate (a plasticizer)

ATBM antitactical ballistic missile

ATC acid-treated coal • air-traffic control • Air Training Command • Air Training Corps • Air Transport Command • *Law* Annotated Tax Cases • Art Teacher's Certificate • *Computing* authorization to copy (of software) • automatic temperature control • automatic train control

ATCC air traffic control centre

ATCE *Astronautics* ablative thrust chamber engine

atchd attached

ATCL Associate of Trinity College of Music, London

ATCO Air Traffic Control Officer

ATCRBS Air Traffic Control Radar Beacon System

ATCSP Association of Teachers of the Chartered Society of Physiotherapists

ATD actual time of departure • advanced technology development • *Physics* alpha-track detector • Art Teacher's Diploma • Australian Tax Decisions

ATDS Association of Teachers of Domestic Science

ATE Amusement Trades Exhibition • Automatic Telephone and Electric Company • *Electronics* automatic test equipment

ATEC Air Transport Electronics Council

ATEE *Chem.* N-acetyl-L-tyrosine ethyl ester

a tem. *Music* a tempo (Italian: in time)

ATF *Nuclear physics* accelerator test facility • *Nuclear physics* advanced toroidal facility

ATFS Association of Track and Field Statisticians

ATG *Med.* antithymocyte globulin

ath. athlete • athletic

ATHE Association of Teachers in Higher Education

Athen. Athenian

athl. athlete • athletic

ATI Associate of the Textile Institute • Association of Technical Institutions

ATII Associate of the Institute of Taxation

Atl. Atlantic

ATL *Insurance* actual total loss • *Med.* adult T-cell leukaemia • *Computing* automated (*or* automatic) tape library

ATLAS ('ætləs) *Astronomy* airborne tunable laser absorption spectrometer • (USA) Argonne tandem-linac accelerator system

ATLB Air Transport Licensing Board

ATLEED *Physics, Chem.* automated tensor low-energy electron diffraction

ATLS *Med.* advanced trauma life support

ATLV adult T-cell leukaemia virus

atm *symbol for* atmosphere (unit of pressure)

atm. atmospheric

ATM *Computing* Adobe Type Manager • air training memorandum • antitank missile • approved testing material • Association of Teachers of Management • Association of Teachers of Mathematics • *Computing* asynchronous transfer mode • *Banking* automated teller machine

ATMS assumption-based truth maintenance system

ATN *Med.* acute tubular necrosis • *Maths.* arc tangent • *Telecom.* augmented transition network

ATNA Australasian Trained Nurses' Association

ATNF Australia Telescope National Facility

at. no. atomic number

ATO (*or* **ato**) Ammunition Technical Officer (bomb-disposal officer in the RAOC) • *Aeronautics* assisted takeoff

A to A air-to-air

A to J (New Zealand) Appendices to Journals (of the House of Representatives or Parliament)

ATOL ('ætɒl) Air Travel Organizers' Licence

ATP *Biochem.* adenosine triphosphate • *Aeronautics* advanced turboprop • *Commerce* aid trade provision • Air Technical Publications • Associated Theatre Properties • Association of Tennis Professionals • *Railways* automatic train protection

ATPAS Association of Teachers of Printing and Allied Subjects

ATPC Association of Tin Producing Countries

ATPG automatic test-pattern generation

ATPL(A) Airline Transport Pilot's Licence (Aeroplanes)

ATPL(H) Airline Transport Pilot's Licence (Helicopters)

ATR *Nuclear physics* advanced test reactor • *Nuclear physics* advanced thermal reactor • *Aeronautics* air turbo-ram jet engine • *Radar* antitransmit–receive (in **ATR tube**) • Association of Teachers of Russian • *Military* automatic target recognition

ATRAN automatic terrain recognition and navigation

a.t.r.i.m.a. *Law* as their respective interests may appear

a.t.s. *Law* at the suit of

ATS (Australia) Amalgamated Television Services • American Temperance Society • American Tract Society • American Transport Service • *Med.* antitetanus serum • (USA) Army Transport Service • Associate of Theological Study • automated trade system • Auxiliary Territorial Service (in World War II)

ATSC Associate of the Tonic Sol-Fa College

ATSDR (USA) Agency for Toxic Substances and Disease Registry

ATSIS *Computing* automated technical-specification information system

ATSS Association for the Teaching of Social Sciences

att. attached • attention • attorney

ATT antitetanus toxoid · Associate of the Association of Tax Technicians

Att-Gen. Attorney-General

attn attention · for the attention of

attrib. attribute · attributed (to) · attribution · attributive(ly)

atty. (*or* **Atty.**) attorney

Atty-Gen. Attorney-General

ATU (USA) Amalgamated Transit Union

ATUC African Trade Union Confederation

ATV all-terrain vehicle · Associated Television

at. wt. atomic weight

Au (*or* **au**) author · *Chem., symbol for* gold (Latin *aurum*)

AU (USA) Actors' Union · *Printing* all up (all set in type) · angstrom unit · arithmetic unit · astronomical unit · *British vehicle registration for* Nottingham

AUA *Finance* agricultural unit of account (in the EU) · American Unitarian Association · American Urological Association

AUBC Association of Universities of the British Commonwealth

AUBTW Amalgamated Union of Building Trade Workers

AUC ab urbe condita *or* anno urbis conditae (Latin: (in the year) from the founding of the city; indicating years numbered from the founding of Rome) · *Chem.* ammonium uranyl carbonate · *Maths.* area under the curve (on a graph) · Association of Underwater Contractors · Australian Universities Commission

AUCAS Association of University Clinical Academic Staff

auct. *Botany* auctorum (Latin: of (other) authors; indicating that the specific name of a plant is that given by authorities other than Linnaeus)

aud. audit · auditor

Aud-Gen. Auditor-General

AUEW Amalgamated Union of Engineering Workers (*see under* AEU)

Aufl. Auflage (German: edition)

AUFW Amalgamated Union of Foundry Workers

aug. *Grammar* augmentative · augmented

Aug. August

augm. *Grammar* augmentative

AUI *Computing* attachment unit interface

AULLA (ˈaʊlə) Australasian Universities Language and Literature Association

AUM air-to-underwater missile

AUMLA Australian Universities Modern Language Association

a.u.n. absque ulla nota (Latin: with no identifying mark)

AUO African Unity Organization

AUP Aberdeen University Press · acceptable use policy

Aur *Astronomy* Auriga

Aus. Australia(n) · Austria(n)

AUS Army of the United States · Assistant Undersecretary · *international vehicle registration for* Australia

AUSA Association of the United States Army

Ausg. Ausgabe (German: edition)

Aust. Australia(n) · Austria(n)

Austral. Australasia · Australia(n)

aut. autograph · autumn

Aut. Autriche (French: Austria)

AUT *Chem.* ammonium uranyl tricarbonate · Association of University Teachers

AUTA Association of University Teachers of Accounting

AUTEC *US Navy* Atlantic Underwater Test Evaluation Center

auth. authentic · author(ess) · authority · authorize(d)

Auth. Ver. Authorized Version (of the Bible)

Autif Association of Unit Trusts and Investment Funds

auto. automatic · automobile · automotive

autobiog. autobiographical · autobiography

autog. autograph

AUT(S) Association of University Teachers (Scotland)

a.u.w. all-up-weight

AUWE Admiralty Underwater Weapons Establishment

aux. auxiliary

AUX *Linguistics* auxiliary verb

av. avenue · average · avoirdupois · avril (French: April)

a.v. (*or* **a/v**) *Finance* ad valorem (Latin: according to value) · annos vixit (Latin: (he *or* she) lived (so many) years) · *Finance* asset value

Av. Avenue · Avocat (French: lawyer)

AV *Chem.* acid value · (*or* **A/V**) *Finance* ad valorem (Latin: according to value) · *Photog.* aperture value · Artillery Volunteers · *Med.* atrioventricular (as in **AV**

bundle, AV node) • audiovisual • *Numismatics* aurum (Latin: gold) • Authorized Version (of the Bible) • average value • *airline flight code for* Avianca • *British vehicle registration for* Peterborough

AVA (Canada) Alberta Veterinary Association • Amateur Volleyball Association of Great Britain • audiovisual aids • Audiovisual Association • Australian Veterinary Association

a.v.c. *Electronics* automatic volume control

av.C. (Italy) avanti Cristo (before Christ; BC)

AVC additional voluntary contribution (in pension schemes) • American Veterans' Committee • *Electronics* automatic volume control

Av. Cert. Aviator's Certificate

AVCM Associate of Victoria College of Music

AVCO *Accounting* average cost

AVD Army Veterinary Department

avdp. avoirdupois

ave. (*or* **Ave.**) avenue • average

AVF *Military* (USA) all-volunteer force

avg. average

AVGAS aviation gasoline (of high octane for piston-type engine)

avge. *Cricket* average

AVI Association of Veterinary Inspectors • audio-video interleaved

avia. aviation

av. J-C (France) avant Jésus-Christ (before Jesus Christ; BC)

AVL *Computing* Adel'son-Vel'skii–Landis (as in **AVL tree**) • *Transport* automatic vehicle location

AVLA Audio Visual Language Association

AVLIS atomic vapour laser isotope separation

AVM Air Vice-Marshal • *Med.* arteriovenous malformation

AVMA Action for the Victims of Medical Accidents • Automatic Vending Machine Association

AVMS administered vertical marketing system

avn aviation

AVN *Med.* atrioventricular node

AVO Administrative Veterinary Officer • *Transport* average vehicle occupancy

avoir. avoirdupois

AVR Army Volunteer Reserve

AVRI Animal Virus Research Institute

AVRO ('ævrəʊ) *Aeronautics* A. V. Roe and Co.

AVRP audiovisual recording and presentation

AVS Anti-Vivisection Society

AVSL assistant venture scout leader

AVTRW Association of Veterinary Teachers and Research Workers

a.w. actual weight • *Shipping* all water • atomic weight

AW added water (in foodstuffs, etc.) • Alfred Wainwright (1907–91, British walker and writer) • *Physics* Alfvén wave(s) • Armstrong Whitworth (aircraft) • Articles of War • atomic warfare • *British vehicle registration for* Shrewsbury

A/W actual weight • airworthy • artwork

AWA Amalgamated Wireless (Australasia) Ltd

AWACS (*or* **Awacs**) ('eɪwæks) airborne warning and control system

AWAM Association of West African Merchants

AWAS Australian Women's Army Service

AWASM Associate of the Western Australia School of Mines

AWB (South Africa) Afrikaner Weerstandsbeweging (right-wing political party) • Agricultural Wages Board • (USA) air waybill • Australian Wool Board

AWBA American World's Boxing Association

AWC Allied Works Council • (USA) Army War College • Australian Wool Corporation

AWE Atomic Weapons Establishment

AWEA American Wind Energy Association

AWeldI Associate of the Welding Institute

AWG American Wire Gauge • Art Workers' Guild

AWH *Physics* Alfvén wave heating

AWHA Australian Women's Home Army

AWJ *Engineering, Building trades* abrasive water jet

AWL (*or* **awl**) absent with leave

AWMC Association of Workers for Maladjusted Children

AWNL Australian Women's National League

AWO American Waterways Operators, Inc • Association of Water Officers

AWOL (*or* **awol**) ('eɪwɒl) absent without (official) leave

AWP amusements with prizes (forms of minor gaming in British Gaming Acts) • annual wood production

AWPR Association of Women in Public Relations

AWR Association for the Study of the World Refugee Problem

AWRA Australian Wool Realization Agency

AWRE Atomic Weapons Research Establishment

aws Graduate of Air Warfare Course

AWS *Geology* acoustic well sounding • Agricultural Wholesale Society • American Welding Society • automatic warning system

AWSA American Water Ski Association

AWU Australian Workers' Union

AWW (South Africa) Afrikaner Weerstandsbeweging (right-wing political party)

ax. axiom

AX *British vehicle registration for* Cardiff

AXAF Advanced X-ray Astrophysical Facility (at NASA)

AY *airline flight code for* Finnair • *British vehicle registration for* Leicester

AYH American Youth Hostels

AYLI (ital.) As You Like It (Shakespeare)

AYM *Freemasonry* Ancient York Mason

Ayr. Ayrshire

az. azimuth • azure

AZ *airline flight code for* Alitalia • *US postcode for* Arizona • *international vehicle registration for* Azerbaijan • *British vehicle registration for* Belfast

Azapo (ə'zɑ:pəʊ) (South Africa) Azanian People's Organization

az. ld. *Printing* azure laid (paper)

Azo. Azores

AZT azidothymidine (drug used in treating Aids)

az. wo. *Printing* azure wove (paper)

B

b *Physics, symbol for* barn • *Physics* bottom (a quark flavour) • *indicating* the second vertical row of squares from the left on a chessboard

b. bag • bale • ball • base • *Music* bass (*or* basso) • bath • batsman • bay • beam • bedroom • before • billion • bis (Latin: twice) • bitch • bloody (euphemism) • book • born • bound • *Med.* bowels • *Cricket* bowled by • breadth • brother • bugger (euphemism) • bust • by • *Cricket* bye

B *symbol for* baht (Thai monetary unit) • *symbol for* balboa (Panamanian monetary unit) • *Irish fishing port registration for* Ballina • *Physics, symbol for* baryon number • *Chem., symbol for* base-catalysed reaction • Baumé (temperature scale) • *British fishing port registration for* Belfast • *international vehicle registration for* Belgium • best (quality of wrought iron) • *postcode for* Birmingham • *Chess, symbol*

for bishop • black (indicating the degree of softness of lead in a pencil; also in **BB** (*or* **2B**), double black, etc.) • *symbol for* bolívar (Venezuelan monetary unit; *see also under* Bs) • (USA) bomber (as in **B-52**) • *Chem., symbol for* boron • breathalyzer (as in **B-test**) • *Photog.* B-setting • *Immunol.* bursa of Fabricius *or* bone marrow (in **B cell** *or* **lymphocyte**) • *international civil aircraft marking for* China *or* Taiwan • *symbol for* eleven (in hexadecimal notation) • (bold ital.) *Physics, symbol for* magnetic flux density • *symbol for* secondary road (as in **B405**, etc.) • (ital.) *Physics, symbol for* susceptance • *indicating* administrative, professional (occupational grade) • *Med., indicating* a blood group or its associated antigen (*see also under* ABO) • *indicating* a musical note or key • *Astronomy, indicating* a spectral type • *indicating* a standard paper size (*see under* B0–B10) • *indicating* something of secondary importance or interest (as in **B-film**, **B-road**)

B. (*or* **B**) Bachelor (in academic degrees) • Baptist • Baron • *Music* bass (*or* basso) • battle • *Cartography* bay • Beatus (Latin: blessed) • Benediction • Bey • Bible • billion • bishop • Blessed • blue • board • boatswain • book • breadth • British • brotherhood • building

B0 *indicating* a standard paper size, 1000 × 1414 mm

B1 *indicating* a standard paper size, 707 × 1000 mm

B2 *indicating* a standard paper size, 500 × 707 mm

B3 *indicating* a standard paper size, 353 × 500 mm

B4 *indicating* a standard paper size, 250 × 353 mm

B5 *indicating* a standard paper size, 176 × 250 mm

B6 *indicating* a standard paper size, 125 × 176 mm

B7 *indicating* a standard paper size, 88 × 125 mm

B8 *indicating* a standard paper size, 62 × 88 mm

B9 *airline flight code for* Caribair • *indicating* a standard paper size, 44 × 62 mm

B10 *indicating* a standard paper size, 31 × 44 mm

3B *international civil aircraft marking for* Mauritius

5B *international civil aircraft marking for* Cyprus

b.a. *Taxation* balancing allowance • blind approach

Ba *Chem.*, *symbol for* barium

BA able-bodied seaman • Bachelor of Arts • *British fishing port registration for* Ballantrae • *Finance* bank (*or* banker's) acceptance • *postcode for* Bath • *Med.* biliary atresia • (ital.) (USA) Biological Abstracts (publication) • Board of Agriculture • Booksellers' Association (of Great Britain and Ireland) • *Chem.* boric acid • British Academy • British Airways • British America • British Association (for the Advancement of Science) • British Association (pitch of screw thread, as in **4BA**) • bronchial asthma • Buenos Aires • *Computing* bus automaton • *British vehicle registration for* Manchester

BAA Bachelor of Applied Arts • Booking Agents' Association of Great Britain •

British Accounting Association • British Airports Authority • British Archaeological Association • British Astronomical Association

BAA & A British Association of Accountants and Auditors

BAAB British Amateur Athletic Board

BA(Admin) Bachelor of Arts in Administration

BAAF British Agencies for Adoption and Fostering

BAAL British Association for Applied Linguistics

BA(Art) Bachelor of Arts in Art

BAAS British Association for the Advancement of Science

Bab. Babylonia(n)

BABIE ('beɪbɪ) British Association for Betterment of Infertility and Education

BABS *Aeronautics* beam (*or* blind) approach beacon system

Bac. (France) baccalauréat (school examination) • *Education* Baccalaureus (Latin: bachelor)

BAc Bachelor of Acupuncture

BAC biologically active compound(s) • blood-alcohol concentration (*or* content) • British Aircraft Corporation • British Association of Chemists • *Med.* bronchoalveolar cell(s) • *Chem.* Brønsted acid centre • Business Archives Council

BACAH British Association of Consultants in Agriculture and Horticulture

BACAT (bæ'kæt) barge aboard catamaran • barge canal traffic

BAcc Bachelor of Accountancy

bach. bachelor

BACIE British Association for Commercial and Industrial Education

BACM British Association of Colliery Management

BACO British Aluminium Company Ltd

BACS Bankers' Automated Clearing System

bact. bacteria(l) • bacteriology

bacteriol. bacteriological • bacteriology

BAD base air depot • British Association of Dermatology

BADA British Antique Dealers' Association

BADGE (bædʒ) base air defence ground environment

BAdmin Bachelor of Administration

BAe British Aerospace

BAE Bachelor of Aeronautical

Engineering • Bachelor of Arts in Education • Badminton Association of England • Belfast Association of Engineers • (USA) Bureau of Agricultural Economics

BAEA British Actors' Equity Association

BAEC Bangladesh Atomic Energy Commission • British Agricultural Export Council

BA(Econ) Bachelor of Arts in Economics

BA(Ed) Bachelor of Arts in Education

BAED Bachelor of Arts in Environmental Design

BAEF Belgian-American Educational Foundation

BAF biological aerated filter • British Athletics Federation

BAFM British Association of Forensic Medicine

BAFMA British and Foreign Maritime Agencies

BAFO British Air Forces of Occupation • British Army Forces Overseas

BAFRA British Antique Furniture Restorers' Association

BAFSC British Association of Field and Sports Contractors

BAFSV British Armed Forces Special Vouchers

BAFTA ('bæftə) British Academy of Film and Television Arts

BAGA ('bɑːgə) British Amateur Gymnastics Association

BAgEc Bachelor of Agricultural Economics

BAgr Bachelor of Agriculture

BAgrSc Bachelor of Agricultural Science

Bah. Bahamas

BAHA British Association of Hotel Accountants

BAHOH British Association for the Hard of Hearing

BAHS British Agricultural History Society

BAI Bachelor of Engineering (Latin *Baccalaureus Artis Ingeniariae*) • *Military* battlefield air interdiction • Book Association of Ireland • *Med.* bronchial arterial infusion

BAIE British Association of Industrial Editors

BA(J) (*or* **BAJour**) Bachelor of Arts in Journalism

bal. *Book-keeping* balance

Bal. Ballarat • Balliol College (Oxford)

BAL blood-alcohol level • British anti-lewisite (dimercaprol; antidote to war gas and metal poisoning) • *Med.* bronchoalveolar lavage

BALH British Association for Local History

ball. ballast • ballistics

Ball. Balliol College (Oxford)

BALPA ('bælpə) British Airline Pilots' Association

bals. balsam

Balt. Baltic • Baltimore

balun ('bælən) *Telecom.* balanced unbalanced (transformer)

BAM Bachelor of Applied Mathematics • Bachelor of Arts in Music

BAMA British Aerosol Manufacturers' Association • British Amsterdam Maritime Agencies • British Army Motoring Association

BAMBI ('bæmbɪ) ballistic missile boost intercept

BAMTM British Association of Machine Tool Merchants

BA(Mus) Bachelor of Arts in Music

BAMW British Association of Meat Wholesalers

Ban. Bangor • Bantu

BAN British Association of Neurologists

Banc. Sup. *Law* Bancus Superior (Latin: higher bench; i.e. Queen's (*or* King's) Bench)

b. and b. (*or* **B & B**) bed and breakfast

B & C *Insurance* building and contents

B & D (*or* **B and D**) bondage and discipline (*or* domination)

b & e beginning and ending

B&FBS British and Foreign Bible Society

B and S (*or* **B & S**) brandy and soda • Brown and Sharpe (wire gauge)

B&W (*or* **b&w**) *Photog.* black and white

B & WE Bristol and West of England

BANS British Association of Numismatic Societies

BANZARE British, Australian, New Zealand Antarctic Research Expedition

BAO Bachelor of (Arts in) Obstetrics • Bankruptcy Annulment Order • British American Oil

BAOD British Airways Overseas Division

BA of E Badminton Association of England

BAOMS British Association of Oral and Maxillo-Facial Surgeons

BAOR British Army of the Rhine

b. à p. billets à payer (French: bills payable)

bap. baptized

Bap. Baptist

BAP *Chem.* 6-benzylaminopurine • *Chem.* bottom agitation process

BAPA British Amateur Press Association

BAPC British Aircraft Preservation Council

BAPCO ('bæp,kəʊ) Bahrain Petroleum Company

BA(PE) Bachelor of Arts in Physical Education

BAPL Bettis Atomic Power Laboratory

BAPLA British Association of Picture Libraries and Agencies

BAPM British Association of Physical Medicine

BAPNA *Chem.* Nα-benzoyl-DL-arginine-*p*-nitroaniline

BAppArts Bachelor of Applied Arts

BAppSc Bachelor of Applied Science

BAppSc(MT) Bachelor of Applied Science (Medical Technology)

BAPS beacon automated processing system (for lighthouses) • British Association of Paediatric Surgeons • British Association of Plastic Surgeons

bapt. baptism • baptized

Bapt. Baptist

BAPT British Association for Physical Training

b. à r. billets à recevoir (French: bills receivable)

bar. barleycorn (obsolete unit of length) • barometer • barometric • barrel (container or unit of measure) • barrister

Bar. baritone • Barrister • *Bible* Baruch

BAR *Computing* base address register • (ital.) Book Auction Records • British Association of Removers • Browning Automatic Rifle

Barb. Barbados

BARB (bɑːb) British Association of Rose Breeders • British Audience Research Bureau • Broadcasters' Audience Research Board

barbie ('bɑːbɪ) (Australia) *Colloquial* barbecue

BARC British Automobile Racing Club

BArch Bachelor of Architecture

BArchE Bachelor of Architectural Engineering

barg. bargain

barit. baritone

BARP British Association of Retired Persons

Barr. Barrister

BARR British Association of Rheumatology and Rehabilitation

BARS *Management* behaviourally anchored rating scales • British Association of Residential Settlements

Bart. Baronet

Bart's St Bartholomew's Hospital, London

BAS Bachelor in Agricultural Science • Bachelor of Applied Science • British Antarctic Survey

BASA British Architectural Students' Association • British Australian Studies Association

BASAF British and South Africa Forum

BASc Bachelor of Agricultural Science • Bachelor of Applied Science

BASC British Association for Shooting and Conservation

BASCA British Academy of Songwriters, Composers, and Authors

BASF Badische Anilin und Soda-Fabrik (German chemical and electronics company)

BASI British Association of Ski Instructors

Basic ('beɪsɪk) (*or* **BASIC**) *Computing* beginners' all-purpose symbolic instruction code (programming language) • (*or* **basic**) British-American scientific international commercial (in **Basic English**)

BASMA ('bæsmə) Boot and Shoe Manufacturers' Association and Leather Trades Protection Society

bass. con. *Music* basso continuo (Italian: continuous bass)

BASW ('bæzwə) British Association of Social Workers

bat. battalion • battery • battle

Bat. Batavia

BAT *Chem.* benzaldehyde–ammonia titration • best available technology • British Aerial Transport (aircraft) • British-American Tobacco Company • *Med.* brown adipose tissue

BATF (USA) Bureau of Alcohol, Tobacco, and Firearms

bath. bathroom

BA(Theol) Bachelor of Arts in Theology

BATO balloon-assisted takeoff

BA(TP) Bachelor of Arts in Town and Country Planning

BATS (bætz) *Ecology* biosphere–atmosphere transfer scheme

batt. (*or* **battn**) battalion • *Military* battery

BAU *Physics* baryon asymmetry of the universe • British Association Unit • business as usual

BAUA Business Aircraft Users' Association

BAUS British Association of Urological Surgeons

b. à v. *Finance* bon à vue (French: good at sight)

Bav. Bavaria(n)

BAWA British Amateur Wrestling Association

BAWLA British Amateur Weight-Lifters' Association

BAYS British Association of Young Scientists

bb books

b.b. ball bearing • bearer bonds • *Nautical* below bridges

Bb. bishops

BB bail bond • balloon barrage • bank book • bed and breakfast • best best (quality of wrought iron) • *postcode for* Blackburn • Blue Book (the HMSO publication *UK National Accounts*) • B'nai B'rith (Jewish international society) • Boys' Brigade • Brigitte Bardot (1934–), French film actress) • (*or* **2B**) double black (indicating a very soft lead pencil) • *British vehicle registration for* Newcastle upon Tyne • *Nautical, indicating* a certificate issued to vessels arriving from foreign destinations to the effect that their papers, etc., are in order • *indicating* a standard size of lead shot, 0.18 in. in diameter

BBA Bachelor of Business Administration • Big Brothers of America • *Obstetrics* born before arrival • British Bankers' Association • British Bee-keepers' Association • British Bloodstock Association • British Board of Agreement • British Bobsleigh Association

BBAC British Balloon and Airship Club

BBB bed, breakfast, and bath • (USA) Better Business Bureau • *Physiol.* blood–brain barrier • (*or* **3B**) treble black (indicating a very soft lead pencil)

BBBC (*or* **BBB of C**) British Boxing Board of Control

BBC British Broadcasting Corporation

BBCM Bandmaster of the Bandsmen's College of Music

BBCMA British Baby Carriage Manufacturers' Association

BBEM bed, breakfast, and evening meal

BBFC British Board of Film Classification (formerly British Board of Film Censors)

BBI British Bottlers' Institute

BBIP (ital.) British Books in Print (publication)

BBIRA British Baking Industries' Research Association

B. Bisc. Bay of Biscay

bbl. barrel (container or unit of measure for oil, etc.)

BBM *Physics* beam-bending magnet • *Histology* brush-border membrane

BBMA British Brush Manufacturers' Association • British Button Manufacturers' Association

9-BBN *Chem.* 9-borobicyclononane

BBO *Chem.* 2,5-bis(4-biphenylyl)oxazole

BBOT *Chem.* 2,5-bis(5-*t*-butyl-2-benzoxazolyl)thiophene

BBQ barbecue

BBS Bachelor of Business Science (*or* Studies) • *Computing* bulletin board system (*or* service)

BBSR Bermuda Biological Station for Research

BBSRC Biotechnology and Biological Sciences Research Council

BBT *Med.* basal body temperature

BBV Banco Bilbao Vizcaya (Spanish bank)

b.c. (*or* **bc**) *Music* basso continuo • blind copy

BC Bachelor of Chemistry • Bachelor of Commerce • Bachelor of Surgery (Latin *Baccalaureus Chirurgiae*) • *Military* (formerly) bad character • badminton club • *Taxation* balancing charge • bank clearing • bankruptcy court • *Chemical engineering* barrels of condensate • basketball club • *Music* basso continuo (Italian: continuing bass) • Battery Commander • battle cruiser • bayonet cap (on electric light bulbs) • (small caps.) before Christ (following a date) • bicycle club • billiards club • (*or* **B/C**) bills for collection • bishop and confessor • *Physics* Bloch curves • *Med.* blood consumption • board of control • boat (*or* boating) club • *Biochem.* body composition • *Astronomy* bolometric correction • Bomber

Command • borough council • *Physics* Bose condensation • bowling (*or* bowls) club • boxing club • boys' club • Bristol Channel • British Coal • British Columbia • British Commonwealth • British Council • *Med.* bronchial carcinoma • *airline flight code for* Brymon Airways • *Chem.* bubble combustion • budgeted cost • *Education* Burnham Committee • *Chem.* butyl chloride • *British vehicle registration for* Leicester

BCA (New Zealand) Bachelor of Commerce and Administration • *Chem.* bifunctional chelating agent • *Physical chem.* binary collision approximation • Boys' Clubs of America • British-Caribbean Association • *Angling* British Casting Association • British Chiropractic Association

BCAA *Biochem.* branched-chain amino acid

BCAB Birth Control Advisory Bureau

BCAC British Conference on Automation and Computation

BC&T (USA) Bakery, Confectionery, and Tobacco Workers International Union

BCAP British Code of Advertising Practice

BCAR British Civil Airworthiness Requirements • British Council for Aid to Refugees

BCAS British Compressed Air Society

BCBC British Cattle Breeders' Club

BCAT ('biː,kæt) Birmingham Centre for Art Therapies

b.c.c. (*or* **bcc**) blind carbon copy • *Crystallog.* body-centred cubic

BCC *Med.* basal-cell carcinoma • British Coal Corporation • British Colour Council • British Copyright Council • British Council of Churches • British Crown Colony

BCCA British Cyclo-Cross Association

BCCG British Cooperative Clinical Group

BCCI Bank of Credit and Commerce International

BCD (USA) bad conduct discharge • *Chem.* betacyclodextrin • *Computing* binary-coded decimal • *Astronomy* blue compact dwarf (type of star)

BCDP *Med.* balloon catheter dilation of the prostate

BCDTA British Chemical and Dyestuffs Traders' Association

BCE Bachelor of Chemical Engineering • Bachelor of Civil Engineering • (small

caps.) before common (*or* Christian) era (following a date; the non-Christian equivalent of BC) • Board of Customs and Excise

BCEAO Banque centrale des états de l'Afrique de l'Ouest (French: Central Bank of West African States)

BCECC British and Central European Chamber of Commerce

BCF battle cruiser force • billion cubic feet • British Chess Federation • British Cycling Federation • *Chem.* bromo-chlorodifluoromethane • *Textiles* bulked continuous filament (indicating yarns in man-made fibres) • (USA) Bureau of Commercial Fisheries

BCFA British-China Friendship Association

BCG *Med.* bacille Calmette–Guérin (anti-tuberculosis vaccine) • *Med.* ballisto-cardiography • *Astronomy* blue compact galaxy

BCGA British Commercial Gas Association • British Cotton Growing Association

BCGF *Biochem.* B-cell growth factor

bch (*or* **Bch**) branch • bunch

b.c.h. *Crystallog.* body-centred hypercubic

BCh (*or* **BChir**) Bachelor of Surgery (Latin *Baccalaureus Chirurgiae*)

BCH *Computing* Bose-Chaudhuri-Hocquenghem (in **BCH code**)

BChD Bachelor of Dental Surgery (Latin *Baccalaureus Chirurgiae Dentalis*)

BChE (*or* **BChemEng**) Bachelor of Chemical Engineering

BCINA British Commonwealth International Newsfilm Agency

BCIRA British Cast-Iron Research Association

BCIS Building Cost Information Service • Bureau central international de séismologie (French: International Central Bureau of Seismology)

BCK *British fishing port registration for* Buckie

BCL Bachelor of Canon Law • Bachelor of Civil Law • *Chemical engineering* brown-coal liquefaction

BCM (USA) Boston Conservatory of Music • British Commercial Monomark • British Consular Mail

BCMA British Colour Makers' Association • British Columbia Medical Association

BCMD biological and chemical munitions disposal (type of bomb disposal)

BCMF British Ceramic Manufacturers' Federation

BCMG Birmingham Contemporary Music Group

BCMS Bible Churchmen's Missionary Society

bcn beacon

BCNZ Broadcasting Corporation of New Zealand

BCO *Med.* bilateral carotid artery occlusion

B.Col.P. British Columbia Pine

BCom (*or* **BComm**) Bachelor of Commerce

BComSc Bachelor of Commercial Science

BCP *Chem.* basic calcium phosphate • Book of Common Prayer • Bulgarian Communist Party (*see under* BSP)

BCPC British Crop Protection Council

BCPIT British Council for the Promotion of International Trade

BCPL Basic Computer (*or* Combined) Programming Language

BCPMA British Chemical Plant Manufacturers' Association

BCR *Chem.* base-catalysed rearrangement • battlefield casualty replacement • *Chem.* branched-chain reaction

BCRA British Carbonization Research Association • British Ceramic Research Association

BCRC British Columbia Research Council

BCRD British Council for the Rehabilitation of the Disabled

BCS Bachelor of Chemical Science • Bachelor of Commercial Science • *Physics* Bardeen–Cooper–Schrieffer (in **BCS theory** of superconductivity) • battle cruiser squadron • Bengal Civil Service • British Calibration Service • British Cardiac Society • British Ceramic Society • British Computer Society • (USA) Bureau of Criminal Statistics

BCSA British Constructional Steelwork Association

bcst broadcast

BCT Belfast Chamber of Trade • bi-CMOS technology • *Med.* body computer tomograph

BCTA British Canadian Trade Association • British Children's Theatre Association

BCTGA British Christmas Tree Growers Association

BCTV *Microbiol.* beet curly top virus

BCU *Films* big close-up • British Canoe Union

BCURA British Coal Utilization Research Association

BCVA British Columbia Veterinary Association

BCWMA British Clock and Watch Manufacturers' Association

BCYC British Corinthian Yacht Club

bd board • bold • *Insurance, Finance* bond • *Bookbinding* bound • broad • bundle

b.d. *Med.* bis die (Latin: twice a day; in prescriptions) • *Commerce* bill(s) discounted

b/d barrels per day • *Book-keeping* brought down

Bd *Computing, symbol for* baud • Band (German: volume) • Board • Boulevard

BD Bachelor of Divinity • *symbol for* Bahrain dinar (monetary unit) • *international vehicle registration for* Bangladesh • battle dress • *Physics* beam deflection • *British fishing port registration for* Bideford • *Med.* bile duct • *Commerce* bill(s) discounted • bomb disposal • *Med.* bone density • *Astronomy* Bonner Durchmusterung (German: Bonn Survey; star catalogue) • boom defence • *postcode for* Bradford • *airline flight code for* British Midland • Bundesrepublik Deutschland (Federal Republic of Germany) • *British vehicle registration for* Northampton

B/D bank draft • *Commerce* bill(s) discounted • *Book-keeping* brought down

BDA Bachelor of Domestic Arts • Bachelor of Dramatic Art • bomb-damage assessment • British Deaf Association • British Dental Association • British Diabetic Association

b.d.c. *Engineering* bottom dead centre

BDC Book Development Council

BDCC British Defence Coordinating Committee

BDCS *Chem.* t-butyldimethylchlorosilane

Bde Bände (German: volumes) • Brigade

BDE *Chem.* bond dissociation energy

BDentSc Bachelor in Dental Science

BDes Bachelor of Design

BDF *Computing* backward differentiation formulae (as in **BDF methods**) • Botswana Defence Force

BDFA British Dairy Farmers' Association

bd.ft. board foot (measure of timber)

BDG *Bookbinding* binding

BDG/ND *Bookbinding* binding, no date can be given

BDH British Drug Houses

b.d.i. bearing deviation indicator • both dates (*or* days) included

BDI (Germany) Bundesverband der Deutschen Industrie (Federal Association of German Industry)

bdl. (*or* **bdle**) bundle

BDL British Drama League

BDM births, deaths, marriages • bomber defence missile • branch delegates' meeting

BDMA British Direct Marketing Association • British Disinfectant Manufacturers' Association

BDMAA British Direct Mail Advertising Association

Bdmr Bandmaster

BDNF *Med.* brain-derived neurotrophic factor

BDO Boom Defence Officer

BDP breakdown pressure

BDPA *Chem.* α,γ-bisdiphenylene-β-phenylallyl

Bdr Bombardier • Brigadier

bdrm bedroom

bds *Bookbinding* boards • bundles

b.d.s. *Med.* bis in die sumendus (Latin: to be taken twice a day; in prescriptions)

BDS Bachelor of Dental Surgery • *international vehicle registration for* Barbados • bomb disposal squad • British Driving Society

BDSA (USA) Business and Defense Services Administration

BDSc Bachelor of Dental Science

BDST British Double Summer Time

BDU *Astronomy* baryon-dominated universe • bomb disposal unit

b.d.v. *Physics* breakdown voltage

BDV *Physics* breakdown voltage • *Accounting* budget day value

Bdx. Bordeaux

be. bezüglich (German: regarding *or* with reference to)

b.e. bill of exchange • binding edge

Be *Chem.*, *symbol for* beryllium • *Astronomy*, *indicating* a spectral type

Bé Baumé (temperature scale)

BE Bachelor of Economics • Bachelor of Education • Bachelor of Engineering •

Bank of England • *Physics* Barkhausen emission • *British fishing port registration for* Barnstaple • best estimate • bill of exchange • *Computing* binary encounter • *Physics* binding energy • (USA) Board of Education • borough engineer • *Physics* Bose–Einstein • British Element • British Embassy • British Empire • *British vehicle registration for* Lincoln

B/E bill of entry • bill of exchange

BEA British East Africa • British Epilepsy Association • British Esperanto Association • (formerly) British European Airways

BEAB British Electrical Approvals Board

BEAC Banque des états de l'Afrique centrale (French: Bank of Central African States)

BEAIRE British Electrical and Allied Industries' Research Association

BEAM (biːm) *Med.* brain electrical activity mapping

BEAMA (Federation of) British Electro-technical and Allied Manufacturers' Associations

bearb. bearbeitet (German: compiled *or* edited)

BEAS British Educational Administration Society

bec. because

BEc Bachelor of Economics

BEC Building Employers' Confederation • Bureau européen du café (French: European Coffee Bureau) • (USA) Bureau of Employees' Compensation

BECA British Exhibition Contractors' Association

Bech. Bechuanaland (now Botswana)

BECO booster-engine cut-off

BEcon Bachelor of Economics

BEcon(IA) Bachelor of Economics in Industrial Administration

BEcon(PA) Bachelor of Economics in Public Administration

BECTU ('bektuː) Broadcasting, Entertainment, Cinematograph, and Theatre Union (formed by amalgamation of BETA with ACTT)

BEd Bachelor of Education

BED *Biochem.* bio-emf device

BEDA British Electrical Development Association • Bureau of European Designers' Associations

BEd(Com) Bachelor of Education in Commerce

BEd(HEc) Bachelor of Education in Home Economics

BEd(N) Bachelor of Education in Nursing

BEd(PE) Bachelor of Education in Physical Education

beds. bedrooms

Beds Bedfordshire

BEd(Sc) Bachelor of Education in Science (*or* Educational Science)

BEE Bachelor of Electrical Engineering

Beeb (biːb) *Colloquial* BBC (British Broadcasting Corporation)

BEEL biological equivalent exposure limit (of radiation)

bef. before

b.e.f. blunt end first

BEF British Equestrian Federation • British Expeditionary Force (in World Wars I and II)

BEFA British Emigrant Families Association

beg. beginning

BEG (USA) Bureau of Economic Geology

BEHA British Export Houses Association

BEI Bachelor of Engineering (Dublin) • *Physics* back-scattered electron imaging • Banque européenne d'investissement (French: European Investments Bank)

Beibl. Beiblatt (German: supplement)

beigeb. beigebunden (German: bound, in with something else)

beil. beiliegend (German: enclosed)

BEIR biological effects of ionizing radiation

Bel. Belgian • Belgium

BEL British Electrotechnical Committee

bel ex. bel exemplaire (French: fine copy; of a book or engraving)

Belf. Belfast

Belfox ('belfɒks) Belgian Futures and Options Exchange

Belg. Belgian • Belgic • Belgium

BEM British Empire Medal • bug-eyed monster

BEMA British Essence Manufacturers' Association • (formerly) Business Equipment Manufacturers' Association (replaced by the EIS)

BEMAC British Exports Marketing Advisory Committee

BEMAS British Education Management and Administration Society

BEME ('biːˈmiː) Brigade Electrical and Mechanical Engineer

BEMSA British Eastern Merchant Shippers' Association

ben. benedictio (Latin: blessing)

BEN *Electronics* broadband electrostatic noise

Bend. Bendigo (Australia)

BenDr Bachelier en droit (French: Bachelor of Law)

benef. benefice

Benelux ('benɪˌlʌks) Belgium, Netherlands, Luxembourg (customs union)

Beng. Bengal(i)

BEng Bachelor of Engineering

BEngr Bachelor of Engraving

BenH Bachelier en humanités (French: Bachelor of Humanities)

BEO Base Engineer Officer

BEPC *Physics* Beijing Electron–Positron Collider • British Electrical Power Convention

beq. bequeath

beqt bequest

Ber. (*or* **Berl.**) Berlin

BERCO British Electric Resistance Company

Berks Berkshire

Berm. Bermuda

Berw. Berwickshire (former Scottish county)

BES Bachelor of Engineering Science • Bachelor of Environmental Studies • Biological Engineering Society • British Ecological Society • (formerly) Business Expansion Scheme (replaced by the EIS)

BèsA Bachelier ès arts (French: Bachelor of Arts)

BESA British Esperanto Scientific Association

BESI bus electronic scanning indicator

BèsL Bachelier ès lettres (French: Bachelor of Letters)

BESO British Executive Service Overseas

BèsS (*or* **BèsSc**) Bachelier ès sciences (French: Bachelor of Science)

BESS Bank of England Statistical Summary • *Oceanog.* bottom environmental sensing system

Best. *Commerce* Bestellung (German: order)

BEST (best) British Expertise in Science and Technology (database)

bet. between

BET British Electric Traction Company • *Chem.* Brunauer–Emmet–Teller (as in

BET adsorption theory) • buildings energy technology

BETA Broadcasting and Entertainment Trades Alliance (*see* BECTU) • Business Equipment Trades' Association

BETAA British Export Trade Advertising Association

BETRO British Export Trade Research Organization

betw. between

BEU *Engineering* batch extraction unit • Benelux Economic Union

BEUC ('beruːk) Bureau européen des unions de consommateurs (French: European Bureau of Consumers' Unions)

bev. bevel

BeV (USA) billion electronvolts

BEVA British Exhibition Venues' Association

BEXA British Exporters Association

bez. bezahlt (German: paid) • bezüglich (German: with reference to)

bezw. beziehungsweise (German: respectively)

bf (*or* **bf.**) brief

b.f. bankruptcy fee • base frequency • beer firkin • (*or* **bf**) *Colloquial* bloody fool • (*or* **bf**) *Printing* bold face • bona fide

b/f bring forward (date on correspondence) • *Book-keeping* brought forward

BF Bachelor of Forestry • *British fishing port registration for* Banff • Banque de France (Bank of France) • *symbol for* Belgian franc • black face (sheep) • blast furnace • *Colloquial* bloody fool • body fat • breathing frequency • British Funds • *international vehicle registration for* Burkina Faso • *British vehicle registration for* Stoke-on-Trent

B/F bring forward (date on correspondence) • *Book-keeping* brought forward

BFA Bachelor of Fine Arts • British Film Academy

BFAP British Forces Arabian Peninsula

BFASS *Nuclear physics* BWR fuel assembly sealing system

BFAWU Bakers', Food, and Allied Workers' Union

BFBPW British Federation of Business and Professional Women

BFBS (*or* **B & FBS**) British and Foreign Bible Society • British Forces Broadcasting Service

BFCA British Federation of Commodity Associations

BFCS British Friesian Cattle Society

BFEBS British Far Eastern Broadcasting Service

BFET *Electronics* ballistic field-effect transistor

BFFA British Film Fund Agency

BFFC British Federation of Folk Clubs

BFG Big Friendly Giant (title character of a Roald Dahl children's book)

BFI British Film Institute

BFIA British Flower Industry Association

BFM *Physics* boson–fermion coupling model

BFMA British Farm Mechanization Association

BFMF British Federation of Music Festivals • British Footwear Manufacturers' Federation

BFMIRA British Food Manufacturing Industries' Research Association

BFMP British Federation of Master Printers

Bfn Bloemfontein

BFN British Forces Network

b.f.o. (*or* **BFO**) *Electronics* beat-frequency oscillator

BFor Bachelor of Forestry

BForSc Bachelor of Forestry Science

BFP Bureau of Freelance Photographers

BFPA British Film Producers' Association

BFPC British Farm Produce Council

BFPO British Forces Post Office

BFr Belgian franc

BFS *Mining* blast-furnace slag • British Fuchsia Society

BFSA British Fire Services' Association

BFSS British and Foreign Sailors' Society • British Field Sports Society

BFT *Physiol., Psychol.* biofeedback training

BFTA British Fur Trade Alliance

BFUW British Federation of University Women

BFV *Military* Bradley fighting vehicle

bg *Commerce* bag

b.g. *Horse racing* bay gelding

b/g bonded goods

BG *airline flight code for* Biman Bangladesh • Birmingham (wire) gauge • blood group • Brigadier General • British Guiana (now Guyana) • *international vehicle registration for* Bulgaria • *British vehicle registration for* Liverpool

BGA (USA) Better Government

Association • British Gliding Association • British Graduates Association

BGB Booksellers' Association of Great Britain and Ireland • (Germany) Bürgerliches Gesetzbuch (code of civil law)

BGC bank giro credit • British Gas Corporation (now British Gas)

BGCS Botanic Gardens Conservation Secretariat

BGEA Billy Graham Evangelistic Association

BGenEd Bachelor of General Education

bGH (or **BGH**) *Biochem.* bovine growth hormone

BGIRA British Glass Industry Research Association

b.g.l. below ground level

BGL Bachelor of General Laws

bglr bugler

BGM Bethnal Green Museum

BGMA British Gear Manufacturers' Association

BGMV *Microbiol.* bean golden mosaic virus

BGS Brigadier General Staff • British Geological Survey • British Geriatrics Society • British Goat Society • Brothers of the Good Shepherd

bgt bought

BGV below-ground vault

bh (or **b.h.**) barrels per hour • *Colloquial* bloody hell

BH base hospital • *Astronomy* black hole • *British fishing port registration for* Blyth • *postcode for* Bournemouth • *Metallurgy* Brinell hardness • British Hovercraft • Burlington House, London (home of the Royal Academy) • *British vehicle registration for* Luton

B/H bill of health

BHA *Mining* bottom-hole assembly • British Homeopathic Association • British Humanist Association

B'ham Birmingham

BHB British Hockey Board

BHC *Med.* benign hereditary chorea • benzene hexachloride (insecticide) • British High Commissioner

bhd beachhead • billhead • bulkhead

BHDF British Hospital Doctors' Federation

BHE Bachelor of Home Economics

B'head Birkenhead

Bhf Bahnhof (German: railway station)

BHF British Hardware Foundation • British Heart Foundation

BHGA British Hang Gliding Association

BHI British Horological Institute • Bureau hydrographique international (French: International Hydrographic Bureau)

Bhm Birmingham

BHMRA British Hydromechanics' Research Association

Bhn (or **BHN**) *Metallurgy* Brinell hardness number

BHort Bachelor of Horticulture

BHortSc Bachelor of Horticultural Science

bhp brake horsepower

BHP (Australia) Broken Hill Proprietary

bhpric bishopric

BHQ Brigade Headquarters

BHRA British Hydromechanics' Research Association

BHRCA British Hotels, Restaurants, and Caterers' Association

BHS boys' high school • British Home Stores • British Horse Society

BHT *Electronics* bipolar heterojunction transistor • *Mining* bottom-hole temperature • *Chem.* butylated hydroxytoluene

BHTA British Herring Trade Association

Bhu. Bhutan

BHy Bachelor of Hygiene

Bi *Chem., symbol for* bismuth

BI background information • Bahama Islands • Balearic Islands • base ignition • *Chem.* benzene-insoluble • Bermuda Islands • *Med.* bone injury • bulk issue • *Irish vehicle registration for* Monaghan • *airline flight code for* Royal Brunei Airlines

BIA British Ironfounders' Association • (USA) Bureau of Indian Affairs

BIAA British Industrial Advertising Association

BIAC Business and Industry Advisory Committee

BIAE British Institute of Adult Education

BIAS Bristol Industrial Archaeological Society

BIATA British Independent Air Transport Association

bib. *Med.* bibe (Latin: drink) • bibliothèque (French: library)

Bib. Bible • Biblical

BIB *Military* baby incendiary bomb • *Electronics* blocked impurity band

BIBA British Insurance Brokers' Association (now British Insurance and Investment Brokers' Association, BIIBA)

BIBC British Isles Bowling Council

BIBF British and Irish Basketball Federation

bibl. (*or* **bibliog.**) bibliographer • bibliographical • bibliography

BIBRA British Industrial Biological Research Association

BIC Bahá'í International Community • *Electronics* biased ion collector • Bureau international du cinéma (French: International Cinema Bureau) • Butter Information Council

BICC Berne International Copyright Convention

BICE Bureau international catholique de l'enfance (French: International Catholic Child Bureau)

BICEMA British Internal Combustion Engine Manufacturers' Association

BICEP British Industrial Collaborative Exponential Programme

BICERI British Internal Combustion Engine Research Institute

BICFET ('bɪk,fɛt) *Electronics* bipolar inversion-channel field-effect transistor

bi-CMOS (baɪ'siːmɒs) *Electronics* (merged) bipolar/complementary metal oxide semiconductor

BICS British Institute of Cleaning Science

b.i.d. *Med.* bis in die (Latin: twice a day; in prescriptions)

BID Bachelor of Industrial Design • Bachelor of Interior Design • *Med.* brought in dead

BIDS British Institute of Dealers in Securities

BIE Bachelor of Industrial Engineering • *Maths.* boundary integral equation • Bureau international d'éducation (French: International Bureau of Education) • Bureau international des expositions (French: International Exhibition Bureau)

BIEE British Institute of Energy Economics

bien. biennial

BIET British Institute of Engineering Technology

BIF *Geology* banded iron formation • *Biochem.* B-cell growth inhibitory factor • *Physics* beta-induced fluorescence • British Industries Fair

BIFFEX ('bɪfɛks) Baltic International Freight Futures Market (formerly Exchange)

BIFU ('bɪfuː) Banking, Insurance, and Finance Union

BIH *Med.* benign intracranial hypertension • *international vehicle registration for* Bosnia-Herzegovina • Bureau international de l'heure (French: International Time Bureau)

BIHA British Ice Hockey Association

BIIBA British Insurance and Investment Brokers' Association

bim. bimestrale (Italian: bimonthly) • bimestre (Italian: a two-month period)

BIM *Maths.* boundary integral method • British Institute of Management • British Insulin Manufacturers

BIMA *Astronomy* Berkeley-Illinois-Maryland Association (in **BIMA array**)

BIMBO buy-in management buyout

BIMCAM British Industrial Measuring and Control Apparatus Manufacturers' Association

bin. *Maths.* binary

BIN (ital.) Bulletin of International News

BINC Building Industries' National Council

bind. binding

BIO (Canada) Bedford Institute of Oceanography

biochem. biochemistry

biog. biographer • biographic(al) • biography

biogeog. biogeography

biol. biological • biologist • biology

BIOS ('baɪɒs) *Computing* basic input-output system • Biological Investigations of Space • (*or* **bios**) *Astronautics* biological satellite • British Intelligence Objectives Subcommittee

BIOT British Indian Ocean Territory

BIP Botswana Independence Party • British Industrial Plastics • British Institute in Paris

BIPCA Bureau international permanent de chimie analytique pour les matières destinées à l'alimentation de l'homme et des animaux (French: Permanent International Bureau of Analytical Chemistry of Human and Animal Food)

BIPL Burmah Industrial Products Limited

BIPM Bureau international des poids et mésures (French: International Bureau of Weights and Measures)

BIPP *Med.* bismuth iodoform paraffin paste (formerly used for treating wounds) • British Institute of Practical Psychology

BIR Board of Inland Revenue • British Institute of Radiology

BIRD Banque internationale pour la reconstruction et le développement (French: International Bank for Reconstruction and Development; IBRD)

BIRE British Institution of Radio Engineers

BIRF Brewing Industry Research Foundation

Birm. Birmingham

BIRMO British Infra-Red Manufacturers' Organization

BIRS British Institute of Recorded Sound

bis. bissextile

BIS Bank for International Settlements • *Physics* beam-induced stripping • British Information Services • British Interplanetary Society • Bureau international du scoutisme (French: Boy Scouts International Bureau)

BISA British International Studies Association

Bisc. Biscayan

BISF British Iron and Steel Federation

BISFA British Industrial and Scientific Film Association

bish. bishop

bis in 7d. *Med.* bis in septem diebus (Latin: twice a week; in prescriptions)

BISPA British Independent Steel Producers' Association

BISRA British Iron and Steel Research Association

BISYNC ('baɪˌsɪŋk) *Computing* binary synchronous communications

bit (bɪt) *Computing, Maths.* binary digit

BIT Bureau international du travail (French: International Labour Office)

BITA British Industrial Truck Association

bitm. bituminous

BITNET ('bɪtˌnɛt) *Computing* Because It's Time Network

BITO British Institution of Training Officers

BITOA British Incoming Tour Operators' Association

bitum. bituminous

BIU Bermuda Industrial Union • Bureau international des universités (French: International University Bureau)

biv. bivouac

BIWF British-Israel World Federation

BIWS Bureau of International Whaling Statistics

Bix (bɪks) Byte magazine's information exchange service

BIZ Bank für Internationalen Zahlungsausgleich (German: Bank for International Settlements)

BJ Bachelor of Journalism • *British vehicle registration for* Ipswich

BJA British Judo Association

BJCEB British Joint Communications Electronics Board

BJJ *Electronics* boundary Josephson junction

BJOS (ital.) British Journal of Occupational Safety

BJP (India) Bharatiya Janata Party

BJSM British Joint Services Mission

BJT *Electronics* bipolar junction transistor

BJTRA British Jute Trade Research Association

BJur (*or* **BJuris**) Bachelor of Jurisprudence

bk backwardation • bank • bark • barrack • black • block • book • break

Bk *Chem.*, *symbol for* berkelium

BK *British fishing port registration for* Berwick-on-Tweed • *Trademark* Burger King • *British vehicle registration for* Portsmouth

BKA (Germany) Bundeskriminalamt (criminal investigations office)

bkble bookable

BKC *Chem.* benzalkonium chloride

bkcy bankruptcy

bkd blackboard • booked

BKD bacterial kidney disease

bkg banking • booking • book-keeping

bkgd background

bkkg book-keeping

bklt booklet

Bklyn (USA) Brooklyn

bkm buckram

BKN *Meteorol.* broken (cloud)

bkpt (*or* **bkrpt**) bankrupt

bks barracks • books

BKSTS British Kinematograph, Sound, and Television Society

bkt basket • bracket

bl barrel (container *or* unit of measure)

bl. bale • black • blue

b.l. (*or* **b/l**) bill of lading • breech-loading (rifle)

Bl. Blat (German: newspaper) • Blessed

BL Bachelor of Law • Bachelor of Literature (*or* Letters) • Barrister-at-Law • base line • bill lodged • bill of lading • *Printing* black letter • *Physics* Bloch lines • *Med.* blood lead • boatswain lieutenant • Bodleian Library (Oxford) • *postcode for* Bolton • *British fishing port registration for* Bristol • (formerly) British Leyland • British Library • *Med.* Burkitt's lymphoma • *airline flight code for* Pacific Airlines • *British vehicle registration for* Reading

B/L bill of lading

BLA Bachelor of Landscape Architecture • Bachelor of Liberal Arts • British Legal Association • British Liberation Army

BLACC British and Latin American Chamber of Commerce

BLAISE (bleɪz) British Library Automated Information Service

BLB *Horticulture* bacterial leaf blight • Boothby, Lovelace, and Bulbulian (in **BLB nasal oxygen mask**)

BLC British Lighting Council

bld *Printing* bold(face)

bldg building

BLE (USA) Brotherhood of Locomotive Engineers (trade union)

BLESMA ('blɛsmə) British Limbless Ex-Servicemen's Association

BLEU Belgo-Luxembourg Economic Union • *Aeronautics* Blind Landing Experimental Unit

BLG (ital.) Burke's Landed Gentry

BLH British Legion Headquarters

BLI British Lighting Industries

BLib (*or* **BLibSc**) Bachelor of Library Science

BLIC Bureau de liaison des industries du caoutchouc de la CE (French: Rubber Industries Liaison Bureau of the EU)

BLIS (USA) Bibliographic Literature Information System

Bliss (blɪs) baby life support systems • *Military* bend, low silhouette, irregular shape, small, secluded (escape technique taught to pilots)

BLit Bachelor of Literature

BLitt Bachelor of Letters (Latin *Baccalaureus Litterarum*)

blk black • blank • block • bulk

BLL Bachelor of Laws

BLM *Biochem.* bilayer lipid membrane • blind landing machine • (USA) Bureau of Land Management

BLMA British Lead Manufacturers' Association

BLMAS Bible Lands Missions' Aid Society

BLMRA British Leather Manufacturers' Research Association

BLO *Meteorol.* below clouds

BLOF British Lace Operatives' Federation

BLOX (blɒks) *Finance* block order exposure system

BLP Barbados Labour Party

b.l.r. breech-loading rifle

BLR *Spectroscopy* broad-line (emission) region

BLRA British Launderers' Research Association

BLRG *Astronomy* broad-line radio galaxy

BLS Bachelor of Library Science • basic life support • benevolenti lectori salutem (Latin: greeting to the well-wishing reader) • Branch Line Society • *Physics* Brillouin light scattering • (USA) Bureau of Labor Statistics

BLSN *Meteorol.* blowing snow

blt built

BLT bacon, lettuce, and tomato (sandwich)

BLV *Microbiol.* bovine leukaemia virus • British Legion Village

Blvd Boulevard

BLWA British Laboratory Ware Association

BLWN *Electronics* band-limited white noise

b.m. bene merenti (Latin: to the well-deserving) • *Horse racing* black mare • board measure (measurement of wood) • *Med.* bowel movement • breech mechanism (of a firearm)

BM *airline flight code for* Air Sicilia • Bachelor of Medicine • Bachelor of Music • bandmaster • *Med.* basal metabolism • base metal • beatae memoriae (Latin: of blessed memory) • Beata Maria (Latin: the Blessed Virgin) • *Surveying* benchmark • Bishop and Martyr • bonae memoriae (Latin: of happy memory) • *Med.* bone marrow • *Computing* Boyer–Moore (in **BM algorithm**) • brigade major • British Monomark • British Museum • *British fishing port registration for* Brixham • Bronze Medallist • (USA) Bureau of Mines • *British vehicle registration for* Luton

BMA Bahrain Monetary Agency • British

Manufacturers' Association • British Medical Association

BM algorithm *Computing* Boyer–Moore algorithm

BMath Bachelor of Mathematics

BMC *Med.* bone marrow cell(s) • *Med.* bone mineral content (*or* concentration) • Book Marketing Council • British Match Corporation • British Medical Council • British Metal Corporation • (formerly) British Motor Corporation • British Mountaineering Council • *Bibliog.* British Museum Catalogue (of 15th-century books)

BMCIS Building Maintenance Cost Information Service

BMD ballistic missile defence • births, marriages, and deaths • *Med.* bone mineral density

BMDM *Med.* bone-marrow-derived macrophage(s) • British Museum Department of Manuscripts

Bmdr *Military* Bombardier

BME Bachelor of Mechanical Engineering • Bachelor of Mining Engineering • Bachelor of Music Education • *Med.* benign myalgic encephalomyelitis

BMEC British Marine Equipment Council

BMed Bachelor of Medicine

BMedSci Bachelor of Medical Science

BMEF British Mechanical Engineering Federation

BMEG Building Materials Export Group

BMEO British Middle East Office

BMEP (*or* **bmep**) *Engineering* brake mean effective pressure

BMet Bachelor of Metallurgy

BMetE Bachelor of Metallurgical Engineering

BMEWS (biː'mjuːz) ballistic missile early-warning system

BMF Builders' Merchants' Federation

BMFA Boston Museum of Fine Arts

BMH British Military Hospital

BMI ballistic missile interceptor • Birmingham and Midland Institute • body-mass index • Broadcast Music Incorporated

BMJ (ital.) British Medical Journal

BML British Museum Library

BMM British Military Mission

BMMA Bacon and Meat Marketing Association

b.m.o. business machine operator

B'mouth Bournemouth

bmp (*or* **b.m.p.**) brake mean power

BMP biochemical methane potential (*or* production) • *Computing* bitmap (format)

BMPA British Metalworking Plantmakers' Association

BMPS British Musicians' Pension Society

BMR *Physiol.* basal metabolic rate

BMRA Brigade Major Royal Artillery

BMRB British Market Research Bureau

BMRMC British Motor Racing Marshals' Club

BMRR (USA) Brookhaven Medical Research Reactor

BMS Bachelor of Marine Science • Baptist Missionary Society • British Mycological Society • building management system • business modelling system

BMSE Baltic Mercantile and Shipping Exchange

BMT basic motion time-study • bone-marrow transplant • borehole-mining tool • British Mean Time

BMTA British Motor Trade Association

BMus Bachelor of Music

BMusEd Bachelor of Music Education

BMV Blessed Mary the Virgin • *Microbiol.* brome mosaic virus

BMW Bayerische Motorenwerke (German: Bavarian Motor Works)

BMWE (USA) Brotherhood of Maintenance of Way Employees (trade union)

BMWS ballistic missile weapon system

BMX bicycle motocross

bn bassoon • battalion • beacon • been • billion • born

Bn Baron • Battalion

BN Bachelor of Nursing • bank note • *Chem.* boron nitride • *British fishing port registration for* Boston • *postcode for* Brighton • Britten-Norman (aircraft) • *British vehicle registration for* Manchester

BNA *Chem.* base-neutralized acid • British Naturalists' Association • British North America • *Insurance* British North Atlantic • British Nursing Association

BNAF British North Africa Force

BNB *Physiol.* blood–nerve barrier • British National Bibliography

BNBC British National Book Centre

BNC *Chem.* base-neutralizing capacity • *Computing* bayonet nut couplers • Brasenose College, Oxford • *Linguistics* British National Corpus

BNCC British National Committee for Chemistry

BNCI (Madagascar) Banque nationale

pour le commerce et l'industrie (French: National Bank for Commerce and Industry)

BNCM (France) Bibliothèque nationale du conservatoire de musique (National Library of Music, Paris)

BNCS British National Carnation Society

BNCSAA British National Committee on Surface-Active Agents

BNCSR British National Committee on Space Research

BND (Germany) Bundesnachrichtendienst (national intelligence service)

B/ND *Bookbinding* binding, no date given

BNDD (USA) Bureau of Narcotics and Dangerous Drugs

Bndr Bandmaster

BNEC British National Export Council • British Nuclear Energy Conference

BNES British Nuclear Energy Society

BNF *Computing* Backus–Naur (*or* Backus normal) form • *Pharmacol.* British National Formulary • (*or* **BNFL**) British Nuclear Fuels (Limited) • British Nutrition Foundation

BNFC British National Film Catalogue

BNFMF British Non-Ferrous Metals Federation

BNGA British Nursery Goods Association

BNGM British Naval Gunnery Mission

BNHQ battalion headquarters

bnkg banking

BNL Banca Nazionale del Lavoro (Italian: National Bank of Labour) • Brookhaven National Laboratory

BNM Bureau national de metrologie (French: National Bureau of Meteorology)

BN object *Astronomy* Becklin–Neugebauer object

BNOC British National Oil Corporation • British National Opera Company

BNP Banque nationale de Paris (French: National Bank of Paris) • Barbados National Party • British National Party

BNS Bachelor of Natural Science • Bathymetric Navigation System • British Numismatic Society • buyer no seller

BNSc Bachelor of Nursing Science

BNSC British National Space Centre

BNTA British Numismatic Trade Association

BNurs Bachelor of Nursing

BNX British Nuclear Export Executive

bnzn *Chem.* benzoin

b.o. back order • blackout • *Med.* bowels open • branch office • broker's order • buyer's option

b/o *Book-keeping* brought over

BO biological oceanography • *Colloquial* body odour • *British fishing port registration for* Borrowstownness • box office • *Med.* bowels opened • *British vehicle registration for* Cardiff

B/O *Book-keeping* brought over • buyer's option

BOA British Olympic Association • British Optical Association • British Orthopaedic Association • broad ocean area

BOAC (formerly) British Overseas Airways Corporation

BOAD Banque ouest-africaine de développement (French: West African Development Bank)

BOA(Disp) British Optical Association, Dispensing Certificate

BOAI *Med.* balloon occluded arterial infusion

BOAT (bəʊt) byroad open to all traffic

BOBA British Overseas Banks' Association

BOBMA British Oil Burner Manufacturers' Association

BOC *Computing* beginning of cycle • *Computing* bimodal optical computer • British Oxygen Corporation • Burmah Oil Company

BOCE Board of Customs and Excise

BOC-ON *Chem.* 2-(*t*-butoxycarbonyl-oxyimino)-2-phenylacetonitrile

Bod. (*or* **Bodl.**, **Bodley**) Bodleian Library (Oxford)

BOD biochemical oxygen demand

BOE Board of Education • *Building trades* brick on edge (in **BOE sill**)

BOF *Chemical engineering* basic oxygen furnace • *Computing* beginning of file • British Orienteering Federation • British Overseas Fairs

B of E Bank of England

B of H Band of Hope Union

BOGMC Bangladesh Oil, Gas, and Minerals Corporation

Boh. Bohemia(n)

BOHA *Med.* balloon occlusion hepatic angiography

BoJ (*or* **BOJ**) Bank of Japan

bol. *Med.* bolus (Latin: large pill)

Bol. Bolívar (1783–1830, South American statesman) • Bolivia(n)

BOL beginning of life • *international vehicle registration for* Bolivia

BOLTOP ('bɒl,tɒp) better on lips than on paper (written on the back of envelopes containing love letters)

b.o.m. bill of materials (in manufacturing)

Bom. Bombay

BOM bill of materials • (USA) Bureau of Mines

Bomb. Bombardier • Bombay

BomCS Bombay Civil Service

BomSC Bombay Staff Corps

bon. bataillon (French: battalion)

BON acid *Chem.* beta-oxynaphthoic acid

BONUS ('bəunəs) *Finance* Borrower's Option for Notes and Underwritten Standby

Boo *Astronomy* Bootes

BOP *Chem.* basic oxygen process • *Mining* blowout preventer • Boy's Own Paper (former magazine)

BOptom Bachelor of Optometry

BOQ *US military* bachelor officers' quarters • base officers' quarters

bor. borough

BOr Bachelor of Orientation

BORAD British Oxygen Research and Development Association

boro. borough

Bos *Computing* business operating system

BOS *Chemical engineering* basic oxygen steelmaking

bos'n boatswain

Bos Pops (USA) Boston Pops Orchestra

BOSS (bɒs) Bioastronautic Orbiting Space Station • (South Africa) Bureau of State Security

Boswash ('bɒs,wɒʃ) (USA) Boston–Washington, DC (indicating the urban area between these cities)

bot. botanic(al) • botanist • botany • bottle • bottom • bought

BOT *Computing* beginning of tape (in **BOT marker**) • (*or* **BoT**) Board of Trade (now part of DTI)

BOTB British Overseas Trade Board

BOT marker *Computing* beginning of tape marker

BOU British Ornithologists' Union

boul. (France) boulevard

BOV brown oil of vitriol (commercial sulphuric acid)

BOWO Brigade Ordnance Warrant Officer

bp (*or* **b.p.**) *Biochem.* base pair(s) (unit of length of nucleic acid) • below proof (of alcohol density) • bills payable • *Chem.* boiling point • bonum publicum (Latin: the public good)

bp. baptized • birthplace

b/p blueprint • bills payable

Bp Bishop

BP *airline flight code for* Air Botswana • Bachelor of Pharmacy • Bachelor of Philosophy • back projection • (*or* **B-P**) (Robert) Baden-Powell (1857–1941, founder of the Boy Scout movement • barometric pressure • *Finance* basis point • *Military* beach party • (small caps.) before present (following a number of years) • *Chem.* benzoyl peroxide • be prepared (motto of the Scout movement) • *Shipping* between perpendiculars • *Biochem.* binding protein • *Med.* blood pressure • Blue Peter (television programme) • (Netherlands) Boerenpartij (Farmers' Party) • *Chem.* boiling point • *Physics* bright point • British Patent • British Petroleum • British Pharmacopoeia • British Public • *Chem.* bromopentane • *Chem.* burnable poison • *British vehicle registration for* Portsmouth

B/P bills payable

BPA Bachelor of Professional Arts • Bahnpostamt (German: railway post office) • (USA) Biological Photographic Association • Bookmakers' Protection Association • British Paediatric Association • British Parachute Association • (USA) Brookhaven (*or* BNL) Plant Analyzer • (USA) Business Publications Audit of Circulation

BPAA British Poster Advertising Association

BPAGB Bicycle Polo Association of Great Britain

BPAO *Biochem.* bovine plasma amine oxidase

BPAS British Pregnancy Advisory Service

b.p.b. bank post bills

BPBF British Paper Box Federation

BPBIRA British Paper and Board Industry Research Association

BPBMA British Paper and Board Makers' Association

BPC Book Prices Current • British Pharmaceutical Codex • British Pharmacopoeia Commission • British Printing Corporation • British Productivity Council • (USA) Business and

Professional Code • *Chem.* butyl pyridium chloride

BPCA British Pest Control Association

b.p.c.d. barrels per calendar day

BPCF British Precast Concrete Federation

BPCR Brakes on Pedal Cycle Regulations

BPCRA British Professional Cycle Racing Association

b.p.d. barrels per day

BPDB Bangladesh Power Development Board

BPDMS *Military* basic point defence missile system

BPE (*or* **BPEd**) Bachelor of Physical Education

BPEA *Chem.* 9,10-bis(phenylethenyl)-anthracene

BPEN *Chem.* 5,10-bis(phenylethenyl)-naphthacene

b.p.f. *Commerce* bon pour francs (French: value in francs)

BPF bottom pressure fluctuation • British Plastics Federation • British Polio Fellowship

BPG Broadcasting Press Guild

BPh Bachelor of Philosophy

BPH Bachelor of Public Health • *Med.* benign prostatic hyperplasia (*or* hypertrophy)

BPharm Bachelor of Pharmacy

BPhil Bachelor of Philosophy

bpi *Computing* bits per inch

BPI Booksellers' Provident Institution • British Pacific Islands

BPICA Bureau permanent international des constructeurs d'automobiles (French: International Permanent Bureau of Motor Manufacturers)

BPIF British Printing Industries' Federation

bpl. birthplace

Bpl. Barnstaple

BPLA *Physics* Beijing Proton Linear Accelerator

bpm (*or* **b.p.m.**) barrels per minute • *Music* beats per minute

BPMA British Premium Merchandise Association • British Pump Manufacturers' Association

BPMF British Postgraduate Medical Federation • British Pottery Manufacturers' Federation

BPO base post office • Berlin Philharmonic Orchestra

BPOE (USA) Benevolent and Protective Order of Elks

BPP Botswana People's Party

BPPMA British Power-Press Manufacturers' Association

BPR business process re-engineering

BPRA Book Publishers' Representatives' Association

BPRO Blind Persons Resettlement Officer

bps *Computing* bits per second

BPs Bachelor of Psychology

BPS *Med.* blood-pool scintigraphy • border patrol sector (*or* station) • British Pharmacological Society • British Psychological Society • (USA) Bureau of Professional Standards (Internal Affairs)

BPsS British Psychological Society

Bp Suff. Bishop Suffragan

BPsych Bachelor of Psychology

bpt *Computing* bits per track

b.pt. boiling point

BPT battle practice target • British Petroleum Tanker

BPV *Microbiol.* bovine papilloma virus

bpy *Chem.* bipyridine (used in formulae)

bq. (*or* **bque**) barque

Bq *Physics, symbol for* becquerel

BQ bene quiescat (Latin: may he (*or* she) rest well)

BQA British Quality Association

BQMS battery quartermaster-sergeant

br. bearing • branch • bridge • brief • brig • bronze • (*or* **br**) brother • brown

b.r. bank rate • (*or* **b/r**) bills receivable

Br *symbol for* birr (Ethiopian monetary unit) • Bombardier • *Chem., symbol for* bromine • *RC Church* Brother • Bugler

Br. Branch • Brazil • Breton • Britain • British

BR *Law* Bancus Reginae *or* Regis (Latin: Queen's (*or* King's) Bench) • bioaccumulation ratio • *Med.* blink reflex • *Education* block release • book of reference • *Physics* branching ratio • *international vehicle registration for* Brazil • breeding ratio • *Physics* bremsstrahlung radiation • *British fishing port registration for* Bridgwater • British Rail • *postcode for* Bromley • (poly)butadiene rubber • butyl rubber • *British vehicle registration for* Newcastle upon Tyne

B/R bills receivable • Bordeaux or Rouen (in the grain trade) • *Insurance* builders' risks

BRA Bee Research Association • Brigadier Royal Artillery • British Records

Association • British Rheumatism and Arthritis Association

braai ('brɑːɪ) (South Africa) *Colloquial* braaivleis (Afrikaans: barbecue)

Brad. Bradford

BRAD (bræd) (ital.) British Rates and Data (publications directory)

Br. Am. British America

Bras. Brasil (Portuguese: Brazil)

BRAS ballistic rocket air suppression

Braz. Brazil(ian)

Brazza. Brazzaville

BRB *Med.* blood retinal barrier

Br. C. British Columbia

BRC (USA) base residence course • Biological Records Centre, Nature Conservancy Council • British Rabbits Council • British Radio Corporation

brch (*or* **Brch**) branch

BRCS British Red Cross Society

BRD *British fishing port registration for* Broadford • (USA) Building Research Division, National Bureau of Standards

BRDC British Racing Drivers' Club

brdcst broadcast

BRE Bachelor of Religious Education • Building Research Establishment

b. rec. bills receivable

BRE(L) British Rail Engineering (Limited)

BREMA British Radio Equipment Manufacturers' Association

Bret. Breton

brev. brevet • breveté (French: patent) • brevetto (Italian: patent)

brew. brewer(y) • brewing

brf *Law* brief

BRF Bible Reading Fellowship • British Road Federation

BRFC *Angling* British Record Fish Committee

brg bearing

br. g. *Horse racing* brown gelding

BRG (formerly) *international vehicle registration for* Guyana (now **GUY**)

Br. I. British India • British Isles

BRI Banque des règlements internationaux (French: Bank for International Settlement) • *Computing* basic-rate ISDN • (USA) Biological Research Institute • (USA) Brain Research Institute

Brig. Brigade • Brigadier

Brig. Gen. Brigadier General

brill. *Music* brillante (Italian: brilliant)

BRIMEC ('brɪmɛk) British Mechanical Engineering Confederation

BRINCO ('brɪŋkəʊ) British Newfoundland Corporation Limited

Brisb. Brisbane

Brist. Bristol

Brit. Britain • Britannia • British • Briton

Brit. Mus. British Museum

Brit. Pat. British Patent

Britt. *Numismatics* Britanniarum (Latin: of Great Britain)

brk brick

brkt bracket

brkwtr breakwater

brl barrel

BRL Ballistic Research Laboratory • (USA) Bible Research Library

BRM *Computing* binary-rate multiplier • *Med.* biological response modifier • British Racing Motors

BRMA Board of Registration of Medical Auxiliaries • British Rubber Manufacturers' Association

BRMCA British Ready-Mixed Concrete Association

BRMF British Rainwear Manufacturers' Federation

brn brown

BRN *international vehicle registration for* Bahrain

BRNC Britannia Royal Naval College

brng *Navigation* bearing • burning

bro. (brəʊ) brother

Bro. Brotherhood

BRO brigade routine order

brom. *Chem.* bromide

bros. (*or* **Bros.**) (brɒs) brothers

BRP biological reclamation process

BRS British Record Society • British Road Services • Building Research Station

BRSA British Rail Staff Association

BRSCC British Racing and Sports Car Club

brt bright

BRT (Belgium) Belgische Radio en Televisie (Dutch broadcasting company) • *Shipping* Brutto-Registertonnen (German: gross register tons)

BRTA British Racing Toboggan Association • British Regional Television Association • British Road Tar Association

BRU *international vehicle registration for* Brunei

Brum. (brʌm) Brummagem (colloquial name for Birmingham)

Brunsw. Brunswick

Brux. Bruxelles (French: Brussels)

BRW British Relay Wireless

bry. (or **bryol.**) bryology

brz. bronze

bs bags • bales

b.s. back stage • balance sheet • bill of sale

Bs symbol for bolivars (see under B) • symbol for boliviano (Bolivian monetary unit)

BS British vehicle registration for Aberdeen • (USA) Bachelor of Science • Bachelor of Surgery • international vehicle registration for Bahamas • battleship • battle squadron • British fishing port registration for Beaumaris • below specification • Chem. benzene-soluble • Bibliographical Society • bill of sale • Computing binary state • Biochemical Society • Blackfriars Settlement • Blessed Sacrament • Boy Scouts • Angling breaking strain of line • Med. breath sounds • postcode for Bristol • Aeronautics Bristol Siddeley • British Standard (indicating the catalogue or publication number of the British Standards Institution) • British Steel plc • Budgerigar Society • building society • Slang bullshit • Computing bus switch • British vehicle registration for Inverness

B/S bill of sale • Commerce bill of store

BSA Bachelor of Science in Agriculture • Bachelor of Scientific Agriculture • Bibliographical Society of America • Birmingham Small Arms Company • Chem. bis(trimethylsilyl)acetamide • Med. body surface area • Med. bovine serum albumin • Boy Scouts' Association • Boy Scouts of America • British School at Athens • British Speleological Association • Building Societies' Association

BSAA Bachelor of Science in Applied Arts • British School of Archaeology at Athens

BSAC British Sub-Aqua Club

BSAdv Bachelor of Science in Advertising

BSAE (or **BSAeEng**) Bachelor of Science in Aeronautical Engineering • (or **BSAgE**) Bachelor of Science in Agricultural Engineering

BSAgr Bachelor of Science in Agriculture

BS & W basic (or bottom) sediment and water

BSAP British Society of Animal Production

BSArch Bachelor of Science in Architecture

BSAS British Ship Adoption Society

BSAVA British Small Animals Veterinary Association

Bsb. Brisbane

BSB British Satellite Broadcasting (see BSkyB) • British Standard brass (type of screw thread)

BSBA Bachelor of Science in Business Administration

BSBC British Social Biology Council

BSBI Botanical Society of the British Isles

BSBus Bachelor of Science in Business

bsc basic

BSc Bachelor of Science

BSC Bachelor of Science in Commerce • Bengal Staff Corps • Bibliographical Society of Canada • Computing binary symmetric channel • Computing binary synchronous communications • Biomedical Sciences Corporation • British Safety Council • British Shoe Corporation • British Standard Channel • British Stationery Council • British Steel Corporation • British Sugar Corporation • British Supply Council • Broadcasting Standards Council

BSCA British Swimming Coaches' Association • (USA) Bureau of Security and Consular Affairs

BScA (or **BSc(Ag)**) Bachelor of Science in Agriculture

BScApp Bachelor of Applied Science

BSCC British Society of Clinical Cytology • British Synchronous Clock Conference

BSCE Bachelor of Science in Civil Engineering

BSChE Bachelor of Science in Chemical Engineering

BSCCO Physics Bi-Sr-Ca-Cu-O superconductor

BScD Bachelor of Dental Science

BSc(Dent) Bachelor of Science in Dentistry

BSc(Econ) Bachelor of Science in Economics

BSc(Ed) Bachelor of Science in Education

BSc(Hort) Bachelor of Science in Horticulture

BScMed Bachelor of Medical Science

BSc(Nutr) Bachelor of Science in Nutrition

BSCP (ital.) British Standard Code of Practice (publication)

BSCRA British Steel Castings Research Association

BScSoc Bachelor of Social Sciences

BSD (or **BSDes**) Bachelor of Science in Design • ballistic system division • British Society of Dowsers • British Space Development

b.s.d.l. boresight datum line

BSD Unix *Computing* Berkeley Systems Distribution UNIX

BSE (or **BSEd**) Bachelor of Science in Education • (or **BSEng**) Bachelor of Science in Engineering • bovine spongiform encephalopathy • *Med.* breast self-examination

BSEc Bachelor of Science in Economics

BSECP Black Sea Economic Cooperation Project

BSEE (or **BSEEng**) Bachelor of Science in Electrical Engineering • Bachelor of Science in Elementary Education

BSEIE Bachelor of Science in Electronic Engineering

BSEM Bachelor of Science in Engineering of Mines

BSES Bachelor of Science in Engineering Sciences • British Schools Exploring Society

BSF Bachelor of Science in Forestry • British Salonica Force • British Slag Federation • British Standard fine (type of screw thread) • British Stone Federation

BSFA British Science Fiction Association • British Steel Founders' Association

BSF(L) British Shipping Federation (Limited)

BSFM Bachelor of Science in Forestry Management

BSFor Bachelor of Science in Forestry

BSFS Bachelor of Science in Foreign Service • British Soviet Friendship Society

BSFT Bachelor of Science in Fuel Technology

BSG British Standard Gauge

b.s.g.d.g. breveté sans garantie du gouvernement (French: patented without government guarantee)

BSGE Bachelor of Science in General Engineering

bsh. bushel

BSH British Society of Hypnotherapists • British Standard Hardness

BSHA Bachelor of Science in Hospital Administration

BSHE (or **BSHEc**) Bachelor of Science in Home Economics

BSHS British Society for the History of Science

BSHyg Bachelor of Science in Hygiene

BSI British Sailors' Institute • British Standards Institution • Building Societies' Institute

BSIA British Security Industry Association

BSIB Boy Scouts International Bureau

BSIC British Ski Instruction Council

BSIE Bachelor of Science in Industrial Engineering

BSIP British Solomon Islands Protectorate

BSIRA British Scientific Instrument Research Association

BSIS Business Sponsorship Incentive Scheme

BSIU British Society for International Understanding

BSJ Bachelor of Science in Journalism

BSJA British Show Jumping Association

bsk. (or **bskt**) basket

BSkyB British Sky Broadcasting (formed by merger of BSB with Sky Television)

Bs/L bills of lading

BSL Bachelor of Sacred Literature • Bachelor of Science in Linguistics • *Med.* bacterial skin lesion • boatswain sublieutenant • British Sign Language

BSLS Bachelor of Science in Library Science

BSM Bachelor of Sacred Music • Bachelor of Science in Medicine • battery sergeant-major • *Chem.* benzene-soluble matter • branch sales manager • British School of Motoring • (USA) bronze star medal

BSMA British Skate Makers' Association

BSME Bachelor of Science in Mechanical Engineering • Bachelor of Science in Mining Engineering

BSMet Bachelor of Science in Metallurgy

BSMetE Bachelor of Science in Metallurgical Engineering

BSMGP British Society of Master Glass-Painters

bsmt basement

BSMT (or **BSMedTech**) Bachelor of Science in Medical Technology

BSMV *Microbiol.* barley stripe mosaic virus

BSN Bachelor of Science in Nursing

BSNE Bachelor of Science in Nuclear Engineering

BSNS Bachelor of Naval Science

BSO base supply officer • *Chem.* benzene-soluble organic • Boston Symphony Orchestra • Bournemouth Symphony Orchestra • Business Statistics Office

BSocSc Bachelor of Social Science

BSOT Bachelor of Science in Occupational Therapy

BSP (*or* **BSPhar, BSPharm**) Bachelor of Science in Pharmacy • Bering Sea Patrol • *Computing* binary space-partitioning (in **BSP tree**) • Birmingham School of Printing • *Printing* bleached sulphite pulp • British Standard pipe (type of screw thread) • Bulgarian Socialist (formerly Communist) Party • business systems planning

BSPA Bachelor of Science in Public Administration

BSPE Bachelor of Science in Physical Education

BSPH Bachelor of Science in Public Health

BSPT (*or* **BSPhTh**) Bachelor of Science in Physical Therapy

BSR Birmingham Sound Reproducers • *Med.* blood sedimentation rate • *Physics* Bragg scattering region • British School at Rome • *Physics* bulk shielding reactor

BSRA British Ship Research Association • British Society for Research on Ageing • British Sound Recording Association

BSRAE British Society for Research in Agricultural Engineering

BSRC (USA) Biological Serial Record Center

BSRT Bachelor of Science in Radiological Technology

BSS Bachelor of Secretarial Science • Bachelor of Social Science • *Med.* balanced salt solution • basic safety standards • Bibliothèque Saint-Sulpice (Montreal) • *Geology* borehole seismic system • British Sailors' Society • British Standard size • British Standards Specification

BSSA Bachelor of Science in Secretarial Administration

BSSc Bachelor of Social Science

BSSE Bachelor of Science in Secondary Education

BSSG (USA) Biomedical Sciences Support Grant

BSSO British Society for the Study of Orthodontics

BSSS Bachelor of Science in Secretarial Studies • Bachelor of Science in Social Science • British Society of Soil Science

BST Bachelor of Sacred Theology • *Agriculture* bovine somatotrophin • British Standard Time (1968–71) • British Summer Time • bulk supply tariff

B/St bill of sight

BSTA British Surgical Trades' Association

BSTC British Student Travel Centre

bstd bastard

BSTFA *Chem.* bis(trimethylsilyl)trifluoro-acetamide

bstr booster

bstr rkt booster rocket

BSU *Engineering* bench scale unit

BSurv Bachelor of Surveying

b.s.w. barrels of salt water

BSW Bachelor of Social Work • British Standard Whitworth (type of screw thread)

BSWB Boy Scouts World Bureau

BSWE British Scouts in Western Europe

BSWIA British Steel Wire Industries' Association

bt beat • benefit • bent • bought

Bt baht (Thai monetary unit) • Baronet • *Military* brevet

BT *airline flight code for* Air Baltic • Bachelor of Teaching • Bachelor of Theology • basic trainer • *Psychol.* behaviour therapy • *postcode for* Belfast • *Med.* benign tumour • bishop's transcript • British Telecom (*or* Telecommunications) • *British vehicle registration for* Leeds

BTA Billiards Trade Association • (USA) Blood Transfusion Association • British Theatre Association • British Tourist Authority (formerly British Travel Association) • British Tuberculosis Association

BTAC *Computing* binary tree algebraic computation

BTASA Book Trade Association of South Africa

BTB *Med.* blood–tumour barrier • *Med.* breakthrough bleeding

BTBA British Ten Pin Bowling Association

BTBS Book Trade Benevolent Society

BTC (USA) Bankers' Trust Company • (USA) basic training center • British

Textile Confederation • British Transport Commission

btca biblioteca (Spanish: library)

BTCC Board of Transportation Commissioners of Canada

BTCh Bachelor of Textile Chemistry

BTCP Bachelor of Town and Country Planning

BTCV British Trust for Conservation Volunteers

b.t.d. bomb testing device

BTDB Bermuda Trade Development Board

BTDC *Engineering* before top dead centre

bté breveté (French: patent)

BTE Bachelor of Textile Engineering

BTEC Business and Technology Education Council

BTech Bachelor of Technology

BTEE *Chem.* N-benzoyl-L-tyrosine ethyl ester

BTEF Book Trade Employers' Federation

BTEMA British Tanning Extract Manufacturers' Association

BTEX *Chem.* benzene, toluene, ethylbenzene, and xylene (solvents)

b.t.f. barrels of total fuel • bomb tail fuse

BTF British Tarpaviors' Federation • British Trawlers Federation • British Turkey Federation

BTG British Technology Group

bth bath(room) • berth

BTh Bachelor of Theology

BTH British Thomson-Houston Company

BTHMA British Toy and Hobby Manufacturers' Association

BThU British thermal unit

BTI (ital.) British Technology Index (publication)

BTIA British Tar Industries' Association

btk buttock

btl. bottle

BTL *Meteorol.* between layers

btm bottom

BTM *Chem.* bromotrifluoromethane

BTMA British Typewriter Manufacturers' Association

BTMSA *Chem.* bis(trimethylsilyl)acetylene

btn baton • button

BTN Brussels Tariff Nomenclature

BTO big time operator • British Trust for Ornithology

BTP Bachelor of Town Planning

BTR *Physics* bimetric theory of relativity •

(ital.) British Tax Review • British Telecommunications Research

BTRA Bombay Textile Industry's Research Association

BTRP Bachelor of Town and Regional Planning

btry *Military* battery

BTS Blood Transfusion Service • British Telecommunications Systems

Btss Baroness

BTTA British Thoracic and Tuberculosis Association

Btu British thermal unit

BTU Board of Trade unit • (USA) British thermal unit

BTUC Bahamas Trade Union Congress

BTW by the way (used in electronic mail)

btwn between

BTX *Chem.* benzene, toluene, and xylene • (Germany) Bildschirmtext (videotext system)

bty *Military* battery

B-type *Psychol.* Basedow type

bu. (*or* **Bu.**) bureau • (*or* **bu**) bushel(s)

b.u. base unit • *Finance* break-up

Bu *Chem.*, *symbol for* butyl group (in formulae)

BU Baptist Union of Great Britain and Ireland • Brown University, Rhode Island • *British fishing port registration for* Burntisland • *British vehicle registration for* Manchester

BuAer (USA) Bureau of Aeronautics

BUAF British United Air Ferries

BUA of E Badminton Umpires' Association of England

BUAV British Union for the Abolition of Vivisection

BUC Bangor University College

buck. buckram

Bucks Buckinghamshire

BUCOP (ital.) British Union Catalogue of Periodicals

bud. budget

Budd. (*or* **Bud.**) Buddhism • Buddhist

BuDocks *US Navy* Bureau of Yards and Docks

Budpst Budapest

BUF British Union of Fascists

BUIC back-up interceptor control

BUJ Bachelor of Canon and Civil Law (Latin *Baccalaureus utriusque juris*)

bul. bulletin

Bulg. Bulgaria(n)

bull. bulla (seal used on a papal bull) • bulletin • *Med.* bulliat (Latin: let it boil)

buloga (bʊˈləʊɡə) business logistics game

BULVA Belfast and Ulster Licensed Vintners' Association

BuMed *US Navy* Bureau of Medicine and Surgery

BUN *Med.* blood urea nitrogen

Buna (ˈbuːnə) *Trademark* butadiene + natrium (i.e. sodium) (type of synthetic rubber)

BUNAC British Universities North America Club

BUNCH (bʌntʃ) Burroughs, Univac, NCR, Control Data, Honeywell (computer manufacturers)

BuOrd *US Navy* Bureau of Ordnance

BUP British United Press

BUPA (ˈbuːpə) British United Provident Association (health-insurance company)

BuPers *US Navy* Bureau of Naval Personnel

Bu. Pub. Aff. (USA) Bureau of Public Affairs

bur. bureau • buried

Bur. Burma • Burmese

BUR *international vehicle registration for* Myanmar (Burma)

BuRec (USA) Bureau of Reclamation

burg. burgess • burgomaster

burl. burlesque

Burm. Burma • Burmese

Burs. Bursar

bus. bushel • business

BuSandA *US Navy* Bureau of Supply and Accounts

BUSF British Universities' Sports Federation

bush. bushel

BuShips *US Navy* Bureau of Ships

bus. mgr business manager

BUSWE British Union of Social Work Employees

but. butter • button

buy. buyer • buying

bv. *Taxonomy* biovar

b.v. balanced voltage • *Accounting* book value

BV Beatitudo Vestra (Latin: Your Holiness) • bene vale (Latin: farewell) • (Netherlands) Besloten Vennootschap (after the name of a private limited company) • Bible Version (of Psalms) • Blessed Virgin (Latin *Beata Virgo*) • *Med.*

blood volume • *Shipping* (France) Bureau Veritas (classification society) • *British vehicle registration for* Preston

BVA British Veterinary Association

BVD *Vet. science* bovine virus diarrhoea

BVetMed Bachelor of Veterinary Medicine

BVetSc Bachelor of Veterinary Science

BVF *Chemical engineering* bulk volume fermenter

BVI British Virgin Islands (abbrev. *or* IVR)

BVJ (ital.) British Veterinary Journal

BVK Bundesverdienstkreuz (German: Federal Cross of Merit)

BVM Bachelor of Veterinary Medicine • Blessed Virgin Mary (Latin *Beata Virgo Maria*)

BVMA British Valve Manufacturers' Association

BVMS Bachelor of Veterinary Medicine and Surgery

BVO Bundesverdienstorden (German: Federal Order of Merit)

BVP British Visitors' Passport • British Volunteer Programme

BVRLA British Vehicle Rental and Leasing Association

BVS Bachelor of Veterinary Surgery

BVSc Bachelor of Veterinary Science

BVSc & AH Bachelor of Veterinary Science and Animal Husbandry

bvt brevet

b.w. bitte wenden (German: please turn over, PTO) • bridleways

b/w *Photog., etc.* black and white

BW *British fishing port registration for* Barrow • Bath and Wells (episcopal see) • biological warfare • *Photog., etc.* black and white • Black Watch • Board of Works • *Med.* body weight • bonded warehouse • British Waterways • business week • *airline flight code for* BWIA • *British vehicle registration for* Oxford

B/W *Photog., etc.* black and white • black to white

BWA backward wave amplifier • Baptist World Alliance • British Waterworks Association • (formerly) British West Africa

B-way (USA) Broadway

BWB British Waterways Board

BWC (USA) Board of War Communications • British War Cabinet

BWCC British Weed Control Conference

bwd backward

BWD *Vet. science* bacillary white diarrhoea (pullorum disease)

BWF British Whiting Federation • British Wool Federation

BWG *Engineering* Birmingham Wire Gauge

BWI British West Indies

BWIA British West Indian Airways

BWIR British West India Regiment

BWISA British West Indies Sugar Association

bwk brickwork • bulwark

BWM British War Medal

BWMA British Woodwork Manufacturers' Association

BWMB British Wool Marketing Board

BWO backward wave oscillator

BWP basic war plan

BWPA backward wave power amplifier • British Waste Paper Association • British Word Preserving Association

BWPUC British Wastepaper Utilization Council

BWR *Nuclear physics* boiling-water reactor

BWS *Med.* battered woman (*or* wife) syndrome • British Watercolour Society

BWSF British Water Ski Federation

BWTA British Women's Temperance Association

BWU Barbados Workers' Union

b.w.v. back water valve

BWV *Music* Bach Werke-Verzeichnis (German: Catalogue of Bach's Works; precedes the catalogue number of a work by J. S. Bach)

BWVA British War Veterans of America

BWWA British Waterworks Association

bx box

BX *US Air Force* Base Exchange • British Xylonite • *British vehicle registration for* Haverfordwest

b.y. billion years

By. Barony

BY *international vehicle registration for* Belarus • *airline flight code for* Britannia Airways • *British vehicle registration for* NW London

BYDV *Microbiol.* barley yellow dwarf virus

Bye. Byelorussia(n)

Byo. Bulawayo

BYO bring your own

BYOB bring your own beer (booze, *or* bottle)

BYOG bring your own girl

byr billion years

BYT *Colloquial* bright young things

BYU (USA) Brigham Young University

Byz. Byzantine

Bz (*or* **bz**) *Chem., symbol for* benzene (in formulae) • *Chem., symbol for* benzoyl group (in formulae)

BZ *international vehicle registration for* Belize • B'nai Zion • British Zone • *British vehicle registration for* Down

B–Z (*or* **BZ**) *Chem.* Belousov–Zhabotinskii (as in **B–Z reaction**)

BZS *Chem.* basic zirconium sulphate

bzw. beziehungsweise (German: respectively)

C

c *symbol for* centi- (prefix indicating 10^{-2}, as in **cm**, centimetre) • *Physics* charm (a quark flavour) • *Chem., symbol for* concentration • *Maths., symbol for* constant • cubic (in **cc**, cubic centimetre) • (ital.) *Physics, symbol for* specific heat capacity • (ital.) *Physics, symbol for* speed of light in a vacuum • *indicating* the third vertical row of squares from the left on a chessboard

c. canine (tooth) • capacity • caput (Latin: chapter) • carat • carbon (paper) • case • *Baseball* catcher • cathode • *Cricket* caught • cent(s) • *Currency* centavo(s) • centime(s) • centre • century (*or* centuries) • chairman • chairwoman • chapter • child • church • *Med.* cibus (Latin: meal) • (ital.) circa (Latin: about; preceding a date) • circiter (Latin: approximately) • circum (Latin: around) • city •

cloudy • cold • colt • compound • consul • contra (Latin: against) • contralto • contrast • convection • copy • copyright • coupon • court • cousin • crowned • *Med.* cum (Latin: with; in prescriptions) • currency • cycle(s)

c/- (Australia, New Zealand) care of (in addresses) • case • coupon • currency

C *Geology* Cambrian • *international civil aircraft marking for* Canada • (ital.) *Physics, symbol for* capacitance • *Chem., symbol for* carbon • *Geology* Carboniferous • (USA) cargo transport (specifying a type of military aircraft, as in **C-5**) • Celsius (in °**C**, degree Celsius; formerly degree centigrade) • *Theatre* centre (of stage) • century • (ital.) *Physics, symbol for* charm quantum number • *Colloquial* cocaine • cold (water) • *Parliamentary procedure* Command Paper (prefix to serial number, 1870–99). *See also under* Cd; Cmd; Cmnd • Companion (in British Orders of Chivalry) • *Immunol., symbol for* complement • *Immunol.* constant region (of an immunoglobulin chain) • (ital.) *Physics, symbol for* compliance • *Pharmacol.* congius (Latin: gallon) • *Irish fishing port registration for* Cork • *Botany, symbol for* corolla (in a floral formula) • *Physics, symbol for* coulomb • crown (a standard paper size; *see under* C4; C8; C16; C32) • *international vehicle registration for* Cuba • *Biochem., symbol for* cytidine • *Biochem., symbol for* cytosine • (ital.) *Physics, symbol for* Euler number • (ital.) *Physics, symbol for* heat capacity • *Roman numeral for* hundred • (ital.) *Chem., symbol for* molecular concentration • *Computing, indicating* a programming language (developed from B) • *indicating* a musical note or key

C. (*or* **C**) Caesar • caldo (Italian: hot) • caliente (Spanish: hot) • calle (Spanish: street) • canon • canto • Cape (on maps) • Captain • Cardinal • catechism • Catholic • Celtic • Chancellor • Chancery • chaud (French: hot) • chief • Christ(ian) • circuit • *Cards* clubs • Commander • commended • Commodore • Confessor • Congregation(al) • Congress • Conservative • contract • Corps • council • Count • *Music* countertenor • county • coupon • cross • curacy • curate

C1 *indicating* supervisory, clerical (occupational grade)

C2 *airline flight code for* Air Caribbean • *international civil aircraft marking for* Nauru • *indicating* skilled manual (occupational grade)

C3 *indicating* a low standard of physical fitness; hence inferior (originally a military grade)

C4 *Television* Channel Four • crown 4to (a standard paper size, $7\frac{1}{2} \times 10$ in)

C5 *Television* Channel Five • *international civil aircraft marking for* Gambia

C6 *international civil aircraft marking for* Bahamas

C8 crown 8vo (a standard paper size, $5 \times 7\frac{1}{2}$ in)

C9 *international civil aircraft marking for* Mozambique

C16 crown 16mo (a standard paper size, $3\frac{3}{4} \times 5$ in)

C32 crown 32mo (a standard paper size, $2\frac{1}{2} \times 3\frac{3}{4}$ in)

3C *international civil aircraft marking for* Equatorial Guinea

ca. *Med.* carcinoma • *Law* cases • centiare (unit of area) • (ital.) circa (*see under* c.)

c.a. *Commerce* capital asset • *Engineering* close annealed • *Music* coll'arco (Italian: with the bow) • *Physics* corriente alterna (Spanish: alternating current) • *Physics* courant alternatif (French: alternating current)

Ca *Chem., symbol for* calcium

Ca. Canada • Canadian • compagnia (Italian: company)

C^a companhia (Portuguese: company; Co.) • compañia (Spanish: company; Co.)

CA *airline flight code for* Air China • *postcode for* California • Canadian army • *Taxation* capital allowances • *Biochem.* carbonic anhydrase • *British fishing port registration for* Cardigan • *postcode for* Carlisle • Caterers' Association • Catholic Association • *Chem.* cellulose acetate • Central America • *Physics* centrifugal atomization • Certificate of Airworthiness • chargé d'affaires • (Scotland) Chartered Accountant • (ital.) Chemical Abstracts • *British vehicle registration for* Chester • Chief Accountant • *Genetics* chromosomal aberration • chronological age • Church Army • Church Assembly • *Chem.* cinnamic aldehyde • citric acid • City Architect • City Attorney • civil affairs •

civil aviation • Classical Association • clean air • coast artillery • College of Arms • commercial agent • Community Association • Companies Act • (USA) Confederate Army • Constituent Assembly • Consular Agent • Consumers' Association • *Med.* contrast angiography • Contributions Agency • controlled atmosphere • Controller of Accounts • Cooperative Agreement • Corps d'Armée (French: Army Corps) • County Alderman • County Architect • Court of Appeal • Croquet Association • Crown Agent • Cruising Association • *Commerce* current assets

C/A (*or* **CA**) capital account • credit account • current account

CAA Canadian Authors' Association • *Commerce* Capital Allowances Act • Central African Airways Corporation • (USA) Civil Aeronautics Administration (*or* Administrator) • Civil Aviation Authority • Clean Air Act • Commonwealth Association of Architects • (USA) Community Action Agency • Concert Artists' Association • *Med.* coronary artery aneurysm • Cost Accountants' Association • County Agricultural Adviser

CAAA Canadian Association of Advertising Agencies

CAADRP civil aircraft airworthiness data recording program

CAAE Canadian Association of Adult Education

CAAIS computer-assisted action information systems

CAAR compressed-air accumulator rocket

CAARC Commonwealth Advisory Aeronautical Research Council

CAAT (Canada) College of Applied Arts and Technology • *Accounting* computer-assisted audit technique

CAAtt Civil Air Attaché

CAAV Central Association of Agricultural Valuers

cab. cabalistic • cabin • cabinet • cable

CAB Canadian Association of Broadcasters • *Chem.* cellulose acetate butyrate • Citizens' Advice Bureau • (USA) Civil Aeronautics Board • Commonwealth Agricultural Bureaux

CABAS City and Borough Architects' Society

CABEI (*or* **Cabei**) Central American Bank for Economic Integration

CABG *Med.* coronary artery bypass graft(ing)

CABM Commonwealth of Australia Bureau of Meteorology

CABMA Canadian Association of British Manufacturers and Agencies

CABS *Med.* coronary artery bypass surgery

cabtmkr cabinetmaker

CAC Canadian Armoured Corps • Central Advisory Committee • Central Arbitration Committee • (USA) Climate Analysis Center • Colonial Advisory Council • (France) Compagnie des agents de change (French stockbrokers' association) • (USA) Consumer Advisory Council • County Agricultural Committee

CACA Canadian Agricultural Chemicals Association

CAC&W continental aircraft control and warning

CACC Civil Aviation Communications Centre • Council for the Accreditation of Correspondence Colleges

CACD computer-aided circuit design

CACDS Centre for Advanced Computing and Decision Support

CACE Central Advisory Council for Education

CACGP Commission on Atmospheric Chemistry and Global Pollution

CACM Central Advisory Council for the Ministry • Central American Common Market

CACSD computer-aided control system design

CACUL Canadian Association of College and University Libraries

c-à-d c'est-à-dire (French: that is to say)

cad. *Med.* cadaver • *Music* cadenza • cadet

Cad. Cádiz

CAD (*or* **c.a.d.**) cash against documents • (USA) civil air defense • comité d'aide au développement (French: development assistance committee) • (kæd) computer-aided design (*or* drawing) • contract award date • *Med.* coronary artery disease • Crown Agent's Department

cadav. *Med.* cadaver

CADC central air-data computer • colour analysis display computer

CADCAM (*or* **CAD/CAM**) ('kæd,kæm) computer-aided design (*or* drawing), computer-aided manufacturing

CADD computer-aided drafting and design

CADE computer-assisted data evaluation

CADF Commutated Antenna Direction Finder

CADIN (USA) continental air defense integration north

CADIS computer-aided design information system

CADMAT ('kæd,mæt) computer-aided design, manufacturing, and testing

CADO (USA) central air documents office

CADPO communications and data-processing operation

CADS computer-aided design system

Cae *Astronomy* Caelum

CAE Canadian Aviation Electronics • cóbrese al entregar (Spanish: cash on delivery) • (Australia) College of Advanced Education • computer-aided education • computer-aided engineering

CAEC County Agricultural Executive Committee

CAEM (France) Conseil d'assistance économique mutuelle (Council for Mutual Economic Aid; COMECON)

CAER Conservative Action for Electoral Reform

Caern. (*or* **Caerns**) Caernarvonshire (former Welsh county)

Caes. Caesar

CAES compressed-air energy storage

CAF *Med.* cardiac assessment factor • Central African Federation • charities aid fund (*or* foundation) • clerical, administrative, and fiscal • (*or* **c.a.f.**) *Commerce* cost and freight • (*or* **c.a.f.**) *Commerce* coût, assurance, fret (French: cost, insurance, freight; c.i.f.)

CAFBC *Chemical engineering* circulating atmospheric fluidized bed combustion (process)

CAFE (kæ'feɪ) (USA) Corporate Average Fuel Economy (standard for minimum fuel consumption by cars)

CAFEA-ICC Commission on Asian and Far Eastern Affairs of the International Chamber of Commerce

CAFIC Combined Allied Forces Information Centre

c.a.f.m. commercial air freight movement

Cafod ('kæfɒd) Catholic Fund for Overseas Development

CAFR *Accounting* comprehensive annual financial report

CAFS *Computing* content-addressable file system

CAFU civil aviation flying unit

CAG Canadian Association of Geographers • *US Navy* carrier air group • (USA) civil air guard • commercial arbitration group • Commercial Artists' Guild • Composers'-Authors' Guild • Concert Artists' Guild

CAGI (USA) Compressed Air and Gas Institute

CAGR civil (*or* commercial) advanced gas-cooled reactor

CAGS (USA) Certificate of Advanced Graduate Study

CAH *Med.* chronic active hepatitis • *Med.* congenital adrenal hyperplasia • *Chem.* cyanacetic hydrazide

Cai. Caithness (former Scottish county) • Caius College (Cambridge)

CAI *Geology* calcium–aluminium-rich inclusion • Canadian Aeronautical Institute • Club Alpino Italiano (Italian Alpine Club) • *Physics* coded-aperture imaging • colour alteration indices • computer-aided (*or* -assisted) instruction • Confederation of Aerial Industries (Limited) • *Chemical engineering* controlled air incineration (*or* incinerator) • *Forestry* current annual increment

CAIB Certified Associate of the Institute of Bankers

CAIRC (USA) Caribbean Air Command

CAIS *Computing* (USA) common APSE interface set (now replaced by **CAIS-A**)

CAISM Central Association of Irish Schoolmistresses

CAISSE computer-aided information system on solar energy

Caith. Caithness (former Scottish county)

cal *symbol for* calorie

cal. (ital.) *Music* calando (Italian: calming) • calendar • calibre

Cal *symbol for* kilocalorie (*or* Calorie)

Cal. Calcutta • Caledonia • Calends • California

CAL computer-aided (*or* -assisted) learning • *Computing* conversational algebraic language • (USA) Cornell (University) Aeronautical Laboratory

CALA Civil Aviation (Licensing) Act

CALANS Caribbean and Latin American News Service

calc. calculate(d) • calculus

Calc. Calcutta

Calç. Calçada (Portuguese: street)

cald calculated

CALE Canadian Army Liaison Executive

calg calculating

Calg. (Canada) Calgary

calibr. calibrate • calibration

Calif. California

Cal. Mac. (Scotland) Caledonian MacBrayne (ferry company)

caln calculation

CALPA Canadian Air Line Pilots' Association

CALS computer-aided (acquisition and) logistics support

Caltech ('kæltɛk) California Institute of Technology

Calv. Calvin(ism)

cam. camber • camouflage

Cam *Astronomy* Camelopardalis

Cam. Cambodia(n) • Cambrian • Cambridge • Cameroon

CAM *international vehicle registration for* Cameroon • (kæm) *Biochem.* cellular adhesion (*or* cell adherence) molecule • *Computing* cellular automata machine • commercial air movement • communication, advertising, and marketing (as in **CAM Foundation**) • computer-aided (*or* -assisted) manufacture • *Computing* content-addressable memory • continuous air monitor • *Botany* crassulacean acid metabolism (as in **CAM plant**)

CAMA Civil Aerospace Medical Association

CAMAL ('kæməl) continuous airborne missile-launched and low-level system

Camb. Cambrian • Cambridge

Cambs Cambridgeshire

CAMC Canadian Army Medical Corps

CAMD *Chem.* computer-aided molecular design

CAMDA Car and Motorcycle Drivers' Association

CAMDS *Military* chemical agent munitions disposal system

CAMM computer-aided (*or* -assisted) maintenance management

cAMP *Biochem.* cyclic AMP (adenosine 3'-5'-phosphate)

CAMRA ('kæmrə) Campaign for Real Ale

CAMRIC ('kæmrɪk) Campaign for Real Ice Cream

CAMS Certificate of Advanced Musical Study • *Chem.* collision activation mass spectrum • *Engineering* computer-aided manipulation system

CaMV *Microbiol.* cauliflower mosaic virus

CAMW Central Association for Mental Welfare

can. canal • cancel • cannon • *Music* canon • canto • canton

Can. Canada • Canadian • Canberra • *Ecclesiast.* Canon(ry) • Cantoris (Latin: the place of the cantor; indicating the side of a choir where the precentor sits, usually the north side)

CAN *Chem.* ceric ammonium nitrate • customs-assigned number

Canad. Canada • Canadian

canc. cancellation • cancelled

Canc. Cancellarius (Latin: Chancellor)

CANCIRCO (kæn'sɜːkəʊ) Cancer International Research Cooperative

cand. candidate

c & b *Cricket* caught and bowled (by)

c & c carpets and curtains

c & d collection and delivery

C & E Customs and Excise

c & f cost and freight

C & G City and Guilds

c & i cost and insurance

C & I commerce and industry • commercial and industrial

C & J *Weightlifting* clean and jerk

c & m care and maintenance

c & p carriage and packing • *Bookbinding* collated and perfect

CANDU ('kændjuː) Canadian Deuterium Uranium Reactor

C & W country and western (music)

C & W Ck caution and warning (system) check

CANEL (USA) Connecticut Advanced Nuclear Engineering Laboratory

Can. Fr. Canadian French

Can. I. Canary Islands

CANO Chief Area Nursing Officer

Can. Pac. Canadian Pacific

CANSG Civil Aviation Navigational Services Group

cant. cantilever

Cant. Canterbury • *Bible* Canticles • Cantonese

Cantab. (kæn'tæb) Cantabrigiensis (Latin: of Cambridge; used with academic awards)

canton. *Military* cantonment

CANTRAN ('kæn,træn) cancel in transmission

Cantuar. ('kæntjuˌɑː) Cantuariensis (Latin: (Archbishop) of Canterbury)

CANUS *Military* Canada–United States

canv. canvas

CAO Chief Accountant Officer • Chief Administrative Officer • County Advisory Officer • County Agricultural Officer • Crimean Astrophysical Observatory

CAORB Civil Aviation Operational Research Branch

CAORG Canadian Army Operational Research Group

CAOT Canadian Association of Occupational Therapy

cap. capacity • *Med.* capiat (Latin: let him (*or* her) take; in prescriptions) • capital • capitalize • capital letter • capitulum *or* caput (Latin: chapter *or* heading) • foolscap

c.a.p. (Italy) codice di avviamento postale (postcode number)

Cap *Astronomy* Capricornus

Cap. Captain

CAP Canadian Association of Physicists • *Genetics* catabolite activator protein • (USA) civil air patrol • Code of Advertising Practice • College of American Pathologists • (USA) combat air patrol • Common Agricultural Policy (in the EU) • (USA) Community Action Program • computer-aided planning • computer-aided production

CAPA Canadian Association of Purchasing Agents

CAPAC Composers', Authors', and Publishers' Association of Canada

CAPCOM ('kæp,kɒm) capsule communicator (in NASA)

CAPD *Med.* continuous ambulant peritoneal dialysis

CAPE Clifton Assessment Procedures for the Elderly

CAPM *Accounting* capital asset pricing model • computer-aided production management

Capn Captain

CAPO Canadian Army Post Office • Chief Administrative Pharmaceutical Officer

CAPP computer-aided process planning

Capric. Capricorn

caps. capital letters • capsule

CAPS (USA) Center for Analysis of Particle Scattering

capt. caption

Capt. Captain

car. carat

c.a.r. compounded annual rate (of interest)

Car *Astronomy* Carina

Car. Carlow

CAR Canadian Association of Radiologists • Central African Republic • Civil Air Regulations • *Aeronautics* cloudtop altitude radiometer • *Chemical engineering* coal–alkali reagent • *Nuclear physics* collisionally activated reaction • (ital.) (Australia) Commonwealth Arbitration Reports • *Finance* compound annual return (*or* rate) • computer-assisted retrieval • *Computing* contents of address register

CARA combat air rescue aircraft

CARAC Civil Aviation Radio Advisory Committee

carb. *Chem.* carbon • *Chem.* carbonate

CARB California Air Resources Board

card. *Maths., Logic* cardinal

Card. *RC Church* Cardinal

CARD (kɑːd) Campaign Against Racial Discrimination • compact automatic retrieval device • computer-augmented road design

CARDE Canadian Armament Research and Development Establishment

Cards Cardiganshire (former Welsh county; now called Ceredigion)

CARE (kɛə) Christian Action for Research and Education • *Psychiatry* communicated authenticity, regard, empathy • computer-aided risk evaluation • continuous aircraft reliability evaluation • Cooperative for American Relief Everywhere • Cottage and Rural Enterprises

CAREC ('kærɛk) Caribbean Epidemiology Centre

Carib. Caribbean

CARIBANK ('kærɪ,bæŋk) Caribbean Investment Bank

CARICAD ('kærɪ,kæd) Caribbean Centre for Administration Development

CARICOM ('kærɪ,kɒm) Caribbean Community and Common Market

CARIFTA (kæ'rɪftə) Caribbean Free Trade Area

Carliol. ('kɑːlɪɒl) Carlioliensis (Latin: (Bishop) of Carlisle)

Carms Carmarthenshire

carn. carnival

Carns Caernarvonshire (former Welsh county)

carp. carpenter • carpentry

Carp. Carpathian mountains

carr. fwd *Commerce* carriage forward

CARS (kɑːz) Canadian Arthritis and Rheumatism Society • *Physics* coherent anti-Stokes Raman spectroscopy

cart. cartage

CART collision avoidance radar trainer

Carth. Carthage

cartog. cartographer • cartography

cas. castle • casual • casualty

Cas *Astronomy* Cassiopeia

CAS Cambridge Antiquarian Society • (Guyana) Carib Advertising Services • *Chem.* cationic and ampholytic surfactants • Centre for Administrative Studies • CERN Accelerator School • (USA) Certificate of Advanced Studies • Chemical Abstracts Service • Chief of Air Staff • Children's Aid Society • close air support • *Aeronautics* collision avoidance system • controlled airspace • *Med.* coronary arterial stenosis • *Physics* criticality alarm system • Fellow of the Connecticut Academy of Arts and Sciences (Latin *Connecticutensis Academiae Socius*)

ca. sa. (keɪ seɪ) *Law* capias ad satisfaciendum (a writ of execution)

CA(SA) Chartered Accountant (South Africa)

CASA (USA) Coal Advisory Service Association • Contemporary Art Society of Australia

CASAC (USA) Clean Air Scientific Advisory Committee

CASB captured anaerobic sludge bed

CASE Centre for Advanced Studies in Environment (of the Architecture Association) • (USA) Committee on Academic Science and Engineering • computer-aided (*or* -assisted) software (*or* system) engineering • Confederation for the Advancement of State Education • Cooperative Awards in Science and Engineering

casevac ('kæsɪ,væk) *Military* casualty evacuation

cash. cashier

CASI Canadian Aeronautics and Space Institute

CASIG Careers Advisory Service in Industry for Girls

CASLE Commonwealth Association of Surveying and Land Economy

Caspar (*or* **CASPAR**) ('kæspɑː) *Physics* Cambridge analog simulator for predicting atomic reactions

Cast. Castile • Castilian

CAST Consolidated African Selection Trust

CASTE Civil Aviation Signals Training Establishment

CASU Cooperative Association of Suez Canal Users

CASW (USA) Council for the Advancement of Scientific Writing

cat. catalogue • catamaran • *Med.* cataplasm (a poultice) • catapult • *Christianity* catechism • category • caterpillar tractor • cattle

Cat. Catalan • Catholic • Catullus (?84–?54 BC, Roman poet)

CAT Centre for Alternative Technology • *Psychol.* Children's Apperception Test • *Genetics* chloramphenicol acetyl transferase (as in **CAT assay**) • Civil Air Transport • cleanup and treatment • *Aeronautics* clear-air turbulence • College of Advanced Technology • compressed-air tunnel • computer-aided (*or* assisted) teaching • computer-aided (*or* assisted) testing • computer-aided (*or* assisted) trading • computer-aided (*or* assisted) training • computer-aided (*or* assisted) translation • computer-aided (*or* assisted) typesetting • (kæt) *Med.* computerized axial (*or* computer-assisted) tomography

Cata ('kætə) Commonwealth Association of Tax Administrators

catachr. catachrestic

Catal. Catalan

CATC Commonwealth Air Transport Commission

CATCC Canadian Association of Textile Colorists and Chemists

cath. cathode

Cath. Cathedral • Catholic • St Catherine's College, Oxford

CATI computer-assisted telephone interviewing

c.atk counterattack

CATOR Combined Air Transport Operations Room

CATRA Cutlery and Allied Trades Research Association

CATS *Education* credit accumulation transfer scheme

CATU Ceramic and Allied Trade Union

CATV cable television • community antenna television

caus. causation • causative

cav. cavalier • *Law* caveat

Cav. Cavaliere (Italian: Knight) • (*or* **cav.**) cavalry

CAV constant air-volume (control system) • *Computing* constant angular velocity • (*or* **c.a.v.**) *Law* curia advisari vult (Latin: the court wishes to consider it; used in law reports when the judgment was given after the hearing) • *Med.* cyclophosphamide, adriamycin, and vincristine (in chemotherapy)

CAVD *Psychol.* Completion, Arithmetic Problems, Vocabulary, following Directions (intelligence test)

CAVI Centre audio-visuel international (French: International Audio-Visual Centre)

CAWU Clerical and Administrative Workers' Union (later amalgamated with APEX, now part of GMB)

Cay. Cayenne • Cayman Islands

c.b. cash book • cast brass • centre of buoyancy (of a boat, etc.) • circuit breaker • compass bearing • continuous breakdown

c/b *Cricket* caught and bowled

Cb *Chem., symbol for* columbium • (ital.) *Meteorol.* cumulonimbus

CB *postcode for* Cambridge • Cape Breton (Canada) • carbon black • carte blanche • cavalry brigade • (USA) Census Bureau • *Theatre* centre back (of stage) • chemical and biological (weapons *or* warfare) • Chief Baron • *Chem.* chlorobromomethane (used for fire extinction) • *Radio* Citizens' Band • Coal Board • *Law* Common Bench • Companion of the (Order of the) Bath • *Physics* conduction band • *Navy* confidential book • confined to barracks • *Military* construction battalion • cost benefit • county borough • *Chem.* covalent bonding • currency bond • *British vehicle registration for* Manchester • *airline flight code for* Suckling Airways

CBA *Physics* colliding-beam accelerator • Commercial Bank of Australia • Commonwealth Broadcasting Association • (USA) Community Broadcasters' Association • cost benefit analysis • Council for British Archaeology

CBAA Canadian Business Aircraft Association

CB & PGNCS circuit breaker and

primary guidance navigation control system

CBAT (USA) College Board Achievement Test

CBB Campaign for Better Broadcasting

CBBC Children's BBC

CBC Canadian Broadcasting Corporation • Caribbean Broadcasting Company • (USA) Children's Book Council • (Australia) Christian Brothers' College • *Computing* cipher block chaining • (*or* **c.b.c.**) *Med.* complete blood count • County Borough Council

CBCRL Cape Breton Coal Research Laboratory

CBCS (Australia) Commonwealth Bureau of Census and Statistics

CBD cash before delivery • central business district • *Anatomy* common bile duct

CBDC Cape Breton Development Corporation

CBE chemical, biological, and environmental • Commander of the Order of the British Empire • Council for Basic Education

CBED *Physics* convergent-beam electron diffraction

CBEL Cambridge Bibliography of English Literature

CBEVE Central Bureau for Educational Visits and Exchanges

CBF Central Board of Finance • *Med.* cerebral blood flow

CBG *Biochem.* corticosteroid-binding (*or* cortisol-binding) globulin

c.b.i. complete background investigation

CBI Cape Breton Island • (USA) Central Bureau of Identification • computer-based information • computer-based instruction • Confederation of British Industry • (USA) Cumulative Book Index

CBIM Companion of the British Institute of Management

CBiol Chartered Biologist

CBIS computer-based information system

CBIV computer-based interactive video-disc

CBJO Coordinating Board of Jewish Organizations

cbk cheque book

cbl. cable

c.b.l. commercial bill of lading

CBL computer-based learning • *Meteorol.* convective boundary layer

CBM Californian Business Machines •

Chem. carbon-bond mechanism • confidence building measure • conveyor belt monitor (*or* monitoring)

CBMIS computer-based management information system

CBMM Council of Building Materials Manufacturers

CBMPE Council of British Manufacturers of Petroleum Equipment

CBMS (USA) Conference Board of Mathematical Sciences

C-BN *Chem.* cubic boron nitride

CBNM (USA) Central Bureau for Nuclear Measurements

CBNS Commander British Navy Staff

Cbo Colombo (Sri Lanka)

CBO Conference of Baltic Oceanographers • (USA) Congressional Budget Office • Counter-Battery Officer

CBOE Chicago Board of Options Exchange

C-bomb cobalt bomb

CBOT Chicago Board of Trade

CBPC Canadian Book Publishers' Council

CBQ civilian bachelor quarters

CBR (USA) Center for Brain Research • chemical, bacteriological, and radiation (weapons *or* warfare) • *Aeronautics* cloud base recorder • *Immunol.* complement-binding reaction • *Computing* constant bit rate • *Astronomy* cosmic background radiation • crude birth rate

CBRI (India) Central Building Research Institute

CBS Canadian Biochemical Society • (Netherlands) Centraal Bureau voor de Statistiek (Central Statistical Bureau) • Church Building Society • *Med.* citrate-buffered saline • *Computing* close binary system • *Physics* coherent Brems-strahlungradiation spectrum (*or* spectra) • (USA) Columbia Broadcasting System • *Med.* computerized bone scanning • *RC Church* Confraternity of the Blessed Sacrament

CBSA Clay Bird Shooting Association

CBSI Chartered Building Societies Institute

CBSM conveyor-belt service machine

CBSO City of Birmingham Symphony Orchestra

CBT Chicago Board of Trade • computer-based teaching (*or* training) • *Physics* conduction-band tail

c.b.u. clustered bomb unit • (*or* **CBU**)

Commerce completely built-up (of goods for immediate use)

CBV *Med.* cerebral blood volume

CBW chemical and biological warfare

CBWA *Physics* counterstreaming electron-beam beta-wave accelerator

CBX company branch (telephone) exchange

CBZ *Chem.* carbobenzyloxy • coastal boundary zone

cc carbon copy (*or* copies) • cruise control (in motor advertisements) • cubic centi-metre(s)

cc. centuries • chapters

c.c. carbon copy (*or* copies) • cash credit • change course • chronometer correction • close control • colour code • (*or* **c/c**) compte courant (French: current account) • (*or* **c/c**) conto corrente (Italian: current account) • contra credit • *Physics* courant continu (French: direct current) • cubic centimetre(s)

Cc (ital.) *Meteorol.* cirrocumulus

CC *British vehicle registration for* Bangor • Caius College (Cambridge) • Cape Colony • Caribbean Commission • central committee • Chamber of Commerce • Charity Commission • chess club • Chief Clerk • *international civil aircraft marking for* Chile • circuit court • City Council • City Councillor • civil commotion • civil court • *Physics* close coupling • closed circuit (transmission) • collision course • colour centre • *Photog.* colour correction (*or* conversion; as in **CC filter**) • community council • *Genetics* compact chromosome • Companion of the Order of Canada • company commander • *Electronics* compensating current • computer code • concave • *Military* confined to camp • Consular Clerk • *Commerce* continuation clause • control computer • *Med.* corpus callosum • *Med.* corpus cardiacum • *Physics* correction coil • *Physics* corriente continua (Spanish: direct current) • *Biology* Coulter counter • Countryside Commission • County Clerk • County Commissioner • County Council • County Councillor • County Court • *Electronics* coupled channel • credit card • cricket club • croquet club • Crown Clerk • cruise control (in motor advertisements) • cruising club • Curate in Charge • cushion craft • cycling club

C/C cruise control (in motor advertisements)

C–C *Chem.* carbon–carbon (as in **C–C bond**)

CCA Canadian Construction Association • *US Navy* carrier-controlled approach • Chief Clerk of the Admiralty • *Chem.* chromated copper arsenate • (USA) Circuit Court of Appeals • *Med.* common carotid artery • Commonwealth Correspondents' Association • (USA) Consumers' Cooperative Association • continental control area • (USA) Council for Colored Affairs • County Councils' Association • (USA) County Court of Appeals • current-cost accounting

CCAB Canadian Circulations Audit Board • Consultative Committee of Accountancy Bodies

CCAFS (USA) Cape Canaveral Air Force Station

CCAHC Central Council for Agricultural and Horticultural Cooperation

CCAM Canadian Congress of Applied Mechanics

CCAMLR Commission for the Conservation of Antarctic Marine Living Resources

c.c.b. cubic capacity of bunkers

CCB (South Africa) Civil Cooperation Bureau

CCBI Council of Churches for Great Britain and Ireland

CCBN Central Council for British Naturism

CCBW Committee on Chemical and Biological Warfare

c.c.c. cwmni cyfyngedig cyhoeddus (Welsh: public limited company; plc)

CCC Canadian Chamber of Commerce • Central Control Commission • Central Criminal Court • (USA) Chemical Control Corporation • Christ's College, Cambridge • (USA) Civilian Conservation Corps • Club Cricket Conference • Commodity Credit Corporation • Conseil de coopération culturelle (French: European Council for Cultural Cooperation) • Corpus Christi College (Oxford or Cambridge) • Council for the Care of Churches • County Cricket Club • cross-country club • *Physics* cryogenic current comparator • Customs Cooperation Council

CCCA Cocoa, Chocolate, and Confectionery Alliance • Corps Commander, Coast Artillery

CCCC Corpus Christi College, Cambridge

CCCI *Military* command, control, communications, and intelligence

CCCM Central Committee for Community Medicine

CCCO Committee on Climatic Changes and the Ocean

CCCP Soyuz Sovietskikh Sotsialisticheskikh Respublik (Russian: Union of Soviet Socialist Republics)

CCCS Commonwealth and Continental Church Society

CCD Central Council for the Disabled • *Electronics* charged-coupled device • Conseil de coopération douanière (French: Customs Cooperation Council)

CCDA (USA) Commercial Chemical Development Association

CCE *Chem.* carbon-chloroform extract • Chartered Civil Engineer • compound-cycle engine • Conseil des communes d'Europe (French: Council of European Municipalities) • (USA) Council of Construction Employers • *Electronics* current-carrying element • *Accounting* current cash equivalent

c.c.e.i. composite cost effectiveness index

CCES *Physics* cold-cathode discharge induced emission spectroscopy

CCETSW Central Council for Education and Training in Social Work

CCETT (France) Centre commun d'études de télédiffusion et de télécommunications (national TV and telecommunications research centre)

CCF central computing facility • Combined Cadet Force • common-cause failure • Common Cold Foundation • concentrated complete fertilizer • *Med.* congestive cardiac failure • (Canada) Cooperative Commonwealth Federation • *Statistics* cross-correlation function

CCFA Combined Cadet Force Association

CCFD *Statistics* complementary cumulative frequency distribution

CCFL *Electronics* counter-current flow limit

CCFM Combined Cadet Forces Medal

CCFP Certificate of the College of Family Physicians

CCG Control Commission for Germany

CCGB Cycling Council of Great Britain

c.c.h. commercial clearing house • cubic capacity of holds

CCHE Central Council for Health Education

CChem Chartered Chemist

CCHF Children's Country Holidays Fund

CCHMS Central Committee for Hospital Medical Services

CCI Calculated Cetane Index • Chambre de commerce internationale (French: International Chamber of Commerce)

CCIA Commission of the Churches on International Affairs • (USA) Consumer Credit Insurance Association

CCIC Comité consultatif international du coton (French: International Cotton Advisory Committee)

CCIR Catholic Council for International Relations • Comité consultatif international des radiocommunications (French: International Radio Consultative Committee)

CCIS command control information system

CCITT Comité consultatif international télégraphique et téléphonique (French: International Telegraph and Telephone Consultative Committee)

CCJ Circuit Court Judge • Council of Christians and Jews • County Court Judge • *Med.* cranio-cervical junction

CCJO Consultative Council of Jewish Organizations

CCK *Biochem.* cholecystokinin

CCL Canadian Congress of Labour • (USA) commodity control list • *Physics* coupled-cavity linac

C.Cls (USA) Court of Claims

CCM *Med.* caffeine clearance measurement • *Physics* constant current modulation • *Physics* controlled carrier modulation

CCMA Canadian Council of Management Association • Commander, Corps Medium Artillery • Contract Cleaning and Maintenance Association

CCMD (USA) Carnegie Committee for Music and Drama • *Physics* continuous-current monitoring device

CCMS Committee on the Challenge of Modern Society (in NATO)

CCN *Meteorol.* cloud condensation nuclei • command control number • contract change notice (*or* notification)

CCNDT Canadian Council for Non-Destructive Technology

CCNR Consultative Committee for Nuclear Research (in the Council of Europe)

CCNSC (USA) Cancer Chemotherapy National Service Center

CCNY Carnegie Corporation of New York • City College of the City University of New York

CCO Central Coding Office • *Physics* current-controlled oscillator

CCOA County Court Officers' Association

CCOFI California Cooperative Oceanic Fisheries Investigations

CCOP *Chemical engineering* Carnot coefficient of performance • *Chem.* chlorine-catalysed oxidative pyrolysis

c.c.p. credit-card purchase • *Crystallog.* cubic close-packed

CCP Chinese Communist Party • Code of Civil Procedure • Committee on Commodity Problems (in the FAO) • Court of Common Pleas • *Nuclear physics* critical compression pressure

CCPE Canadian Council of Professional Engineers

CCPF Comité central de la propriété forestière de la CE (French: Central Committee on Forest Property for the EU)

CCPI Cyanamid Canada Pipeline Inc

CCPIT China Committee for the Promotion of International Trade

CCPL (USA) Computer Center Program Library

CCPO Comité central permanent de l'opium (French: Permanent Central Opium Board)

CCPR Central Council of Physical Recreation

CCPS Consultative Committee for Postal Studies

CCR cassette camera recorder • (USA) Commission of Civil Rights • Common Centre of Research • *Chemical engineering* continuous catalyst regeneration • *Law* contract change request • *Med.* cranial cavity ratio

CCRA Commander Corps of Royal Artillery

CCRE Commander Corps of Royal Engineers

CCREME Commander Corps of Royal Electrical and Mechanical Engineers

CCRSigs Commander Corps of Royal Signals

CCRU (USA) Common Cold Research Unit

CCS *Computing* calculus of communicating systems • Canadian Cancer Society • Canadian Ceramic Society • casualty clearing station • child-care service • collective call sign • (USA) Combined Chiefs of Staff • controlled combustion system • *Chem.* controlled component synthesis

CCSA Canadian Committee on Sugar Analysis

CCSATU Coordinating Council of South African Trade Unions

CCSD *Computing* coupled-cluster single and double

CCSEM computer-controlled scanning electron microscopy

CCSS centrifugally cast stainless steel

CCST (USA) Center for Computer Sciences and Technology

CCSU Council of Civil Service Unions

CCT *Med.* cerebral circulation time • (USA) clean coal technology • common customs tariff (in the EU) • *Electronics* contact charge-transfer • *Physical chem.* continuous cooling transformation • *Nuclear engineering* continuous-current tokamak • correct corps time • *Med.* cranial computed tomography

CCTA Central Computer and Telecommunications Agency • Commission de coopération technique pour l'Afrique (French: Commission for Technical Cooperation in Africa) • Coordinating Committee of Technical Assistance

CCTG constant concentration tracer gas (system)

CCTR *Nuclear physics* cassette compact toroid reactor

CCTS (USA) Canaveral Council of Technical Societies • Combat Crew Training Squadron

CCTV closed-circuit television

CCU *Med.* coronary care unit

CCUS Chamber of Commerce of the United States

CCV *Aeronautics* control-configured vehicle

ccw. counterclockwise

CCW *Nuclear physics* component cooling water • Curriculum Council for Wales

CCWS *Nuclear physics* component cooling water system

cd *Physics, symbol for* candela • cord • could

c.d. (*or* **c/d**) *Book-keeping* carried down • cash discount • *Finance* cum dividend (i.e. with dividend)

Cd *Chem., symbol for* cadmium • *Military* Command • *Parliamentary procedure* Command Paper (prefix to serial number, 1900–18). *See also under* C; Cmd; Cmnd • *Military* Commissioned

CD *British vehicle registration for* Brighton • Canadian Forces Decoration • *Electronics* carrier density • certificate of deposit • *Law* Chancery Division • *Chem.* chemiluminescence detection • *Physical chem.* circular dichroism • Civil Defence • civil disobedience • closing date • *Immunol., Med.* cluster of differentiation (preceding a number, as in **CD4**) • coal dust • College Diploma • (Jamaica) Commander of the Order of Distinction • commercial dock • *Electronics* compact disc • Conference on Disarmament (of the UN) • confidential document • (USA) Congressional District • *Med.* contagious disease • *Image technol.* contrast detail • *Physics* convective-dispersive • *Astronomy* Córdoba Durchmusterung (Star Catalogue) • *Nuclear engineering* core damage • Corps Diplomatique (French: Diplomatic Corps) • count down • *Freemasonry* Court of Deliberation • *Electronics* current density

C/D consular declaration • customs declaration

CDA Canadian Dental Association • Centre de Donnés Astronomiques (French: Centre for Astronomical Data) • (South Africa) Christian Democratic Alliance • Civil Defence Act • College Diploma in Agriculture • Colonial Dames of America • *Computing* compound document architecture • (Canada) Conference of Defence Associations • Copper Development Association

CDAAA Committee to Defend America by Aiding the Allies

CD&G compact disc and graphics

Cd Armn Commissioned Airman

CDAS Civil Defence Ambulance Service

Cd B Commissioned Boatswain

CDB *Computing* comprehensive database

cdbd cardboard

Cd Bndr Commissioned Bandmaster

CDC Canada Development Corporation • canister decontamination cell (*or* chamber) • *Chem.* carbon derived from coal • Caribbean Defence Command •

(USA) Center(s) for Disease Control • (USA) Combat Development Command • *Computing* command and datahandling console • Commissioners of the District of Columbia • *Computing* common development cycle • Commonwealth Development Corporation • Control Data Corporation (computer manufacturers) • cost determination committee • *Electronics* critical density of current

Cd CO Commissioned Communication Officer

Cd Con Commissioned Constructor

CDD (USA) certificate of disability for discharge • *Electronics* charge-density distribution • *Chem.* chlorinated dibenzo-*p*-dioxin

CDDI *Computing* copper distributed data interface

CDE chemical defence ensemble • (*or* **CD-E**) compact disc erasable • Conference on Confidence- and Security-Building and Disarmament in Europe (as in **CDE treaty**)

CDEE Chemical Defence Experimental Establishment

C de G Croix de Guerre (French military decoration)

CDEM crop disease environment monitor

Cd Eng Commissioned Engineer

c.d.f. *Statistics* cumulative distribution function

CDF *Computing* central database facility • Collider Detector at Fermilab • *Nuclear physics* core-damage frequency

CDFC Commonwealth Development Finance Company

cd fwd *Book-keeping* carried forward

Cdg. Cardigan • Cardiganshire (former Welsh county)

Cd Gr Commissioned Gunner

CDH College Diploma in Horticulture • *Med.* congenital dislocation of the hip

CDHS California Department of Health Services

CDI (*or* **CD-I**) compact-disc interactive

CDIC (USA) Carbon Dioxide Information Center

CDIF *Computing* CASE data interchange format

Cd In O Commissioned Instructor Officer

CDipAF Certified Diploma in Accounting and Finance

c. div. *Finance* cum dividend (i.e. with dividend)

Cdl Cardinal

CDL Central Dockyard Laboratory (in the Ministry of Defence) • central door locking (in motor advertisements) • *Chem.* coal-derived liquid • Council of the Duchy of Lancaster • (Australia) County and Democratic League

CDLD chronic diffuse liver disease

CDM *Astronomy* cold dark matter (as in **CDM theory**) • *Physics* critical damping mode

Cd MAA Commissioned Master-at-Arms

CD-Mo compact disc magneto optical

Cdn Canadian

CDN *international vehicle registration for* Canada • (ital.) Chicago Daily News

cDNA *Biochem.* complementary DNA

Cdo Commando

Cd O Commissioned Officer

Cd Obs Commissioned Observer

Cd OE Commissioned Ordnance Engineer

CDOI Colorado Department of Institutions

Cd OO Commissioned Ordnance Officer

CDOS ('siː,dɒs) *Computing* concurrent disc operating system

CDP *Biochem.* collagenase-digestible protein • Committee of Directors of Polytechnics • *Physical chem.* condensed dispersed phase • *Biochem.* cytidine 5'-diphosphate

CDPE continental daily parcels express

CDPS *Computing* compound document protocol specification

Cdr *Military* Commander • Conductor

CDR carbon dioxide research • (Cuba) Committees for the Defence of the Revolution • (*or* **CD-R**) compact-disc recordable • *Computing* contents of decrement register • critical design review • crude death rate

CDRA Committee of Directors of Research Associations

Cd Rad O Commissioned Radio Officer

CDRB Canadian Defence Research Board

CDRC Civil Defence Regional Commissioner

CDRD (USA) Carbon Dioxide Research Division

Cdre *Military* Commodore

CDRF Canadian Dental Research Foundation

CDRH (USA) Center for Devices and Radiological Health

CDRI (India) Central Drug Research Institute

CD-ROM (ˌsiː.diːˈrɒm) *Computing* compact disc read-only memory

CD-ROM XA *Computing* CD-ROM extended architecture

CDRS Civil Defence Rescue Service

CDS Chief of the Defence Staff • Civil Defence Services • *Physics* coded Doppler sonography

CDSE computer-driver simulation environment

Cd Sh Commissioned Shipwright

Cd SO Commissioned Stores Officer • Commissioned Supply Officer

CDSO Companion of the Distinguished Service Order

Cdt Cadet • Commandant

CDT Carnegie Dunfermline Trust • (USA and Canada) Central Daylight Time • Craft, Design, and Technology (a subject on the GCSE syllabus)

CDTA *Chem.* 1,2-cyclohexylenedinitro-tetraacetic acid

Cdt Mid Cadet Midshipman

CDTV *Computing* Commodore Dynamic Total Vision • compact-disc television

CDU Christlich-Demokratische Union (Christian Democratic Union; German political party)

CDUCE Christian Democratic Union of Central Europe

c.d.v. carte-de-visite (French: visiting card)

CDV CD-video (compact-disc player) • Civil Defence Volunteers • current domestic value

c.d.w. chilled (*or* cold) drinking water

CDW *Physics* charge density wave • (USA) collision damage waiver (in car insurance) • *Physics* continuum distorted wave

CD-Wo compact disc write once

Cd Wdr Commissioned Wardmaster

CDWR California Department of Water Resources

CDWS Civil Defence Wardens' Service

Cdz Cádiz

c.e. *Law* caveat emptor (Latin: let the buyer beware) • compass error • critical examination

Ce *Chem., symbol for* cerium

CE Canada East • carbon equivalent • *Architect.* centre of effort • *Chem.* centrifugal elutriation • *Med.* cerebral embolism • Chancellor of the Exchequer • Chemical Engineer • Chief Engineer • *Chem.* cholesteryl ester • Christian Endeavour • (small caps.) Christian Era (following a date) • Church of England • *Computing* circular error • Civil Engineer • *Med.* clinical evaluation • *British fishing port registration for* Coleraine • *Chem.* combustion efficiency • (USA) Combustion Engineering (Inc.) • *Education* Common Entrance • (small caps.) Common Era (following a date) • Communauté européenne (French: European Community) • community education (as in **CE Centre**) • compression engine • *Computing* computing efficiency • contrast-enhanced (of picture images) • Corps of Engineers • *Physics* correlation energy • *Chem.* coulomb explosion • Council of Europe • counter-espionage • *Electronics* current efficiency • *British vehicle registration for* Peterborough

CEA Canadian Electrical Association • *Med.* carcino-embryonic antigen • Central Electricity Authority • Cinematograph Exhibitors Association • Combustion Engineering Association • Comité européen des assurances (French: European Insurance Committee) • *Commerce* commodity exchange authority • Confédération européenne de l'agriculture (French: European Confederation of Agriculture) • Conference of Educational Associations • control electronics assembly • (USA) Council of Economic Advisers • Council of Educational Advance

CEAA (USA) Center for Editions of American Authors • Council of European-American Associations

CEAC (USA) Citizens Energy Advisory Committee • Commission européenne de l'aviation civile (French: European Civil Aviation Commission)

CEB Central Electricity Board

CEBAR (ˈsiːbɑː) chemical, biological, radiological warfare

CEC California Energy Commission • Canadian Electrical Code (of standardization) • Catholic Education Council • *Chem.* cation exchange capacity • Church Education Corporation • Civil Engineering Corps • Clothing Export Council •

Commission of the European Communities • Commonwealth Economic Committee • Commonwealth Education Conference • Commonwealth Engineering Conference • Community Education Centre • Council for Exceptional Children

CECA Communauté européenne du charbon et de l'acier (French: European Coal and Steel Community)

CECD Confédération européenne du commerce de détail (French: European Confederation of Retail Trades)

CECE Committee for European Construction Equipment

CECG Consumers in the European Community Group

CECLES Conseil européen pour la construction de lanceurs d'engins spatiaux (French: European Launching Development Organization)

CECS Church of England Children's Society • civil engineering computing system • Communications Electronics Coordination Section

CED Committee for Economic Development • computer entry device • (USA) Council for Economic Development

CEDA Committee for Economic Development of Australia

CEDAR ('siːdə) coupling, energetics, and dynamics of atmospheric regions

CEDEL Centrale de Livraison de Valeurs Mobilières (French; eurobond settlement service in Luxembourg)

CEDI Centre européen de documentation et d'information (French: European Documentation and Information Centre)

CEDIC Church Estates Development and Improvement Company

CEDO Centre for Educational Development Overseas

CEDR ('siːdə) Centre for Dispute Resolution

CEE Central Engineering Establishment • Certificate of Extended Education • Commission économique pour l'Europe (French: Economic Commission for Europe) • Commission internationale de réglementation en vue de l'approbation de l'équipement électrique (French: International Commission on Rules for the Approval of Electrical Equipment) • Common Entrance Examination • Communauté économique européenne (French: European Economic

Community) • Council of Environment Education

CEEA Communauté européenne de l'énergie atomique (French: European Atomic Energy Community)

CEEB (USA) College Entrance Examination Board

CEEC Council for European Economic Cooperation

CEED Centre for Economic and Environment Development

CEEL (siːl) *Physical chem.* core electron energy loss

CEEP Centre européen d'études de population (French: European Centre for Population Studies)

CEF Canadian Expeditionary Force • Chinese Expeditionary Force • *Chem.* crystalline electric field

CEFTA ('sɛftə) Central European Free Trade Area

CEFTRI (India) Central Food Technological Research Institute

CEG (France) collège d'enseignement général (college of general education)

CEGB Central Electricity Generating Board

CEGGS Church of England Girls' Grammar School

CEGS Church of England Grammar School

CEI (Switzerland) Centre d'études industrielles (French: Centre for Industrial Studies) • Commission électrotechnique internationale (French: International Electrotechnical Commission) • communications-electronics instructions • cost-effectiveness index • Council of Engineering Institutions

CEIF Council of European Industrial Federations

CEIR Corporation for Economic and Industrial Research

cel. celebrate(d) • celebration • celery • *Music* celesta • celibate

Cel. (*or* **Cels.**) Celsius

CEL (USA) Constitutional Educational League

CELA Council for Exports to Latin America

CELC Commonwealth Education Liaison Committee

CELEX Communitatis Europeae Lex (Latin: European Community Law; a computerized documentation system for community law)

CELJ Conference of Editors of Learned Journals

CELSS controlled ecological life-support system

Celt. Celtic

cem. cement · cemetery

CEM Companhia Electricidade de Macáu (Portuguese: Electricity Company of Macao) · *Physics* continuous emission monitoring · cost and effectiveness method · *Chem.* coulomb explosion method · *Astronautics* crew-escape module

CEMA Canadian Electrical Manufacturers' Association · Catering Equipment Manufacturers' Association · (USA) Conveyor Equipment Manufacturers' Association · Council for Economic Mutual Assistance · Council for the Encouragement of Music and the Arts

CEMAC Committee of European Associations of Manufacturers of Active Electronic Components · Communauté économique et monétaire de l'afrique central (French: Central African Economic and Monetary Community)

CEMAP Commission européenne des méthodes d'analyse des pesticides (French: Collaborative Pesticides Analytical Committee)

CEMF *Electronics* counter-electromotive force

CEMLA Centro de Estudios Monetarios Latino-Americanos (Spanish: Latin-American Centre for Monetary Studies)

CEMR Council of European Municipalities and Regions

CEMS Church of England Men's Society · *Chem.* conversion electron Mössbauer spectroscopy

cen. central · centre · century

Cen *Astronomy* Centaurus

CEN (sɛn) Comité européen de normalisation (French: European Standardization Committee)

CEND Civil Engineers for Nuclear Disarmament · *Physics* Compton-emission neutron detector

CENEL European Electrical Standards Coordinating Committee

CENELEC ('sɛnə,lɛk) Comité européen normalisation électrotechnique (French: European Electrotechnical Standardization Committee)

CEng Chartered Engineer

cens. censor(ship)

cent. *Currency* centavo · *Currency* centesimo · centigrade · centime · central · centrifugal · centum (Latin: hundred) · century

CENTAG ('sɛntæg) Central (European) Army Group (in NATO)

centig. centigrade

CENTO (*or* **Cento**) ('sɛntəʊ) Central Treaty Organization

CEO Chief Education Officer · Chief Executive Officer · Confederation of Employee Organizations

CEOA Central European Operating Agency (in NATO)

Cep *Astronomy* Cepheus

CEP *Computing* circular error probability

CEPCEO Comité d'études des producteurs de charbon d'Europe occidentale (French: Western European Coal Producers' Association)

CEPES Comité européen pour le progrès économique et social (French: European Committee for Economic and Social Progress)

CEPIS Council of European Professional Informatics Societies

CEPO Central European Pipeline Office

CEPS Central European Pipeline System · Cornish Engine Preservation Society

CEPT Conférence européenne des administrations des postes et des télécommunications (French: European Conference of Postal and Telecommunications Administrations)

CEQ (USA) Council on Environmental Quality

CEQA (USA) California Environmental Quality Act

cer. (*or* **ceram.**) ceramic

CER carbon dioxide exchange (*or* evolution) rate · *Physics* ceramic electrochemical reactor · *Electronics* charge-exchange recombination · (Australia and New Zealand) Closer Economic Relations

CERC (USA) Center for Energy Research Computation

CERCA Commonwealth and Empire Radio for Civil Aviation

CERCLA ('sɜːklə) (USA) Comprehensive Environmental Response, Compensation, and Liability Act

Cerdip ('sɜːˌdɪp) *Computing* ceramic dual in-line package

CERES ('sɪəriːz) (USA) Coalition for Environmentally Responsible Economies

CERG *Politics* Conservative European Reform Group

CERI Centre for Educational Research and Innovation (of the OECD)

CERL (sɜːl) Central Electricity Research Laboratories

CERN (sɜːn) Conseil européen pour la recherche nucléaire (French: European Organization for Nuclear Research; now called European Laboratory for Particle Physics)

CERP Centre européen des relations publiques (French: European Centre of Public Relations)

cert. certificate(d) • certification • certified • certify

CERT Charities Effectiveness Review Trust • computer emergency response team

CertCAM Certificate in Communication, Advertising, and Marketing

Cert Ed Certificate in Education

CertHE Certificate in Higher Education

certif. certificate

cert. inv. certified invoice

CertITP (USA) Certificate of International Teachers' Program

cerv. *Med.* cervical

CES (USA) Center for Energy Studies • Christian Evidence Society • (France) collège d'enseignement secondaire (college of secondary education) • community energy system • cost-estimating system

CESAR (USA) Center for Engineering Systems Advanced Research

CESE (USA) Center for Earth Science and Engineering

CESLS *Chem.* constant energy synchronous luminescence spectrometry

CESR (USA) Cornell electron storage ring

CESSAC Church of England Soldiers', Sailors', and Airmen's Clubs

CESSI Church of England Sunday School Institution

CEST Centre for Exploitation of Science and Technology

Cestr. Cestrensis (Latin: (Bishop) of Chester)

Cet *Astronomy* Cetus

CET Central European Time • (France) collège d'enseignement technique (college of technical education) • Common External Tariff • Council for Educational Technology

CETA ('siːtə) (USA) Comprehensive Employment and Training Act

CETEX ('siːtɛks) Committee on Extra-Terrestrial Exploration

CETHV Council for the Education and Training of Health Visitors

CETI communications with extraterrestrial intelligence

CETO Centre for Educational Television Overseas

cet. par. ceteris paribus (Latin: other things being equal)

CETS chemical energy transmission systems • Church of England Temperance Society

CEU Christian Endeavour Union

CEUS Central and Eastern United States

CEUSA Committee for Exports to the United States of America

CEVS (USA) controlled environment vitrification system

CEWMS Church of England Working Men's Society

Cey. Ceylon

CEYC Church of England Youth Council

Ceyl. Ceylon

CEZMS Church of England Zenana Missionary Society

cf. *Bookbinding* calfskin • compare (Latin *confer*)

c.f. *Music* cantus firmus (Latin: fixed song) • (*or* **c/f**) *Book-keeping* carried forward • *Baseball* center fielder • *Sport* centre field • *Sport* centre forward • chemin de fer (French: railway) • communication factor • context free • (*or* **c/f**) cost and freight • cubic feet

Cf *Chem., symbol for* californium

Cf. *RC Church* Confessions

CF calibration factor • *international civil aircraft marking for* Canada • Canadian Forces • capacity factor • Cardiff (postcode *or* British fishing port registration) • *Commerce* carriage forward • centre of flotation • Chaplain to the Forces • charcoal-filtered • Comédie Française (French national theatre) • Commonwealth Fund • *symbol for* Comorian franc (monetary unit of Comoros) • compensation fee • *Med.* conventional fractionation radiotherapy • Corresponding Fellow •

cost and freight • *Chem.* crystal field (as in **CF theory**) • *Med.* cystic fibrosis • *airline flight code for* Faucett Peruvian Airlines • *British vehicle registration for* Reading

CFA Canadian Federation of Agriculture • Canadian Field Artillery • Canadian Forestry Association • cash-flow accounting • (USA) Chartered Financial Analyst • Commission of Fine Arts • Commonwealth Forestry Association • Communauté financière africaine (French: African Financial Community, as in **CFA franc**) • *Med.* complete Freund's adjuvant • Contract Flooring Maintenance (Limited) • Cookery and Food Association • Council for Acupuncture • Council of Foreign Affairs • *Electrical engineering* cross-field amplifier • *Med.* cryptogenic fibrosing alveolitis

CFAF CFA (Communauté financière africaine) franc

CFAL Current Food Additives Legislation

CFAM cogeneration feasibility analysis model

CFAP Canadian Foundation for the Advancement of Pharmacy

CFAR constant false alarm rate

CFAT Carnegie Foundation for the Advancement of Teaching

CFB *Computing* cipher feedback • *Chemical engineering* circulating fluid bed • (USA) Consumer Fraud Bureau • Council of Foreign Bondholders

CFBC *Chemical engineering* circulating fluid-bed combustion

CFBS Canadian Federation of Biological Sciences

CFC carbon-fibre composite • *Chem.* chlorofluorocarbon (*or* chlorinated fluorocarbon) • *Chemical engineering* circulating fluidized combustion • *Microbiol.* colony-forming cell • Common Fund for Commodities (in the UN) • Congregatio Fratrum Christianorum (Latin: Congregation of Christian Brothers) • consolidation freight classification

CFCE Conseil des fédérations commerciales d'Europe (French: Council of European Commercial Federations)

c.f.d. cubic feet per day

CFD *Computing* compact floppy disc • *Engineering* computational fluid dynamics

CFDC Canadian Film Development Corporation

CFDT (France) Confederation française

démocratique du travail (French Democratic Federation of Labour)

CFE Central Fighter Establishment • College of Further Education • *Microbiol.* colony-forming efficiency • *Physics* control fuel element • *Military* Conventional Forces in Europe (as in **CFE treaty**)

CFF Chemins de fer fédéraux Suisses (Swiss national railway) • *Computing* critical flicker frequency • *Physics* critical fusion frequency

CFFLS (USA) Consortium for Fossil Fuel Liquefaction Science

c.f.h. cubic feet per hour

CFHT Canada–France–Hawaii Telescope (Mauna Kea, Hawaii)

CFI Chief Flying Instructor • (*or* **c.f.i.**) cost, freight, and insurance

CFL Canadian Football League • ceasefire line • Central Film Library

cfm confirm(ation)

c.f.m. cubic feet per minute

CFM Cadet Forces Medal • *Chem.* chlorofluoromethane • Council of Foreign Ministers

c.f.o. calling for orders • channel for orders • *Nautical* coast for orders

CFO *Meteorol.* Central Forecasting Office • Chief Financial Officer • Chief Fire Officer

CFOA Chief Fire Officers' Association

CFOD Catholic Fund for Overseas Development

CFP Common Fisheries Policy (of the EU) • Communauté financière du Pacifique (French: Pacific Financial Community) • Compagnie Française des Pétroles (French Petroleum Company)

CFPF CFP (Communauté financière du Pacifique) franc

CFPP coal-fired power plant

cfr. chauffeur • confronta (Italian: compare)

CFR (USA) Code of Federal Regulations • Commander of the Order of the Federal Republic of Nigeria • *Nuclear engineering* commercial fast reactor • *Engineering* Co-operative Fuel Research (Committee) (as in **CFR engine**) • *Med.* coronary flow reserve • *Metallurgy* corrosion-fatigue resistance • Council on Foreign Relations

CFRI (USA) Central Fuel Research Institute

CFRP carbon-fibre reinforced plastic

c.f.s. cubic feet per second

CFS Central Flying School • *Med.* chronic fatigue syndrome • Clergy Friendly Society • *Telecom.* combination frequency signal • *Computing* common file system

CFSAN (USA) Center for Food Safety and Applied Nutrition

CFSE *Chem.* crystal-field stabilization energy

CFSP common foreign and security policy (in the EU)

CFSTI (USA) Clearinghouse for Federal Scientific and Technical Information

cft craft

CFT Compagnie française de télévision (French television company) • *Med.* complement fixation test • *Engineering* cross-flow turbine

CFTB Commonwealth Forestry and Timber Bureau

CFTC (USA) Commodity Futures Trading Commission • Commonwealth Fund for Technical Cooperation

CF theory *Chem.* crystal-field theory

cftmn craftsman

CFU *Microbiol.* colony-forming unit

CFV continuous-flow ventilation

CFWI County Federation of Women's Institutes

CFX *RC Church* Congregatio Fratrum Xaverianorum (Latin: Congregation of Xaverian Brothers)

cg *symbol for* centigram(s) • (*or* **c.g.**) centre of gravity

CG *British vehicle registration for* Bournemouth • Captain-General • Captain of the Guard • centre of gravity • *Aviation* cloud-to-ground • coastguard • Coldstream Guards • *US Army* Commanding General • Commissary-General • computer graphics • concentration guide • conjugate gradient • Consul-General • Croix de Guerre (French military decoration)

C-G Chaplain-General

CGA *Commerce* cargo's proportion of general average • Certified General Accountant • *Chemical engineering* coal-gasification atmosphere • (USA) Coast Guard Academy • (USA) Coast Guard Auxiliary • *Computing* colour graphics adapter • Community of the Glorious Ascension • Country Gentlemen's Association

CGBR Central Government Borrowing Requirement

CGC *Chem.* capillary gas chromatography • *Chemical engineering* coal-gas condensate • (USA) Coast Guard cutter • (France) Confédération générale des cadres (General Confederation of Executive Staff) • Conspicuous Gallantry Cross

CGD *Med.* chronic granulomatous disease

CGDK Coalition Government of Democratic Kampuchea

cge carriage • charge

CGE *Politics* Conservative Group for Europe

CGG compensated geothermal gradient

CGH Cape of Good Hope • computer-generated hologram

c.g.i. corrugated galvanized iron

CGI Chief Ground Instructor • Chief Gunnery Instructor • City and Guilds Institute • commercial grade item • *Computing* common gateway interface • computer graphics interface

CGIA City and Guilds of London Insignia Award

CGIAR Consultative Group on International Agricultural Research

CGIL (Italy) Confederazione Generale Italiana del Lavoro (General Italian Confederation of Labour)

CGL *Electronics* charge-generation layer • corrected geomagnetic latitude

CGLI City and Guilds of London Institute

cgm centigram(s)

CGM computer graphics metafile • Conspicuous Gallantry Medal

CGMW Commission for the Geological Map of the World

cgo cargo • *Finance* contango

CGOU (USA) Coast Guard Oceanographic Unit

CGP College of General Practitioners

CGPM Conférence générale des poids et mesures (French: General Conference of Weights and Measures) • Conseil général des pêches pour la Méditerranée (French: General Fisheries Council for the Mediterranean)

CGPS Canadian Government Purchasing System

CGRI (India) Central Glass and Ceramic Research Institute

CGRM Commandant-General of the Royal Marines

CGRO Compton Gamma Ray Observatory

CGRT *Physics* carbon–glass resistance thermometer

cgs (*or* **c.g.s.**) centimetre, gram, second (in **cgs units**)

CGS central gunnery school • Chief of General Staff • Coast and Geodetic Survey • *US Army* Commissary General of Subsistence

CGSB Canadian Government Specifications Board

CGSC (USA) Command and General Staff College

CGSS (USA) Command and General Staff School

CGSUS Council of Graduate Schools in the United States

CGT capital-gains tax • ceramic gas turbine • closed gas turbine • Compagnie générale transatlantique (French shipping company) • compressed-gas turbomotor • (Argentina) Confederación general del trabajo (Spanish: General Federation of Workers) • (France) Confédération générale du travail (General Confederation of Labour)

CGTB Canadian Government Travel Bureau

CGT-FO (France) Confédération générale du travail–force ouvrière (General Confederation of Labour–Workers' Force)

c.g.u. ceramic glazed units

ch *Maths.* cosh

ch. chain (unit of measure; crochet stitch) • chaldron • chambre (French: room) • chaplain • chapter • charge(s) • chart • *Horse racing* chase • *Chess* check • chemical • chemistry • *Horse racing* chestnut • cheval-vapeur (French: horsepower) • chief • child(ren) • choir • choke • church

c.h. candle hour(s) • central heating • *Sport* centre half • clearing house • compass heading • court house • custom(s) house

Ch. Chairman • Chaldean • Chaldee • Chamber • Champion • Chancellor • Chancery • Chapter • Chile(an) • China • Chinese • Chirurgiae (Latin: of surgery; in academic degrees) • Christ • Church

CH *Military* Captain of the Horse • *Freemasonry* Captain of the Host • Carnegie Hall • *Med.* cerebral haemorrhage • chapter house • Chester (postcode *or* British fishing port registration) • Christ's Hospital • clearing house • *Physics* collision hypothesis • Companion of Honour • *Med.* contact hypersensitivity •

Astronomy coronal hole • corporate hospitality • custom(s) house • *British vehicle registration for* Nottingham • *international vehicle registration for* Switzerland (French *Confédération Helvétique*)

C/H central heating

Cha *Astronomy* Chamaeleon

CHA Catholic Hospital Association • Chest and Heart Association • Community Health Association • Country-wide Holidays Association

chacom ('tʃeɪkɒm) chain of command

chal. chaldron (unit of capacity) • chaleur (French: heat) • challenge

Chal. (*or* **Chald.**) Chaldaic (*or* Chaldee) • Chaldean

Chamb. Chamberlain • Chambers

chan. channel

Chan. (*or* **Chanc.**) Chancellor • Chancery

CHANCOM ('tʃæn,kɒm) Channel Committee (in NATO)

chap. chapel • (*or* **Chap.**) chaplain • chaplaincy • (*or* **Chap.**) chapter

CHAPS (tʃæps) *Chem.* 3-[(3-cholamidopropyl)dimethylammonio]-1-propanesulphonate • Clearing House Automatic Payments System

char. character • characteristic • characterize • charity • charter

CHAR Campaign for Homeless People (originally Campaign for the Homeless and Rootless)

charact. characterize

charc. charcoal

Chas. Charles

Chauc. (Geoffrey) Chaucer (?1340–1400, English poet)

Chb. Cherbourg

ChB Bachelor of Surgery (Latin *Chirugiae Baccalaureus*) • (USA) Chief of the Bureau

CHB Companion of Honour of Barbados

CHC child health clinic • choke coil • Clerk to the House of Commons • Community Health Council • (USA) Confederate High Command • *Chem.* cyclohexylamine carbonate

Ch. Ch. Christ Church (Oxford)

Ch. Clk Chief Clerk

Ch. Coll. Christ's College (Cambridge)

ChD Doctor of Chemistry • Doctor of Surgery (Latin *Chirugiae Doctor*)

Ch. D. *Law* Chancery Division

CHD congenital heart disease • coronary heart disease

CHDL computer hardware description language

ChE Chemical Engineer • Chief Engineer

CHE Campaign for Homosexual Equality

CHEAR Council on Higher Education in the American Republics

CHEC Commonwealth Human Ecology Council

Cheka ('tʃɛkə) Chrezvychainaya Comissiya (Russian: Extraordinary Commission (for Fighting Counter-Revolution and Sabotage); forerunner of KGB)

CHEL (ital.) Cambridge History of English Literature

Chelm. Cheltenham

chem. chemical(ly) • chemist(ry)

CHEM chemical health effects (assessment) methodology

ChemE Chemical Engineer

Ches. Cheshire

CHES *Chem.* 2-(cyclohexylamino)-ethanesulphonic acid

CHESS (tʃes) (USA) Cornell high-energy synchrotron source

chev. chevron

Chev. Chevalier

chf chief

ChF Chaplain of the Fleet

CHF Carnegie Hero Fund • *Biochem.* combined enzymatic hydrolysis and fermentation • *Med.* congestive heart failure • *Physics* coupled Hartree–Fock (mathematical technique)

ch. fwd charges forward

chg. change • (*or* **chge**) *Commerce, Finance* charge

chgd charged

chgph. choreographer • choreographic • choreography

Ch. hist. Church history

Chi. (*or* **Chic.**) Chicago • China • Chinese

Chich. Chichester

CHILL (tʃɪl) *Computing* CCITT high-level language

Chin. China • Chinese

CHIPS (tʃɪps) Clearing House Inter-Bank Payments System

CHIRP (tʃɜːp) *Civil aviation* Confidential Human Incidence Reporting Programme (pilots' comments on safety trends)

Chi. Trib. (ital.) Chicago Tribune

chiv. chivalry

Ch. J. Chief Justice

chk check

Ch. K. Christ the King

chkr checker

chl. chloride • chloroform

Chl *Botany, Biochem.* chlorophyll (as in **Chl a**)

Ch. Lbr. Chief Librarian

ChLJ Chaplain of the Order of St Lazarus of Jerusalem

chlo. chloride • chloroform

CHLW *Nuclear engineering* commercial high-level waste

chm. (*or* **Chm.**, **chmn**) chairman • checkmate • (*or* **Chm.**) choirmaster

ChM Master of Surgery (Latin *Chirurgiae Magister*)

CHM *Chem.* Cape Horn methanol • *see* ARCO(CHM); FRCO(CHM)

CHMC (USA) Children's Hospital Medical Center

CHMR (USA) Center for Hazardous Materials Research

CHNT Community Health Nurse Tutor

cho. choral • chorister • chorus

CHO Crop Husbandry Officer

choc. chocolate

Ch. of S. Chamber of Shipping

Ch. of the F. Chaplain of the Fleet

chor. choral • chorister • chorus

CHP chemical heat pump • combined heat and power (energy use) • (Turkey) Cumhuriyet Halk Partisi (Republican People's Party)

ch. pd charges paid

Ch. ppd. charges prepaid

chq. cheque

CHQ Commonwealth Headquarters (central office of the Girl Guides Association for the UK and Commonwealth) • *Military* Corps Headquarters

Chr. Christ • Christian(ity) • Christmas • *Bible* Chronicles

chrm. (*or* **Chrm.**) chairman

chron. chronicle • (*or* **chronol.**) chronological(ly) • (*or* **chronol.**) chronology • chronometry

Chron. *Bible* Chronicles

Chrs Chambers

chs chapters

CHS *Physics* calibration heat source • Canadian Hydrographic Service • Church Historical Society • *Computing* cylinders, heads, and sectors

CHSA Chest, Heart, and Stroke Association

CHSC Central Health Services Council

ch'ship championship

Ch. Skr. Chief Skipper

ChStJ Chaplain of the Order of St John of Jerusalem

c.h.t. cylinder-head temperature

chtg charting

ch. v. check valve

c.h.w. constant hot water

chwdn churchwarden

chyd churchyard

Chy Div. *Law* Chancery Division

c.i. cast iron

Ci (ital.) *Meteorol.* cirrus • *Physics, symbol for* curie

CI *Med.* cardiac index • *Physics* cascade ionization • *Med.* cerebral infarction • Channel Islands • *Freemasonry* Chapter of Instruction • chemical inspectorate • chemical ionization • Chief Inspector • Chief Instructor • *airline flight code for* China Airlines • *Chem.* collisional ionization • *Astronomy* Colour Index • *Med.* combined injury • *Chem.* combustion index • Commonwealth Institute • Communist International • compression-ignition (as in **CI engine**) • *Statistics* confidence interval • *Chem.* configuration interaction • consular invoice • continuous improvement (in management) • *Photog.* contrast index • *Physics* Copenhagen interpretation • *Astronomy* coronal index (of solar activity) • *Engineering* corrosion inhibitor • *international vehicle registration for* Côte d'Ivoire (Ivory Coast) • counter-intelligence • (Imperial Order of the) Crown of India • *Irish vehicle registration for* Laois

C³I *Military* command, control, communications, and intelligence

Cia compagnia (Italian: company; Co.) • companhia (Portuguese: company; Co.) • compañia (Spanish: company; Co.)

CIA cash in advance • (USA) Central Intelligence Agency • Chemical Industries' Association • Chief Inspector of Armaments • Conseil international des archives (French: International Council on Archives) • Corporation of Insurance Agents • Culinary Institute of America

CIAA Centre international d'aviation agricole (French: International Agricultural Aviation Centre) • Coordinator of Inter-American Affairs

CIAB (USA) Coal Industry Advisory Board • Conseil international des agences bénévoles (French: International Council of Voluntary Agencies)

CIAgrE Companion of the Institution of Agricultural Engineers

CIAI Commerce and Industry Association Institute

CIAL Corresponding Member of the International Institute of Arts and Letters

CIAPG Confédération internationale des anciens prisonniers de guerre (French: International Confederation of Former Prisoners of War)

CIArb Chartered Institute of Arbitrators

CIAS Changi International Airport Services • Conference of Independent African States

CIB Central Intelligence Board • Chartered Institute of Bankers • Corporation of Insurance Brokers • (New Zealand) Criminal Investigation Branch

CIBSE Chartered Institution of Building Services Engineers (formerly Chartered Institute of Building Services, **CIBS**)

Cic. (Marcus Tullius) Cicero (106–43 BC, Roman consul, orator, and writer)

CIC Capital Issues Committee • Chemical Institute of Canada • *Immunol.* circulating immune complexes • (USA) Combat Information Center • Commander-in-Chief • (USA) Command Information Center • Commonwealth Information Centre • (USA) Counterintelligence Corps • (USA) Critical Issues Council

CICA Canadian Institute of Chartered Accountants

CICADA (sɪ'kɑːdə) central instrumentation control and data acquisition

CICAR Cooperative Investigations of the Caribbean and Adjacent Regions

CICB Criminal Injuries Compensation Board

CICC *Electronics* cable-in-conduit conductor • Conférénce internationale des charités catholiques (French: International Conference of Catholic Charities)

Cicestr. Cicestrensis (Latin: (Bishop) of Chichester)

CICG Centre international du commerce de gros (French: International Centre for Wholesale Trade)

CICHE Committee for International Cooperation in Higher Education

CICI Confederation of Information Communication Industries

CICP (USA) Committee to Investigate Copyright Problems

CICR Comité international de la Croix-Rouge (French: International Committee of the Red Cross)

CICRC Commission internationale contre le régime concentrationnaire (French: International Commission Against Concentration Camp Practices)

CICRIS Cooperative Industrial and Commercial Reference and Information Service

CICS *Computing* customer information control system

CICT Conseil international du cinéma et de la télévision (French: International Film and Television Council)

CID *Electronics* charge injection device • *Chem.* collision-induced dissociation • Committee for Imperial Defence • computer-assisted imaging device • Council of Industrial Design • Criminal Investigation Department • *Physics* current-image diffraction

CIDA Canadian International Development Agency • Comisión interamericano de desarrollo agrícola (Spanish: Inter-American Committee for Agricultural Development) • Comité inter-gouvernemental du droit d'auteur (French: Intergovernmental Copyright Committee)

CIDADEC Confédération internationale des associations d'experts et de conseils (French: International Confederation of Associations of Experts and Consultants)

CIDE (*or* **Cide**) (Uruguay) Comisión de inversion y desarrollo económico (Spanish: Commission for Investment and Economic Development)

CIDESA Centre international de documentation économique et sociale africaine (French: International Centre for African Social and Economic Documentation)

CIDNP chemically induced dynamic nuclear polarization

CIDOC (Mexico) Centre for Intercultural Documentation

Cie Compagnie (French: Company; Co.)

CIE captain's imperfect entry • Centre international de l'enfance (French: International Children's Centre) • *Chem.* collisional ionization equilibrium • Commission internationale de l'éclairage (French: International Commission on Illumination; ICI) • Companion of the Order of the Indian Empire (former honour) • Confédération internationale des étudiants (French: International Confederation of Students) • (Ireland) Córas Iompair Éireann (Irish Gaelic: Transport Organization of Ireland)

CIEC Centre international d'études criminologiques (French: International Centre of Criminological Studies) • Commission internationale de l'état civil (French: International Commission on Civil Status)

CIEE Companion of the Institution of Electrical Engineers

CIEO Catholic International Education Office

CIEPS Conseil international de l'éducation physique et sportive (French: International Council of Sport and Physical Education)

c.i.f. (*or* **CIF**) *Commerce* cost, insurance, and freight (as in **c.i.f. contract**)

CIF Canadian Institute of Forestry • Clube internacional de futebol (Portuguese: International Football Club) • Conseil international des femmes (French: International Council of Women) • *Physics, Electronics* contrast improvement factor

CIFA Corporation of Insurance and Financial Advisers

c.i.f.c. *Commerce* cost, insurance, freight, and commission

c.i.f.c.i. *Commerce* cost, insurance, freight, commission, and interest

c.i.f.e. *Commerce* cost, insurance, freight, and exchange

CIFE Colleges and Institutes for Further Education • Conseil des fédérations industrielles d'Europe (French: Council of European Industrial Federations) • Conseil international du film d'enseignement (French: International Council for Educational Films)

c.i.f.i. *Commerce* cost, insurance, freight, and interest

CIFJ Centre international du film pour la jeunesse (French: International Centre of Films for Children)

c.i.f.L.t. *Commerce* cost, insurance, and freight, London terms

CIG *Chemical engineering* coal–iron gasification • Comité international de géophysique (French: International Geophysical Committee)

CIGA (Italy) Compagnia Italiana dei Grandi Alberghi (hotel group)

CIGAR (sɪˈgɑː) *Nuclear physics* channel inspection and gauging apparatus of reactors

CIGasE Companion of the Institute of Gas Engineers

CIGR Commission internationale du génie rural (French: International Commission of Agricultural Engineering)

CIGS Chief of the Imperial General Staff

CIH Certificate in Industrial Health • *Med.* chronic inactive hepatitis

CIHA Comité international d'histoire de l'art (French: International Committee on the History of Art)

CII Centre for Industrial Innovation (of the University of Strathclyde) • Chartered Insurance Institute • Conseil international des infirmières (French: International Council of Nurses)

CIIA Canadian Institute of International Affairs • Commission internationale des industries agricoles (French: International Commission for Agricultural Industries)

CIIR Catholic Institute for International Relations

CIJ Commission internationale de juristes (French: International Commission of Jurists)

CIL Confederation of Irish Labour

CILB Commission internationale de lutte biologique contre les ennemis des plantes (French; International Commission for Biological Control)

CILECT Central international de liaison des écoles de cinéma et de télévision (French; International Association of National Film Schools)

CILG Construction Industry Information Liaison Group

CIM Canadian Institute of Mining • China Inland Mission • Commission for Industry and Manpower • CompuServe Information Manager • computer input on microfilm • computer-integrated manufacturing • Conférence islamique mondial (French: World Muslim Conference) • Conseil international de la musique (French: International Music Council) • Cooperative Investment Management

CIMA Chartered Institute of Management Accountants • (USA) Construction Industry Manufacturers' Association

CIMarE Companion of the Institute of Marine Engineers

CIMAS continuous iron-making and steel-making

CIME Comité intergouvernemental pour les migrations européennes (French: Intergovernmental Committee for European Migration) • Council of Industry for Management Education

CIMechE Companion of the Institution of Mechanical Engineers

CIMEMME Companion of the Institute of Mining Electrical and Mining Mechanical Engineers

CIMGTechE Companion of the Institution of Mechanical and General Technician Engineers

CIMM Canadian Institute of Mining and Metallurgy

CIMPM Comité international de médecine et de pharmacie militaires (French: International Committee of Military Medicine and Pharmacy)

CIMS (USA) chemical information management systems • *Physical chem.* chemical ionization mass spectrometry (*or* spectroscopy)

CIMTP Congrès international de médecine tropicale et de paludisme (French: International Congress of Tropical Medicine and Malaria)

CIN *Med.* cervical intraepithelial neoplasia (in grading cervical smears) • Commission internationale de numismatique (French: International Numismatic Commission)

CINAA *Physics* cyclic instrumental neutron activation analysis

C-in-C (*or* **C in C**) Commander-in-Chief

CINCAFMED Commander-in-Chief Allied Forces Mediterranean

CINCEASTLANT Commander-in-Chief Eastern Atlantic Area

CINCENT Commander-in-Chief Allied Forces Central Europe

CINCEUR Commander-in-Chief Europe

CINCHAN Commander-in-Chief Channel

CINCLANT Commander-in-Chief Atlantic

CINCLANTFLT *US Navy* Commander-in-Chief, Atlantic Fleet

CINCMED Commander-in-Chief British Naval Forces in the Mediterranean

CINCNELM Commander-in-Chief US

Naval Forces in Europe, the East Atlantic, and the Mediterranean

CINCNORTH Commander-in-Chief Allied Forces Northern Europe

CINCPAC Commander-in-Chief Pacific

CINCPACFLT *US Navy* Commander-in-Chief, Pacific Fleet

CINCSOUTH Commander-in-Chief Allied Forces Southern Europe

CINCWESTLANT Commander-in-Chief Western Atlantic Area

CINFO Chief of Information

Cinn. Cincinnati

CINO Chief Inspector of Naval Ordnance

CINR (USA) Central Institute for Nuclear Research

CINS (sɪnz) (USA) Child(ren) In Need of Supervision

CInstR Companion of the Institute of Refrigeration

CInstRE(Aust) Companion of the Institution of Radio Engineers (Australia)

CIO *Physical chem.* catalogue of infrared observations • Church Information Office • Comité international olympique (French: International Olympic Committee) • Commission internationale d'optique (French: International Commission for Optics) • (USA) Congress of Industrial Organizations

CIOB Chartered Institute of Building

CIOMS Council for International Organizations of Medical Sciences

CIOS Combined Intelligence Objectives Subcommittee • Conseil international pour l'organisation scientifique (French: International Committee of Scientific Management)

CIP *Mining* carbon-in-pulp • Cataloguing-in-Publication • (Belgium) Centre d'information de la presse (Catholic news agency) • *Engineering* cold isostatic pressing • Common Industrial Policy (of the EU) • *Engineering* constant-injection pressure

CIPA Canadian Industrial Preparedness Association • Chartered Institute of Patent Agents

CIPFA (*or* **Cipfa**) Chartered Institute of Public Finance and Accountancy

CIPL Comité international permanent de linguistes (French: Permanent International Committee of Linguists)

CIPM Commission internationale des poids et mesures (French: International

Committee on Weights and Measures) • Companion of the Institute of Personnel Management • (USA) Council for International Progress in Management

CIPO Comité international pour la préservation des oiseaux (French: International Committee for Bird Preservation)

CIPP Conseil indo-pacifique des pêches (French: Indo-Pacific Fisheries Council)

CIPR Commission internationale de protection contre les radiations (French: International Commission on Radiological Protection)

CIPS Central Illinois Public Service (Company) • Choice in Personal Safety (organization)

CIPSH Conseil international de la philosophie et des sciences humaines (French: International Council for Philosophy and the Humanities)

CIPW *Geol.* Cross, Iddings, Pirsson, Washington (rock classification; named after its devisers)

cir. (ital.) circa (*see under* c.) • circle • circuit • circular • circulation • circumference • circus

Cir *Astronomy* Circinus

CIR Canada India reactor • Commission (*or* Council) on Industrial Relations • Commissioners of Inland Revenue • cost information report • *Computing* current instruction register

CIRA Conference of Industrial Research Associations

circ. (ital.) circa (*see under* c.) • circle • circuit • circular • circulation • circumcision • circumference • circus

CIRCCE Confédération internationale de la représentation commerciale de la communauté européenne (French: International Confederation of Commercial Representation in the European Community)

circum. circumference

CIRF Centre international d'information et de recherche sur la formation professionelle (French: International Vocational Training Information and Research Centre) • Corn Industries Research Foundation

CIRIA Construction Industry Research and Information Association

CIRIEC Centre international de recherches et d'information sur l'économie collective (French: International

Centre of Research and Information on Collective Economy)

CIRP Collège internationale pour recherche et production (French; International Institution for Production Engineering Research)

CIRRPC (USA) Committee on Interagency Radiation Research and Policy Coordination

CIRS *Computing* cross-interleaved Reed-Solomon (as in **CIRS scheme**)

CIRT Community Initiative Research Trust

cis *Maths.* cos + i sin

CIS *Med.* carcinoma in situ • (*or* **c.i.s.**) cataloguing in source • (USA) Catholic Information Society • (USA) Center for International Studies • Central Information Service on Occupational Health and Safety (of ROSPA) • Chartered Institute of Secretaries (now Institute of Chartered Secretaries and Administrators) • Coal Industry Society • *Med.* combined-injury syndrome • Commonwealth of Independent States (the former Soviet republics) • CompuServe Information Service • Cooperative Insurance Society

CISA Canadian Industrial Safety Association

CISAC Centre for International Security and Arms Control • Confédération internationale des sociétés d'auteurs et compositeurs (French: International Confederation of Societies of Authors and Composers)

CISBH Comité international de standardisation en biologie humaine (French: International Committee for Standardization in Human Biology)

CISC complex instruction-set computer (*or* chip)

CISCO ('sɪskəʊ) City Group for Smaller Companies • Civil Service Catering Organization

CIS–COBOL *Computing* compact interactive standard COBOL

CISF Confédération internationale des sages-femmes (French: International Confederation of Midwives)

CISL Confédération internationale des syndicats libres (French: International Confederation of Free Trade Unions) • (Italy) Confederazione italiana sindacati lavoratori (Italian Confederation of Workers' Trade Unions)

CISM Conseil international du sport militaire (French: International Military Sports Council)

CISPR Comité international spécial des perturbations radioélectriques (French: International Special Committee on Radio Interference)

CISS Conseil international des sciences sociales (French: International Social Science Council)

Cist. Cistercian

CISTI Canadian Institute for Scientific and Technological Information

CISV Children's International Summer Village Association

cit. citadel • citation • cited • citizen • citrate

c.i.t. compression in transit

CIT (USA) California Institute of Technology • (USA) Carnegie Institute of Technology • (New Zealand) Central Institute of Technology • Chartered Institute of Transport • *Astronomy* circumstellar imaging telescope • Comité international des transports par chemins de fer (French: International Railway Transport Committee) • *Nuclear engineering* compact ignition tokamak

CITB Construction Industry Training Board

CITC Canadian Institute of Timber Construction • Construction Industry Training Centre

CITCE Comité international de thermodynamique et de cinétique électrochimiques (French: International Committee of Electrochemical Thermodynamics and Kinetics)

CITEL Committee for Inter-American Telecommunications

CITES ('sartiz) Convention on International Trade in Endangered Species

CITI Confédération internationale des travailleurs intellectuels (French: International Confederation of Professional and Intellectual Workers)

cito disp. *Med.* cito dispensetur (Latin: let it be dispensed quickly)

CITV Children's ITV

CIU Club and Institute Union

CIUS Conseil international des unions scientifiques (French: International Council of Scientific Unions)

CIUSS Catholic International Union for Social Service

civ. civil • civilian • civilization • civilize

CIV City Imperial Volunteers • Commission internationale du verre (French: International Glass Commission) • Convention internationale concernant le transport des voyageurs et des bagages par chemin de fer (French: International Convention Concerning the Carriage of Passengers and Baggage by Rail) • *Physics* critical ionization velocity

CivE Civil Engineer

CIW (USA) Carnegie Institute of Washington

CIWF Compassion in World Farming

CIWO Companion of the Institute of Welfare Officers

Cix Compulink Information eXchange

CIX commercial Internet exchange

cj. conjectural

CJ Chief Justice • *British vehicle registration for* Gloucester

CJA Commonwealth Journalists' Association • Criminal Justice Act

CJCC Commonwealth Joint Communications Committee

CJD Creutzfeldt–Jakob disease

CJM Code de justice militaire (French: Code of Military Justice) • *RC Church* Congregation of Jesus and Mary

ck cask • check • cook

CK *Trademark* Calvin Klein • *British fishing port registration for* Colchester • *Biochem.* creatine kinase • *airline flight code for* Gambia Airways • *British vehicle registration for* Preston

CKD *Computing* centre for key distribution • (*or* **c.k.d.**) *Commerce* completely knocked down (i.e. in parts; of goods for sale)

ckpt cockpit

ckw. clockwise

cl *symbol for* centilitre(s)

cl. claim • clarinet • (*or* **Cl.**) class • classical • classics • classification • clause • clearance • clergy(man) • clerk • climb • close • closet • closure • cloth • clove • council

c.l. carload • centre line • cum laude (Latin: with praise) • cut lengths

c/l (*or* **C/L**) cash letter • craft loss

Cl *Chem.*, *symbol for* chlorine

CL calendar line • *British fishing port registration for* Carlisle • cathode luminescence • central locking (in motor advertisements) • *Chem.* chain length • *Electronics* Child–Langmuir • civil law •

Civil Lord • (Belgium) Commander of the Order of Leopold • common law • Communication Lieutenant • confidence limits • craft loss • critical list • *Med.* cruciate ligament • *British vehicle registration for* Norwich • *international vehicle registration for* Sri Lanka (formerly Ceylon)

Cla. (*or* **Clack.**) Clackmannan

CLA Canadian Library Association • Canadian Lumbermen's Association • Copyright Licensing Agency • Country Landowner's Association

CL(ADO) Contact Lens Diploma of the Association of Dispensing Opticians

CLAM (klæm) chemical ramjet low-altitude missile

CLAN (klæn) *Computing* cordless local-area network

CLAPA Cleft Lip and Palate Association

clar. *Printing* clarendon type • clarinet

Clar. *Heraldry* Clarenceux (King of Arms)

Clasc (klæsk) Confederación latino-americana de sindicalistas cristianos (Spanish: Latin-American Central of Workers; originally Latin-American Federation of Christian Trade Unionists, CLAT)

class. classic(al) • classification • classified • classify

CLASS (klɑːs) Computer-based Laboratory for Automated School Systems

CLAT *see* Clasc

clav. *Music* clavier

CLB (USA) Cape Lookout Bight (North Carolina) • Church Lads' Brigade

CLC Canadian Labour Congress • Chartered Life Underwriter of Canada • Commonwealth Liaison Committee

CLCB Committee of London Clearing Banks (*or* Bankers)

CLCert(BOA) Supplementary Contact Lens Certificate of the British Optical Association

CLCert(SMC) Supplementary Contact Lens Certificate of the Worshipful Company of Spectacle Makers

CLCr Communication Lieutenant-Commander

cld *Stock exchange* called • cancelled • cleared • coloured • cooled • could

CLD chronic liver disease • *Biochem.* cross-linked dextran • Doctor of Civil Law

CLE Council of Legal Education

CLEA Council of Local Education Authorities

CLEAPSE Consortium of Local Education Authorities for the Provision of Science Equipment

CLEM *Maths.* closed loop eigenvector matrix

cler. clerical

CLFT *Physics* classical local field theories

cl.gt. *Bookbinding* cloth gilt

CLH Croix de la Légion d'Honneur (French: Cross of the Legion of Honour)

CLI *Computing* command-line interface · (*or* **cli**) cost-of-living index

CLIBS (klɪbz) *Physical chem.* colinear laser ion-beam spectroscopy

CLIC Cancer and Leukaemia in Children

clim. climate · climatic

CLIMAP ('klaɪmæp) Climate: Long-range Interpretation, Mapping, and Prediction (international project)

clin. clinic(al)

CLIP (klɪp) *Biochem.* corticotrophin-like intermediate lobe peptide

CLit (*or* **CLitt**) Companion of Literature

CLJ (ital.) Cambridge Law Journal · Commander of the Order of St Lazarus of Jerusalem

clk clerk · clock

clkw. clockwise

cl. L. classical Latin

CLL chronic lymphocytic leukaemia

Cllr Councillor

clm column

CLM coal–liquid mixture · conservation and load management

CLML (ital.) Current List of Medical Literature

CLNS *Computing* connectionless network service

clo. closet · clothing

CLO chief liaison officer · cod liver oil

CLP China Light and Power (Company Limited) · *Chemical engineering* coal liquefaction process · *Computing* concurrent logic programming · Constituency Labour Party · *Computing* constraint logic programming

CLPA Common Law Procedure Acts

clr clear · colour · cooler

CLR City of London Rifles · computer-language recorder · (USA) Council on Library Resources

CLRAE Conference of Local and Regional Authorities of Europe

clrm classroom

CLRU Cambridge Language Research Unit

CLS Certificate in Library Science · Christian Literature Society · *Statistics* classical least-squares · *Statistics* constrained least-squares · Courts of London Sessions

CLSES *Chem.* closed loop solvent extraction system

CLT (ital.) Canadian Law Times · computer-language translator

CLTech(ADO) Contact Lens Diploma of the Association of Dispensing Opticians

CLU Chartered Life Underwriter · *Computing* colour lock-up table

CLV *Microbiol.* carnation latent virus · *Microbiol.* cassava latent virus · *Computing* constant linear velocity

clvd clavichord

cm *symbol for* centimetre(s)

c.m. carat métrique (French: metric carat) · causa mortis (Latin: by reason of death) · *Music* common metre · court martial

cM *Genetics, symbol for* centimorgan

Cm *Chem., symbol for* curium

CM Canadian Militia · carboxymethyl (in **CM-cellulose**) · Catholic Mission · (USA) Central Maine · central meridian · *Physics* centre of mass · Certificated (*or* Certified) Master · Certificate of Merit · *postcode for* Chelmsford · church mission(ary) · circulation manager · (USA) Cleveland Museum (of Art) · combustion modification · command module (spacecraft) · Common Market · *Music* common metre · complex mixture · composite material · condition monitoring · *Computing* configuration management · *RC Church* Congregation of the Mission · *Radiology* contrast medium · *Nuclear physics* core meltdown · corporate membership · Corresponding Member · *Microbiol.* culture medium · *Cell biology* cytoplasmic membrane · *British vehicle registration for* Liverpool · Master of Surgery (Latin *Chirurgiae Magister*) · Member of the Order of Canada

C-M *Astronomy* colour-magnitude (in **C-M diagram**)

CMa *Astronomy* Canis Major

CMA Cable Makers' Association · *Chem.* calcium magnesium acetate · Canadian Medical Association · Catering

Managers' Association • (USA) Certificate of Management Accounting • Church Music Association • civil–military affairs • coal-mineral analysis • (USA) Colorado Mining Association • Commonwealth Medical Association • Communication Managers' Association • (New Zealand) cost and management accountant • Court of Military Appeals • *Optics* cylindrical mirror analyser

CMAC Catholic Marriage Advisory Council

CMACP Conseil mondial pour l'assemblée constituante des peuples (French: World Council for the People's World Convention)

CMAS Clergy Mutual Assurance Society • Confédération mondiale des activités subaquatiques (French: World Underwater Federation)

CMB Central Medical Board • Central Midwives Board (superseded by UKCC) • Chase Manhattan Bank • *Chem.* chemical mass balance • coastal motor boat • *Astronomy* cosmic microwave background

CMBHI Craft Member of the British Horological Institute

CMBI Caribbean Marine Biological Institute

CMBS *Chemical engineering* countercurrent moving-bed separator

cmbt combat

CMC Canadian Marconi Company • Canadian Meteorological Centre • Canadian Music Council • *Chem.* carboxymethyl cellulose • Central Manpower Committee • ceramic-matrix composite • (USA) certified management consultant • Collective Measures Committee (of the UN) • Commandant of the Marine Corps • *Physical chem.* compact molecular cloud • Conservation Monitoring Centre

CMChM Master of Surgery

CMCW Calvinistic Methodist Church of Wales

Cmd *Parliamentary procedure* Command Paper (prefix to serial number, 1919–56). *See also under* C; Cd; Cmnd

CMD *Astronomy* colour-magnitude diagram • *Music* common metre double • conventional munitions disposal (type of bomb disposal) • *Physics* cryogenic magnetic detector

cmdg commanding

Cmdr *Military* Commander

Cmdre Commodore

Cmdt Commandant

CME *Chem.* chemically modified electrode • Chicago Mercantile Exchange • Conférence mondiale de l'énergie (French: World Power Conference) • *Astronomy* coronal mass ejection • (USA) cost and manufacturability expert

CMEA Council for Mutual Economic Assistance (i.e. COMECON)

CMF Cement Makers' Federation • Central Mediterranean Force • (Australia) Citizen Military Forces • *Physics* coherent memory filter • Commonwealth Military Forces • *Physics* compressed magnetic field • *RC Church* Cordis Mariae Filii (Latin: Missionary Sons of the Immaculate Heart of Mary)

CMG Commission on Marine Geology • Companion of the Order of St Michael and St George • Computer Management Group

CMH Campaign for the Mentally Handicapped • combined military hospital • (USA) Congressional Medal of Honor

CMHA Canadian Mental Health Association

CMHC (Canada) Central Mortgage and Housing Corporation

CMi *Astronomy* Canis Minor

CMI *Med.* cell-mediated immunity • Comité maritime international (French: International Maritime Committee) • Commission mixte internationale pour la protection des lignes de télécommunication et des canalisations (French: Joint International Committee for the Protection of Telecommunication Lines and Ducts) • Commonwealth Mycological Institute • computer-managed instruction

CMIA Coal Mining Institute of America

CMIPS *Computing* common management information protocol/service

CMIR (USA) Center for Medical Imaging Research

CMJ Church's Ministry among the Jews

cml commercial • *Computing* current mode logic

CML Central Music Library • chronic myelocytic (*or* myelogenous, myeloid) leukaemia

CMLA Chief Martial Law Administrator

CMLJ Commander of Merit of the Order of St Lazarus of Jerusalem

CMM *Computing* Capability and Maturity Model • (Canada) Commander of the Order of Military Merit

CMMA Concrete Mixer Manufacturers' Association

CMMC (USA) Coal Mining Management College

cmn commission

Cmnd *Parliamentary procedure* Command Paper (prefix to serial number since 1957). *See also under* C; Cd; Cmd

cmnr commissioner

CMO (Zambia) Central Merchandising Office • Chief Medical Officer • (USA) collateralized mortgage-backed obligation • *Astronomy* compact massive object

CMOPE Confédération mondiale des organisations de la profession enseignante (French: World Confederation of Organizations of the Teaching Profession)

CMOS (*or* **C/MOS**) ('siːmɒs) *Electronics* complementary metal oxide semiconductor (*or* silicon)

cmp. compromise

CMP (USA) Central Maine Power Company • *Astronautics* command module pilot • Commissioner of the Metropolitan Police • *Chem.* correlation of molecular pairs • cost of maintaining project • *Biochem.* cytidine monophosphate

cmpd compound

CMPDI (USA) Central Mine Planning and Design Institute Ltd

cm. pf. cumulative preference (*or* preferred) shares

Cmpn Companion

CMR Cape Mounted Rifles • central meter reading • *Electronics* common mode rejection

CMRA Chemical Marketing Research Association

CMRC (USA) Coal Mining Research Company

CMRO County Milk Regulations Officer

CMRR *Electronics* common mode rejection ratio

CMRS (USA) Central Mining Research Station

c.m.s. *Med.* cras mane sumendus (Latin: to be taken tomorrow morning; in prescriptions)

CMS *Chem.* carbon molecular sieve •

(USA) Center for Measurement of Science • central materials supply • Certificate in Management Studies • Church Missionary Society • *Computing* conversational monitor system

CMSER (USA) Commission on Marine Science, Engineering, and Resources

CM/SM command module service module (spacecraft)

cmt cement

CMV *Med.* cerebral major vessel • *Med.* conventional mechanical ventilation • *Microbiol.* cucumber mosaic virus • *Med., Microbiol.* cytomegalovirus

CMY *Computing* cyan, magenta, yellow

CMYK *Computing* cyan, magenta, yellow, black

CMZS Corresponding Member of the Zoological Society

c.n. *Med.* cras nocte (Latin: tomorrow night)

Cn. *Ecclesiast.* Canon

CN *British fishing port registration for* Campbeltown • Canadian National (Railway) • *Chem.* cellulose nitrate • Chinese Nationalists • *Chem.* chloracetophenone • Code Napoléon (French: Napoleonic Code) • common network • *Physics* compound nucleus • *Physics* condensation nucleus • Confederate Navy • *Chem.* coordination number • *Biochem.* cyclic nucleotide • *international civil aircraft marking for* Morocco • *British vehicle registration for* Newcastle upon Tyne

C/N *Biochem., Ecology* carbon–nitrogen (in **C/N ratio**) • circular note • consignment note • contract note • *Insurance* cover note • *Commerce* credit note

CNA Canadian Nuclear Association • *US Navy* Center for Naval Analyses • (Taiwan) Central News Agency • (USA) Chemical Notation Association • *Astronomy* cosmic noise absorption

CNAA Council for National Academic Awards

CNADS Conference of National Armaments Directors (in NATO)

CNAR *Finance* compound net annual rate

CNAS Chief of Naval Air Services

Cnc *Astronomy* Cancer

CNC computerized numerical control

Cncl Council

Cnclr Councillor

CNCMH Canadian National Committee for Mental Hygiene

cncr. concurrent

CND Campaign for Nuclear Disarmament

CNDC China Nuclear Data Centre

CNE (Spain) Comisión nacional de energía (National Energy Commission)

CNEAF coal, nuclear, electric, and alternate fuels

CN(Eng)O Chief Naval Engineering Officer

CNES Centre national d'études spatiales (French space agency)

CNF Challis National Forest • Commonwealth Nurses' Federation • *Maths.* conjunctive normal form

CNFD (USA) Commercial Nuclear Fuel Division

CNG compressed (*or* consolidated) natural gas

CNH *Chem.* carbon–nitrogen hydrogenolysis

CNI Chief of Naval Information • Companion of the Nautical Institute

CNIPA Committee of National Institutes of Patent Agents

cnl cancel

CNL Canadian National Library • (Australia) Commonwealth National Library • (USA) Crocker Nuclear Laboratory

CNLA (USA) Council of National Library Associations

CNLS *Statistics* complex nonlinear least square

CNM (USA) Certified Nurse-Midwife

CNN Cable News Network • Certified Nursery Nurse

CNO Chief Nursing Officer • Chief of Naval Operations

CNP Chief of Naval Personnel

CNPS *Physics* compact nuclear power source

cnr corner

CNR Canadian National Railway • Civil Nursing Reserve • (Burkina Faso) Conseil National de la Révolution (French: National Revolutionary Council) • Council of National Representatives

CNRE (USA) cooperative networks on rural energy

CNRS (France) Centre national de la recherche scientifique (National Centre for Scientific Research)

CNS *Anatomy* central nervous system • (USA) Cherokee Nuclear Station • Chief of the Naval Staff • *Physics* cold-neutron

source • Community Nursing Services • Congress of Neurological Surgeons

CNSI (USA) Chem-Nuclear Systems, Inc.

CNSLD chronic nonspecific lung disease

CNSSO Chief Naval Supply and Secretariat Officer

CNT Canadian National Telegraphs • celestial navigation trainer • (USA) Center for Neighborhood Technology • (Spain) Confederación nacional del trabajo (National Confederation of Labour)

cntn contain

cntr. container • contribute • contribution

cnvt convict

c.o. *Med.* complains of

c/o care of • *Book-keeping* carried over • change over

Co *Chem., symbol for* cobalt

Co. Coalition • Colorado • Company • County

CO Cabinet Office • *British fishing port registration for* Caernarvon • *Med.* cardiac output • careers officer • central office • Chief Officer • Clerical Officer • *postcode for* Colchester • *international vehicle registration for* Colombia • *US postcode for* Colorado • *Military* combined operations • Commanding Officer • command order • Commissioner of Oaths • Commonwealth Office (now part of the FCO) • *Finance* compte ouvert (French: open account) • conscientious objector • *airline flight code for* Continental Airlines • criminal offence • Crown Office • *Electronics* current oscillations • *Biochem.* cyclo-oxygenase • *British vehicle registration for* Exeter

C/O case oil • cash order • certificate of origin

CoA *Biochem.* coenzyme A

COA change of address • College of Aeronautics • condition on admission

coad. coadjutor

Coal. *Politics* Coalition

COAS Council of the Organization of American States

c.o.b. close of business

COBCCEE Comité des organisations de la boucherie et charcuterie de la CE (French: Committee of the Meat Trade Organizations of the EU)

COBE *Astronomy* Cosmic Background Explorer

Cobol (*or* **COBOL**) ('kəʊˌbɒl) *Computing* common business-oriented language

c.o.b.q. cum omnibus bonis quiescat (Latin: may he or she rest with all good souls)

Cobra ('kəʊbrə) Cabinet Office Briefing Room

coc. cocaine

COC Chamber of Commerce • Clerk of the Chapel • *Chem.* colloidal organic carbon • (USA) combat operations center • *Med.* combined oral contraceptive • Corps of Commissionaires

COCA (USA) consent order and compliance agreement

COCAST Council for Overseas Colleges of Arts, Science, and Technology

COCEMA Comité des constructeurs européens de matériel alimentaire (French: Committee of European Machinery Manufacturers for the Food Industries)

coch. (*or* **cochl.**) *Med.* cochleare (Latin: spoonful)

coch. amp. *Med.* cochleare amplum (Latin: heaped spoonful)

coch. mag. *Med.* cochleare magnum (Latin: tablespoonful)

coch. med. *Med.* cochleare medium (Latin: dessertspoonful)

coch. parv. *Med.* cochleare parvum (Latin: teaspoonful)

COCI Consortium on Chemical Information

COCOM (*or* **CoCom**) ('kəʊˌkɒm) Coordinating Committee for Multinational Export Controls

COCOMO (*or* **CoCoMo**) (kɒ'kəʊməʊ) *Computing* constructive cost model

cod. codicil • codification

c.o.d. cargo on deck • cash on delivery

Cod. (*or* **cod.**) codex

COD cash on delivery • Chamber of Deputies • chemical oxygen demand • (USA) collect on delivery • (ital.) Concise Oxford Dictionary

CODAG *Engineering* combined diesel and gas turbine

CODAN ('kəʊdæn) *Telecom.* carrier-operated device anti-noise

CODASYL ('kəʊdəsɪl) *Computing* Conference on Data Systems Languages

CODATA Confederation of Design and Technology Associations

CODC cuttings-oil discharge content

codd. codices

codec ('kəʊdɛk) *Computing* coder-decoder

Codesh ('kəʊdɛʃ) Convention for a Democratic South Africa

CODIPHASE *Computing* coherent digital phased array system

CODOG *Engineering* combined diesel or gas turbine

CODOT Classification of Occupations and Directory of Occupational Titles

COE Conseil œcuménique des églises (French: World Council of Churches) • (USA) Corps of Engineers • cost of electricity • cost of energy

co-ed (ˌkəʊˈɛd) coeducational

COED computer-operated electronic display • (ital.) Concise Oxford English Dictionary

coeff. coefficient

CoEnCo Committee for Environmental Conservation

COESA (USA) Committee on Extension to the Standard Atmosphere

COF Comité olympique française (French Olympic Committee)

C of A Certificate of Airworthiness • College of Arms

COFACE Compagnie française pour l'assurance du commerce exterieur (French export credit guarantee company)

C of B confirmation of balance

C of C Chamber of Commerce

C of E Church of England • Council of Europe

C of ECS Church of England Children's Society

C of F Chaplain of the Fleet • chief of finance • *Mechanical engineering* coefficient of friction

c of g centre of gravity

C of GH Cape of Good Hope

C of I Church of Ireland

COFI Committee on Fisheries (of the FAO)

C of L City of London

C of M *Aeronautics* Certificate of Maintenance

C of S Chief of Staff • Church of Scotland • conditions of service

cog. cognate • cognisant • cognomen

CoG (*or* **c.o.g.**) centre of gravity

COG Cleansing Officers' Guild • *Chemical engineering* coke oven gas

COGAG *Engineering* combined gas and gas turbine

COGB certified official government business

COGECA Comité général de la coopération agricole des pays de la CE (French: General Committee for Agricultural Cooperation in the EU Countries)

COGLA Canada Oil and Gas Lands Administration

COGS *Astronautics* continuous orbital guidance system • *Accounting* cost of goods sold

coh. coheir

c.o.h. cash on hand

COHO ('kəu,həu) *Physics* coherent oscillator

COHSE (*or* **Cohse**) ('kəuzɪ) (formerly) Confederation of Health Service Employees (merged with NALGO and NUPE to form Unison)

COI Central Office of Information • certificate of origin and interest • Commission océanographique intergouvernementale (French: Intergovernmental Oceanographic Commission) • (USA) cost of illness

COIC Canadian Oceanographic Identification Centre • Careers and Occupational Information Centre

CoID Council of Industrial Design (now Design Council)

COIL (kɔɪl) chemical oxygen–iodine laser

COIN (kɔɪn) (USA) counter-insurgency

COINS (USA) Committee on Improvement of National Statistics

COJO Conference of Jewish Organizations

col. collect(ed) • collection • collector • college • collegiate • colon • colonial • colony • colour(ed) • column

Col *Astronomy* Columba

Col. Colombia(n) • Colonel • Colorado • *Bible* Colossians

COL computer-oriented language • cost of living

COLA ('kəulə) (USA) cost of living adjustment (clause in employment contracts)

Col Comdt Colonel Commandant

coll. collateral • colleague • collect(ed) • collection • collector • (*or* **Coll.**) college • collegiate • colloquial(ism) • *Med.* collyrium (eyewash)

collab. collaborate • (in) collaboration (with) • collaborator

collat. collateral(ly)

collect. collective(ly)

Coll of FE College of Further Education

colloq. colloquial(ly) • colloquialism

coll'ott. (*or* **col 8ᵛᵃ**) *Music* coll'ottava (Italian: in octaves)

collr collector

colly. colliery

Colo. Colorado

colog. cologarithm

Coloss. *Bible* Colossians

Col. P. *Advertising* colour page

cols columns

COLS *Computing* communications for online systems

Col-Sergt (*or* **Col-Sgt**) Colour-Sergeant

com. comedy • comic • comma • commentary • commerce • commercial • commission(er) • committee • common(ly) • commoner • commune • communicate(d) • communication(s) • community

Com *Astronomy* Coma Berenices

Com. Commander • Commissary • Commissioner • Committee • Commodore • Commonwealth • Communist

COM coal–oil mixture • *US Navy* Commander • *Chem.* complex organic mixture • computerized operations management • computer output on microfilm (*or* microfiche)

COMA (*or* **Coma**) ('kəumə) Committee on Medical Aspects of Food Policy

COMACA (formerly) Corresponding Member of the Academy of Arts of the USSR

COMAF Comité des constructeurs de matériel frigorifique de la CE (French: Committee of Refrigerating Plant Manufacturers of the EU)

COMAIRCHAN (,kɒmɛəˈtʃæn) Maritime Air Commander Channel

COMAIREASTLANT (,kɒmɛəiːstˈlænt) Air Commander Eastern Atlantic Area

COMAIRLANT (,kɒmɛəˈlænt) (USA) Commander Air Force, Atlantic

COMAL ('kəumæl) *Computing* common algorithmic language

COMAR ('kəumɑː) (USA) Code of Maryland Air Regulations

COMARE ('kəumɛə) (USA) Committee on Medical Aspects of Radiation in the Environment

COMART ('kæmɑːt) Commander, Marine Air Reserve Training

comb. combination • combine(d) • combining • combustible • combustion

combu. combustion

COMCRULANT (ˌkɒmˌkruːˈlænt) (USA) Commander Cruisers, Atlantic

comd command

Comd. Commander

COMDEV Commonwealth Development Finance Company

Comdg *Military* Commanding

Comdr *Military* Commander

Comdt *Military* Commandant

COME Chief Ordnance Mechanical Engineer

COMECON (ˈkɒmɪˌkɒn) Council for Mutual Economic Assistance

Comet (*or* **COMET**) (ˈkɒmɪt) Committee for Middle East Trade • computer-operated management evaluation technique

COMEX (ˈkəʊˌmɛks) (USA) Commodity Exchange (New York)

COMEXO Committee for Exploitation of the Oceans

Com-Gen Commissary-General

COMIBOL Corporación Minera de Bolivia (Bolivian state mining company)

Cominform (ˈkɒmɪnˌfɔːm) Communist Information Bureau

COMINT (ˈkɒmɪnt) communications intelligence

Comintern (ˈkɒmɪnˌtɜːn) Communist International

COMISCO (ˈkɒmɪsˌkəʊ) Committee of International Socialist Conference

coml commercial

COMLOGNET (ˌkɒmˌlɒgˈnɛt) *US Air Force* combat logistics network

comm. commentary • commerce • commercial(ly) • committee • commonwealth • communication

Comm. Commander • Commodore

Commd *Military* Commissioned

commem. commemoration • commemorative

Commiss. Commissary

Commissr (*or* **Commr**) Commissioner

Commn Commission

Commnd *Military* Commissioned

commun. communication • community

Commy Commissary

COMNAVNORTH (ˌkɒmˌnævˈnɔːθ) Commander Allied Naval Forces Northern Europe

COMO Committee of Marketing Organizations

comp. companion • comparative •

compare • comparison • compass • compensation • compete • competitive • competitor • compilation • compiled • compiler • complete • compose(r) • composite • composition • compositor • compound(ed) • comprehensive • compression • comprising

Comp Companion (of an institution; as in **ComplMechE**, Institution of Mechanical Engineers)

COMPAC (ˈkɒmpæk) Commonwealth Trans-Pacific Telephone Cable

compar. comparative • comparison

compd compound

Comp-Gen Comptroller-General

compl. complement • complete • compliment • complimentary

complt complainant • complaint

compo. composition

compr. compressive

compt compartment

Compt. (*or* **Comptr**) Comptroller

Comr Commissioner

Com. Rom. Common Romance (language)

comsat (ˈkɒmsæt) communications satellite

ComSec (ˈkɒmˌsɛk) Commonwealth Secretariat

COMSER Commission on Marine Science and Engineering Research (in the UN)

COMSUBEASTLANT (ˌkɒmˌsʌbˌiːstˈlænt) Commander Submarine Force Eastern Atlantic

Com. Teut. Common Teutonic (language)

Com. Ver. Common Version (of the Bible)

Com. W Ger. Common West Germanic (language)

Comy-Gen Commissary-General

con. concentration • concerning • concerto • conclusion • confidence • conics • conjunx (Latin: wife) • connection • consolidate(d) • consols • continue(d) • contra (Latin: against) • convenience • conversation

Con. Conformist • Conservative • *Finance* Consols • *Military* Constructor • Consul • *Music* contralto

Con A *Immunol.* concanavalin A

CONAC (ˈkɒnæk) *US Air Force* Continental Air Command

CONAD (ˈkɒnæd) (USA) Continental Air Defense Command

CONARC ('kɒnɑːk) (USA) Continental Army Command

conbd contributed

conc. concentrate(d) • concentration • concerning • concerto

Con. C Constructor Captain

concd concentrated

concg concentrating

conch. conchology

concn concentration

concr. concrete

con. cr. *Book-keeping* contra credit

Con. Cr Constructor Commander

cond. condense(r) • condition(al) • conduct(ed) • conductivity • (*or* **condr**) conductor

con esp. (*or* **con espr.**) *Music* con espressione (Italian: with expression)

conf. *Med.* confection • confectionery • confer (Latin: compare) • conference • confessor

Confed. Confederacy • Confederate • Confederation

cong. *Pharmacol.* congius (Latin: gallon) • congregation(ist)

Cong. Congregational(ist) • Congress • Congressional

Cong. R. (*or* **Cong. Rec.**) (USA) Congressional Records

CONGU Council of National Golf Unions

CONI Comitato olimpico nazionale italiano (Italian National Olympic Committee)

con. inv. consular invoice

conj. conjugation • conjunction • conjunctive

Con. L Constructor Lieutenant

CONLAN ('kɒnlæn) *Computing* consensus language

Con. LCr Constructor Lieutenant-Commander

conn. connect(ed) • connection • connotation

Conn. Connecticut

conq. conquer(or)

cons. consecrate(d) • consecration • consecutive • consequence • conservation • conservative • *Med.* conserve • consigned • consignment • consolidated • consonant • constable • constitution(al) • construction • consul • consult(ing)

Cons. Conservative • Conservatoire (*or* Conservatorium, Conservatory) • Constable • Constitution • Consul

CONS *Computing* connection-oriented network service

con. sec. conic section

Conserv. Conservatoire (*or* Conservatorium, Conservatory)

cons. et prud. consilio et prudentia (Latin: by counsel and prudence)

consgt consignment

consid. consideration

Con. SL Constructor Sub-Lieutenant

Consols ('kɒnsɒlz) consolidated annuities (*or* stock)

const. constable • constant • constituency • (*or* **Const.**) constitution • (*or* **constl**) constitutional • construction

constr. construct(ion) • construe

cont. container • containing • contents • (*or* **Cont.**) continent(al) • continue(d) • (*or* **cont**) *Music* continuo • continuum • contra (Latin: against) • contract • contraction • control(ler)

contag. contagious

contbd contraband

contbg contributing

cont. bon. mor. contra bonos mores (Latin: contrary to good manners)

contd contained • continued

contemp. contemporary

contg containing

contn continuation

contr. contract(s) • contracted • contraction • contralto • contrary • contrast(ed) • control(ler)

contr. bon. mor. contra bonos mores (Latin: contrary to good manners)

cont. rem. *Med.* continuantur remedia (Latin: let the medicines be continued; in prescriptions)

contrib. contributed • contributing • contribution • contributor

CONUS continental United States

conv. convenient • convent • convention(al) • conversation • converter • convertible

Conv. Convocation

convce conveyance

COOC *Insurance* contact with oil or other cargo

co-op. (*or* **coop.**) cooperative

COORS communications outage restoral section

cop. copper • copulative • copyright(ed)

Cop. Copernican • Coptic

COP (New Zealand) Certificate of

Proficiency (a pass in a university subject) · *Thermodynamics* coefficient of performance · (USA) community-oriented policing · custom of port

COPA Comité des organisations professionnelles agricoles de la CE (French: Committee of Agricultural Organizations in the EU)

COPAL ('kəʊˌpæl) Cocoa Producers' Alliance

COPD chronic obstructive pulmonary disease

Copec ('kɒpɛk) Conference on Christian Politics, Economics, and Citizenship

COPPSO Conference of Professional and Public Service Organizations

copr. copyright

COPS Council of Polytechnic Secretaries

Copt. Coptic

coptr copartner

COPUS (*or* **Copus**) ('kəʊpəs) Committee on the Public Understanding of Science

cor. corner · cornet · coroner · corpus (Latin: the body) · correct(ed) · correction · correlative · correspondence · correspondent · corresponding · corrupt

Cor. *Bible* Corinthians

Corat ('kɔːræt) Christian Organizations Research and Advisory Trust

CORBA ('kɔːbə) *Computing* common object request broker architecture

CORD chronic obstructive respiratory disease

CORE (kɔː) (USA) Congress of Racial Equality

CORES *Commerce* (Japan) Computerized Order Routing and Execution System

CORGI ('kɔːgɪ) Council for Registration of Gas Installers

Cor. Mem. Corresponding Member

Corn. Cornish · Cornwall

Cornh. (ital.) Cornhill Magazine

corol. (*or* **coroll.**) corollary

Corp. (*or* **Corpl**) Corporal · (*or* **Corpn**) Corporation · Corpus Christi College, Oxford

corr. correct(ed) · correction · corrective · correlative · corrente (Italian: current) · correspond(ent) · correspondence · corresponding · (*or* **Corr.**) corrigenda · corrugated · corrupt(ed) · corruption

CORRA Combined Overseas Rehabilitation Relief Appeal

correl. correlative

corresp. correspondence · corresponding

Corresp. Mem. Corresponding Member

corrupt. corruption

Cors. Corsica(n)

CORS (kɔːz) *Chem.* catalytic oxygen removal system · Chief of the Regulating Staff

Cor. Sec. Corresponding Secretary

CORSO ('kɔːsəʊ) (New Zealand) Council of Organizations for Relief Services Overseas

cort. cortex

cos (kɒz) *Maths.* cosine

c.o.s. *Commerce* cash on shipment

Cos. (*or* **cos.**) Companies · Counties

COS *Commerce* cash on shipment · Chamber of Shipping · Charity Organization Society · (*or* **CoS**) Chief of Staff

co. sa. come sopra (Italian: as above)

COSA Colliery Officials and Staffs Association · *Accounting* cost of sales adjustment

COSAG ('kəʊsæg) combined steam turbine and gas turbine (ship propulsion)

COSAR ('kəʊsɑː) compression scanning array radar

COSATI (kəʊˈsɑːtɪ) Committee on Scientific and Technical Information

COSATU (*or* **Cosatu**) (kəʊˈsɑːtuː) Congress of South African Trade Unions

COSBA (kɒsbɑː) Computer Services and Bureaux Association

Cose *Computing* common open software environment

cosec ('kəʊsɛk) *Maths.* cosecant

COSEC ('kəʊsɛk) Coordinating Secretariat of the National Union of Students

cosech ('kəʊʃɛk) *Maths.* hyperbolic cosecant

COSFPS Commons, Open Spaces, Footpaths Preservation Society

cosh (kɒʃ) *Maths.* hyperbolic cosine

COSHH Control of Substances Hazardous to Health

CoSIRA (kəʊˈsaɪrə) Council for Small Industries in Rural Areas

COSLA (*or* **Cosla**) ('kɒzlə) Convention of Scottish Local Authorities

COSMD Combined Operations Signals Maintenance Division

Cosmo ('kɒzməʊ) (ital.) Cosmopolitan (magazine)

cosmog. cosmogony · cosmographical · cosmography

COSMOS ('kɒsˌmɒs) Coast Survey Marine Observation System

co. so. come sopra (Italian: as above)

COSPAR ('kəʊˌspɑː) Committee on Space Research

Coss. Consules (Latin: Consuls)

COSSAC Chief of Staff to Supreme Allied Commander

COST (kɒst) Committee for Overseas Science and Technology (of the Royal Society)

cot (kɒt) *Maths.* cotangent

CoT college of technology

COT *Physics* change of temperature

cotan ('kəʊˌtæn) *Maths.* cotangent

COTC Canadian Officers' Training Corps

COTE (*or* **Cote**) Committee for the Accreditation of Teacher Education

coth (kɒθ) *Maths.* hyperbolic cotangent

COTR (USA) contracting officers' technical representative

COTT (South Africa) Central Organization for Technical Training

couch. *Heraldry* couchant

Coun. Council(lor) • Counsellor

cour. courant (French: instant (this month); in correspondence)

cov. covenant

Cov. Coventry

COV *Statistics* coefficient of variation • *Chem.* concentrated oil of vitriol • *Electronics* corona onset voltage • (*or* **Cov**) *Statistics* covariance • *Genetics* crossover value

covers ('kəʊvɜːs) *Maths.* coversed sine

cov. pt. *Cricket* cover point

Cow character-oriented windows

COWAR Committee on Water Research

COWPS (USA) Council on Wages and Price Stability

Coy. *Military* company

COZI communications zone indicator

cp *Chem.* cyclopentadienyl (used in formulae)

cp. compare

c.p. candlepower • carriage paid • *Aeronautics* centre of pressure • condensation product • constant pressure

Cp. *RC Church* compline

CP *international civil aircraft marking for* Bolivia • *Med.* Campylobacter pylori (cause of gastric infections) • *airline flight code for* Canadian Airlines International • Canadian Pacific (Railway) • Canadian Press • (formerly) Cape Province (South Africa) • (Isle of Man) Captain of the Parish • *Navigation* cardinal point • casella postale (Italian: post office box) • *Computing* central processor • centrifugal pump • *Med.* cerebral palsy • *Surveying* change point • *Physics* charge (conjugation) parity (as in **CP invariance**) • charterparty • chemically pure • Chief of Police • Chief Patriarch • *Chem.* chlorophenol • *Med.* chronic pancreatitis • *Med.* cisplatin (anticancer drug) • civil power • civil procedure • Clarendon Press • Clerk of the Peace • code of procedure • Codice Penale (Italian: penal code) • College of Preceptors • colour printing (as in **CP filter**) • *Military* Command Post • commercial paper • Common Pleas • Common Prayer • Communist Party • community physician • Community Programme (employment scheme) • (Portugal) Companhia dos Caminhos de Ferro Portugueses (railway company) • *Stock exchange* concert party • conference paper • conference proceedings • *RC Church* Congregatio Passionis (Latin: Congregation of the Passion) • (India) Congress Party • (South Africa) Conservative Party • convict prison • corporal punishment • (Australia) Country Party (now the National Party) • Court of Probate • *Biochem.* creatine phosphate • *Physics* cross polarization • current paper • *British vehicle registration for* Huddersfield

C/P charterparty

CPA Canadian Pharmaceutical Association • Canadian Psychological Association • central planning area • (USA) Certified Public Accountant • Chartered Patent Agent • Chick Producers' Association • *Insurance* claims payable abroad • Clyde Port Authority • Commonwealth Parliamentary Association • Communist Party of Australia • Construction Plant Association • Contractors' Plant Association • contract price adjustment • cost planning and appraisal • Council of Provincial Associations • critical path analysis

CPAA *Physics* charged-particle activation analysis

CPAC Collaborative Pesticides Analytical Committee • Consumer Protection Advisory Committee

CPAG Child Poverty Action Group • (USA) Collision Prevention Advisory Group

CPAI Canvas Products Association International

C. Pal. Crystal Palace

CPAM Committee of Purchasers of Aircraft Material

CPAS Catholic Prisoners' Aid Society • Church Pastoral Aid Society

CPB *Book-keeping* casual payments book • Central Planning Bureau • Communist Party of Britain

CPC (USA) City Planning Commission • City Police Commissioner • Clerk of the Privy Council • *Med.* clinicopathological conference • Communist Party of China • Conservative Political Centre

CPCIZ Comité permanent des congrès internationaux de zoologie (French: Permanent Committee of International Zoological Congresses)

CPCU (USA) Chartered Property and Casualty Underwriter

cpd compound

c.p.d. *Commerce* charterers pay dues

CPD *Astronomy* Cape Photographic Durchmüsterung (star catalogue) • chronic pulmonary disease • *Electronics* contact potential difference • continuing professional development

CPDL Canadian Patents and Developments Limited

CPDM Centre for Physical Distribution Management

CPE Certificate of Physical Education • (USA) Certified Property Exchanger • *Med.* chronic pulmonary emphysema • College of Physical Education • *Law* Common Professional Examination • Congrès du peuple européen (French: Congress of European People) • contractor performance evaluation • *Microbiol.* cytopathic effect

CPEA Catholic Parents' and Electors' Association • (USA) Cooperative Program for Educational Administration

c. pén. code pénal (French: penal code)

CPEQ Corporation of Professional Engineers of Quebec

CPF contributory pension fund

CPFF cost plus fixed fee

CPFS Council for the Promotion of Field Studies

CPGB Communist Party of Great Britain

c.p.h. cycles per hour

CPH Certificate in Public Health

CPHA Canadian Public Health Association

CPhys Chartered Physicist

cpi *Printing* characters per inch

CPI *Biochem.* carboxypeptidase inhibitor • chief pilot instructor • Communist Party of India • (USA) consumer price index

CPJI Cour permanente de justice internationale (French: Permanent Court of International Justice)

Cpl *Military* Corporal

CPL central public library • Colonial Products Laboratory • *Computing* combined programming language • Commercial Pilot's Licence

cpm cycles per minute

CPM (USA) Certified Property Manager • Colonial Police Medal • *Music* common particular metre • Communist Party of Malaya • computer program module • critical path method

CP/M *Computing, trademark* Control Program for Microcomputers

CPMEE&W Council for Postgraduate Medical Education in England and Wales

CPMG Carr-Purcell-Meiboom-Gill (in **CPMG sequence**)

cpn coupon

Cpn Copenhagen

CPN Communistische Partij van Nederland (Netherlands Communist Party) • Community Psychiatric Nurse

CPNA Council of Photographic News Agencies

Cpnhgn Copenhagen

CPNZ Communist Party of New Zealand

CPO cancel previous order • *Chem.* catalytic partial oxidation • Chief Petty Officer • command pay office • Commonwealth Producers' Organization • compulsory purchase order • County Planning Officer

c.p.p. controllable pitch propeller

CPP chemical processing plant • *Chemical engineering* condensate purification plant • (Ghana) Convention People's Party • *Computing* critical path planning • current purchasing power (in **CPP accounting**)

CPPA Canadian Pulp and Paper Association

CPPCC Chinese People's Political Consultative Conference

CPPO *Chem.* bis(2-carbopentyloxy-3,5,6-trichlorophenyl) oxalate

CPPS *Computing* critical path planning and scheduling

CPR Canadian Pacific Railway • *Med.* cardiopulmonary resuscitation

CPRC Central Price Regulation Committee

CPRE Council for the Protection of Rural England

CPRS Central Policy Review Staff

CPRW Council for the Protection of Rural Wales

cps *Computing* characters per second • *Physics* cycles per second

CPS Centre for Policy Studies • cents per share • (USA) Certified Professional Secretary • Church Patronage Society • Clerk of Petty Sessions • Commonwealth Public Service • Communist Party of Syria • Congregational Publishing Society • Crown Prosecution Service • Custos Privati Sigilli (Latin: Keeper of the Privy Seal)

CPSA Civil and Public Services Association (clerical civil servants' union) • Clay Pigeon Shooting Association

CPSC (USA) Consumer Product Safety Commission

CPSS Certificate in Public Service Studies

CPSU Communist Party of the Soviet Union

CPsychol Chartered Psychologist

cpt cockpit • counterpoint

Cpt. Captain

CPT Canadian Pacific Telegraphs • *Physics* charge (conjugation) parity–time (reversal) (as in **CPT theorem**) • cost per thousand • *Computing* critical path technique

CPTB Clay Products Technical Bureau

CPU central packaging unit • (*or* **c.p.u.**) *Computing* central processing unit • collective protection unit • Commonwealth Press Union

CPUSA Communist Party of the United States of America

CPVE Certificate of Pre-vocational Education

CPWC (Burma) Central People's Workers' Council

CQ *Military* charge of quarters • conditionally qualified

CQM Chief Quartermaster • Company Quartermaster

CQMS Company Quartermaster-Sergeant

CQR (ital.) Church Quarterly Review

CQS Court of Quarter Sessions

CQSW Certificate of Qualification in Social Work

cr. created • creation • (*or* **Cr.**) credit(or) • creek • *Music* crescendo • crew • crimson • crown • cruise • crux (Latin: cross)

c.r. con riserva (Italian: with reservations) • *Finance* cum rights (i.e. with rights)

c/r company's risk

Cr *Chem., symbol for* chromium • Commander • Councillor • cruiser

CR Carolina Regina (Latin: Queen Caroline) • Carolus Rex (Latin: King Charles) • carriage return • carrier's risk • cash receipts • central railway • central registry • Chief Ranger • *Med.* chronic rejection • Civis Romanus (Latin: Roman citizen) • (USA) Commendation Ribbon • *Anglican Church* Community of the Resurrection • company's risk • *Med.* complete regression (*or* remission) • *Engineering* compression ratio • *Med.* computed radiography • *Chem.* concentration ratio • *Psychol.* conditioned reflex (*or* response) • conference report • Congo red (dye and chemical indicator) • (USA) consciousness raising • control relay • cosmic rays • Costa Rica (abbrev. *or* IVR) • credit • credit rating • credit report • *Statistics* critical ratio • *postcode for* Croydon • current rate • Custos Rotulorum (Latin: Keeper of the Rolls) • *British vehicle registration for* Portsmouth • *international civil aircraft marking for* Portugal

CrA *Astronomy* Corona Australis

CRA California Redwood Association • Canadian Rheumatism Association • (Northern Ireland) Civil Rights Association • Coal Research Association of New Zealand • Commander of the Royal Artillery • Commercial Rabbit Association • composite research aircraft • corrosion-resistant alloy

CrAA Commander-at-Arms

CRAC Careers Research and Advisory Centre • Central Religious Advisory Committee • Construction Research Advisory Council

CRAD Committee for Research into Apparatus for the Disabled

CRAeS Companion of the Royal Aeronautical Society

CRAF (USA) Civil Reserve Air Fleet

CRAMRA Convention on the Regulation of Antarctic Mineral Resource Activities

cran. (*or* **craniol.**) craniology

craniom. craniometry

CRASC Commander of the Royal Army Service Corps

CrB *Astronomy* Corona Borealis

CRB Central Radio Bureau

CRC *Printing* camera-ready copy • Cancer Research Campaign • child-resistant closure • Civil Rights Commission • coal rank code (for classifying coals) • Community Relations Council • *Printing* composing room chapel • (USA) Coordinating Research Council • Cycle Racing Club • *Computing* cyclic redundancy check (*or* code)

CRCC Canadian Red Cross Committee

CRCH central register and clearing house

CRCP Certificant of the Royal College of Physicians

CRCS Certificant of the Royal College of Surgeons

CRD chronic respiratory disease • Crop Research Division

CRDEC (USA) Chemical Research, Development, and Engineering Center

CRDF cathode-ray direction-finding

CRE (USA) Coal Research Establishment • Commander of the Royal Engineers • Commercial Relations and Exports • Commission for Racial Equality

CREFAL Centro Regional de Educación Fundamental para la América Latina (Spanish: Regional Centre of Fundamental Education for Latin America)

cres. (*or* **cresc.**) *Music* crescendo

Cres. Crescent (in street names)

CREST (krɛst) *Med.* calcinosis, Raynaud's phenomenon, (o)esophageal malfunction, sclerodactyly, telangiectasia (in **CREST syndrome**)

CRF *Finance* capital recovery factor • chronic renal failure • *Med.* coagulase reacting factor • *Biochem.* corticotrophin-releasing factor

crg. carriage

CRH corticotrophin-releasing hormone

CRI Caribbean Research Institute • *Med.* chronic renal insufficiency • *Photog.* colour reversal intermediate • (USA) Cray Research Incorporated • Croce Rossa Italiana (Italian Red Cross)

CRIB Current Research in Britain (publication)

CRIC *RC Church* Canons Regular of the Immaculate Conception

crim. criminal

crim. con. *Law* criminal conversation (i.e. adultery)

criminol. criminology

Crimp. *Textiles* Crimplene

CRIS *Med.* clinical radiology imaging system • *Computing* command retrieval information system • current research information system

crit. criterion • critic • critical(ly) • criticism

Crk Cork (Ireland)

CRL *RC Church* Canons Regular of the Lateran • Certified Record Librarian • Certified Reference Librarian • Chemical Research Laboratory

CRM (USA) Central Rocky Mountains • certified reference material • *Physics* chemical remanent magnetization • counter-radar missile • count-rate meter • cruise missile

CRMA Cotton and Rayon Merchants' Association

CRMF Cancer Relief Macmillan Fund

CRMP Corps of Royal Military Police

crn. crane • crown

CRNA Clinical Research Nurses' Association

CRNCM Companion of the Royal Northern College of Music

CRNSS Chief of the Royal Naval Scientific Service

CRO cathode-ray oscilloscope (*or* oscillograph) • chief recruiting officer • Commonwealth Relations Office (now part of the FCO) • community relations officer • compulsory rights order • Criminal Records Office

Croat. Croatia(n)

CRP Calendarium Rotulorum Patentium (Latin: Calendar of the Patent Rolls) • *RC Church* Canons Regular of Prémontré • capacity requirement plan • (India) Central Reserve Police • coordinated research programme • *Immunol.* C-reactive protein • *Biochem.* cyclic-AMP receptor protein

CRPL (USA) Central Radio Propagation Laboratory

CRPPH (USA) Committee on Radiation Protection and Public Health

CRR constant ratio rule • Curia Regis Roll

CRS Catholic Record Society • cold-rolled steel • Cooperative Retail Society • cosmic radio source

CRSA Cold Rolled Sections Association • Concrete Reinforcement Steel Association

CRSI (USA) Concrete Reinforcing Steel Institute

Crt Court • *Astronomy* Crater

CRT cathode-ray tube • *Maths.* Chinese remainder theorem • combat readiness training • composite rate tax

CRTC Canadian Radio-Television Commission

crtkr caretaker

CRTS Commonwealth Reconstruction Training Scheme

Cru *Astronomy* Crux

CRU civil resettlement unit • *Computing* control register user

CRUDESPAC ('kruːdɛs,pæk) *US Navy* Cruiser-Destroyer Forces, Pacific

CRULANT ('kruːlænt) *US Navy* Cruiser Forces, Atlantic

Crv *Astronomy* Corvus

crypto. cryptographic • cryptography

cryst. crystal • crystalline • (*or* **crystallog.**) crystallography

crystd crystallized

crystn crystallization

CRZZ (Poland) Centralna Rada Zwiazkow Zawadowych (Central Council of Trade Unions)

cs. case • census • consul

c.s. capital stock • come sopra (Italian: as above)

c/s cases • cycles per second

Cs *Chem., symbol for* caesium • (ital.) *Meteorol.* cirrostratus

CS *Chem.* calcium silicate • capital stock • carbon steel • *Biochem.* casein • cast steel • Certificate in Statistics • Chartered Surveyor • Chemical Society • Chief of Staff • Chief Secretary • *Computing* chip select • Christian Science (*or* Scientist) • City Surveyor • civil servant • Civil Service • Clerk of Session • Clerk to the Signet • *Med.* clinical stage • close shot • close support • College of Science • *Mining* colliery screened • Common Serjeant • *Physics* composite superconductor • *Physics* Compton scattering • *Psychol.* conditioned stimulus • (USA) Confederate States • *RC Church* Congregation of Salesians • *Physics* coolant

system • Cooperative Society • cotton seed • County Surveyor • Court of Session • *British fishing port registration for* Cowes • credit sales • cruiser squadron • Custos Sigilli (Latin: keeper of the seal) • *British vehicle registration for* Glasgow • *international civil aircraft marking for* Portugal • *see* CS gas

C/S channel shank (buttons) • cycles per second

CSA Canadian Standards Association • Casualty Surgeons' Association • Child Support Agency • Common Services Agency • (USA) Community Services Administration • (Canada) Computer Science Association • Confederate States Army • Confederate States of America

CSAA Child Study Association of America

CSAE Canadian Society of Agricultural Engineering

CSAP Canadian Society of Animal Production

CSAR Communication Satellite Advanced Research

CSB Bachelor of Christian Science • *Engineering* calcium silicate brick • Central Statistical Board • *Med.* chemical stimulation of the brain

CSBGM Committee of Scottish Bank General Managers

CSBM *Military* Confidence- and Security-Building Measures

csc *Maths.* cosecant

CSC Civil Service Commission • Commonwealth Science Council • Comprehensive Schools Committee • (Belgium) Confédération des syndicats chrétiens (French: Federation of Christian Trade Unions) • Congregation of the Holy Cross • Conspicuous Service Cross (replaced by the DSC)

CSCBS Commodore Superintendent, Contract-Built Ships

CSCC Civil Service Commission of Canada • Council of Scottish Chambers of Commerce

CSCE Conference on Security and Cooperation in Europe

CSCW computer-supported cooperative working

CSD Chartered Society of Designers • Civil Service Department • Commonwealth Society for the Deaf • *Engineering* constant speed drive • Cooperative Secretaries Diploma • Doctor of Christian Science

CSDE *RAF* Central Servicing Development Establishment

cse course

CSE Central Signals Establishment • Certificate of Secondary Education (replaced by General Certificate of Secondary Education, GCSE) • *Computing* cognitive systems engineering • Council of the Stock Exchange

CSEA (USA) Civil Service Employees Association

CSED *US Navy* coordinated ship electronics design

CSEU Confederation of Shipbuilding and Engineering Unions

CSF *Med.* cerebrospinal fluid • Coil Spring Federation • *Med.* colony-stimulating factor (as in **G-CSF**, granulocyte CSF)

CSFA Canadian Scientific Film Association

CSFE Canadian Society of Forest Engineers

CSG Catholic Social Guild • Companion of the Order of the Star of Ghana • *Computing* constructive solid geometry

CSGA Canadian Seed Growers' Association

CS gas *indicating* a tear gas (surname initials of its US inventors, Ben Carson and Roger Staughton)

CSH *Chem.* calcium silicate hydrate

CSI Chartered Surveyors' Institution • *Chem.* chlorosulphonyl isocyanate • Church of South India • *Image technol.* compact source iodide (as in **CSI lamp**) • Commission sportive internationale (French: International Sporting Commission) • Companion of the Order of the Star of India • (USA) Construction Specifications Institute

CSICC Canadian Steel Industries Construction Council

CSIEJB Certificate of the Sanitary Inspectors Examination Joint Board

CSIP Committee for the Scientific Investigation of the Paranormal

CSIR Council for Scientific and Industrial Research

CSIRA Council for Small Industries in Rural Areas

CSIRO Commonwealth Scientific and Industrial Research Organization (Australia)

CSJ (ital.) Christian Science Journal

csk cask • countersink

CSL (Australia) Commonwealth Serum Laboratories • Communication Sub-Lieutenant • computer simulation language • *Computing* control and simulation language • cub scout leader

CSLATP Canadian Society of Landscape Architects and Town Planners

CSLO Canadian Scientific Liaison Office • Combined Services Liaison Officer

CSLT Canadian Society of Laboratory Technologists

CSM cerebrospinal meningitis • (ital.) Christian Science Monitor • Christian Socialist Movement • *Astronautics* command service module • Commission for Synoptic Meteorology • Committee on Safety of Medicines • Company Sergeant-Major • (USA) corn, soya, milk (food supplement)

CSMA Chemical Specialities Manufacturers' Association • Civil Service Motoring Association

CSMA/CD *Computing* carrier sense multiple access, collision detection (network protocol)

CSMMG Chartered Society of Massage and Medical Gymnastics

CSMTS Card Setting Machine Tenters' Society

CSN Confederate States Navy

CSNI (USA) Committee on the Safety of Nuclear Installations

CSO Caltech Submillimeter Observatory • Central Selling Organization • Central Statistical Office • Chief Scientific Officer • Chief Signal Officer • Chief Staff Officer • Colonial Secretary's Office • *Image technol.* colour separation overlay • Command Signals Officer • Commonwealth Scientific Office • community service order

CSP Chartered Society of Physiotherapists (*or* Physiotherapy) • Civil Service of Pakistan • *Computing* communicating sequential processes • Congregation of Saint Paul • Council for Scientific Policy

CSPAA Conférence de solidarité des pays afro-asiatiques (French: Afro-Asian People's Solidarity Conference)

CSPCA Canadian Society for the Prevention of Cruelty to Animals

CSPR *Chem.* chlorosulphonated polyethylene rubber

CSR (Australia) Colonial Sugar Refining Company • combat-stress reaction • Czechoslovak Socialist Republic

CSS *Computing* centralized structure store • Certificate in Social Service • computer systems simulator • (*or* **CSSp**) (member of the) Congregation of the Holy Ghost (Sanctus Spiritus); Holy Ghost Father • Council for Science and Society

CSSA Civil Service Supply Association

CSSB Civil Service Selection Board

CSSDA (USA) Council for Social Science Data Archives

CSSR *RC Church* Congregatio Sanctissimi Redemptoris (Latin: Congregation of the Most Holy Redeemer; Redemptorists)

CSSS Canadian Soil Science Society

CST (USA) Central Standard Time • College of Science and Technology

CSTA Canadian Society of Technical Agriculturists • (New Zealand) Canterbury Science Teachers' Association

CSTI Council of Science and Technology Institutes

CStJ Commander of the Most Venerable Order of the Hospital of St John of Jerusalem

CSU *Med.* catheter specimen of urine • Central Statistical Unit • Christlich-Soziale Union (Christian Social Union; German political party) • Civil Service Union • Colorado State University • *Engineering* constant speed unit

CSV *Computing* comma-separated variables • community service volunteer • Community Service Volunteers

CSW (USA) Certified Social Worker • continuous seismic wave

CSYS (Scotland) Certificate of Sixth Year Studies

ct carat • caught • cent • circuit • courant (French: instant (this month); in correspondence) • court • crate • credit • current

ct. centum (Latin: hundred) • certificate

Ct. Canton • Connecticut • Count • (*or* **Ct**) *Music* countertenor • Court

CT cable transfer • Candidate in Theology • *postcode for* Canterbury • Cape Town • *British fishing port registration for* Castletown • *Med.* cell therapy • (USA) Central Time • certificated (*or* certified) teacher • Civic Trust • code telegrams • college of technology • commercial traveller • (Italy) commissario tecnico (sports coach) • *Med.* computerized (*or* computed) tomography (as in **CT scanner**) • *US postcode for* Connecticut •

corporation tax • counter trade • *Electrical engineering* current transformer • cycle time • *British vehicle registration for* Lincoln

C/T Californian Terms (in the grain trade)

c.t.a. *Law* cum testamento annexo (Latin: with the will annexed)

CTA Camping Trade Association • Canadian Tuberculosis Association • Caribbean Technical Assistance • Caribbean Tourist Association • Catering Teachers' Association • *Chem.* cellulose triacetate • Channel Tunnel Association • Chaplain Territorial Army • Chicago Transit Authority • Commercial Travellers' Association • (USA) commodities trading adviser • *Med.* computerized (*or* computed) tomographic angiography (*or* arteriography)

CTAB *Chem.* cetyltrimethylammonium bromide

CTAU Catholic Total Abstinence Union

CTB Commonwealth Telecommunications Board • comprehensive test ban

CTBT comprehensive test ban treaty

CTC Canadian Transport Commission • *Chem.* carbon tetrachloride • centralized traffic control • Central Training Council • (USA) Citizen's Training Corps • city technology college • Civil Technical Corps • Commando Training Centre • Confederación de Trabajadores Cubanos (Spanish: Confederation of Cuban Workers) • Congrès du travail du Canada (French: Canadian Labour Congress) • corn trade clauses • crushing, tearing, and curling (machine) • Cyclists' Touring Club

ctd crated

CTD central training depot • *Electronics* charge-transfer device • classified telephone directory

ctDNA *Biochem.* chloroplast DNA

Cte Comte (French: Count)

CTE *Physics* coefficient of (linear) thermal expansion

CTEB Council of Technical Examining Bodies

Ctesse Comtesse (French: Countess)

CTETOC Council for Technical Education and Training for Overseas Countries

CText Chartered Textile Technologist

ctf. certificate • certify

CTF Chaplain to the Territorial Forces • coal-tar fuels

ctge cartage • cartridge • cottage
CTH Corporation of Trinity House
CTI computer telephony integration
CTIO Cerro Tololo Inter-American Observatory (Chile)
CTK (formerly) Ceskoslovenská Tisková Kancelář (Czechoslovak Press Bureau)
ctl central
CTL constant tensile load • *Insurance* constructive total loss • *Med.* cytotoxic T lymphocyte
CTM *Med.* computerized (*or* computed) tomographic myelography
ctn carton • *Maths.* cotangent
CTO *Philately* cancelled to order (of postage stamps) • Central Telegraph Office • Central Treaty Organization • Chief Technical Officer
CTOL ('siːtɒl) conventional takeoff and landing
CTP Confederación de Trabajadores del Peru (Spanish: Peruvian Confederation of Labour) • *Biochem.* cytidine 5′-triphosphate
ctpt *Music* counterpoint
ctptal *Music* contrapuntal
ctptst *Music* contrapuntist
ctr. centre • contribution • contributor
CTR certified test record • controlled thermonuclear reaction • controlled thermonuclear research
CTRA Coal Tar Research Association
Ctrl *Computing* control
CTRP Confederación de Trabajadores de la República de Panama (Spanish: Confederation of Workers of the Republic of Panama)
CTRU Colonial Termite Research Unit
cts centimes • cents • certificates • crates
CTS *Med.* carpal tunnel syndrome • Catholic Truth Society • characteristic time scale • *Computing* clear to send • *Med.* computerized (*or* computed) tomographic scanner • Consolidated Tin Smelters
CTSA Crucible and Tool Steel Association
CTT capital-transfer tax
CTTB Central Trade Test Board (of the RAF)
Cttee (*or* **C'ttee**) committee
CTTSC Certificate in the Teaching and Training of Subnormal Children
CTUS Carnegie Trust for the Universities of Scotland
CTV cable television • Canadian

Television Network Limited • Confederación de Trabajadores de Venezuela (Spanish: Confederation of Venezuelan Workers)
CTZ control traffic zone (around an aerodrome or airport)
cu. cubic
Cu *Chem.*, *symbol for* copper (Latin *cuprum*) • (ital.) *Meteorol.* cumulus
CU Cambridge University • Christian Union • Church Union • *Photog.* close-up • Congregational Union of England and Wales • *Computing* control unit • Cooperative Union • (USA) Cornell University • *international civil aircraft marking for* Cuba • *airline flight code for* Cubana Airlines • Customs Union • *British vehicle registration for* Newcastle upon Tyne
CUA Canadian Underwriters' Association • Colour Users' Association • *Computing* common user access
CUAC Cambridge University Athletic Club
CUAFC Cambridge University Association Football Club
CUAS Cambridge University Agricultural Society • Cambridge University Air Squadron
cub. cubic
CUBC Cambridge University Boat Club • Cambridge University Boxing Club
CUC Canberra University College • Coal Utilization Council
CUCC Cambridge University Cricket Club
CUDAT Community Urban Development Assistance Team
CUDS Cambridge University Dramatic Society
CUEW Congregational Union of England and Wales
CUF Catholicarum Universitatum Foederatio (Latin: Federation of Catholic Universities) • common university fund
CUGC Cambridge University Golf Club
CUHC Cambridge University Hockey Club
cuis. cuisine
CUKT Carnegie United Kingdom Trust
cul. culinary
CUL Cambridge University Library
CULS convertible unsecured loan stock
CULTC Cambridge University Lawn Tennis Club
cum. *Finance* cumulative

CUM Cambridge University Mission

Cumb. Cumberland (former county)

cum div. *Finance* cum dividend (i.e. with dividend)

cum. pref. *Finance* cumulative preference (shares)

CUMS Cambridge University Musical Society

CUNA Credit Union National Association

CUNY City University of New York

CUOG Cambridge University Opera Group

CUP Cambridge University Press • *symbol for* Cuban peso (monetary unit)

cur. currency • current

CURAC ('kjʊəræk) (Australia) Coal Utilization Research Advisory Committee

cur. adv. vult *Law* curia advisari vult (*see under* CAV)

CURE (kjʊə) Care, Understanding, and Research (an organization for the welfare of drug addicts)

curt. current (i.e. this month)

CURUFC Cambridge University Rugby Union Football Club

CURV *US Navy* cable-controlled undersea recovery vehicle

CUS Catholic University School

CUSO ('kjuːsəʊ) Canadian University Services Overseas

CUSRPG Canada–United States Regional Planning Group (in NATO)

custod. custodian

CUTF Commonwealth Unit Trust Fund

CUTS (kʌts) *Computing* cassette users' tape specification • Computer Users' Tape System

cv. *Botany* cultivar

c.v. cheval-vapeur (French: horsepower) • chief value • *Med.* cras vespere (Latin: tomorrow evening) • curriculum vitae • cursus vitae (Latin: course of life)

CV calorific value • *Crystallog.* carbon vacancy • *Med.* cardiovascular • *Chem.* carrier vehicle (in chromatography) • *Astronomy* cataclysmic variable • cavallo vapore (Italian: horsepower) • cavalos vapor (Portuguese: horsepower) • *Statistics* coefficient of variation • *Maths.* collective vector • common valve • Common Version (of the Bible) • *Med.* contrast ventriculography • *Finance* convertible • *postcode for* Coventry • (Canada) Cross of Valour • *Chem.* crystal violet •

curriculum vitae • *British vehicle registration for* Truro

2CV 2 cheval-vapeur (deux chevaux; two-horsepower French car)

CVA *Med.* cerebrovascular accident • (USA) Columbia Valley Authority

CVC current-voltage (*or* capacitance-voltage) characteristics

CVCP Committee of Vice-Chancellors and Principals (of the Universities of the United Kingdom)

c.v.d. *Commerce* cash versus documents

CVD *Physics* Cerenkov viewing device • *Electronics* chemical vapour deposition • common valve development

CVE Certificate of Vocational Education • Council for Visual Education

CVEsc *symbol for* escudo (monetary unit of Cape Verde)

CVI chemical vapour infiltration • *Med.* common variable immunodeficiency

CVJ constant velocity joint (in vehicles)

CVK centre vertical keel

CVM Company of Veteran Motorists

CVn *Astronomy* Canes Venatici

CVN *Engineering* Charpy V-notch

CVO Commander of the Royal Victorian Order

C.voc. *Music* colla voce (Italian: with the voice; instruction to accompanist)

CVP *Med.* central venous pressure • (Belgium) Christelijk Volkspartif (Flemish Christian Social Party)

CVS *Med., Zoology* cardiovascular system • *Med.* chorionic villus sampling • Council of Voluntary Service

CVSNA Council of Voluntary Service National Association

cvt. *Finance* convertible

CVT *Maths.* canonical variational theory • *Chem.* chemical vapour transport • constant variable transmission (in vehicles)

Cvt Gdn Covent Garden

CVWS combat vehicle weapons system

cw. clockwise

c.w. *Physics* carrier wave • continuous weld

c/w chainwheel (in cycling)

CW *airline flight code for* Air Marshall Islands • Canada West • cavity wall • chemical weapons (*or* warfare) • child welfare • clerk of works • Cockcroft–Walton (as in **CW generator**) • cold-worked (of metals) • commercial weight •

Commissions and Warrants Department of the Admiralty • complete with • continuous wave(s) (as in **CW radar**; also indicating Morse code, as in **CW speed**) • *postcode for* Crewe • *Physics* Curie–Weiss (in **CW law**) • *British vehicle registration for* Preston

CWA Catering Wages Act • chemical warfare agent • (USA) Civil Works Administration • (USA) Clean Water Act • (Australia) Country Women's Association • Crime Writers' Association

CWB Canadian Wheat Board • Central Wages Board

CWBW chemical warfare–bacteriological warfare

CWC Catering Wages Commission • Commonwealth of World Citizens

CWD civilian war dead

CWDE Centre for World Development Education

C'wealth Commonwealth

CWF coal–water fuel

CWGC Commonwealth War Graves Commission

CWINC (India) Central Waterways, Irrigation, and Navigation Commission

CWIS (kwɪz) *Computing* campus-wide information service

CWL Catholic Women's League

Cwlth Commonwealth

CWME Commission on World Mission and Evangelism (of the World Council of Churches)

CWNA Canadian Weekly Newspapers Association

c.w.o. cash with order

CWO Chief Warrant Officer

CWOIH Conference of World Organizations Interested in the Handicapped

CWP (Malta) Christian Workers' Party • (USA) Communist Workers' Party

CWR (*or* **c.w.r.**) continuous welded rail

CW radar continuous-wave radar

CWS coal–water slurry • *Physics* continuous-wave spectroscopy • Cooperative Wholesale Society • Court Welfare Service

cwt *symbol for* hundredweight

CWT central war time

CWU Chemical Workers' Union • Communication Workers' Union (formed from merger of UCW and NCU)

cx *Med.* cervix • convex

CX *airline flight code for* Cathay Pacific Airways • *British vehicle registration for* Huddersfield • *international civil aircraft marking for* Uruguay

CXR chest X-ray

CXT Common External Tariff of the European Community

cy capacity • currency • *Chem., symbol for* cyanide (in formulae) • cycle(s)

CY calendar year • *British fishing port registration for* Castlebay • *international vehicle registration for* Cyprus • *airline flight code for* Cyprus Airways • *British vehicle registration for* Swansea

cyath. *Med.* cyathus (Latin: glassful)

cyath. vin. *Med.* cyathus vinarius (Latin: wineglassful)

cyber. cybernetics

cyc. cycle(s) • cycling • cyclopedia • cyclopedic

CYCA Clyde Yacht Clubs Association

cyclo. cyclopedia

CYEE Central Youth Employment Executive

Cyg *Astronomy* Cygnus

cyl. cylinder • cylindrical

Cym. Cymric

CYMS Catholic Young Men's Society

CYO (USA) Catholic Youth Organization

Cyp. Cyprian • Cyprus

Cys (*or* **cys**) *Biochem.* cysteine

Cz *Geology* Cenozoic

CZ *British vehicle registration for* Belfast • Canal Zone • *airline flight code for* China Southern Airlines • *international vehicle registration for* Czech Republic

CZI *Cytology* chlor-zinc iodide

CZMA (USA) Coastal Zone Management Act

D

d *symbol for* day • *symbol for* deci- (prefix indicating 0.1, as in **dB**, decibel) • *Biochem., symbol for* deoxyribonucleoside (preceding the nucleoside symbol, as in **dT**, thymidine) • *Physics, symbol for* deuteron • (ital.) *Chem.* dextrorotatory (as in **d-tartaric acid**) • (ital.) *symbol for* diameter • *Music* doh (in tonic sol-fa) • *Physics* down (a quark flavour) • (ital.) *symbol for* relative density • (ital.) *symbol for* thickness • *Maths., indicating* a small increment in a given variable or function (as in **d**y/**d**x) • *Physics, Chem., indicating* the electron state $l=2$, where l is the orbital angular momentum quantum number • *indicating* the fourth vertical row of squares from the left on a chessboard

d. dam (in animal pedigrees) • damn • date • daughter • day • deacon • dead • deceased • *Dentistry* deciduous • decision • decree • degree • delete • deliver(y) • delta • density • depart(s) • depth • deputy • desert • deserter • destro (Italian: right) • died • dime • *Currency* dinar(s) (*see also under* D; DA; Din; JD) • discharge • distance • dividend • dollar(s) (symbol **$**) • *Med.* dose • *Currency* drachma(s) (*see also under* Dr) • drama • droite (French: right) • dump • (formerly) penny *or* pennies (Latin *denarius*)

D (ital.) *Physics, Med., symbol for* absorbed dose (of radiation) • *symbol for* dalasï (Gambian monetary unit) • *Geology, symbol for* darcy • defence (in **D notice**) • *Government* Department • *Chem., symbol for* deuterium • *Geology* Devonian • (ital.) *symbol for* diameter • (ital.) *Chem., symbol for* diffusion coefficient • dimension *or* dimensional (preceded by a number) • *symbol for* dinar (Tunisian monetary unit) • *Optics, symbol for* dioptre • (ital.) *Physics, symbol for* dispersion • *Immunol.* diversity region (of an immunoglobulin chain) • *symbol for* dong (Vietnamese monetary unit) • (ital.) *Aeronautics,* *symbol for* drag • *Irish fishing port registration for* Dublin • (bold ital.) *Physics, symbol for* electric flux density (*or* displacement) • *Roman numeral for* five hundred • *international vehicle registration or civil aircraft marking for* Germany • *Maths., indicating* the first derivative of a function (in **D operator**) • (small cap.) *Chem., indicating* an optically active compound having a configuration related to dextrorotatory glyceraldehyde (as in **D-glucose**) • *indicating* a musical note or key • *Med., indicating* a rhesus antigen (as in **anti-D**) • *indicating* semiskilled *or* unskilled (occupational grade)

D. (*or* **D**) Damen (German: ladies) • December • (USA) Democrat(ic) • demy (paper size) • destroyer • Deus (Latin: God) • Deutschland (German: Germany) • *Cards* diamonds • Director • Diretto (Italian: slow train) • distinguished • doctor • dogana (Italian: customs) • Dom (monastic title) • Dominus (Latin: God *or* Christ) • Don (Spanish title) • douane (French: customs) • Dowager • Duchess • Duke • Dutch

D2 *international civil aircraft marking for* Angola

D2T2 *Computing* dye diffusion thermal transfer

D4 *international civil aircraft marking for* Cape Verde Islands

D6 *airline flight code for* Inter Air

2,4-D 2,4-dichlorophenoxyacetic acid (herbicide)

3D *airline flight code for* Palair Macedonian • *international civil aircraft marking for* Swaziland

3-D three-dimensional

4D *airline flight code for* Air Sinai

da *symbol for* deca- (prefix indicating 10, as in **dam**, decametre)

d/a *Commerce* days after acceptance • *Commerce* documents against acceptance

Da *Biochem., symbol for* dalton(s)

Da. Danish

D/a deposit account • discharge afloat

DA *airline flight code for* Air Georgia • *British vehicle registration for* Birmingham • *postcode for* Dartford • Daughters of America • deed of arrangement • Defence Act • delayed action (bomb) • (USA) Department of Agriculture • Deputy Advocate • deputy assistant • design automation • destructive analysis • *symbol for* dinar (Algerian monetary unit) • Diploma in Anaesthesia (*or* Anaesthetics) • Diploma in Art • direct action • *Chem.* dissolved acetylene • (USA) District Attorney • Doctor of Arts • doesn't answer • *Med.* dopamine • *Irish fishing port registration for* Drogheda • duck's arse (hairstyle)

D/A *Commerce* days after acceptance • *Commerce* delivery on acceptance • deposit account • (*or* **D-A**) *Computing, Electronics* digital-to-analogue (converter) • *Commerce* documents against acceptance (as in **D/A bill**)

DAA défense anti-aérienne (French: anti-aircraft defence) • *Chem.* diacetone acrylamide • *Chem.* diacetone alcohol • Diploma of the Advertising Association

DAA&QMG Deputy Assistant Adjutant and Quartermaster-General

DAAG Deputy Assistant Adjutant-General

DA&QMG Deputy Adjutant and Quartermaster-General

DAAS data acquisition and analysis system

DAB daily audience barometer • Deutsches Apothekerbuch (German Pharmacopoeia) • (ital.) Dictionary of American Biography • digital audio broadcasting

DABCO *Chem.* 1,4-diazobicyclooctane

DABN *Med.* diffuse acute bacterial nephritis

d.a.c. *Insurance* deductible average clause • direct air cycle

DAc Doctor of Acupuncture

DAC data analysis and control • Development Assistance Committee (of the OECD) • *Computing* digital-to-analogue converter

DACC Dangerous Air Cargoes Committee

DACG Deputy Assistant Chaplain-General

dacr. dacron (fabric)

DAD Deputy Assistant Director (as in **DADMS**, Deputy Assistant Director of Medical Services; **DADOS**, Deputy Assistant Director of Ordnance Services;

DADQ, Deputy Assistant Director of Quartering; **DADST**, Deputy Assistant Director of Supplies and Transport)

DADG Deputy Assistant Director General (as in **DADGMS**, Deputy Assistant Director General of Medical Services)

DAdmin Doctor of Administration

DAE (USA) Department of Atomic Energy • (ital.) Dictionary of American English • differential algebraic equation • Diploma in Advanced Engineering • Director of Army Education

DAEP Division of Atomic Energy Production

DAER Department of Aeronautical and Engineering Research

daf (dæf) (Netherlands) Doorn Automobielfabriek (Dutch gearless car)

d.a.f. described as follows

DAF (USA) Department of the Air Force • dissolved air flotation • dry ash-free (coal)

DAFS Department of Agriculture and Fisheries for Scotland

dag *symbol for* decagram(s)

DAG Deputy Adjutant-General • (Germany) Deutsche Angestellten-Gewerkschaft (trade union of salaried staff) • development assistance group

DAGMAR ('dægmɑː) *Commerce* defining advertising goals for measured advertising results

DAgr Doctor of Agriculture

DAgrSc Doctor of Agricultural Science

DAH *Med.* disordered action of the heart

d.a.i. death from accidental injuries

DAI disease activity index • distributed artificial intelligence

DAJAG Deputy Assistant Judge Advocate General

Dak. Dakota

dal *symbol for* decalitre(s)

DAL direct acid leaching

DALR *Meteorol.* dry adiabatic lapse rate

dal S. *Music* dal segno (Italian: (repeat) from the sign)

dam *symbol for* decametre(s)

DAM Diploma in Ayurvedic Medicine

DAMN *Chem.* diaminomaleonitrile

DAMS (USA) defense against missiles system(s) • Deputy Assistant Military Secretary

Dan. *Bible* Daniel • Danish

D&AD Designers and Art Directors Association

D&B discipline and bondage

D & C dean and chapter • *Med.* dilatation and curettage (of the uterus)

d and d drunk and disorderly

D & D death and dying

D & G Dolce e Gabbana

D & HAA Dock and Harbour Authorities' Association

d and p development and printing

d & s demand and supply

D and V *Med.* diarrhoea and vomiting

DAO district advisory officer

DAOT Director of Air Organization and Training

d.a.p. do anything possible • documents against payment

DAP *Chem.* diammonium phosphate • Director of Administrative Planning • *Computing* distributed array processor • *Electronics* donor–acceptor pair • Draw a Person (psychological test)

DAP&E Diploma in Applied Parasitology and Entomology

DAPM Deputy Assistant Provost-Marshal

DAppSc Doctor of Applied Science

DAPS Director of Army Postal Services

DAQMG Deputy Assistant Quartermaster-General

DAR Daughters of the American Revolution • (USA) Defense Aid Reports • direct absorption receiver (of solar radiation) • (Canada) Directorate of Atomic Research

DArch Doctor of Architecture

DARD Directorate of Aircraft Research and Development

DARE demand and resource evaluation

DARPA ('dɑːpə) (USA) Defense Advanced Research Projects Agency (formerly ARPA)

DArt Doctor of Art

d.a.s. *Commerce* delivered alongside ship

DAS *Computing* data-acquisition system • development advisory service • Director of Armament Supply • double algebraic sum • Dramatic Authors' Society

DASA (USA) Defense Atomic Support Agency

DASc Doctor of Agricultural Science(s)

DASC (USA) Direct Air Support Center

DASD *Computing* direct-access storage device • Director of Army Staff Duties

DASH drone antisubmarine helicopter

DASM delayed-action space missile

DAST *Chem.* diethylaminosulphur trifluoride

dat. dative

DAT *Med.* dementia of the Alzheimer type • digital audio tape

DATEC ('deɪtɛk) Art and Design Committee, Technician Education Council

Datel ('deɪ,tɛl) *Trademark* data telex (data transmission service)

DATV digitally assisted television

dau. daughter

DAV Disabled American Veterans

DAvMed Diploma in Aviation Medicine

DAW *Nuclear engineering* dry active waste

DAWS Director of Army Welfare Services

DAX *Stock exchange* Deutsche Aktienindex (German share price index)

DAyM Doctor of Ayurvedic Medicine

d.b. *Music* double bass • double bed • double-breasted • drawbar (on tractors, etc.)

dB *symbol for* decibel(s)

Db *symbol for* dobra (monetary unit of São Tomé and Príncipe)

DB Bachelor of Divinity (Latin *Divinitatis Baccalaureus*) • *airline flight code for* Brit Air • *Chem.* dangling bond • database • (or **D/B**) *Book-keeping* daybook • deals and battens (timber) • *Radio, Television* delayed broadcast • (Germany) Deutsche Bundesbahn (German Federal Railway) • (Germany) Deutsche Bundesbank (German Federal Bank) • Domesday Book • double-barrelled • *British vehicle registration for* Manchester

2,4-DB *Chem.* 4-(2,4-dichlorophenoxy)-butyric acid

dBA (or **dBa**) *symbol for* decibel A (decibels above reference noise, adjusted: unit for measuring noise)

DBA *Computing* database administrator (or administration) • dihydrodimethylbenzo-pyranbutyric acid (sickle-cell anaemia treatment) • Doctor of Business Administration • (or **d.b.a.**) doing business as (or at)

DBB deals, battens, and boards (timber) • (Germany) Deutscher Beamtenbund (civil servants' trade union) • (or **DB&B**) dinner, bed, and breakfast (in accommodation advertisements)

DBC Deaf Broadcasting Council

DBE Dame Commander of the Order of the British Empire • design-basis event

DBEATS dispatch payable both ends all time saved

DBELTS dispatch payable both ends on lay time saved

Dbh (*or* **DBH**) *Forestry* diameter at breast height

DBib Douay Bible

DBIU (Canada) Dominion Board of Insurance Underwriters

dbk debark • drawback

dbl. (*or* **dble**) double

DBM Diploma in Business Management

DBMS *Computing* database management system

DBMT displacement bone-marrow transplantation

Dbn Durban (South Africa)

DBO Diploma of the British Orthoptic Board

dbre diciembre (Spanish: December)

DBS *Physics* diffraction back-scattering • direct broadcast(ing by) satellite

DBST double British summer time

DBT *Chem.* dual-buffer titration

d.c. dead centre • *Physics* direct current • *Printing* double column • *Music* double crochet • *Navigation* drift correction

d.C. depois de Cristo (Portuguese: after Christ, i.e. AD) • dopo Cristo (Italian: after Christ, i.e. AD)

DC *Music* da capo (Italian: from the head; i.e. repeat from the beginning) • Daughters of Charity of St Vincent de Paul • death certificate • (*or* **DC.**) *Botany* de Candolle (indicating the author of a species, etc.) • decimal currency • (Italy) Democrazia Cristiana (Christian Democratic Party) • (USA) Dental Corps • depth charge • Deputy Chief • Deputy Commissioner • Deputy Consul • Deputy Counsel • Detective Constable • *Computing* device coordinates • diagnostic centre • diplomatic corps • *Physics* direct current • Disarmament Conference • Disciples of Christ • District Commissioner • district council • district court • District of Columbia (abbrev. *or* postcode) • Doctor of Chiropractic • *Commerce* documents (against) cash • Douglas Commercial (aircraft, as in **DC10**) • *Theatre* down centre (of stage) • *Physics* drift chamber • *British vehicle registration for* Middlesbrough

D/C *Marine insurance* deviation clause

DCA (USA) Defense Communications Agency • (Australia) Department of Civil Aviation • *Computing* document content architecture

DCAe Diploma of the College of Aeronautics

DCAO Deputy County Advisory Officer

d.cap. double foolscap (standard paper size)

DCAS Deputy Chief of the Air Staff

DCB Dame Commander of the Order of the Bath • double cantilever beam

DCBE *Med.* double-contrast barium enema

DCC Deputy Chief Constable • digital compact cassette • Diocesan Consistory Court • Diploma of Chelsea College

DCD (USA) Department of Community Development • Diploma in Chest Diseases • *Physics* double-crystal (X-ray) diffraction • *Electronics* drain-current drift

DCE *Computing* data-communications equipment • design and construction error • *Chem.* dichloroethane • Diploma in Chemical Engineering • Doctor of Civil Engineering • domestic credit expansion

DCEP Diploma of Child and Educational Psychology

DCF *Accounting* discounted cash flow

DCG Deputy Chaplain-General • direct coal gasification

DCGS Deputy Chief of the General Staff

DCh Doctor of Surgery (Latin *Doctor Chirurgiae*)

DCH Diploma in Child Health

DChD Doctor of Dental Surgery

DChE Doctor of Chemical Engineering

DCI desorption chemical ionization • Detective Chief Inspector • direct computer interviewing • *Computing* display control interface • (*or* **d.c.i.**) double column inch (in advertisements) • ductile cast iron

DCJ (USA) district court judge

dcl. declaration

DCL Distillers' Company Limited • Doctor of Civil Law

DCLI Duke of Cornwall's Light Infantry

DCLJ Dame Commander of the Order of St Lazarus of Jerusalem

DCISc Doctor of Clinical Science

DCM Diploma in Community Medicine • *Military* Distinguished Conduct Medal • district court-martial • Doctor of Comparative Medicine

DCMG Dame Commander of the Order of St Michael and St George

DCMS Deputy Commissioner Medical Services

DCnL Doctor of Canon Law

DCNS Deputy Chief of Naval Staff

DCO Duke of Cambridge's Own (regiment)

DC of S Deputy Chief of Staff

d.col. double column (in advertisements)

DComL Doctor of Commercial Law

DComm Doctor of Commerce

DCompL Doctor of Comparative Law

DCP Diploma in Conservation of Paintings • Diploma in Clinical Pathology • Diploma of Clinical Psychology • *Physics* direct-current plasma

DCPA (USA) Defense Civil Preparedness Agency

DCPath Diploma of the College Pathologists

DCR Diploma of the College of Radiographers (as in **DCR(MU)**, medical ultrasound; **DCR(NM)**, nuclear medicine; **DCR(R)**, diagnostic radiography; **DCR(T)**, radiotherapy)

DCrim Doctor of Criminology

DCS Deputy Chief of Staff • Deputy Clerk of Sessions • *Physics* differential cross section • digital camera system • Doctor of Christian Science • Doctor of Commercial Sciences • *Med.* dorsal column stimulator

DCSO Deputy Chief Scientific Officer

DCST Deputy Chief of Supplies and Transport

dct document

DCT *Computing* discrete cosine transform • Doctor of Christian Theology • *Med.* dynamic computerized tomography

DCU *Chem.* N,N-dichlorourethane

DCV *Accounting* direct charge voucher

DCVO Dame Commander of the Royal Victorian Order

DCW dead carcass weight • domestic cold water

dd dated • dedicated • delivered

d.d. days after date • delayed delivery • *Commerce* delivered dock • demand draft • dono dedit (Latin: given as a gift) • dry dock • due date (*or* day) • today's date (Latin *de dato*)

D.d. Deo dedit (Latin: gave to God)

D/d days after date • *Commerce* delivered

DD *Insurance* damage done • *Chem.* D

dimer • *Banking* demand draft • (USA) Department of Defense • Deputy Director (*or* Directorate) • dichloropropylene and dichloropropane (soil fumigant) • Diploma in Dermatology • *Finance* direct debit • direttissimo (Italian: fast train) • discharged dead • dishonourable discharge • Doctor of Divinity • double demy (standard paper size) • *Computing* double-density • *Med.* Duchenne dystrophy • *postcode for* Dundee • *British vehicle registration for* Gloucester

D/D delivered at docks • demand draft • dock dues

DDA Dangerous Drugs Act • Disabled Drivers' Association

D-Day Day Day (the specified day, i.e. 6 June 1944, for the Allied invasion of Europe)

DDBMS distributed database management system

ddC (*or* **DDC**) *Med.* dideoxycytidine (drug for treating Aids)

DDC Dewey Decimal Classification (of library books) • *Computing* direct digital control • *Computing* display data channel

DDCMP digital data communication message protocol

DDD dat, dicat, dedicat (Latin: gives, devotes, and dedicates) • deadline delivery date • *Chem.* dichlorodiphenyldichloroethane • dono dedit dedicavit (Latin: gave and consecrated as a gift)

DDDA *Chem.* dodecadien-1-yl acetate

DDDOL *Chem.* dodecadien-1-ol

DDDS Deputy Director of Dental Services

DDE *Computing* direct data entry • Dwight David Eisenhower (1890–1969, US president 1953–61) • dynamic data exchange

DDG Deputy Director-General

DDGAMS Deputy Director-General, Army Medical Services

DDH Diploma in Dental Health

ddI (*or* **DDI**) dideoxyinosine (drug used in treating Aids)

DDI Divisional Detective Inspector

d.d. in d. *Med.* de die in diem (Latin: from day to day)

DDL *Computing* data definition (*or* description) language(s) • Deputy Director of Labour

DDM Diploma in Dermatological Medicine • Doctor of Dental Medicine

DDME Deputy Director of Mechanical Engineering

DDMI Deputy Director of Military Intelligence

DDMOI Deputy Director of Military Operations and Intelligence

DDMS Deputy Director of Medical Services

DDMT Deputy Director of Military Training

DDNI Deputy Director of Naval Intelligence

DDO Diploma in Dental Orthopaedics · District Dental Officer

DDOL *Chem.* dodecen-1-ol

DDOS Deputy Director of Ordnance Services

DDP *Chem.* dichlorodiammine-platinum(II) · *Computing* distributed data processing

DDPH Diploma in Dental Public Health

DDPR Deputy Director of Public Relations

DDPS Deputy Director of Personal Services · Deputy Director of Postal Services

DDR Deutsche Demokratische Republik (German Democratic Republic *or* East Germany; now part of Germany) · Diploma in Diagnostic Radiology

DDRA Deputy Director, Royal Artillery

DDRD Deputy Directorate of Research and Development

dd/s delivered sound

DDS deep diving system · Deputy Directorate of Science · Dewey Decimal System (library-book classification) · *Med.* diaminodiphenyl sulphone (dapsone; leprosy treatment) · digital data storage · Director of Dental Services · (*or* **DDSc**) Doctor of Dental Science · Doctor of Dental Surgery

DDSD Deputy Director of Staff Duties

DDSM (USA) Defense Distinguished Service Medal

DDSR Deputy Director of Scientific Research

DDST Deputy Director of Supplies and Transport

DDT dichlorodiphenyltrichloroethane (insecticide)

DDVP dimethyldichlorovinyl phosphate (dichlorvos; insecticide)

DDVS Deputy Director of Veterinary Services

DDWE&M Deputy Director of Works, Electrical and Mechanical

d.e. deckle edge · diesel-electric · *Fencing* direct elimination · *Book-keeping* double entry

DE Dáil Éireann (lower chamber of the Irish parliament) · deflection error · *US postcode for* Delaware · Department of Employment · *postcode for* Derby · *Military* destroyer escort · destruction efficiency · diesel engine · direct electrolysis · Doctor of Engineering · Doctor of Entomology · double elephant (standard paper size) · *British fishing port registration for* Dundee · *British vehicle registration for* Haverfordwest

Dea. Deacon · Dean

DEA *Computing* data encryption algorithm · Department of Economic Affairs (former UK government department) · *Chem.* diethanolamine · (USA) Drug Enforcement Administration (*or* Agency)

DEAD *Chem.* diethyl azodicarboxylate

deb. *Finance* debenture · debit

deb. stk *Finance* debenture stock

dec *Astronomy* declination

dec. deceased · décembre (French: December) · decimal · declaration · declare(d) · *Grammar* declension · decoration · decorative · decrease · *Music* decrescendo (Italian; decrease loudness)

Dec. Decani (Latin: of the dean; indicating the dean's side (usually the south side) of a choir) · Decanus (Latin: dean) · (*or* **Dec**) December

DEc Doctor of Economics

DEC dental examination centre · (USA) Department of Environmental Conservation · (USA) Digital Equipment Corporation · Dollar Export Council

decd deceased

decid. deciduous

decl. *Grammar* declension

decn decontamination

DEcon Doctor of Economics

DEconSc Doctor of Economic Science

DECR *Meteorol.* decrease

decresc. *Music* decrescendo (Italian; decrease loudness)

ded. dedicate(d) · dedication · deduce · deduct(ion)

DEd Doctor of Education

de d. in d. (ital.) *Med.* de die in diem (Latin: from day to day)

DEE Diploma in Electrical Engineering

DEED (USA) Department of Energy and Economic Development

DEEP Directly Elected European Parliament

def. defecate • defecation • defect • defection • defective • defector • defence • defendant • deferred • deficit • define • definite • definition • deflagrate • deflect(ion) • defoliate • defrost • defunct • defunctus (Latin: deceased)

def. art. *Grammar* definite article

DEFCON ('def,kɒn) *Military* defence readiness condition

defl. deflate (*or* deflation) • deflect(ion)

deft. defendant

deg. degree

DEG *Meteorol.* degrees

DEHA di-2-ethylhexyl adipate (plasticizer in clingfilm)

DEI Dutch East Indies (former name for Indonesia)

del. delegate • delegation • delete • delineavit (Latin: he (*or* she) drew it) • deliver(ed)

Del *Astronomy* Delphinus

Del. Delaware • Delhi

deld. delivered

deleg. delegate • delegation

DElo Doctor of Elocution

delv. deliver(ed)

dely. delivery

dem. demand • demerara • democracy • demurrage

Dem. (USA) Democrat(ic)

DEME Directorate of Electrical and Mechanical Engineering

demon. demonstrate • demonstrative

DEMS defensively equipped merchant ships

DemU (Northern Ireland) Democratic Unionist

demur. demurrage

den. denier • denotation • denote(d) • dental • dentist(ry)

Den. Denbighshire (former Welsh county) • Denmark • Denver (USA)

DEn (*or* **D.En.**) Department of Energy • Doctor of English

DEN District Enrolled Nurse

dend. (*or* **dendrol.**) dendrology

DenD Docteur en droit (French: Doctor of Laws)

DEng (*or* **DEngg**) Doctor of Engineering

DEngS Doctor of Engineering Science

DenM (*or* **DenMed**) Docteur en médecine (French: Doctor of Medicine)

denom. *Religion* denomination

DENR (USA) Department of Energy and Natural Resources

dens. density

dent. dental • dentist(ry) • denture

DEnt Doctor of Entomology

DEOVR Duke of Edinburgh's Own Volunteer Rifles

dep. department • depart(s) • departure • dependant • dependency • dependent • deponent • depose(d) • deposit • depositor • depot • (*or* **Dep.**) deputy

dép. département (administrative division in France) • député (French: deputy)

DEP Department of Employment and Productivity • (USA) Department of Environmental Protection • *Chem.* diethyl pyrocarbonate

dept department • deponent

dept. deputy

der. derecha (Spanish: right) • derivation • derivative • derive(d)

DER (USA) Department of Environmental Resources

Derby. Derbyshire

deriv. derivation • derivative • derive(d)

DERL derived emergency reference level (of radiation)

derm. dermatitis • (*or* **dermat.**, **dermatol.**) dermatology

DERR Duke of Edinburgh's Royal Regiment

derv (dɜːv) diesel-engined road vehicle (diesel oil)

des. desert • design(er) • designate(d) • designation • desire • dessert

Des. Deaconess • Designer

DES *Computing* data encryption standard • Department of Education and Science (former government department, replaced by the DfEE) • *Med.* diethylstilboestrol

desc. descend • descendant • descent • describe

desid. desideratum (Latin: something wanted)

desig. designate

DèsL Docteur ès lettres (French: Doctor of Letters)

desp. despatch(ed)

DesRCA Designer of the Royal College of Art

DèsS (or **DèsSc**) Docteur ès sciences (French: Doctor of Science)

DèsScPol Docteur ès science politique (French: Doctor of Political Science)

dest. destroyer

destn destination

det. detach · detachment · detail · determine · *Med.* detur (Latin: let it be given)

Det Detective (as in **Det Con**, Detective Constable; **Det Insp**, Detective Inspector; **Det Sgt**, Detective Sergeant; **Det Supt**, Detective Superintendent)

DET *Linguistics* determiner · diethyltryptamine (hallucinogenic drug) · direct energy transfer

detn determination

Deut. *Bible* Deuteronomy

dev. develop(ed) · developer · development · deviate · *Navigation* deviation

Dev. Devon (or Devonshire)

Devon. Devonshire

devp develop

devpt development

devs. devotions

DEW (dju:) directed-energy weapon · *Military* distant early-warning (as in **DEW line**; network of radar stations)

DEXA dual-energy X-ray absorptiometry

dez. dezembro (Portuguese: December)

Dez. Dezember (German: December)

df. draft

d.f. *Commerce* dead freight

DF Dean of Faculty · *Nuclear engineering* decontamination factor · Defender of the Faith · (or **D/F**) *Telecom.* direction finder (or finding) · *symbol for* Djibouti franc (monetary unit) · Doctor of Forestry · double foolscap (paper) · *British vehicle registration for* Gloucester

DFA (USA) Department of Foreign Affairs · Diploma in Foreign Affairs · Doctor of Fine Arts

DFAA *Biochem.* dissolved free amino acids

DFC Distinguished Flying Cross

DFD *Computing* dataflow diagram

DFDL dual-frequency dye laser

DfEE Department for Education and Employment

DFHom Diploma of the Faculty of Homeopathy

Dfl Dutch florin

DFLS Day Fighter Leaders' School

DFM Diploma in Forensic Medicine · Distinguished Flying Medal

dfndt defendant

DFP *Chem.* diisopropyl fluorophosphate

DFR Dounreay fast reactor

DFS disease-free survival

dft defendant · draft

DFT *Maths.* discrete Fourier transform · dual-flow turbine

DFW director of fortifications and works

dg *symbol for* decigram(s)

DG Dei gratia (Latin: by the grace of God) · Deo gratias (Latin: thanks be to God) · dependence graph · Deutsche Grammophon (German record company) · *Biochem.* D-glucose · differential geometry · *Navigation* directional gyro · Director-General · Dragoon Guards · *postcode for* Dumfries · *British vehicle registration for* Gloucester

DGA Director-General, Aircraft · Directors' Guild of America

DGAA Distressed Gentlefolks Aid Association

DGAMS Director-General, Army Medical Services

DGB (Germany) Deutscher Gewerkschaftsbund (German Trade Union Federation)

DGC *Biology* density-gradient centrifugation · Diploma in Guidance and Counselling

DGCA Director-General of Civil Aviation

DGCE Directorate-General of Communications Equipment

DGCStJ Dame Grand Cross of the Order of Saint John of Jerusalem

DGD Director, Gunnery Division

DGD&M Director-General, Dockyards and Maintenance

DGE Directorate-General of Equipment

DGEME Director-General, Electrical and Mechanical Engineering

DGI Director-General of Information · Director-General of Inspection

DGLP(A) Director-General, Logistic Policy (Army)

DGM Diploma in General Medicine · Director-General of Manpower

DGMS Director-General of Medical Services

DGMT Director-General of Military Training

DGMW Director-General of Military Works

Dgn Dragoon

DGNPS Director-General of Naval Personal Services

DGO Diploma in Gynaecology and Obstetrics

DGP *Biochem.* D-glucose 6-phosphate · Director-General of Personnel · Director-General of Production

DGPS Director-General of Personal Services

DGR Director of Graves Registration

DGS Diploma in General Studies · Diploma in General Surgery · Diploma in Graduate Studies · Directorate-General of Signals · Director-General, Ships

DGSRD Directorate-General of Scientific Research and Development

DGT Director-General of Training

DGW Director-General of Weapons · Director-General of Works

d.h. das heisst (German: that is; i.e.) · dead heat

Dh *symbol for* dirham (monetary unit of United Arab Emirates)

DH *British fishing port registration for* Dartmouth · *Sport* dead heat · De Havilland (aircraft) · Department of Health · *Baseball* designated hitter · *symbol for* dirham (Moroccan monetary unit) · district heating · Doctor of Humanities · *Electronics* double heterostructure · *British vehicle registration for* Dudley · *postcode for* Durham

DHA District Health Authority

DHBA *Chem.* 3,4-dihydroxybenzylamine

Dhc Doctor honoris causa (honorary doctorate)

DHHS (USA) Department of Health and Human Services

DHL Doctor of Hebrew Letters (*or* Literature) · Doctor of Humane Letters

DHMSA Diploma in the History of Medicine (Society of Apothecaries)

DHO *Electronics* damped harmonic oscillator

DHQ district (*or* divisional) headquarters

DHR *Nuclear engineering* decay heat removal

DHS Diploma in Horticultural Science · Doctor of Health Sciences

DHSA Diploma of Health Service Administration

DHSS Department of Health and Social Security (former government department, now split into the DoH and DSS)

DHumLit Doctor of Humane Letters

DHW domestic hot water

DHyg Doctor of Hygiene

di. diameter

d.i. daily inspection · das ist (German: that is; i.e.) · de-ice · diplomatic immunity · document identifier

Di *Chem., symbol for* didymium

DI Defence Intelligence · Department of the Interior · Detective Inspector · *airline flight code for* Deutsche AB · direct injection · Director of Infantry · district (*or* divisional) inspector · *Med.* donor insemination · double imperial (standard paper size) · drill instructor · *Irish vehicle registration for* Roscommon

dia. diagram · dialect · diameter

DIA (USA) Defense Intelligence Agency · Design and Industries Association · Diploma in International Affairs · Driving Instructors' Association

DIAD *Chem.* diisopropyl azodicarboxylate

diag. diagnose · diagonal · diagram

dial. dialect(al) · dialectic(al) · dialogue

diam. diameter

diamat ('daɪə,mæt) *Philosophy, Economics* dialectical materialism

DIANE (daɪ'æn) Direct Information Access Network for Europe

diap. *Music* diapason

diaph. diaphragm

diars *Chem.* o-phenylenebis(dimethylarsine) (used in formulae)

DIAS Dublin Institute of Advanced Sciences

DIBAL-H *Chem.* diisobutylaluminium hydride

dic. dicembre (Italian: December) · diciembre (Spanish: December)

DIC Diploma of Membership of the Imperial College of Science and Technology (London) · *Med.* disseminated intravascular coagulation · *Chem.* dissolved inorganic carbon

DIChem Diploma in Industrial Chemistry

dict. dictate(d) · dictation · dictator · dictionary

dicta. dictaphone

DICTA Diploma of Imperial College of Tropical Agriculture

DIDS *Electronics* donor-impurity density of states

DIE Designated Investment Exchange · Diploma in Industrial Engineering · Diploma of the Institute of Engineering

dieb. alt. *Med.* diebus alternus (Latin: on alternate days)

DIEL (Advisory Committee on Telecommunications for) Disabled and Elderly People

DIEME Directorate of Inspection of Electrical and Mechanical Equipment

dien *Chem.* diethylenetriamine (used in formulae)

diet. dietary • dietetics • dietician

dif. differ • differential

Dif data interchange format

DIF *Meteorol.* diffuse • District Inspector of Fisheries

diff. differ • difference • different • differential

dig. digest (book or summary) • digestion • digestive • digit(al)

DIG Deputy Inspector-General • disablement income group

digraph ('daɪgræf) *Computing* directed graph

DIH Diploma in Industrial Health

DIIR digital image-intensifier radiography

Dij. Dijon (France)

dil. dilute(d) • dilution

DIL *Electronics* dual in-line (as in **DIL switch**)

diln dilution

dim. dimanche (French: Sunday) • dimension • dimidium (Latin: one-half) • diminish • *Music* diminuendo • diminutive

DIM Diploma in Industrial Management

dimin. diminutive

Dimm (dɪm) *Electronics, Computing* dual in-line memory module

DIMS *Computing* Data and Information Management System

din. dining car (*or* room) • dinner

Din *symbol for* dinar (Yugoslavian monetary unit)

Din. Dinsdag (Dutch: Tuesday)

DIN Deutsches Institut für Normung (German national standards organization; as in **DIN connector**, etc.) • *Computing* digital imaging network

DIng Doctor of Engineering (Latin *Doctor Ingeniariae*)

dinky (*or* **Dink, DINK**) ('dɪŋkɪ *or* dɪŋk) *Colloquial* double (*or* dual) income, no kids (used of couples)

DInstPA Diploma of the Institute of Park Administration

dio. diocese

DIO district intelligence officer

dioc. diocesan • diocese

dioc. syn. diocesan synod

DIOP *Chem.* di-iso-octyl phthalate

dip. diploma

Dip (dɪp) Diploma (in degrees and qualifications) • *Computing* document image processing

DIP *Physics* deep inelastic processes • (dɪp) *Electronics* dual in-line package (*or* pin)

DipAD Diploma in Art and Design

DipAe Diploma in Aeronautics

DipAgr Diploma in Agriculture

DipALing Diploma in Applied Linguistics

DipAM Diploma in Applied Mechanics

DipAppSc Diploma of Applied Science

DipArch Diploma in Architecture

DipArts Diploma in Arts

DipASE Diploma in Advanced Study of Education, College of Preceptors

DipAvMed Diploma of Aviation Medicine, Royal College of Physicians

DipBA Diploma in Business Administration

DipBac Diploma in Bacteriology

DipBMS Diploma in Basic Medical Sciences

DipBS Diploma in Fine Art, Byam Shaw School

DipCAM Diploma in Communication, Advertising, and Marketing of CAM Foundation

DipCC Diploma of the Central College

DipCD Diploma in Child Development • Diploma in Civic Design

DipCE Diploma in Civil Engineering

DipChemEng Diploma in Chemical Engineering

DipCom Diploma in Commerce

DipDHus Diploma in Dairy Husbandry

DipDP Diploma in Drawing and Painting

DipDS Diploma in Dental Surgery

DipEcon Diploma in Economics

DipEd Diploma in Education

DipEl Diploma in Electronics

DipEng Diploma in Engineering

DipESL Diploma in English as a Second Language

DipEth Diploma in Ethnology

DipFA Diploma in Fine Arts

DipFD Diploma in Funeral Directing

DipFE Diploma in Further Education

DipFor Diploma in Forestry

DipGSM Diploma in Music, Guildhall School of Music and Drama
DipGT Diploma in Glass Technology
DipHA Diploma in Hospital Administration
DipHE Diploma in Higher Education · Diploma in Highway Engineering
DipHum Diploma in Humanities
diphos *Chem.* 1,2-bis(diphenylphosphino)ethane (used in formulae)
DipHSc Diploma in Home Science
diphth. diphthong
DipJ Diploma in Journalism
dipl. diploma · diplomacy · diplomat(ic)
DipL Diploma in Languages
DipLA Diploma in Landscape Architecture
DipLib Diploma of Librarianship
DipLSc Diploma of Library Science
DipM Diploma in Marketing
DipMechE Diploma of Mechanical Engineering
DipMet Diploma in Metallurgy
DipMFOS Diploma in Maxial, Facial, and Oral Surgery
DipMusEd Diploma in Musical Education
DipN Diploma in Nursing
DipNEd Diploma in Nursery School Education
DipO&G Diploma in Obstetrics and Gynaecology
DipOL Diploma in Oriental Learning
DipOrth Diploma in Orthodontics
DipPA Diploma in Public Administration
DipP&OT Diploma in Physiotherapy and Occupational Therapy
DipPharmMed Diploma in Pharmaceutical Medicine
DipPhysEd Diploma in Physical Education
DipQS Diploma in Quantity Surveying
DipRADA Diploma of the Royal Academy of Dramatic Art
DipREM Diploma in Rural Estate Management
DIPS digital image-processing system
DipS&PA Diploma in Social and Public Administration
DipSMS Diploma in School Management Studies
DipSoc Diploma in Sociology
DipSpEd Diploma in Special Education
DipSS Diploma in Social Studies

DipSW Diploma in Social Work
DipT Diploma in Teaching
DipTA Diploma in Tropical Agriculture
DipT&CP Diploma in Town and Country Planning
DipTech Diploma in Technology
DipTEFL Diploma in the Teaching of English as a Foreign Language
DipTh Diploma in Theology
DipTP Diploma in Town Planning
DipTPT Diploma in Theory and Practice of Teaching
dir. direct(ed) · direction · director · *Currency* dirham (*see also under* Dh; DH)
DIR *Photog.* developer inhibitor release
Dir-Genl Director-General
dis. discharge · disciple · discipline · disconnect · discontinue · discount · dispense · distance · distant · distribute
DIS (USA) Defense Intelligence School · (the) Disney Channel
disab. (*or* **disabl.**) disability
disb. disbursement
disc. disciple · discipline · discount · discover(ed) · discoverer · discovery
disch. discharge
DISH *Med.* diffuse idiopathic skeletal hyperostosis
dishon. dishonourable · dishonourably
dismac ('dɪs,mæk) *Military* digital scene-matching area correlation sensors
disp. dispensary · dispensation · dispense · disperse · dispersion
displ. displacement
diss. dissenter · dissertation · dissolve
dist. distance · distant · distilled · distinguish(ed) · district
Dist. Atty (USA) District Attorney
distr. distribution · distributor
distrib. distributive
DistTP Distinction in Town Planning
DIT *Med.* desferrioxamine infusion test · Detroit Institute of Technology · double income tax (in **DIT relief**)
div (dɪv) *Maths.* divergence
div. diversion · divide(d) · dividend · divine · division · divisor · divorce(d)
Div. Divine · Divinity
div. in par. aeq. *Med.* dividatur in partes aequales (Latin: divide into equal parts; in prescriptions)
divn division
divnl divisional
DIY (*or* **d.i.y.**) do-it-yourself

d.J. der Jüngere (German: junior) • dieses Jahres (German: of this year)

DJ (*or* **d.j.**) dinner jacket • Diploma in Journalism • (*or* **d.j.**) disc jockey • (USA) District Judge • Divorce Judge • Doctor of Law (Latin *Doctor Juris*) • *Finance* Dow Jones • dust jacket (of a book) • *British vehicle registration for* Liverpool

DJAG Deputy Judge Advocate-General

DJI *Finance* (USA) Dow Jones Index

DJIA *Finance* (USA) Dow Jones Industrial Average

DJS Doctor of Juridical Science

DJT Doctor of Jewish Theology

DJTA *Finance* (USA) Dow Jones Transportation Average

DJUA *Finance* (USA) Dow Jones Utilities Average

DJur Doctor of Law (Latin *Doctor Juris*)

dk dark • deck • dock • duck

DK *international vehicle registration for* Denmark and Greenland • Dorling Kindersley • *Irish fishing port registration for* Dundalk • *British vehicle registration for* Manchester

DKB *Computing* distributed knowledge base

DKNY *Trademark* Donna Karan New York

Dkr *symbol for* krone (monetary unit of Denmark and Greenland)

DKS Deputy Keeper of the Signet

dkt docket

dkyd dockyard

dl (*or* **dL**) *symbol for* decilitre(s)

DL *postcode for* Darlington • *British fishing port registration for* Deal • *airline flight code for* Delta Airlines • Deputy Lieutenant • *Physics* detection limit • diesel • *Psychol.* difference limen • *Electronics* distribution line • Doctor of Laws (Latin *Doctor Legum*) • *Physics* dose level • *Physical chem.* double layer • *Book-keeping* double ledger • *Theatre* down left (of stage) • driving licence • *British vehicle registration for* Portsmouth

D/L data link • demand loan

DLA diffusion limited aggregation

d.l.c. direct lift control

DLC Diploma of Loughborough College • divisional land commissioner • Doctor of Celtic Literature • *Theatre* down left centre (of stage)

dld. delivered

d.l.d. deadline date

DLE *Med.* discoid lupus erythematosus

DLES Doctor of Letters in Economic Studies

DLett Docteur en lettres (French: Doctor of Letters)

DLG David Lloyd George (1863–1945, British statesman, prime minister 1916–22)

DLI Durham Light Infantry

D-Lib. (USA) Liberal Democrat

DLit Doctor of Literature

DLitt Doctor of Letters (Latin *Doctor Litterarum*)

DLittS Doctor of Sacred Letters

DLJ Dame of Justice of the Order of St Lazarus of Jerusalem

DLL *Computing* dynamic link library

DLM *Genetics* dominant lethal mutation

DLO dead letter office (*see under* RLO) • Diploma in Laryngology and Otology (*or* Otorhinolaryngology) • (*or* **d.l.o.**) dispatch (money payable) loading only

DLOY Duke of Lancaster's Own Yeomanry

DLP (Australia) Democratic Labor Party (former political party) • Democratic Labour Party (of Barbados, Trinidad and Tobago, etc.) • *Chem.* dicalcium lead propionate

dlr dealer

DLR Docklands Light Railway

DLS debt liquidation schedule • (*or* **DLSc**) Doctor of Library Science • Dominion Land Surveyor

DLTP dilauryl thiodipropionate (antioxidant for rubber)

dlvd delivered

dlvy delivery

DLW Diploma in Labour Welfare

dly daily

dm *symbol for* decimetre(s)

d.M. dieses Monats (German: this month; inst.)

DM *British vehicle registration for* Chester • (ital.) Daily Mail • *Astronomy* dark matter • Deputy Master • design manual • *symbol for* Deutschmark (German monetary unit) • *Physics* dipole moment • direct mail • *Chem.* direct methanation • Director of Music • dispersion measures • *Computing* distributed memory • district manager • Docteur en médecine (French: Doctor of Medicine) • Doctor of Mathematics • Doctor of Medicine • Doctor of Music

DMA Diploma in Municipal Administration • *Computing* direct memory access

DMAC *Computing* direct memory access control • *Television* duobinary multiplexed analogue component

DM&CW Diploma in Maternity and Child Welfare

DMAP *Chem.* dimethylaminopyridine

D-mark Deutschmark

DMath Doctor of Mathematics

DMC direct manufacturing costs • district medical committee

DMD *Computing* digital micromirror device • Doctor of Mathematics and Didactics • Doctor of Medical Dentistry (*or* Dental Medicine) • *Med.* Duchenne muscular dystrophy

DME *Chem.* 1,2-dimethoxyethane • Diploma in Mechanical Engineering • *Computing* direct machine environment • *Aeronautics* distance-measuring equipment • *Computing* distributed management environment • *Electronics* dropping mercury electrode

DMed Doctor of Medicine

DMet Doctor of Metallurgy • Doctor of Meteorology

DMF *Chem.* N,N-dimethylformamide • *Electronics* dose-modifying (*or* -modification) factor

DMFOS Diploma in Maxillofacial and Oral Surgery

dmg. damage

DMGO Divisional Machine-Gun Officer

DMGT *Genetics* DNA-mediated gene transfer

DMHS Director of Medical and Health Services

DMI *Computing* desktop management interface • Director of Military Intelligence

DMin Doctor of Ministry

DMJ Diploma in Medical Jurisprudence

DMJ(Path) Diploma in Medical Jurisprudence (Pathology)

DMK (India) Dravida Munnetra Kazgham (political party)

dml. demolish

DML *Computing* data-manipulation language • (formerly, USSR) Defence Medal for Leningrad • Doctor of Modern Languages

DMLJ Dame of Merit of the Order of St Lazarus of Jerusalem

DMLS Doppler microwave landing system

DMLT Diploma in Medical Laboratory Technology

DMM (formerly, USSR) Defence Medal for Moscow • *Med.* diffuse malignant mesothelioma

dmn. dimension(al)

DMN (*or* **DMNA**) *Chem.* dimethyl-nitrosamine

DMO (formerly, USSR) Defence Medal for Odessa • *Astronomy* diffuse massive object • Director of Military Operations • district medical officer

DMO&I Director, Military Operations and Intelligence

DMP Diploma in Medical Psychology • Director of Manpower Planning

DMPB Diploma in Medical Pathology and Bacteriology

DMPM *Physics* differential mechanical-properties microprobe

DMPO *Chem.* 5,5-dimethyl-l-pyrroline-N-oxide

DMPP *Chem.* 1,1-dimethyl-4-phenyl-piperazinium iodide

DMPS *Chem.* 2,3-dimercapto-1-propane-sulphonic acid

DMPU *Chem.* N,N'-dimethyl- propylene-urea

DMR Diploma in Medical Radiology • Director of Materials Research

DMRD Diploma in Medical Radiological Diagnosis • Directorate of Materials Research and Development

DMRE (*or* **DMR&E**) Diploma in Medical Radiology and Electrology

DMRT Diploma in Medical Radiotherapy

DMs ('di:'emz) Dr Martens (boots)

DMS *Computing* data-management system • *Mining* dense-medium separation • *Chem.* dimethyl sulphide • Diploma in Management Studies • Directorate of Military Survey • Director of Medical Services • Dis manibus sacrum (Latin: consecrated to the souls of the departed) • Doctor of Medical Science • Doctor of Medicine and Surgery • *Physics* drift magnetic separator

DMSA *Med.* dimercaptosuccinic acid (diagnostic aid)

DMSO *Med.* dimethylsulphoxide (used in ointments)

DMSSB Direct Mail Services Standard Board

dmstn (*or* **dmst.**) demonstration

dmstr. demonstrator

DMSV (formerly, USSR) Defence Medal for Sevastopol

DMT *Chem.* dimethyl terephthalate • dimethyltryptamine (hallucinogenic drug) • Director of Military Training

DMTF *Computing* desktop management task force

DMU *Commerce* decision-making unit (within an organization) • directly managed unit (of NHS hospitals)

DMus Doctor of Music

DMV (USA) Department of Motor Vehicles • Docteur en médecine vétérinaire (French: Doctor of Veterinary Medicine)

DMZ demilitarized zone

dn down • dozen

Dn Deacon • Dragoon

D^n Don (Spanish: Mr)

DN (*or* **D/N**) debit note • de novo (Latin: from the beginning) • Diploma in Nursing • Diploma in Nutrition • Dominus noster (Latin: our Lord) • *postcode for* Doncaster • *British vehicle registration for* Leeds

D^{na} Doña (Spanish: Mrs)

DNA (USA) Defense Nuclear Agency • *Genetics* deoxyribonucleic acid • Det Norske Arbeiderpartiet (Norwegian Labour Party) • Deutscher Normenausschuss (German Committee of Standards) • *Med.* did not attend • Director of Naval Accounts • District Nursing Association

DNAD Director of Naval Air Division

DNase (*or* **DNAase**) *Biochem.* deoxyribonuclease

DNB (ital.) Dictionary of National Biography

DNC *Nuclear engineering* delayed-neutron counting • Director of Naval Construction • *Computing* distributed numerical control

DND Director of Navigation and Direction

dne. douane (French: customs)

DNE Diploma in Nursing Education • Director of Naval Equipment • Director of Nurse Education • Dounreay Nuclear Establishment

DNES Director of Naval Education Service

DNF (*or* **dnf**) did not finish (of competitors in a race)

DNH Department of National Heritage

DNHW (USA) Department of National Health and Welfare

DNI Director of Naval Intelligence

DNJC Dominus noster Jesus Christus (Latin: Our Lord Jesus Christ)

DNMS Director of Naval Medical Services

DNO Director of Naval Ordnance • district naval officer • district nursing officer • divisional nursing officer

DNOC dinitro-*o*-cresol (pesticide)

D notice defence notice

DNP declared national programme • *Immunol.* dinitrophenyl • dynamic nuclear polarization

DNPDE Dounreay Nuclear Power Development Establishment

DNPP Dominus Noster Papa Pontifex (Latin: Our Lord the Pope)

DNR (USA) Department of Natural Resources • Director of Naval Recruiting • *Med.* do not resuscitate

dns *Cartography* downs

DNS Department for National Savings • *Chem.* 5-dimethylamino-1-naphthalenesulphonic acid • *Computing* domain name server (*or* system)

DNSA Diploma in Nursing Service Administration

DNT Director of Naval Training

DNWS Director of Naval Weather Service

do. ditto

d/o (*or* **DO**) *Commerce* delivery order

DO defence order • *Finance* deferred ordinary (shares) • Diploma in Ophthalmology • Diploma in Osteopathy • (*or* **d.o.**) *Grammar* direct object • (*or* **D/O**) *Commerce* direct order • dissolved oxygen • district office (*or* officer) • divisional office (*or* officer) • Doctor of Optometry • Doctor of Oratory • Doctor of Osteopathy • *British fishing port registration for* Douglas • drawing office • *British vehicle registration for* Lincoln

DOA date of availability • *Med.* dead on arrival • (USA) Department of the Army • dissolved-oxygen analyser

DOAE Defence Operational Analysis Establishment (of the Ministry of Defence)

d.o.b. (*or* **DOB**) date of birth

DObstRCOG Diploma of the Royal College of Obstetricians and Gynaecologists

doc. document(s)

Doc. Doctor

DOC Denominazione di Origine Controllata (Italian: name of origin controlled; Italian wine classification) • (USA) Department of Commerce • direct operating cost • *Chem.* dissolved organic carbon • District Officer Commanding

d.o.c.a. date of current appointment

d.o.c.e. date of current enlistment

DocEng Doctor of Engineering

DOCG Denominazione di Origine Controllata Garantita (Italian: name of origin guaranteed controlled; Italian wine classification)

docu. document • documentary • documentation

d.o.d. date of death • died of disease

DOD (USA) Department of Defense

DOE (USA) Department of Energy • (*or* **DoE**) Department of the Environment • (*or* **d.o.e.**) depends on experience (referring to salary in job advertisements) • (*or* **DoE**) Director of Education

DOF *Chem., Physics* degrees of freedom

D. of Corn. LI Duke of Cornwall's Light Infantry

D of H degree of honour

D of L Duchy of Lancaster

D of S director of stores

DoH (*or* **DOH**) Department of Health

DOHC double (*or* dual) overhead camshaft

DOHS (USA) Department of Health Services

DOI (USA) Department of the Interior • died of injuries • Director of Information

DOJ (USA) Department of Justice

dol. *Music* dolce (Italian: sweetly; i.e. gently) • dollar

DOL (USA) Department of Labor • Doctor of Oriental Learning

dolciss. *Music* dolcissimo (Italian: very sweetly)

dom. domain • domenica (Italian: Sunday) • domestic • domicile • dominant • domingo (Spanish, Portuguese: Sunday) • dominion

Dom. Dominica • Dominical • *RC Church* Dominican • Dominus (Latin: Lord)

DOM Deo Optimo Maximo (Latin: to God, the best, the greatest) • *Colloquial* dirty old man • dissolved (*or* dispersed) organic material • *international vehicle registration for* Dominican Republic • Dominus omnium magister (Latin: God the Lord of all)

Dom. Proc. *Law* Domus Procerum (Latin: House of Lords)

DOMS Diploma in Ophthalmic Medicine and Surgery

don. donec (Latin: until)

Don. (*or* **Doneg.**) Donegal

DON (USA) Department of the Navy • Diploma in Orthopaedic Nursing • dissolved organic nitrogen

DOPC *Biochem.* dioleoylphosphatidylcholine

DOPS *Chem.* DL-*threo*-3,4-dihydroxyphenylserine

DOpt Diploma in Ophthalmic Optics

Dor *Astronomy* Dorado

Dor. Dorian • Doric

DOr Doctor in Orientation

DOR Director of Operational Requirements • *Chem.* direct oxide reduction

DORA ('dɔːrə) Defence of the Realm Act (1914)

Dors. Dorset

DOrth Diploma in Orthodontics • Diploma in Orthoptics

DOS day of sale • *Physics* density of states • (USA) Department of State • Diploma in Orthopaedic Surgery • Directorate of Overseas Surveys • Director of Ordnance Services • (dᴏs) *Computing* disk operating system • Doctor of Ocular Science

DOSV deep ocean survey vehicle

dot. *Law* dotation (endowment)

DoT Department of Transport

DOT Department of Overseas Trade • (USA) Department of Transportation • *Finance* (USA) designated order turnaround (on the New York Stock Exchange) • *Computing* digital optical tape • Diploma in Occupational Therapy • *Med.* directly observed therapy • *Physiol.* dissolved oxygen tension

dott. dottore (Italian: doctor)

DOV double oil of vitriol (commercial sulphuric acid)

DOVAP *Physics* Doppler velocity and position

dow. (*or* **Dow.**) dowager

DOW (*or* **d.o.w.**) died of wounds

doz. dozen

dp deep • *Baseball* double play

d.p. damp-proof(ing) • deep penetration • departure point • *Finance* depreciation percentage • *Med.* directione propria

(Latin: with proper direction; in prescriptions) • direct port • double paper • dry powder • dual purpose

DP *airline flight code for* Air 2000 • data processing • *Chem.* degree of polymerization • delivery point • Democratic Party • diametral pitch • Diploma in Psychiatry • (by) direction of the president • Director of Photography • disabled person • displaced person • Doctor of Philosophy • domestic prelate • *Law* Domus Procerum (Latin: House of Lords) • durable press • duty paid • *Computing* dynamic programming • *British vehicle registration for* Reading

D/P delivery on payment • *Commerce* documents against payment (*or* presentation)

2,4-DP *Chem.* 2-(2,4-dichlorophenoxy)-propionic acid

d.p.a. deferred payment account

DPA Deutsche-Presse Agentur (German news agency) • Diploma in Public Administration • discharged prisoners' aid • (USA) Doctor of Public Administration

DPath Diploma in Pathology

DPB *Finance* deposit pass book • *Chem.* 1,4-diphenyl-1,3-butadiene

DPC Defence Planning Committee (in NATO) • *Biochem.* DNA–protein crosslinks (*or* complex)

DPCM *Telecom.* differential pulse code modulation

DPCP Department of Prices and Consumer Protection

DPCS data personal communications service

DPD Data Protection Directive • *Botany* diffusion pressure deficit • Diploma in Public Dentistry • *Statistics* discrete probability distributions

DPE Diploma in Physical Education

DPEc Doctor of Political Economy

DPed Doctor of Pedagogy

DPh Doctor of Philosophy

DPH *Chem.* 1,6-diphenylhexatriene • Diploma in Public Health • Director of Public Health • (USA) Doctor of Public Health

DPharm Doctor of Pharmacy

DPHD Diploma in Public Health Dentistry

DPhil ('diː 'fɪl) Doctor of Philosophy

DPHN Diploma in Public Health Nursing

DPhysMed Diploma in Physical Medicine

dpi *Computing* dots per inch

dpl. diploma • (*or* **DPL**) diplomat • duplex

DPM *Chem.* bis(diphenylphosphino)-methane • data-processing manager • Deputy Prime Minister • Deputy Provost-Marshal • Diploma in Psychological Medicine

DPMI *Computing* DOS/Protected Mode Interface

DPMS *Computing* display power management system

DPN *Biochem.* diphosphopyridine nucleotide (now called NAD)

DPO distributing post office • district pay office

d.p.o.b. date and place of birth

DPolSc Doctor of Political Science

DPP *Insurance* deferred payment plan • Diploma in Plant Pathology • Director of Public Prosecutions • *Finance* (USA) direct participation program

DPPA *Chem.* diphenylphosphoryl azide

DPPH *Chem.* 2,2-di(4-t-octylphenyl)-1-picrylhydrazyl

DPR Data Protection Register • Director of Public Relations

DPRK Democratic People's Republic of Korea (North Korea)

DPS *Chem.* trans-4,4′-diphenylstilbene • Director of Personal Services • Director of Postal Services • dividend per share • Doctor of Public Service

DPSA Diploma in Public and Social Administration

DPSK *Computing* differential phase shift keying

DPSN Diploma in Professional Studies in Nursing

DPsSc Doctor of Psychological Science

DPsy Docteur en psychologie (French: Doctor of Psychology)

DPsych Diploma in Psychiatry • Doctor of Psychology

dpt department • deponent • deposit • depot

DPT *Med.* dental pantomogram • diphtheria, pertussis (whooping cough), tetanus (in **DPT vaccine**)

dpty deputy

DPW Department of Public Works

dpx duplex

d.q. direct question

DQ dispersion quotient • *Sport* disqualify • *international civil aircraft marking for* Fiji

DQDB *Computing* distributed queue dual bus

DQMG Deputy Quartermaster-General

DQMS Deputy Quartermaster-Sergeant

dr debtor · *symbol for* dram · *Meteorol.* drizzle and rain

dr. debit · door · drachm · drachma (*see also under* Dr) · drama · draw(n) · (*or* **Dr.**) drawer · dresser · driver · drum(mer)

d.r. deficiency report · design requirements · development report · document report

Dr Director · Doctor · *symbol for* drachma (Greek monetary unit) · Drive (in street names)

DR (USA) Daughters of the Revolution · *Navigation* dead reckoning · *Physics* decay ratio · defence regulation · dining room · Diploma in Radiology · discount rate · dispatch rider · dispersion relation · district railway · district registry · double royal (standard paper size) · *British fishing port registration for* Dover · *Theatre* down right (of stage) · dry riser (pipe with attachment for fireman's hose) · dynamic relaxation · *British vehicle registration for* Exeter

D/R deposit receipt

Dra *Astronomy* Draco

Dr^a Doctora (Spanish: (female) doctor) · Doutora (Portuguese: (female) doctor)

DRA de-rating appeals

DRAC Director, Royal Armoured Corps

DrAgr Doctor of Agriculture

dram. drama · dramatic · dramatist

DRAM (dræm) *Computing* dynamic random-access memory

dram. pers. dramatis personae (Latin: characters in a play)

DRAO Dominion Radio Astrophysical Observatory

Drav. Dravidian (language)

DRAW (drɔː) *Computing* direct read after write

DrBusAdmin Doctor of Business Administration

DRC Diploma of the Royal College of Science and Technology, Glasgow · *Theatre* down right centre (of stage) · Dutch Reformed Church

DrChem Doctor of Chemistry

DRCOG Diploma of the Royal College of Obstetricians and Gynaecologists

DRCPath Diploma of the Royal College of Pathologists

DRCS *Computing* dynamically redefinable character set

DRD Diploma in Restorative Dentistry

DRDOS Digital Research's disk-operating system

DRDW *Computing* direct read during write

DRE *Med.* digital rectal examination · Director of Religious Education · (*or* **DRelEd**) Doctor of Religious Education

DrEng Doctor of Engineering

DRG *Med.* diagnosis-related group

Dr hc Doctor honoris causa (Latin: honorary doctor)

Dr ing Doctor of Engineering

Dr jur Doctor of Laws

DRK Deutsches Rotes Kreuz (German Red Cross)

DRLS dispatch-rider letter service

DRM Diploma in Radiation Medicine · Diploma in Resource Management

DrMed Doctor of Medicine

drn drawn

DrNatSci Doctor of Natural Science

DRO daily routine order · disablement resettlement officer · divisional routine order

DrOecPol Doctor of Political Economics (Latin *Doctor Œconomiae Politicae*)

DRP dividend reinvestment plan

DrPH Doctor of Public Health

DrPhil Doctor of Philosophy

DrPolSci Doctor of Political Science

Dr rer. nat. Doctor of Natural Science

DRS Diploma in Religious Studies

DRSAMD Diploma of the Royal Scottish Academy of Music and Drama

DRSE *Med.* drug-related side effects

DRSN *Meteorol.* drifting snow

DRT *Psychiatry* diagnostic rhyme test

DrTheol Doctor of Theology

Dr. und (*or* **u**) **Vrl.** Druck und Verlag (German: printed and published (by))

DRurSc Doctor of Rural Science

DRW *Computing* draw format

drx. drachma

ds. destro (Italian: right)

d.s. date of service · daylight saving · days after sight · day's sight · document signed

Ds Deus (Latin: God) · Dominus (Latin: Lord *or* Master)

DS *airline flight code for* Air Senegal · *Music* dal segno (Italian: (repeat) from the sign) · *Finance* debenture stock · defect score · *Physics* defect size · *Chem.*

dehydroepiandrosterone sulphate • dental surgeon • Department of State • Deputy Secretary • Detective Sergeant • *Physics* diffractive system • Directing Staff • Doctor of Science • Doctor of Surgery • *Computing* double-sided (floppy disk) • Down's syndrome • driver seated (in **DS vehicle**) • *British fishing port registration for* Dumfries • *British vehicle registration for* Glasgow

DSA *Med.* digital subtraction angiography • Diploma in Social Administration • Docteur en sciences agricoles (French: Doctor of Agriculture) • Driving Standards Agency

DSAC Defence Scientific Advisory Council

DSAO Diplomatic Service Administration Office

DSASO Deputy Senior Air Staff Officer

DSB (*or* **d.s.b.**) double sideband • Drug Supervisory Body (of the UN)

DSc Doctor of Science

DSC *Physics* differential scanning calorimetry • (the) Discovery Channel • *Military* Distinguished Service Cross • *Chem.* N,N′-disuccinimidyl carbonate • Doctor of Surgical Chiropody

DScA Docteur en sciences agricoles (French: Doctor of Agricultural Sciences) • Docteur en sciences appliquées (French: Doctor of Applied Sciences)

DSc(Agr) Doctor of Science in Agriculture

DScEng Doctor of Science in Engineering

DScFor Doctor of Science in Forestry

DSCHE Diploma of the Scottish Council for Health Education

DScMil Doctor of Military Science

DScTech Doctor of Technical Science

DSD Director of Signals Division • Director of Staff Duties

dsDNA *Genetics* double-stranded DNA

DSDP Deep-Sea Drilling Project

DSE Doctor of Science in Economics

dsgn. design(er)

DSIR (New Zealand) Department of Scientific and Industrial Research

DSL district scout leader • Doctor of Sacred Letters

DSM (*or* **d.s.m.**) deputy stage manager • (ital.) Diagnostic and Statistical Manual (of the American Psychiatric Association) • *Computing* digital storage

medium • Directorate of Servicing and Maintenance • *Military* Distinguished Service Medal • Doctor of Sacred Music

dsmd dismissed

DSN *Astronautics* Deep Space Network

DSO *Military* (Companion of the) Distinguished Service Order • District Staff Officer

DSocSc Doctor of Social Science(s)

d.s.p. decessit sine prole (Latin: died without issue)

DSP Democratic Socialist Party • *Electronics* digital signal processing (*or* processor) • (Canada) Doctor of Political Science (French *Docteur en sciences politiques*) • dye sublimation printing

dspl disposal

d.s.p.l. decessit sine prole legitima (Latin: died without legitimate issue)

d.s.p.m. decessit sine prole mascula (Latin: died without male issue)

d.s.p.m.s. decessit sine prole mascula superstite (Latin: died without surviving male issue)

dspn disposition

d.s.p.s. decessit sine prole superstite (Latin: died without surviving issue)

d.s.p.v. decessit sine prole virile (Latin: died without male issue)

d.s.q. discharged to sick quarters

DSR *Commerce* debt service ratio • *Med.* digital subtraction radiography • Director of Scientific Research • *Med.* dynamic spatial reconstructor

DSRD Directorate of Signals Research and Development

Dss Deaconess

DSS *Computing* decision-support system • Department of Social Security • Director of Social Services • Doctor of Holy (*or* Sacred) Scripture (Latin *Doctor Sacrae Scripturae*) • Doctor of Social Science • duplex stainless steel

DSSc Diploma in Sanitary Science • Doctor of Social Science

DST Daylight Saving Time • Director of Supplies and Transport • Doctor of Sacred Theology • Double Summer Time • *Mining* drill-stem test

DStJ Dame of Justice (*or* of Grace) of the Order of St John of Jerusalem

dstn destination

D. Surg. Dental Surgeon

DSW (New Zealand) Department of Social

Welfare • Doctor of Social Welfare (*or* Work)

d.t. delirium tremens • double throw

dT *Biochem., symbol for* thymidine

DT (ital.) Daily Telegraph • damage-tolerant • daylight time • *Med.* dead from tumour • delirium tremens • dental technician • (USA) Department of Transportation • (USA) Department of Treasury • destructive testing • *Colloquial* (USA) detective • (*or* **D–T**) *Nuclear engineering* deuterium–tritium • Director of Transport • Doctor of Divinity (*or* Theology) (Latin *Doctor Theologiae*) • *postcode for* Dorchester (Dorset) • *British vehicle registration for* Sheffield • *airline flight code for* TAAG-Angola Airlines

DTA *Physics* differential thermal analysis • Diploma in Tropical Agriculture • Distributive Trades' Alliance

DTAM data transfer, access, and manipulation

d.t.b.a. *Commerce* date to be advised

DTC (USA) Depository Trust Company • Diamond Trading Company • *Med.* differentiated thyroid carcinoma • Diploma in Textile Chemistry • Docklands Transportation Consortium

DTCD (USA) Department of Technical Cooperation for Development • Diploma in Tuberculosis and Chest Diseases

dtd dated

d.t.d. *Med.* detur talis dosis (Latin: let such a dose be given)

DTD Diploma in Tuberculous Diseases • Director of Technical Development

DTDP ditridecyl phthalate (plasticizer for PVC)

DTE *Computing* data terminal equipment • *Chem.* dithioerythritol

DTech Doctor of Technology

Dtg Dienstag (German: Tuesday)

DTh (*or* **DTheol**) Doctor of Theology

DTH *Immunol.* delayed-type hypersensitivity • Diploma in Tropical Hygiene

DThPT Diploma in Theory and Practice of Teaching

DTI Department of Trade and Industry

DTIC dacarbazine (anticancer drug)

DTL *Electronics* diode-transistor logic • down the line (in shooting) • *Physics* drift-tube linac (linear accelerator)

DTM Diploma in Tropical Medicine

DTMH (*or* **DTM&H**) Diploma in Tropical Medicine and Hygiene

DTNB *Chem.* 5,5'-dithiobis-(2-nitrobenzoic acid)

dt° direito (Portuguese: right)

DTOD Director of Trade and Operations Division

DTp Department of Transport

DTP desktop publishing

DTPA *Med.* diethylenetriaminepentaacetic acid (diagnostic aid)

DTPH Diploma in Tropical Public Health

DTR *Photog.* diffusion-transfer reversal • Diploma in Therapeutic Radiology • double taxation relief

DTRP Diploma in Town and Regional Planning

DTRT *Meteorol.* deteriorate (*or* deteriorating)

DT's (ˌdiːˈtiːz) *Colloquial* delirium tremens

DTS deep-sleep therapy • *Med.* dual tracer scintigraphy

DTT *Chem.* dithiothreitol

DTV digital television

DTVM Diploma in Tropical Veterinary Medicine

Du. Ducal • Duchy • Duke • Dutch

DU *British vehicle registration for* Coventry • *Physics* depleted uranium • died unmarried • Doctor of the University • *Med.* duodenal ulcer

DUART (ˈdjuːˌɑːt) *Electronics, Computing* dual universal asynchronous receiver transmitter

dub. dubious • dubitans (Latin: doubting)

Dub. (*or* **Dubl.**) Dublin

DUC *Meteorol.* dense upper cloud

DUI (USA) driving under the influence (of alcohol or drugs)

Dumf. Dumfries and Galloway Region

Dun. Dundee

Dunelm. Dunelmensis (Latin: (Bishop) of Durham)

DUniv Doctor of the University

duo. duodecimo (standard paper size)

dup. duplicate

DUP (Northern Ireland) Democratic Unionist Party • Docteur de l'Université de Paris (French: Doctor of the University of Paris)

Dur. Durban • Durham

DUS Diploma of the University of Southampton

DUSC *US Air Force* deep underground support center

Dut. Dutch

DUV damaging ultraviolet (radiation)

DV defective (*or* double) vision • Deo volente (Latin: God willing) • Diploma in Venereology • direct vision • distinguished visitor • district valuer • Douay Version (of the Bible) • *British vehicle registration for* Exeter

DVA *Med.* digital video angiography • Diploma of Veterinary Anaesthesia

DV&D Diploma in Venereology and Dermatology

DVC *Microbiol.* direct viable count

DVD *Computing* digital versatile disc

DVD-R *Computing* digital versatile disc-recordable

DVH Diploma in Veterinary Hygiene

DVI *Computing* digital video imaging (*or* interactive)

DVLA Driver and Vehicle Licensing Authority

DVLC Driver and Vehicle Licensing Centre (Swansea)

d.v.m. decessit vita matris (Latin: died in the lifetime of the mother)

DVM digital voltmeter • Doctor of Veterinary Medicine

DVMS Doctor of Veterinary Medicine and Surgery

DVO District Veterinary Officer

d.v.p. decessit vita patris (Latin: died in the lifetime of the father)

DVPH Diploma in Veterinary Public Health

dvr driver

DVR Diploma in Veterinary Radiology • *Computing* discrete-variable representation

DVS Doctor of Veterinary Surgery

DVSc (*or* **DVSci**) Doctor of Veterinary Science

DVT *Med.* deep-vein thrombosis

d.w. dead weight • delivered weight

d/w dust wrapper

DW *British vehicle registration for* Cardiff •

Chem. deionized water • (*or* **D/W**) *Commerce* dock warrant

DWA driving without awareness

d.w.c. deadweight capacity

DWEM *Colloquial* (USA) dead white European male

dwg drawing • (*or* **dwel.**) dwelling

DWI (USA) driving while intoxicated • Dutch West Indies (former name for Netherlands Antilles)

dwr drawer

DWR Duke of Wellington's Regiment

dwt pennyweight

d.w.t. (*or* **DWT**) deadweight tonnage

DWU (USA) Distillery, Wine, and Allied Workers International Union

DX *Photog.* daylight exposure (indicating that the speed of a film can be set automatically in a suitably equipped camera) • *British vehicle registration for* Ipswich • *Telecom.* long-distance (*or* -distant)

DXF data exchange format

DXR *Med.* deep X-ray

DXRT deep X-ray therapy

DXS diagnostic X-ray spectrometry

dy. delivery • demy (standard paper size)

Dy *Chem., symbol for* dysprosium

DY *international vehicle registration for* Benin (formerly Dahomey) • *British vehicle registration for* Brighton • dockyard • *postcode for* Dudley

Dyd. Dockyard

dyn *Physics, symbol for* dyne

dyn. dynamics • dynamite • dynamo • dynasty

DYS Duke of York's Royal Military School

dz. dozen

DZ *international vehicle registration for* Algeria • *British vehicle registration for* Antrim • Doctor of Zoology • *Meteorol.* drizzle • *Military* drop zone

DZool Doctor of Zoology

D-Zug Durchgangszug (German: express train)

E

e *Maths., symbol for* the base of natural (Napierian) logarithms • (ital.) *Maths., symbol for* eccentricity (of an ellipse or other conic) • (*or* **e.**) electromotive • (*or* **e⁻**) *Physics, symbol for* electron • (ital.) *Physics, symbol for* electron (*or* proton) charge • (ital.) *Chem., symbol for* equatorial conformation (of molecules) • *Physics, symbol for* positron (in **e⁺**) • *Maths., symbol for* the transcendental number 2.718.282... • (bold ital.) *Maths., symbol for* unit coordinate vectors (in **e**$_x$, **e**$_y$, **e**$_z$) • *Meteorol., symbol for* wet air • *indicating* the fifth vertical row of squares from the left on a chessboard

e. edition • educated • elder • eldest • electric • electricity • engineer(ing) • *Telephony, symbol for* Erlang (unit of traffic intensity) • *Baseball* error • excellence • excellent

E earth (indicating the terminal in an electrical circuit) • east(ern) • *postcode for* east London • *Colloquial* Ecstasy (hallucinogenic drug) • (ital.) *Physics, symbol for* electric field strength (light ital. in non-vector equations; bold ital. in vector equations) • (ital.) *Chem., symbol for* electrode potential • *Chem., symbol for* electromeric effect • (ital.) *Physics, symbol for* electromotive force • *Chem., symbol for* elimination reaction • *Astronomy, symbol for* elliptical galaxy (followed by a number) • *symbol for* emalangeni (sing. lilangeni; monetary unit of Swaziland) • (ital.) *Physics, symbol for* energy • *Numismatics* English shilling • E-number (EU-approved code number of a food additive, as in **E200**) • *symbol for* exa- (prefix indicating 10^{18}, as in **EJ**, exajoule) • *British fishing port registration for* Exeter • (ital.; *or* **E**$_v$) *Physics, symbol for* illuminance • (ital.; *or* **E**$_e$) *Physics, symbol for* irradiance • *symbol for* second-class merchant ship (in Lloyd's Register) • *international vehicle registration for* Spain • (ital.) *Physics, symbol for* Young modulus •

indicating casual workers (occupational grade) • *indicating* a musical note or key • *Logic, indicating* a universal negative categorical proposition

E. (*or* **E**) Earl • earth (planet) • east(ern) • Easter (in the Book of Common Prayer) • Edinburgh • efficiency • Egypt(ian) • elocution • eminence • enemy • engineer(ing) • England • English • envoy • equator • España (Spanish: Spain) • evening • evensong (in the Book of Common Prayer)

E2 *airline flight code for* Everest Air

E9 *airline flight code for* Compagnie Africaine D'Aviation

7E *airline flight code for* Nepal Airways

8E *airline flight code for* Bering Air

ea. each

EA *British vehicle registration for* Dudley • early antigen • East Anglia • economic adviser • educational age • effective action • effective agent (in **EA content**) • electrical artificer • Ente Autonomo (Italian: Autonomous Corporation) • *Freemasonry* Entered Apprentice • enterprise allowance • environmental assessment • *Chem.* ethylacrylate • Evangelical Alliance • experimental area • exposure age

E/A enemy aircraft • experimental aircraft

EAA Edinburgh Architectural Association • Electrical Appliance Association • (USA) Engineer in Aeronautics and Astronautics • *Chem.* ethylene (*or* ethene) acrylic acid

EAAA European Association of Advertising Agencies

EAAC European Agricultural Aviation Centre

EAAFRO East African Agriculture and Forestry Research Organization (Kenya)

EAAP European Association for Animal Production

EAB *Med.* extra-anatomic bypass

EAC East African Community •

Educational Advisory Committee • Engineering Advisory Council

EACA East Africa Court of Appeal Reports • *Med.* epsilon aminocaproic acid

EACC East Asia Christian Conference

EACN *Chem.* equivalent alkane carbon number

EACSO (iː'ɑːksəu) East African Common Services Organization

EAEG European Association of Exploration Geophysicists

EAES European Atomic Energy Society

e.a.f. emergency action file

EAF electric-arc furnace

EAFFRO (*or* **EAFRO**) East African Freshwater Fisheries Research Organization

EAG Economists Advisory Group

EAGGF European Agricultural Guidance and Guarantee Fund (in the EU)

EAK *international vehicle registration for* (East Africa) Kenya

EAM *Physics* embedded-atom method • Ethniko Apelentherotiko Metopo (Greek: National Liberation Front; in World War II)

EAMF European Association of Music Festivals

EAMFRO East African Marine Fisheries Research Organization

EAMTC European Association of Management Training Centres

EAN *Chem.* effective atomic number • *Computing* European Academic Network

e. & e. each and every

E & O *Accounting* errors and omissions

E & OE errors and omissions excepted (on invoice forms)

e.a.o.n. except as otherwise noted

EAP East Africa Protectorate • Edgar Allan Poe (1809–49, US writer) • (USA) employee-assistance program • English for academic purposes

EAPR European Association for Potato Research

EAPROM ('iːprɒm) *Computing* electrically alterable programmable read-only memory

EAR employee attitude research • *Engineering* energy-absorbing resin

EARCCUS East African Regional Committee for Conservation and Utilization of Soil

EARN (ɜːn) *Computing* European Academic and Research Network

EAROM ('ɪərɒm) *Computing* electrically alterable read-only memory

EAROPH East Asia Regional Organization for Planning and Housing

EAS *Med.* endotoxin-activated serum • *Aeronautics* equivalent air speed • *Physics* extensive air (*or* atmospheric) shower (of cosmic rays)

EASA Entertainment Arts Socialist Association

EASEP *Astronautics* Early Apollo Scientific Experiments Package

EASHP European Association of Senior Hospital Physicians

east. eastern

EASTROPAC ('iːstrəu,pæk) Eastern Tropical Pacific

e.a.t. earliest arrival time

EAT *Med.* Ehrlich ascites tumour • Employment Appeals Tribunal • *Med.* experimental autoimmune thyroiditis • *international vehicle registration for* (East Africa) Tanzania (*see also* EAZ)

EATF *Physics* externally applied thermal field

EATRO East African Trypanosomiasis Research Organization

EAU *international vehicle registration for* (East Africa) Uganda

EAVRO East African Veterinary Research Organization

EAW Electrical Association for Women • equivalent average words

EAX *Telecom.* electronic automatic exchange

EAZ *international vehicle registration for* (East Africa) Tanzania (Zanzibar; *see also under* EAT)

Eb *symbol for* exabyte(s)

EB electricity board • electron beam • electronic book • (ital.) Encyclopaedia Britannica • Epstein–Barr (in **EB virus**) • Evans blue (dye) • *British vehicle registration for* Peterborough

EBA English Bowling Association

e.b.a.r. edited beyond all recognition

EBC English Benedictine Congregation • European Billiards Confederation • European Brewery Convention

EBCDIC ('ɛpsɪ,dɪk) *Computing* extended binary-coded decimal-interchange code

e-beam electron beam

EBIC ('ɛbɪk) *Physics* electron-beam induced (*or* induction) current

EBICON ('ɛbɪkɔn) *Physics* electron-bombardment-induced conductivity

EBIS ('ɛbɪs) *Physics* electron-beam ion source

EBIT ('ɛbɪt) *Accounting* earnings before interest and tax • *Physics* electron-beam ion trap

EBL electron-beam lithography • European Bridge League

EBM electron-beam machined (*or* machining) • *Physics* energy-balance model • *Med.* expressed breast milk

EBMC English Butter Marketing Company

EbN east by north

EBNF *Computing* extended BNF (Backus normal form)

E-boat Enemy War Motorboat (German torpedo boat in World War II)

Ebor. ('iːbɔː) Eboracensis (Latin: (Archbishop) of York)

EBP *Physics* electron-beam process • *Physics* electron-beam-pumped

EBR electron-beam recording • experimental breeder reactor

EBRA (*or* **EB & RA**) Engineer Buyers' and Representatives' Association

EBRD European Bank for Reconstruction and Development

EbS east by south

EBS Emergency Bed Service • emergency broadcast system • engineered barrier system • English Bookplate Society

EBT *Med.* electron-beam therapy • *Physics* Elmo Bumpy Torus

EBU English Bridge Union • European Badminton Union • European Boxing Union • European Broadcasting Union

EBV *Physics* electron-beam vaporization • *Med.* Epstein–Barr virus

EBW electron-beam welding

EBWR *Nuclear physics* experimental boiling-water reactor

e.c. earth closet • enamel-coated (*or* -covered) • error correction • exempli causa (Latin: for example) • extended coverage • extension course

Ec. Ecuador

EC East Caribbean • *postcode for* East Central London • east coast • Eastern Command • Ecclesiastical Commissioner • *international vehicle registration for* Ecuador • *Physics* eddy current • education (*or* educational) committee • *Chem.* effective concentration • electricity council • electrolytic corrosion • *Physics* electron capture • *Physics* electron cyclotron • electronic computer • emergency commission • *Med.* endothelial cell • Engineering Corps • *Biochem.* Enzyme Commission (preceding the code number of an enzyme) • Episcopal Church • Established Church • *Chem.* ethylene (*or* ethene) carbonate • (Canada) Etoile du Courage (French: Star of Courage) • European Commission • European Community (*see also under* EU) • executive committee • *British vehicle registration for* Preston • *international civil aircraft marking for* Spain

ECA Early Closing Association • Economic Commission for Africa (UN agency) • (USA) Economic Cooperation Administration • Educational Centres Association • Electrical Contractors' Association • European Confederation of Agriculture • European Congress of Accountants

ECAC Engineering College Administrative Council • European Civil Aviation Conference

ECAFE Economic Commission for Asia and the Far East (*see* ESCAP)

ECAPO (ɪ'kɑːpəʊ) *Chem.* ethoxy carbonyl amino pyridine N-oxide

ECAS Electrical Contractors' Association of Scotland

ECB electronic codebook • electronic components board • European Central Bank

ECBS engineering of computer-based systems

ecc. eccetera (Italian: et cetera; etc.)

Ecc. Eccellenze (Italian: Excellency)

ECC *Electrical engineering* earth continuity conductor • *Chem.* electrochemical concentration cell • *Physics* electron capture to the continuum • *Physics* emergency core cooling (*or* coolant) • energy-conscious construction • *Computing* error-correction code • European Cultural Centre

ECCD electron-cyclotron current drive

Ecc. Hom. Ecce Homo (Latin: Behold the Man)

eccl. (*or* **eccles.**) ecclesiastic(al)

Eccles. (*or* **Eccl.**) *Bible* Ecclesiastes

ecclesiast. ecclesiastical

Ecclus. *Bible* Ecclesiasticus

ECCM *Military* electronic counter-countermeasure(s)

ECCP European Committee on Crime Problems

ECCS *Physics* emergency core-cooling system

ECCU English Cross Country Union

ECD early closing day • *Physics* electron-capture detector (*or* detection) • *Electronics* electrostatic charge decay • *Computing* enhanced colour display • estimated completion date • *Chem.* ethyl cysteinate dimer

ECDR electrostatic corona discharge reactor

ECE Economic Commission for Europe (UN agency) • *Physics* electron-cyclotron emission

ECF *Med.* (USA) extended-care facility

ECFA European Committee for Future Accelerators

ECFMG Educational Council for Foreign Medical Graduates

ECFMS Educational Council for Foreign Medical Students

ECG *Med.* electrocardiogram (*or* electrocardiograph) • Export Credit Guarantee

ECGB East Coast of Great Britain

ECGC Empire Cotton Growing Corporation

ECGD Export Credits Guarantee Department

ech. echelon

ECH *Physics* electron-cyclotron heating

ECHP (USA) Environmental Compliance and Health Protection

ECHR European Court of Human Rights

ECI East Coast of Ireland • energy-cost indicator

ECIA European Committee of Interior Architects

ECITO European Central Inland Transport Organization

ECJ European Court of Justice

ECL *Physics* emergency cooling limit • *Electronics* emitter-coupled logic

ECLAC Economic Commission for Latin America and the Caribbean (UN agency)

eclec. eclectic(ism)

ecli. eclipse • ecliptic

ECLO *Electronics* emitter-coupled logic operator

ECLOF Ecumenical Church Loan Fund

ECLSS environmental control and life-support systems

ECM electric coding machine • electrochemical machining • *Physics*

electron-cyclotron maser • *Military* electronic countermeasure(s) (for jamming enemy signals, destroying guided missiles, etc.) • energy conservation measure • environmental corrosion monitor

ECMA ('ɛkmə) *Geology* East Coast Magnetic Anomaly • European Computer Manufacturers' Association (Geneva)

ECMO ('ɛkməʊ) *Med.* extracorporeal membrane oxygenation

ECMF Electric Cable Makers' Federation

ECMRA European Chemical Market Research Association

ECMT European Conference of Ministers of Transport

ECMWF European Centre for Medium-range Weather Forecast

ECN epoxy–cresol–novolak (synthetic resin)

ECNSW Electricity Commission of New South Wales

ECO energy conservation opportunity • English Chamber Orchestra

ECODU European Control Data Users

ECoG *Med.* electrocochleography

ecol. ecological • ecology

Ecol. Soc. Am. Ecological Society of America

econ. economical • economics • economist • economy

Econ. J. (ital.) Economic Journal

Econ. R. (ital.) Economic Review

ECOR (USA) Engineering Committee on Ocean Resources

ECOSOC ('ɛkəʊ,sɒk) Economic and Social Council (of the UN)

ECOVAST ('ɛkəʊ,væst) European Council for the Village and Small Town

ECOWAS (ɛ'kəʊəs) Economic Community of West African States

ECP *Finance* Euro-commercial paper • European Committee on Crop Protection • Evangelii Christi Praedicator (Latin: Preacher of the Gospel of Christ)

ECPA Electric Consumers Protection Act (1986)

ECPD (USA) Engineers' Council for Professional Development

ECPS European Centre for Population Studies

ECQAC Electronic Components Quality Assurance Committee

ECR *Physics* electron-cyclotron resonance • electronic cash register

ECRH *Physics* electron-cyclotron-resonance heating

ECS *Physics* electron-capture spectroscopy • *Physics* electronically collimated system • *Physics* emergency cooling (*or* coolant) system • environmental control system • European Communications Satellite • European Components Service

ECSC European Coal and Steel Community

ECT *Med.* electroconvulsive therapy • *Med.* emission-computerized (*or* -computed) tomography

ECTA Electrical Contractors' Trading Association

ECTD (USA) Emission Control Technology Division

ECTF Enterprise Computer Telephony Forum

ecu (*or* **ECU**) ('eɪkju: *or* 'ɛkju:) European Currency Unit (in the EU)

ECU English Church Union • environmental control unit • European Chiropractic Union • *Photog.* extreme close-up

Ecua. Ecuador

ECUK East Coast of the United Kingdom

ECV *Obstetrics* external cephalic version

ECWA Economic Commission for Western Asia (UN agency)

ECWEC European Community Wind Energy Conference and Exhibition

ECY European Conservation Year

ECYO European Community Youth Orchestra

ed. edile (*or* edilizia) (Italian: building) • edited • edition • editor • educated • education

éd. édition (French: edition)

e.d. edge distance • enemy dead • error detecting • excused duty • extra duty

Ed. Edinburgh • Editor

ED (USA) Department of Education • (USA) Doctor of Engineering • economic dispatch • Education Department • *Pharmacol.* effective dose (as in **ED$_{50}$**, mean effective dose) • Efficiency Decoration • election district • *Physics* electromagnetic dissociation • *Electronics* electron device • *Physics, Chem.* electron diffraction • *Med.* embryonic day • Employment Department • *Med.* end-diastole (*or* -diastolic) • entertainments duty • equilibrium dialysis • *Physics*

equivalent dose (of radiation) • European Democrat • *Finance* ex dividend • existence doubtful • experimental detector • *Med.* extensive disease • *Finance* extra dividend • *British vehicle registration for Liverpool*

e.d.a. early departure authorized

EDA (USA) Economic Development Administration • Electrical Development Association • electronic design automation • (Greece) Eniea Dimokratiki Aristera (Union of the Democratic Left; political party)

ED & S (ital.) English Dance and Song (publication of the English Folk Dance and Song Society)

EDAS ('i:dæs) *Computing* enhanced data-acquisition system

EDAX ('i:dæks) *Physics* energy-dispersive analysis of X-rays

EdB Bachelor of Education

EDB ethene dibromide (antiknock agent and soil fumigant)

EDBS *Computing* expert database system

e.d.c. error detection and correction • extra dark colour

EDC Economic Development Committee • *Chem.* effective dielectric constant • *Physics* electron-distribution curve • Engineering Design Centre • *Computing* error-detection code • European Defence Community • *Obstetrics* expected date of confinement

edd. ediderunt (Latin: published by) • editiones (Latin: editions)

EdD Doctor of Education

EDD exactly delayed detonator • *Obstetrics* expected date of delivery

EDE *Physics* effective dose equivalent (of radiation)

EDF Electricité de France (French electricity corporation) • (USA) Environmental Defense Fund • European Development Fund

EDG emergency diesel generator • European Democratic Group

EDHE experimental data handling equipment

EDI *Computing* electronic data interchange

Edin. Edinburgh

Ed. in Ch. Editor in Chief

EDIP European Defence Improvement Programme (in NATO)

EDIS Engineering Data Information System

edit. edited • edition • editor • editore (Italian: publisher) • editorial

e.d.l. edition de luxe

EDL economic discard limits • *Chem.* electrical double layer • electrodeless discharge lamp

EdM Master of Education

EDM electrical-discharge machined (*or* machining) • *Chem.* electric dipole moment • *Surveying* electronic distance measurement • *Physics* energy-density model

EDMA European Direct Marketing Association

Edm. and Ipswich (Bishop of) St Edmundsbury and Ipswich

Edmn Edmonton (Canada)

edn edition

Ednbgh Edinburgh

EDNS expected demand not supplied

EDO *Computing* extended data out

e.d.o.c. effective date of change

EDO DRAM *Computing* extended data out dynamic random-access memory

EDP *Physics* electron-diffraction pattern • (*or* **e.d.p.**) electronic data processing • Emergency Defence Plan • *Med.* end-diastolic pressure

EDPS *Computing* electronic data-processing system

EDPT *Computing* enhanced-drive parameter table

EDR electronic decoy rocket • (Japanese) Electronic Dictionary Research • European Depository Receipts • except during rain (following a pollen count)

EDRF *Med.* endothelium-derived relaxing factor

EDRP European Demonstration Reprocessing Plant (for nuclear fuel)

EdS Education Specialist

EDS Electronic Data Systems Corporation • *Physics* energy-dispersive spectrometry (*or* spectroscopy) • *Physics* energy-dispersive spectrum (*or* spectra) • English Dialect Society • *Computing* exchangeable disk store

EDSAC ('ɛdsæk) *Computing* Electronic Delay Storage Automatic Calculator

EDSAT ('ɛdsæt) Educational Television Satellite

EDSP Exchange Delivery Settlement Price

EDSS expert decision-support system

EDT (USA, Canada) Eastern Daylight Time • energy design technique

EDTA (*or* **edta**) *Chem.* ethylenediaminetetraacetic acid • European Dialysis and Transplant Association

EDTN *Chem.* 1-ethoxy-4-(dichloro-1,3,5-triazinyl)naphthalene

EDU European Democratic Union

educ. educated • (*or* **educn**) education • educational

EDV *Med.* end-diastolic volume

EDVAC ('ɛdvæk) *Computing* Electronic Discrete Variable Automatic Computer

Edw. Edward

EDX *Physics* energy-dispersive X-ray

EDXA *Physics* energy-dispersive X-ray analysis (*or* analyser)

EDXD *Physics, Crystallog.* energy-dispersive X-ray diffraction

EDXF *Physics, Chem.* energy-dispersive X-ray fluorescence

EDXS *Physics, Chem.* energy-dispersive X-ray spectrometer • *Physics, Chem.* energy-dispersive X-ray spectroscopy

e.e. errors excepted • eye and ear

EE Early English • Eastern Electricity • edge-to-edge • electrical engineer(ing) • electron emission • electronic engineer(ing) • employment exchange • environmental education • environmental engineering • Envoy Extraordinary • errors excepted • (New Zealand) ewe equivalent • explosive emission • *British vehicle registration for* Lincoln

EEA Electronic Engineering Association • European Economic Area

EEAIE Electrical, Electronic, and Allied Industries, Europe

EE & MP Envoy Extraordinary and Minister Plenipotentiary

EEB European Environmental Bureau

EEC *Physics* energy–energy correlation • English Electric Company • European Economic Community (replaced by the EC) • *Electronics* explosive-emission cathode

EECA European Electronic Component Manufacturers' Association

EECS electrical-energy conversion system

EED *Physics* effective equivalent dose (of radiation) • electro-explosive device

EEDC Economic Development Committee for the Electronics Industry

EEDQ *Chem.* 1-ethoxycarbonyl-2-ethoxy-1,2-dihydroquinoline

EEF Egyptian Expeditionary Force •

Engineering Employers' Federation • equivalent-energy function

EEG electroencephalogram (*or* electroencephalograph) • Essence Export Group

EEI (USA) Edison Electric Institute • (USA) Environmental Equipment Institute

EEIBA Electrical and Electronic Industries Benevolent Association

EEL *Physics* electron energy loss

EELS *Physics* electron energy loss spectroscopy

EEMJEB Electrical and Electronic Manufacturers' Joint Education Board

EEMS *Computing* enhanced expanded memory specification

EEMUA Engineering Equipment and Materials Users Association

E Eng. Early English

EENT *Med.* eye, ear, nose, and throat

EEO equal employment opportunity

EEOC (USA) Equal Employment Opportunities Commission

EEPLD *Computing* electrically erasable programmable logic device

EEPROM ('i:prɒm) *Computing* electrically erasable programmable read-only memory

EER energy–efficiency ratio • *Physics* equilibrium equivalent radon

EERI (USA) Earthquake Engineering Research Institute

EEROM ('ɪərɒm) *Computing* electrically erasable read-only memory

EES *Physics* electron-energy spectroscopy (*or* spectrum) • European Exchange System

EET Eastern European Time

EETPU Electrical, Electronic, Telecommunications, and Plumbing Union (*see* AEEU)

EETS Early English Text Society

EEZ exclusive economic zone

EF edge-to-face • education(al) foundation • elevation finder • *Genetics* elongation factor • emergency fleet • *Physics* enrichment factor • expectant father • expeditionary force • experimental flight • extra fine • *British vehicle registration for* Middlesbrough

EFA engine fault analysis • *Biochem.* essential fatty acid • Eton Fives Association • European Fighter Aircraft

EFB energy from biomass

EFC European Federation of Corrosion • European Forestry Commission

EFCE European Federation of Chemical Engineering

EFCT European Federation of Conference Towns

EFD *Engineering* early fault detection

EFDSS English Folk Dance and Song Society

eff. effetto (Italian: bill *or* promissory note) • efficiency • effigy

EFF Electronic Frontier Foundation • European Furniture Federation

EFG electric-field gradient

EFGF *Electronics* epitaxial ferrite-garnet film

EFI electronic fuel injection (of car engines)

EFIS *Aeronautics* electronic flight-information system

EFL English as a foreign language • external financial limit

EFM *Med.* electronic fetal monitor • European Federalist Movement

EFNS Educational Foundation for Nuclear Science

EFP *Maths.* Einstein–Fokker–Planck (differential equation) • *Photog.* electronic field production • *Commerce* exchange of futures for physicals • explosively formed projectile (*or* penetrator)

EFPD effective full-power day

EFPH effective full-power hour

EFPW European Federation for the Protection of Waters

EFPY effective full-power year

EFR *Astronomy* emerging flux region • European Fast Reactor • experimental fast reactor

EFRC (USA) Edwards Flight Research Center

EFSA European Federation of Sea Anglers

EFSC European Federation of Soroptimist Clubs

EFT electronic funds-transfer

EFTA ('ɛftə) European Free Trade Association

EFTC Electrical Fair Trading Council

EFTPOS ('ɛftpɒs) electronic funds-transfer at point of sale

EFTS electronic funds-transfer system • elementary flying training school

EFTU Engineering and Fastener Trade Union

EFU energetic feed unit • European Football Union

EFVA Educational Foundation for Visual Aids

EFW electric front windows (in motor advertisements) • energy from waste

e.g. ejusdem generis (Latin: of a like kind) • exempli gratia (Latin: for example)

Eg. Egypt(ian) • Egyptology

EG Engineers' Guild • *Chem.* ethylene glycol • *airline flight code for* Japan Asia Airways • *British vehicle registration for* Peterborough

EGA Elizabeth Garrett Anderson (Hospital for Women) • *Computing* enhanced graphics adapter • European Golf Association

EGARD ('εgɑːd) environmental gamma-ray and radon detector

EGAS Educational Grants Advisory Service

EGBPS *Med.* ECG gated blood pool scintigraphy

EGCI Export Group for the Construction Industries

EGCS English Guernsey Cattle Society

EGD epithermal gold deposit

EGEAS Electric Generating Expansion Analysis System

EGF *Med.* epidermal growth factor

EGFR *Med.* epidermal-growth-factor receptor

EGIFO Edward Grey Institute of Field Ornithology (Oxford)

EGL Engineers' Guild Limited

EGM Empire Gallantry Medal (replaced by the George Cross) • European Glass Container Manufacturers' Committee • extraordinary general meeting

EGmbH (Germany) Eingetragene Gesellschaft mit beschränkter Haftung (registered limited company)

EGO eccentric (orbit) geophysical observatory

EGR *Finance* earned growth rate • exhaust gas recirculation (as in **EGR valve**)

EGSP electronics glossary and symbol panel

e.g.t. exhaust gas temperature

EGT *Physics* Einstein-invariant gauge theory (of gravitation)

EGU English Golf Union • *Chem.* external gelation of uranium

Egypt. Egyptian

Egyptol. Egyptologist • Egyptology

eh. ehrenhalber (German: honorary)

EH *postcode for* Edinburgh • *Music* English horn (cor anglais) • *Med.* essential hypertension • *British vehicle registration for* Stoke-on-Trent

EHC effective heat capacity (of a building) • European Hotel Corporation • *Med.* external heart compression

EHE *Physics* extraordinary Hall effect

EHF *Physics* European Hadron Facility • European Hockey Federation • experimental husbandry farm • *Radio* extremely high frequency

EHG electro-hydraulic governor

EHL *Physics* effective half-life • *Physics* electron-hole liquid

EHMO *Chem.* extended Hückel molecular orbital

EHO Environmental Health Officer

ehp effective horsepower

EHP electric and hybrid propulsion • *Physics* electron-hole plasma

EHR Environmental Hazard Ranking

EHS (USA) Environmental Health Services • European hybrid spectrometer • extra-high strength

EHT (*or* **e.h.t.**) *Electronics* extra-high tension

EHV electric and hybrid vehicle • extra-high voltage

EHWS extreme high water-level spring tides

EI *airline flight code for* Aer Lingus • earth (atmosphere) interface • East Indian • East Indies • electrical insulation • electromagnetic interaction • electron impact • electron ionization • *Finance* endorsement irregular • energy intake • environmentally-induced illness • *Photog.* exposure index • external irradiation • *international civil aircraft marking for* Republic of Ireland • *Irish vehicle registration for* Sligo

EIA East Indian Association • economic impact assessment • (USA) Electrical Industries Association • (USA) Electronic Industries Association • Engineering Industries Association • environmental impact analysis (*or* assessment) • Environmental Investigation Agency • *Med.* enzyme immunoassay • *Med.* exercise-induced asthma

EIB European Investment Bank • Export–Import Bank

EIC East India Company • Electrical Industries' Club • *Physics* electrostatic ion cyclotron • Engineering Institute of Canada

EICM employer's inventory of critical manpower

EICS East India Company's Service

EICW *Physics* electrostatic ion cyclotron wave

EID East India Dock • *Military* Electrical Inspection Directorate (for electrical, electronic, and optical equipment) • *Physics* electron-induced disordering

EIDCT Educational Institute of Design, Craft, and Technology

E-IDE enhanced integrated-drive electronics

EIE *Chem.* equilibrium isotope effect

EIEE *Med.* early infantile epileptic encephalopathy

EIEMA Electrical Installation Equipment Manufacturers' Association

EIF Elderly Invalids Fund

EIFAC European Inland Fisheries Advisory Committee (of the FAO)

E-in-C Engineer-in-Chief

E Ind. East Indies

einschl. einschliesslich (German: including *or* inclusive)

Einw. Einwohner (German: inhabitant)

EIO extended interaction oscillator

EIPC European Institute of Printed Circuits

EIR *Taxation* earned income relief • *Computing* error-indicating recording

EIRMA European Industrial Research Management Association

EIS economic information system • Educational Institute of Scotland • effluent information system • energy information system • Enterprise Investment Scheme (replaced the BES) • environmental impact statement • epidemic intelligence service

EISA *Computing* enhanced (*or* extended) industry standard architecture

EITB Engineering Industry Training Board

EITF *Accounting* (USA) Emerging Issues Task Force

EIU Economist Intelligence Unit

EIVT European Institute for Vocational Training

ej. ejemplo (Spanish: example)

EJ *Physics, symbol for* exajoule(s) • *British vehicle registration for* Haverfordwest

EJC (USA) Engineers' Joint Council

EJMA English Joinery Manufacturers' Association

ejusd. ejusdem (Latin: of the same)

eK etter Kristi (Norwegian: after Christ; AD)

EK *international civil aircraft marking for* Armenia • East Kilbride • *airline flight code for* Emirates Airlines • (Greece) Enosis Kentron (Centre Union; political party) • *British vehicle registration for* Liverpool

EKD Evangelische Kirche in Deutschland (German: Protestant Church in Germany)

EKE *Physics* equatorial Kerr effect

EKG (USA, Germany) electrocardiogram *or* electrocardiograph (German *Elektrokardiogramme*

el. elect(ed) • electric(ity) • element • elevated • elevated railway • elevation • elongation

EL *British vehicle registration for* Bournemouth • electrical laboratory • electroluminescent • electronics laboratory • *Physics* electron lattice • Engineer Lieutenant • epitaxial layer • Everyman's Library • explosive limit • *Med.* extracorporeal lithotripsy • *international civil aircraft marking for* Liberia

ELA electronic learning aid • Eritrean Liberation Army

elas. elasticity

ELAS Ethnikos Laikos Apeleutherotikos Stratos (Greek: Hellenic People's Army of Liberation; in World War II)

ELB Bachelor of English Literature

ELBS English Language Book Society

ELC *Physics* electron loss to continuum • Environment Liaison Centre (Nairobi)

eld. elder • eldest

ELD economic load distribution

ELDC economic load dispatching centre • equivalent load duration curve

ELDO ('ɛldəʊ) *Astronautics* European Launcher Development Organization (now part of ESA)

elec. (*or* **elect.**) election • elector(al) • electric(al) • electrically • electrician • electricity • electron • electronic • *Med.* electuary

ELEC European League for Economic Cooperation

electr. electrical

Electra (ɪ'lɛktrə) Electrical, Electronics, and Communications Trades Association

electron. electronics

elem. element(s) • elementary

elev. elevation • elevator

e.l.f. early lunar flare

ELF Eritrean Liberation Front • European Landworkers' Federation • *Radio* extremely low frequency

ELG *Astronomy* emission-line galaxy

ELH *Biochem.* egg-laying hormone

Eli *Sport* electronic line indicator

Eli. Elias • Elijah

Elien. Eliensis (Latin: (Bishop) of Ely)

Elint ('ɛlɪnt) *Military* electronic intelligence

ELISA (ɪ'laɪzə) *Med.* enzyme-linked immunosorbent assay

elix. *Med.* elixir

Eliz. Elizabethan

ELLA European Long Lines Agency

ellipt. elliptical

ELM edge-localized mode

ELMA electromechanical aid

Elmint ('ɛlmɪnt) *Military* electromagnetic intelligence

elo. elocution • eloquence

ELOISE (ˌɛləʊ'iːz) European Large Orbiting Instrumentation for Solar Experiments

elong. elongate(d) • elongation

E. Long. east longitude

ELP equivalent local potential

ELR *Meteorol.* environment lapse rate • exceptional leave to remain (for four years in the UK; granted to asylum-seekers) • export licensing regulations

ELS Electronic Lodgement Service • *Physics* energy-loss spectroscopy

elsewh. elsewhere

ELSIE ('ɛlsɪ) electronic speech information equipment

ELT *Physics* effective Lagrangian theory • English language teaching (for foreign students) • European letter telegram

ELU English Lacrosse Union

ELV expendable launch vehicle • extra-low voltage

ELWS extreme low water-level spring tides

ely. easterly

em. emanation • embargo • eminent

e.m. (*or* **em**) electromagnetic • electron microscope • emergency maintenance • expanded metal • external memorandum

Em. Eminence

EM Earl Marshal • Edward Medal • *Physics* effective mass • Efficiency Medal • electrical and mechanical • electromagnetic • electromotive • electronic mail • electron microscope (*or* microscopy) • emission measure • engineering model • Engineer of Mines • *Genetics* enhanced mutagenesis • enlisted man • environmental modelling • equipment module • Equitum Magister (Latin: Master of the Horse) • European Movement • evaluation model • expectation maximization • *British vehicle registration for* Liverpool

EMA effective medium approximation • *Immunol.* epithelial membrane antigen • European Marketing Association • European Monetary Agreement • Evangelical Missionary Alliance

EMAD engine maintenance and disassembly

email (*or* **e-mail**, **E-mail**) ('iːmeɪl) electronic mail

EMAS *Computing* Edinburgh multiaccess system • Employment Medical Advisory Service

emb. embargo • embossed

Emb. Embankment (in London) • Embassy

EMB (USA) Energy Mobilization Board

EMBD *Meteorol.* embedded in cloud

EMBL European Molecular Biology Laboratory (Heidelberg)

EMBO European Molecular Biology Organization

embr. embroider(y)

embryol. embryology

EMC (USA) Einstein Medical Center • *Computing* electromagnetic compatibility • Energy Management Centre (*or* Company) • (USA) Engineering Manpower Commission • environmental monitoring and compliance • *Physics* European muon collaboration

EMCB earth-mounded concrete bunker

EMCC European Municipal Credit Community

EMCCC European Military Communications Coordinating Committee

EMCOF ('ɛmkɒf) European Monetary Cooperation Fund

EMCS energy monitoring and control system

EMD *Chem.* electrolytic manganese dioxide

EME East Midlands Electricity

emer. emergency

Emer. Emeritus

EMet Engineer of Metallurgy

EMEU East Midlands Educational Union

EMF electromagnetic force • (*or* **emf, e.m.f.**) electromotive force • European Metalworkers' Federation • European Motel Federation

EMFI (USA) Energy and Minerals Field Institute

EMG *Med.* electromyogram (*or* electromyograph)

EMI (USA) Earth Mechanics Institute • Electric and Musical Industries • *Computing* electromagnetic interference • European Monetary Institute • *Physics* external muon identifier

EMIC emergency maternity and infant care

EMK elektromotorische Kraft (German: electromotive force) • *Chem.* ethyl methyl ketone

EML Environmental Measurement(s) Laboratory • Everyman's Library

EMLA *Med.* eutectic mixture of local anaesthetics (in **EMLA cream**)

Emm. Emmanuel College, Cambridge

EMMS *Computing* expanded memory manager

E. Mn. E. Early Modern English

EMNRD Energy, Minerals, and Natural Resources Department

e.m.o.s. earth's mean orbital speed

emp. *Med.* emplastrum (Latin: a plaster)

Emp. Emperor • Empire • Empress

EMP ecological monitoring programme • electromagnetic pulse • electronic materials programme • environmental management (*or* monitoring) plan (*or* programme) • *Physics* exciton-magnetic polaron

emp. agcy employment agency

EMPC *Chem.* electrostatic molecular potential contour

empld employed

EMR Eastern Mediterranean Region • (*or* **emr**) electromagnetic radiation

EMRIC (USA) Educational Media Research Information Center

EMRS East Malling Research Station

EMS emergency management system • emergency medical service • energy management system • *Chem.* ethyl methane sulphonate • European Monetary System • *Computing* expanded memory specification (*or* support)

EMSA Electron Microscopy Society of America

EMSC (USA) Electrical Manufacturers' Standards Council

EMT *Physics* effective-mass theory • emergency medical technician • *Physics* energy-momentum tensor

EMTA Electro-Medical Trade Association

emu (*or* **e.m.u.**) *Physics* electromagnetic unit

EMU electrical multiple unit • European monetary union (within the EU) • *Astronautics* extravehicular mobility unit

EMV *Accounting* expected monetary value

EMW energy from municipal waste

en *Chem.* ethylenediamine (used in formulae)

en. enemy

En. Engineer • English

EN *airline flight code for* Air Dolomiti • *Meteorol.* El Niño • (Portugal) Emissora Nacional (national broadcasting station) • *postcode for* Enfield • Enrolled Nurse • (Portugal) Estrada Nacional (national highway) • Euro Norm (European standard) • exceptions noted • extrapolation number • *British vehicle registration for* Manchester

ENA Eastern North America • (*or* **Ena**; 'iːnə) (France) École Nationale d'Administration (graduate college for top civil servants) • English Newspaper Association

ENAA *Physics* epithermal neutron activation analysis

ENAB Evening Newspaper Advertising Bureau

enam. enamel(led)

ENB English National Ballet • English National Board for Nursing, Midwifery, and Health Visiting

enc. enclosed • enclosure

ENC equivalent noise charge

ENCA European Naval Communications Agency

Enc. Brit. (*or* **Ency. Brit., Encyc. Brit.**) Encyclopaedia Britannica

encl. enclosed • enclosure

ency. (*or* **encyc., encycl.**) encyclopedia • encyclopedic • encyclopedism • encyclopedist

END European Nuclear Disarmament

ENDO Ethiopian National Democratic Organization

ENDOR ('ɛndɔ:) *Physics* electron-nuclear double resonance

endow. endowment

endp. endpaper

ENDS Euratom Nuclear Documentation System

ENE east-northeast

ENEA European Nuclear Energy Agency

ENF European Nuclear Force

eng. engine • engineer(ing) • engraved • engraver • engraving

Eng. England • English

ENG electronic news gathering (in TV broadcasting)

EN(G) Enrolled Nurse (General) (formerly SEN)

EngD Doctor of Engineering

engg engineering

Eng. hn *Music* English horn

engin. engineer(ing)

Engl. England • English

engr engineer • engraver

engr. engrave(d) • engraving

EngScD Doctor of Engineering Science

EngTech Engineering Technician

ENIAC ('ɛnɪæk) Electronic Numerical Integrator and Calculator (first electronic calculator)

ENIT (Italy) Ente Nazionale Industrie Turistiche (state tourist office)

enl. enlarge(d) • enlargement • enlisted

EN(M) Enrolled Nurse (Mental)

EN(MH) Enrolled Nurse (Mental Handicap)

en° enero (Spanish: January)

ENO English National Opera

ENR Energy and Natural Resources

Ens. (*or* **ens.**) *Music* ensemble • Ensign

ENS empty nest syndrome

ENSA ('ɛnsə) Entertainments National Service Association (in World War II)

ENSDF Evaluated Nuclear Structure Data File

ENSO *Meteorol.* El Niño southern oscillation

ent. entomology • entertainment • entrance

Ent. enter (stage direction)

ENT *Med.* ear, nose, and throat

entom. (*or* **entomol.**) entomological • entomology

Ent. Sta. Hall entered (registered) at Stationers' Hall (requirement to secure copyright on books before 1924)

env. envelope • environs

Env. Envoy

Env. Ext. (*or* **Env. Extra.**) Envoy Extraordinary

e.o. ex officio

E.o. Easter offerings

EO *Christianity* Eastern Orthodox • Education Officer • *Physics* electro-optic(al) • emergency operation • employers' organization • Engineer Officer • Entertainments Officer • Equal Opportunities • *Chem.* ethoxylated sulphonates • *Chem.* ethylene oxide • Executive Officer • executive order • Experimental Officer • *British vehicle registration for* Preston • *airline flight code for* Zaïre Express

EOA (USA) Essential Oil Association • examination, opinion, and advice

EOARDC (USA) European Office of the Air Research and Development Command

EOB *Computing* end of block • Executive Office Building

Eoc *Geology* Eocene

EOC electron-optical camera • (USA) Emergency Operations (*or* Operating) Center • *Physics* end of charge • end of cycle • Equal Opportunities Commission

e.o.d. entry on duty • every other day

EOD end of data (computing code) • *Military* explosive ordnance disposal (*or* demolition)

EOE enemy-occupied Europe • equal opportunity employer • errors and omissions excepted (on invoice forms) • European Options Exchange

EOF (USA) Emergency Operations (*or* Operating) Facility • *Maths.* empirical orthogonal function • *Computing* end of file

EOG *Med.* electrooculogram (*or* electrooculography)

e.o.h.p. except otherwise herein provided

EOJ *Computing* end of job

Eoka (*or* **EOKA**) (ɪ' əʊkə) Ethniki Organosis Kypriakou Agonos (Greek: National Organization of Cypriot Struggle; for union with Greece)

EOL end of life

EOLM electro-optical light modulator

e.o.m. *Commerce* end of the month • every other month

EOM Egyptian Order of Merit • extractable organic material (*or* matter)

EONR European Organization for Nuclear Research

e.o.o.e. *Commerce* erreur ou omission exceptée (French: errors and omissions excepted; on invoice forms)

EOP *Microbiol.* efficiency of plating • emergency operating procedure

EOPH Examined Officer of Public Health

EOQ *Accounting* economic order quantity

EOQC European Organization for Quality Control

EOQL end of qualified life

EOR *Astronautics* earth-orbit rendezvous • *Computing* end of record • enhanced oil recovery

EORI (USA) Economic Opportunity Research Institute

EORTC European Organization for Research on Treatment of Cancer

EOS *Physics* equation of state • *Computing* erasable optical storage • European Orthodontic Society

e.o.t. enemy-occupied territory

EOT *Computing* end of tape (as in **EOT marker**) • end of terrace (in property advertisements) • *Computing* end of transmission

EOTP European Organization for Trade Promotion

e.p. easy projection • editio princeps (Latin: first edition) • electrically polarized • endpaper • engineering personnel • *Chess* en passant • *Navigation* estimated position • extreme pressure

Ep. Episcopus (Latin: Bishop) • *Bible* Epistle

EP early picture • (USA) educational psychologist • *Electronics* electrically conducting polymer • *Metallurgy* electroplate(d) • electrostatic precipitator • end point • environmental protection • *Navigation* estimated position • *Physics* evoked potentials • expanded polystyrene • extended-play (gramophone record) • extraction procedure • Extraordinary and Plenipotentiary • *international civil aircraft marking for* Iran • *British vehicle registration for* Swansea

EPA educational priority area • *Med.* eicosapentaenoic acid (cholesterol-reducing fatty acid) • *Physics* electron–positron annihilation • *Physics* electron probe analyser • Emergency Powers Act • Employment Protection Act • energy-performance assessment • (USA) Environmental Protection Agency • European Productivity Agency

EPAA (USA) Emergency Petroleum Allocation Act

EPACCI Economic Planning and Advisory Council for the Construction Industries

EPAQ electronic parts of assessed quality

EPB equivalent pension benefit

EPC easy-processing channel (in **EPC black,** filler in rubber compounding) • Economic and Planning Council • Educational Publishers' Council • *Foundry* evaporative pattern-casting

EPCA (USA) Energy Policy and Conservation Act • energy production and consumption account • European Petro-Chemical Association

EPCOT Experimental Prototype Community of Tomorrow (Florida)

EPD earliest practicable date • *Crystallog.* etch-pit density

EPDA Emergency Powers Defence Act

EPDC (USA) Electric Power Development Company

EPDM *Chem.* ethylene–propylene diene monomer

EPEA Electrical Power Engineers' Association

EPF emulsified petroleum fuel • European Packaging Federation

EPFM elastic-plastic fracture mechanics

EPG (USA) Electronic Proving Ground • *Med.* electropalatogram (*or* electropalatology) • Emergency Procedure Guideline(s) • Eminent Persons Group

EPGS electric power generating system

Eph. (*or* **Ephes.**) *Bible* Ephesians

EPI echo planar imaging • *Physics* effective-pair interaction • electronic position indicator • Eysenck Personality Test

EPIC Engineering and Production Information Control • (*or* **Epic;** ˈɛpɪk) European Prospective Investigation into Cancer

Epict. Epictetus (*c.* 55–*c.* 110 AD, Greek Stoic philosopher)

epid. epidemic

epil. epilogue

Epiph. Epiphany

EPIRB *Navigation* emergency position indicator radio beacon

Epis. (*or* **Episc.**) Episcopal(ian) • (*or* **Epist.**) *Bible* Epistle

epit. epitaph • epitome

EPL exact power law

EPLD *Computing* erasable programmable logic device

EPLF Eritrean People's Liberation Front

EPM (*or* **EPMA**) electron-probe microanalysis • extract of particulate matter

EPMI Esso Production Malaysia Inc.

EPN elemental point number

EPNdB effective perceived noise decibels

EPNG El Paso Natural Gas Company

EPNS electroplated nickel silver • English Place-Name Society

EPO *Med.* erythropoietin

EPOC Eastern Pacific Oceanic Conference

EPOS ('iːpɒs) *Commerce* electronic point of sale (as in **EPOS terminal, EPOS system**)

e.p.p. (*or* **EPP**) *Physiol.* endplate potential

EPP *Computing* enhanced (*or* extended) parallel port • European People's Party • executive pension plan

EPPO European and Mediterranean Plant Protection Organization

EPPS *Chem.* 4-(2-hydroxyethyl)-1-piperazinepropanesulphonic acid

EPPT *Computing* European printer performance test

EPPV electropneumatic proportional valve

EPR electron paramagnetic resonance (as in **EPR spectroscopy**) • ethylene–propylene rubber

EPRDF Ethiopian People's Revolutionary Democratic Front

EPRI (USA) Electric Power Research Institute

EPROM ('iːprɒm) *Computing* erasable programmable read-only memory

e.p.s. earnings per share

EPS electric(al) power system • electrochemical photocapacitance spectroscopy • *Computing* Encapsulated PostScript • (USA) Environmental Protection Service

e.p.s.p. (*or* **EPSP**) *Physiol.* excitatory postsynaptic potential

EPSRC Engineering and Physical Sciences Research Council

EPSS electrical power supervision system

EPT *Med.* early pregnancy test • ethylene–propylene terpolymer (synthetic rubber) • excess profits tax

EPTA Expanded Programme of Technical Assistance (in the UN)

Ep. tm. *Law* Epiphany term

EPU European Payments Union

Epus. Episcopus (Latin: Bishop)

EPW earth-penetrator weapon • enemy prisoner of war

eq. equal • equate • equation • equator(ial) • equipment • equitable • equity • equivalent

Eq. Equerry

EQ educational quotient • *Electronics* equalize • equipment qualification • *Electronics* equivalence (as in **EQ gate**)

EQA *Commerce* European Quality Award

EQC external quality control

EQD Electrical Quality Assurance Directorate

EQDB equipment qualification data bank

EQ gate *Electronics* equivalence gate

EQI exhaust quality index

eqn equation

eqpt equipment

Equ *Astronomy* Equuleus

equil. equilibrium

equip. equipment

EQUIP (ɪ'kwɪp) equipment usage information programme

equiv. equivalent

er elder

e.r. echo ranging • effectiveness report • electronic reconnaissance • emergency request • emergency rescue • established reliability • external resistance

Er *Chem.*, *symbol for* erbium

ER Eastern Region (British Rail) • East Riding (Yorkshire) • Eduardus Rex (Latin: King Edward) • efficiency report • Elizabeth Regina (Latin: Queen Elizabeth) • *Med.* emergency room • *Cytology* endoplasmic reticulum • engine room • *Chem.* enhanced reactivation • *Computing* entity relationship • evaporation residue • *international civil aircraft marking for* Moldova • *British vehicle registration for* Peterborough

ERA *Baseball* earned run average • (USA) Economic Regulatory Administration • Education Reform Act (1988) • Electrical Research Association • electronic reading automation • Electronic Rentals Association • (USA) Emergency Relief Administration • engine-room artificer • *Computing* entity relationship attribute (as in **ERA diagram, ERA model**) • (USA) Equal Rights Amendment • European Ramblers' Association • Evangelical Radio Alliance

ERAB Energy Research Advisory Board

ERAMS Environmental Radiation Ambients Monitoring System

Eras. (Desiderius) Erasmus (1466–1536, Dutch scholar and humanist)

ERBM extended-range ballistic missile

ERBS Earth Radiation Budget Satellite

ERC Economic Research Council • Electronics Research Council • Employment Rehabilitation Centre • *Med.* endoscopic retrograde cholangiography • (USA) Energy Research Corporation

ERC & I Economic Reform Club and Institute

ERCB (USA) Energy Resources Conservation Board

ERCH *Physics* electron-resonance cyclotron heating

ERCP *Med.* endoscopic retrograde cholangiopancreatography

ERCS emergency response computer system

ERD elastic recoil detection • Emergency Reserve Decoration • emergency return device • environmental radiation data • *Physics* equivalent residual dose (of radiation) • extrapolated response dose

ERDA ('ɜːdə) *Electronics* elastic recoil detection analysis • (Australia) Electrical and Radio Development Association • (USA) Energy Research and Development Administration

ERDAF (USA) Energy Research and Development in Agriculture and Food

ERDE Engineering Research and Development Establishment • Explosives Research and Development Establishment

ERDF European Regional Development Fund

ERDIC Energy Research, Development, and Information Centre

ERDL Engineering Research and Development Laboratory

ERDS emergency response data system

ERE (Greece) Ethniki Rizospastiki Enosis (National Radical Union; political party) • extent of reaction effect

erec. erection

Erf (ɜːf) *Chem.* electrorheological fluid

ERFA European Radio Frequency Agency

ERG electrical resistance gauge • *Med.* electroretinogram

ERGOM European Research Group on Management

ergon. ergonomics

Eri *Astronomy* Eridanus

ERI (*or* **ER(I)**, **ER et I**) Edwardus Rex et Imperator (Latin: Edward King and Emperor)

ERIC (USA) Educational Resources Information Center • energy rate input controller

ERIM Environmental Research Institute of Michigan

ERIS Emergency Response Information System • *Astronautics* Exo-atmospheric Re-entry Vehicle Interceptor Subsystem

ERISA (e'rɪsə) (USA) Employee Retirement Income Security Act (1974)

Erit. Eritrea

Erl. Erläuterung (German: explanatory note)

ERL Energy Research Laboratory

ERLL *Computing* enhanced run length limited

erm. ermine

ERM *Finance* Exchange Rate Mechanism

Ernie (*or* **ERNIE**) ('ɜːnɪ) electronic random number indicating equipment (premium-bond computer)

ERO European Regional Organization of the International Confederation of Free Trade Unions

EROPA Eastern Regional Organization for Public Administration

EROS ('ɪərɒs) earth resources observation satellite • *Astronautics* experimental reflection orbital shot

e.r.p. effective radiated power

ERP *Med.* endoscopic retrograde pancreatography • *Biochem.* enzyme-releasing peptide • European Recovery Programme

ERPF *Med.* effective renal plasma flow

ERR energy release rate

erron. erroneous(ly)

ERS earnings-related supplement • earth resources satellite • emergency radio service • emergency response system • engine repair section • Ergonomics Research Society • (USA) Experimental Research Society

ERT (Greece) Elliniki Radiophonia Tileorasis (Hellenic National Radio and Television Institute) • excess retention tax

ERTS Earth Resources Technology Satellite • European Rapid Train System

ERU English Rugby Union

ERV English Revised Version (of the Bible) • *Med.* exercise radionuclide ventriculography

erw. (*or* **erweit.**) erweitert (German: enlarged *or* extended)

ERW enhanced radiation weapon

es. esempio (Italian: example)

e.s. eldest son • electrical sounding • electric starting • *Med.* enema saponis (Latin: soap enema)

e/s en suite (bathroom) (in property or accommodation advertisements)

Es *Chem.*, *symbol for* einsteinium

ES *British vehicle registration for* Dundee • Econometric Society • Education Specialist • *Physics* elastic scattering • *Chem.* electronic structure • *Physics* electron synchrotron • electrostatic • *international vehicle registration for* El Salvador • *Med.* endoscopic sphincterotomy • *Med.* endsystole (*or* -systolic) • energy spectrum (*or* spectra) • engine-sized (paper) • Entomological Society • *international civil aircraft marking for* Estonia • *Mining* exploratory shaft

ESA Ecological Society of America • electrostatic sector analyser • *Physics* energy sensor array • energy system analysis • Entomological Society of America • environmentally sensitive area • European Space Agency • Euthanasia Society of America • *Physics* excited state absorption

ESAAB (USA) Energy System Acquisition Advisory Board

ESANZ Economic Society of Australia and New Zealand

ESAR electronically steerable array radar

ESB electrical stimulation of the brain • (USA) Electricity Supply Board • electric storage battery • Empire State Building • English Speaking Board • environmental specimen bank

ESBA English Schools' Badminton Association

ESBBA English Schools' Basket Ball Association

ESBTC (USA) European Space Battery Test Center

esc. escompte (French: discount) • escutcheon

Esc *Computing* escape key • *symbol for* escudo (Portuguese monetary unit)

ESC Economic and Social Council (of the UN) • electronic structural correlator • (USA) Energy Security Corporation • Energy Study Centre (Netherlands) • English Stage Company • English Steel Corporation • Entomological Society of Canada • European Space Conference • *Computing* extended core storage

ESCA *Physical chem.* electron spectroscopy for chemical analysis • English Schools' Cricket Association • English Schools' Cycling Association

ESCAP Economic and Social Commission for Asia and the Pacific (UN agency; formerly ECAFE)

ESCC *Engineering* external stress corrosion cracking

ESCCD electrical short-circuit current decay

eschat. eschatology

ESCO ('ɛskəʊ) Educational, Scientific, and Cultural Organization (of the UN)

ESCOM ('ɛskɒm) Electricity Supply Commission of South Africa

ESCR external standard channel ratio

Esd. *Bible* Esdras

ESD echo-sounding device • electromagnetic shower detector • *Computing* electrostatic discharge • (USA) Environmental Sciences Division • Euratom Safeguards Directorate

Esda ('ɛzdə) electrostatic deposition (*or* document) analysis (as in **Esda test**)

ESDAC ('ɛsdæk) European Space Data Centre

ESDI ('ɛzdɪ) enhanced small-device interface

ESE east-southeast • *Physics* electron spin echo • engineers stores establishment

ESECA (USA) Energy Supply and Environmental Coordination Act

ESEF Electrotyping and Stereotyping Employers' Federation

ESES environmentally sound energy system

ESF European Science Foundation

ESG Education Support Grant • *Engineering* English Standard Gauge

ESH Environmental Safety and Health • equivalent standard hours • European Society of Haematology

ESI Electricity Supply Industry • environment sensitivity index

ESITB Electricity Supply Industry Training Board

Esk. Eskimo

e.s.l. expected significance level

ESL English as a second language

ESLAB ('ɛslæb) European Space Research Laboratory

ESMA Electrical Sign Manufacturers' Association

esn. (*or* **esntl**) essential

ESN educationally subnormal

ESNS educationally subnormal, serious

ESNZ Entomological Society of New Zealand

ESO Energy Services Operator • European Southern Observatory (Chile)

ESOC European Space Operations Centre (Germany)

ESOL ('iːsɒl) English for speakers of other languages

ESOMAR ('esəʊˌmɑː) European Society for Opinion Surveys and Market Research

ESOP ('iːsɒp) employee share-ownership plan • (USA) employee stock-option plan

ESOT ('iːsɒt) employee share-ownership trust

esp. especially • *Music* espressivo (Italian: expressively)

Esp. Espagne (French: Spain) • España (Spanish: Spain) • Esperanto

ESP electric submersible pump (*or* pumping) • *Physics* electron spin polarization • *Commerce* emotional selling proposition • English for specific (*or* special) purposes • *Computing* enhanced serial port • extrasensory perception

espg. espionage

espr. (*or* **espress.**) *Music* espressivo (Italian: expressively)

ESPRIT ('esprɪt) European strategic programme for research and development in information technology

Esq. esquire (in correspondence)

ESQA English Slate Quarries Association

esq° esquerdo (Portuguese: left)

ESR electric sunroof (in motor advertisements) • *Physics* electron-spin resonance • *Med.* erythrocyte sedimentation rate • *Physics* experimental storage ring

ESRC Economic and Social Research Council (formerly SSRC) • Electricity Supply Research Council

ESRD *Med.* end-stage renal disease

ESRF *Med.* end-stage renal failure

ESRIN ('ezrɪn) European Space Research Institute (Italy)

ESRO ('ezrəʊ) European Space Research Organization (now part of ESA)

ESRS European Society for Rural Sociology

ess. essence • *Pharmacol.* essentia (Latin: essence) • essential

Ess. Essex

ESS energy storage system

ESSA English Schools' Swimming Association • (USA) Environmental Science Services Administration (of the Department of Commerce)

ESSI ('esɪ) European Systems and Software Initiative

Esso ('esəʊ) Standard Oil (phonetic spelling of SO)

est. establish(ed) • *Law* estate • estimate(d) • estimation • estimator • estuary

Est. Established • Estonia(n)

EST earliest start time • (USA, Canada) Eastern Standard Time • electroshock therapy (*or* electric-shock treatment) • *Nuclear engineering* ellipsoidal shell tokamak • *international vehicle registration for* Estonia

estab. establish(ed) • establishment

ESTEC ('eztek) European Space Technology Centre (Netherlands)

estg estimating

Esth. *Bible* Esther

ESTI European Space Technology Institute

estn estimation

e.s.u. (*or* **ESU**) *Physics* electrostatic unit

ESU English-Speaking Union

ESV *Astronautics* earth satellite vehicle • *Engineering* emergency shutdown valve • *Med.* end-systolic (left ventricular) volume

ESWL *Med.* extracorporeal shock-wave lithotripsy

e.t. educational therapy • electric telegraph • engineering time • English text (*or* translation) • entertainment tax • exchange telegraph

Et *Chem., symbol for* ethyl (used in formulae)

ET (USA) Eastern Time • *international vehicle registration for* (Arab Republic of) Egypt • *Chem.* electron transfer • *Physics, Chem.* electron transition • *Nuclear engineering* elongated tokamak • *Med.* embryo transfer • emerging technology (in NATO) • Employment Training (for the unemployed) • emptying time • ephemeris time • *Physics* equation of time • *international civil aircraft marking for* Ethiopia • *airline flight code for* Ethiopian

Airlines • *Electronics* excitation transport • extraterrestrial • *British vehicle registration for* Sheffield

ETA *Physics* emanation thermal analysis • Entertainment Trades' Alliance • estimated time of arrival • European Teachers' Association • (*or* **Eta**; ˈɛtə) Euzkadi ta Askatsuna (Basque: Basque Nation and Liberty; nationalist organization)

Étab. Établissement (French: business establishment)

ETAC Education and Training Advisory Council

et al. et alibi (Latin: and elsewhere) • et alii (Latin: and others)

ETB English Tourist Board

etc. et cetera (Latin: and other things)

ETC Eastern Telegraph Company • European Translation Centre

ETCTA Electrical Trades Commercial Travellers' Association

ETD estimated time of departure • *Telecom.* extension trunk dialling

ETE estimated time en route • evacuation-time estimate • Experimental Tunnelling Establishment

ETF *Banking* electronic transfer of funds

eth. ether • ethical • ethics

Eth. Ethiopia(n) • Ethiopic

ETH (Switzerland) Eidgenössische Technische Hochschule (German: Federal Institute of Technology) • *international vehicle registration for* Ethiopia

ethnog. ethnography

ethnol. ethnological • ethnology

e.t.i. elapsed time indicator

ETI estimated time of interception • extraterrestrial intelligence

ETJC Engineering Trades Joint Council

e.t.k.m. every test known to man

ETMA English Timber Merchants' Association • European Television Magazine Association

e.t.o. estimated time off

ETO European Theatre of Operations (in World War II) • European Transport Organization

e.t.p. estimated turnaround (*or* turning) point

Etr. Etruscan

ETR engineering test reactor • estimated time of return • *Nuclear engineering* experimental thermonuclear (*or* test) reactor

ETS earth thermal storage • (USA) Educational Testing Service • *Biochem.* electron-transport system • estimated time of sailing • *Military* (USA) estimated time of separation (i.e. discharge from service) • *Med.* exercise thallium scintigraphy • expiration of time of service

ETSA Electricity Trust of South Australia • *Chem.* ethyl (trimethylsilyl)acetate

et seq. et sequens (Latin: and the following) • (*or* **et seqq.**) et sequentia (Latin: and those that follow)

ETSI European Telecommunications Standards Institute

e.t.s.p. entitled to severance pay

et sqq. et sequentes *or* sequentia (Latin: and the following)

ETSU Energy Technology Support Unit (of the Department of Energy)

ETTA English Table Tennis Association

ETTU European Table Tennis Union

ETU experimental test unit

ETUC European Trade Union Confederation

ETUI European Trade Union Institute

et ux. et uxor (Latin: and wife)

ETV educational television • engine test vehicle

ety. (*or* **etym., etymol.**) etymological • etymologist • etymology

Eu (ital.) *Physics, symbol for* Euler number • *Chem., symbol for* europium

EU *British vehicle registration for* Bristol • European Union (now often used in place of EC) • Evangelical Union • *Med.* excretory urography • experimental unit • (France, Spain) United States (French *États-Unis*; Spanish *Estados Unidos*)

EUA *Finance* European unit of account • *Med.* examination under anaesthesia • (France, Portugal, Spain) United States of America (French *États-Unis d'Amérique*; Spanish and Portuguese *Estados Unidos da* (Portuguese) or *de* (Spanish) *América*)

Euc. (*or* **Eucl.**) Euclid (*c.* 330–*c.* 275 BC, Greek mathematician)

EUCOM (ˈjuːkɒm) *Military* (USA) European Command

EUDISED (ˈjuːdɪˌsɛd) European Documentation and Information Service for Education

EUFTT European Union of Film and Television Technicians

eugen. eugenics

EUI energy utilization index · energy utilization intensity

EUL Everyman's University Library

EUM (Mexico, Spain) Mexico (Spanish *Estados Unidos Mexicanos*)

EUMETSAT ('juːmɛt,sæt) European Meteorological Satellite System

EUP English Universities Press

Eur. Europe(an)

EUR Esposizione Universale di Roma (Italian: Roman Universal Exhibition)

Euratom (juəˈrætəm) European Atomic Energy Community

Eureca (juəˈriːkə) *Astronautics* European retrievable carrier

Eur Ing European Engineer

Eurip. Euripides (*c.* 480–406 BC, Greek dramatist)

EURO ('juərəu) European Regional Office (of the FAO)

EUROCAE European Organization for Civil Aviation Electronics

EUROCEAN European Oceanographic Association

EUROCHEMIC European Company for the Chemical Processing of Irradiated Fuels

EUROCOM ('juərəu,kɒm) European Coal Merchants' Union

EUROFINAS Association of European Finance Houses

EuroJazz ('juərəu,jæz) European Community Youth Jazz Orchestra

EUROM ('juərɒm) European Federation for Optics and Precision Mechanics

EURONET ('juərəu,nɛt) European data-transmission network

EUROP European Railway Wagon Pool

EUROSPACE ('juərəu,speɪs) European Industrial Space Study Group

EUROTOX ('juərəu,tɒks) Comité européen permanent de recherches sur la protection de populations contre les risques de toxicité à long terme (French: European Standing Committee for the Protection of Populations against the Risks of Chronic Toxicity)

Eus. (*or* **Euseb.**) Eusebius of Caesarea (*fl.* 4th century AD, churchman and historian)

EUS Eastern United States

EUV *Physics* extreme ultraviolet

EUVE *Astronomy* Extreme Ultraviolet Explorer

EUW European Union of Women

ev. evangelisch (German: Protestant)

e.v. efficient vulcanization

eV eingetragener Verein (German: registered society) · *Physics, symbol for* electronvolt

EV *British vehicle registration for* Chelmsford · electric vehicle · English Version (of the Bible) · *Astronautics* entry vehicle · *Med.* equilibrium venography · *Med.* erythrocyte volume · *Accounting* expected value · *Image technol.* exposure value

EVA Electric Vehicle Association of Great Britain · Engineer Vice-Admiral · *Chem.* ethene and vinyl acetate (copolymers) · *Astronautics* extravehicular activity

evac. evacuate(d) · evacuation

eval. evaluate(d) · evaluation

evan. (*or* **evang.**) evangelical · evangelist

evap. evaporate(d) · evaporation · evaporator

EVBA *Physics* equivalent vector-boson approximation

evce evidence

EVCS *Astronautics* extravehicular communications system

EVFU *Computing* electronic vertical format unit

evg evening

EVGA *Computing* enhanced video graphics array

evid. evidence

EVN *Astronomy* European VLBI Network

evng evening

evol. evolution(ary) · evolutionist

EVR electronics video broadcasting system

EVT Educational and Vocational Training

EVW European Voluntary Workers

evy every

e.w. each way (betting)

EW *international civil aircraft marking for* Belarus · early warning · electric windows (in motor advertisements) · electronic warfare · (USA) enlisted woman · *British vehicle registration for* Peterborough

EWA (USA) Education Writers' Association

EWF Electrical Wholesalers' Federation

EWICS European Workshop for Industrial Computer Systems

EWL evaporative water loss

EWO Educational Welfare Officer · essential work order · European Women's Orchestra

EWP emergency war plan

EWR early-warning radar

EWRC European Weed Research Association

EWS emergency water supply • emergency welfare service

EWSF European Work Study Federation

ex. examination • examine(d) • examiner • example • excellent • except(ed) • exception • excess • exchange • exclude • excluding • exclusive • excursion • *Literary* excursus • execute(d) • executive • executor • exempt • exercise • export • express • extension • extra

Ex. Exeter • Exeter College, Oxford • *Bible* Exodus

EX *postcode for* Exeter • *international civil aircraft marking for* Kyrgyzstan • *British vehicle registration for* Norwich

EXAFS extended X-ray absorption fine structure

exag. exaggerate • exaggeration

exam. examination • examine(r)

examd examined

examg examining

examn examination

ex aq. *Med.* ex aqua (Latin: from water)

ex b. *Finance* ex bonus (without bonus)

exc. excellent • except(ed) • exception • exchange • excommunication • excudit (Latin: (he *or* she) engraved it; after an engraver's name) • excursion

Exc. Excellency

ex cap. *Finance* ex capitalization (without capitalization)

exch. exchange • (*or* **Exch.**) exchequer

excl. (*or* **exclam.**) exclamation • (*or* **exclam.**) exclamatory • exclude • excluding • exclusive

ex cp. *Finance* ex coupon (without the interest on the coupon)

exd examined

ex div. *Finance* ex dividend (without dividend)

exec. execute • execution • executive • executor

execx executrix

EXELFS electron energy-loss fine-structure spectroscopy

exempl. exemplaire (French: copy of a printed work)

exes expenses

Exet. Exeter College, Oxford

ex. g. (*or* **ex. gr.**) exempli gratia (Latin: for example)

exh. exhaust • exhibition

exhib. exhibit • (*or* **exhbn**) exhibition • exhibitioner

Ex.-Im. (*or* **Eximbank**) (USA) Export-Import Bank

ex int. *Banking* ex interest (without interest)

ex lib. ex libris (Latin: from the books (*or* library) of)

ex n. *Stock exchange* ex new (of shares)

EXNOR ('ɛksnɔ:) *Computing, Electronics* exclusive-NOR (as in **EXNOR gate**)

Exod. *Bible* Exodus

ex off. ex officio (Latin: by right of office)

Exon. Exonia (Latin: Exeter) • Exoniensis (Latin: (Bishop) of Exeter)

exor executor

EXOR ('ɛksɔ:) *Computing, Electronics* exclusive-OR (as in **EXOR gate**)

ex p. *Law* ex parte (Latin: on behalf of one party only)

exp *Maths., symbol for* exponential

exp. expand • expansion • expedition • expenses • experience • experiment(al) • expiration • expire(d) • export(ed) • exportation • exporter • express • expression • expurgated

EXP Exchange of Persons Office (in UNESCO) • *Meteorol.* expect(ed)

expdn expedition

exper. experimental

expl. explain • explanation • explanatory • explosion • explosive

exploit. exploitation

expn exposition

exp. o. experimental order

expr. express

ex-Pres. ex-President

expt experiment

exptl experimental

exptr exporter

expurg. expurgate

exr executor

exrx executrix

exs expenses

ext. extend • extension • extent • exterior • external(ly) • extinct • extra • extract • extraction • *Med.* extractum (Latin: extract) • extreme

EXTD *Meteorol.* extend
EXTEL ('ɛkstɛl) Exchange Telegraph (news agency)
EXTEND (ɪk'stɛnd) Exercise Training for the Elderly and/or Disabled
ext. liq. extractum liquidum (Latin: liquid extract)
extn extension
extr. extraordinary
extrad. extradition
exx examples • executrix

EY *British vehicle registration for* Bangor • *international civil aircraft marking for* Tajikistan
EYC European Youth Campaign
EYR East Yorkshire Regiment
Ez. *Bible* Ezra
EZ *British vehicle registration for* Belfast • easy • *international civil aircraft marking for* Turkmenistan
Ezek. *Bible* Ezekiel
Ezr. *Bible* Ezra

F

f *Numismatics* face value • *Music* fah (in tonic sol-fa) • *symbol for* femto- (prefix indicating 10^{-15}, as in **fm**, femtometre) • (*or* **fl**, **f:**) *Photog.* f-number (ratio of the focal length of a lens to its aperture, as in **f8**) • (ital.) *Physics, symbol for* focal length • foreign • (ital.) *Music* forte (Italian; loudly) • (ital.) *Physics, symbol for* frequency • (ital.) *Physical chem., symbol for* fugacity • (ital.) *Maths., symbol for* function (as in $f(x)$) • (ital.) *Biochem., symbol for* furanose • *Chem., Physics, indicating* electron state $l=3$ (where l is orbital angular momentum quantum number) • *indicating* the fifth vertical row of squares from the left on a chessboard

f. facing • fair • farthing • father • fathom(s) • feet • female • *Grammar* feminine • *Horse racing* filly • *Metallurgy* fine • flat • fluid • folio • following (page) • foot • for • forecastle • *Botany* form (in plant taxonomy) • formula • *Sport* foul • founded • franc(s) • freehold • from • furlong • furlough • *symbol for* guilder (monetary unit of the Netherlands)

F Fahrenheit (in °**F**, degree Fahrenheit) • *Logic, Maths.* false • *symbol for* farad(s) • (ital.) *Physics, symbol for* Faraday constant • fast (on a clock or watch regulator) • *British fishing port registration for* Faversham • (USA) fighter (specifying a type of military aircraft, as in **F-106**) • *Genetics, symbol for* filial generation (as in **F₁**, first generation) • *Chem., symbol for*

fluorine • (*or* **Fl**, **F:**) *Photog.* f-number (as in **F4**; *see under* f) • Fokker (aircraft) • (ital.) *Physics, symbol for* force • *symbol for* franc (monetary unit of various countries) • *international vehicle registration or international civil aircraft marking for* France • (ital.) *Thermodynamics, symbol for* Helmholtz function • *indicating* a musical note or key • *Astronomy, indicating* a spectral type

F. (*or* **F**) family • Father • fathom(s) • February • Federation • Fellow • female • *Grammar* feminine • ferrovia (Italian: railway) • *Med.* fiat (Latin: let it be made) • fiction • *Horse racing* filly • finance • *Metallurgy* fine • fleet • folio • foolscap • *Sport* foul • Frauen (German: women) • freddo (Italian: cold) • French • *Ecclesiast.* Frère (French: Brother) • Friday • frio (Portuguese *or* Spanish: cold)

5F *airline flight code for* Arctic Circle Air
6F *airline flight code for* Laker Airways
fª factura (Spanish: invoice)
f.a. *Slang* Fanny Adams (*or* fuck all; i.e. nothing) • fire alarm • first aid • *Commerce* free alongside • *Optics* free aperture • freight agent • friendly aircraft • fuel–air (ratio)
Fa. Faeroes • Firma (German: firm *or* business) • Florida
FA *Statistics* factor analysis • Factory Act • Faculty of Actuaries • family allowance • *Slang* Fanny Adams (*or* fuck all; i.e. nothing) • farm adviser • *Chem.* fatty

acid • field activities • field allowance • field ambulance • *Military* field artillery • filtered air • Finance Act • financial adviser • fine art • *Immunol.* fluorescent antibody • *Building trades* fly (*or* fuel) ash • *Biochem.* folic acid • *Biochem.* folinic acid • Football Association • freight agent • *Nuclear engineering* fuel assembly • *Metallurgy* furnace annealing • *British vehicle registration for* Stoke-on-Trent

F-A *Chem.* fenitrothion in acetone

f.a.a. *Marine insurance* free of all averages

FAA (USA) Federal Aviation Administration • Fellow of the Australian Academy (of Science) • Film Artistes' Association • Fleet Air Arm • *Chem.* formalin acetic alcohol • *Biochem.* free amino acid

FAAAS Fellow of the American Academy of Arts and Sciences • Fellow of the American Association for the Advancement of Science

FAAC Food Additives and Contaminants Committee

FAAO Fellow of the American Academy of Optometry

FAAP Fellow of the American Academy of Pediatrics

FAARM Fellow of the American Academy of Reproductive Medicine

FAAV Fellow of the Central Association of Agricultural Valuers

fab. fabric • fabricate • *Commerce* fabrication (French: make)

f.a.b. first aid box

f. à b. *Commerce* franco à bord (French: free on board)

FAB *Physics, Electronics* fast-atom bombardment (*or* beam) • Flour Advisory Bureau • French–American–British • fuel-air bomb

fabbr. fabbrica (Italian: factory)

FABMDS (USA) field army ballistic missile defense system

FABP *Biochem.* fatty-acid-binding protein

fabr. fabricate • fabrication

Fab. Soc. Fabian Society

fabx. fire alarm box

fac. façade • facial • facility • facsimile • factor • factory • (*or* **Fac.**) faculty

f.a.c. (as) fast as can (be) • *Marine insurance* franc d'avarie commune (French: free of general average)

FAC Federation of Agricultural Cooperatives • *Military* forward air controller

FACA Fellow of the American College of Anesthesiologists

FACC Fellow of the American College of Cardiology • (USA) Ford Aerospace and Communication Corporation

FACCA Fellow of the Association of Certified and Corporate Accountants (now the Chartered Association of Certified Accountants; *see* FCCA)

FACCP Fellow of the American College of Chest Physicians

FACD Fellow of the American College of Dentistry

FACE Fellow of the Australian College of Education • (feis) field artillery computer equipments

FACEM Federation of the Associations of Colliery Equipment Manufacturers

FACerS Fellow of the American Ceramic Society

facet. facetious

FACG Fellow of the American College of Gastroenterology

facil. facility

FACMTA (USA) Federal Advisory Council on Medical Training

FACOG Fellow of the American College of Obstetricians and Gynecologists

FACOM Fellow of the Australian College of Occupational Medicine

FACP Fellow of the American College of Physicians

FACR Fellow of the American College of Radiology

FACRM Fellow of the Australian College of Reproductive Medicine

facs. facsimile

FACS Fellow of the American College of Surgeons • *Immunol.* fluorescence-activated cell sorter

facsim. facsimile

fact. (*or* **fact°**) factura (Spanish: invoice)

Fact (*or* **FACT**) (fækt) Federation Against Copyright Theft • fully automatic compiler-translator

f.a.d. free air delivered

FAD *Biochem.* flavin adenine dinucleotide

FADEC *Computing* fully authorized digital engine control

FADO Fellow of the Association of Dispensing Opticians

FAE *Military* fuel-air explosive(s)

Faer. Faeroe Islands

FAeSI Fellow of the Aeronautical Society of India

FAGO Fellowship in Australia in Obstetrics and Gynaecology

FAGS Federation of Astronomical and Geophysical Services • Fellow of the American Geographical Society

Fah. (*or* **Fahr.**) Fahrenheit

FAHA Fellow of the Australian Academy of the Humanities

FAI Fédération aéronautique internationale (French: International Aeronautical Federation) • Football Association of Ireland • fresh-air inlet

FAIA Fellow of the American Institute of Architects • Fellow of the Association of International Accountants • Fellow of the Australian Institute of Advertising

FAIAA Fellow of the American Institute of Aeronautics and Astronautics

FAIAS Fellow of the Australian Institute of Agricultural Science

FAIB Fellow of the Australian Institute of Builders

FAIBiol Fellow of the Australian Institute of Biology

FAIC Fellow of the American Institute of Chemists

FAIE Fellow of the Australian Institute of Energy

FAIEx Fellow of the Australian Institute of Export

FAIFST Fellow of the Australian Institute of Food Science and Technology

FAII Fellow of the Australian Insurance Institute

FAIM Fellow of the Australian Institute of Management

FAIP Fellow of the Australian Institute of Physics

Fak. Faktura (German: invoice)

FAK *Shipping* freights all kinds

Falk. I. (*or* **Falk. Is.**) Falkland Islands

FALN (Puerto Rico) Fuerzas Armadas de Liberación Nacional (Spanish: Armed Forces of National Liberation)

fam. familiar • family

FAM Free and Accepted Masons

FAMA Fellow of the American Medical Association • Fellow of the Australian Medical Association • Foundation for Mutual Assistance in Africa

FAME *Chem.* fatty-acid methyl ester

FAMEME Fellow of the Association of Mining Electrical and Mechanical Engineers

FAMI Fellow of the Australian Marketing Institute

FAmNucSoc Fellow of the American Nuclear Society

FAMS Fellow of the Ancient Monuments Society • Fellow of the Indian Academy of Medical Sciences

f & a fore and aft

F & AP *Insurance* fire and allied perils

F & C full and change (tides)

f & d fill and drain • (*or* **F & D**) freight and demurrage

f & f fixtures and fittings

F and Gs (*or* **F&Gs**) *Bookbinding* folded and gathered pages

F & M foot and mouth disease

f & t (*or* **F & T**) *Insurance* fire and theft

FANY (*or* **Fany, Fanny**) ('fænɪ) First Aid Nursing Yeomanry

f.a.o. finish all over • for the attention of

FAO Fleet Accountant Officer • Food and Agriculture Organization (of the UN)

FAP (USA) Family Assistance Program • first aid post • *Ethology* fixed action pattern • Força Aérea Portuguesa (Portuguese Air Force) • (*or* **f.a.p.**) *Marine insurance* franc d'avarie particulière (French: free of particular average)

FAPA Fellow of the American Psychiatric Association • Fellow of the American Psychological Association • Fellow of the Association of Authorized Public Accountants

FAPHA Fellow of the American Public Health Association

FAPS Fellow of the American Phytopathological Society

f.a.q. *Commerce* fair average quality • *Commerce* free alongside quay

FAQ (fæk) *Computing* frequently asked question

f.a.q.s. fair average quality of season

far. farriery • farthing

FAR (USA) Federal Aviation Regulation(s) • *Insurance* free (of claim) for accident reported • front arm rests (in motor advertisements)

FArborA Fellow of the Arboricultural Association

FARE Federation of Alcoholic Rehabilitation Establishments

FARELF Far East Land Forces

f.a.s. firsts and seconds • *Commerce* free alongside ship

FAS Federation of American Scientists •

Fellow of the Anthropological Society • Fellow of the Antiquarian Society • *Med.* fetal alcohol syndrome • *Commerce* free alongside ship

FASA Fellow of the Australian Society of Accountants

FASB (USA) Financial Accounting Standards Board

fasc. *Anatomy, Printing* fasciculus

FASc Fellow of the Indian Academy of Sciences

FASCE Fellow of the American Society of Civil Engineers

FASE Fellow of the Antiquarian Society, Edinburgh

FASEB Federation of American Societies for Experimental Biology

FASI Fellow of the Architects' and Surveyors' Institute

FASS Federation of the Associations of Specialists and Subcontractors

FASSA Fellow of the Academy of Social Sciences in Australia

FAST (fɑːst) factor analysis system • fast automatic shuttle transfer • Federation Against Software Theft • first atomic ship transport • forecasting and assessment in science and technology (in the EU)

fastnr fastener

FAT *Computing* file allocation table

fath. fathom

FATIS Food and Agriculture Technical Information Service (of the OEEC)

FAU Friends' Ambulance Unit

FAusIMM Fellow of the Australasian Institute of Mining and Metallurgy

fav. favour • favourite

FAVO Fleet Aviation Officer

FAWA Federation of Asian Women's Associations

fax (*or* **FAX**) (fæks) facsimile transmission • fuel-air explosion

f.b. flat bar • fog bell • freight bill • *Sport* fullback

FB *British vehicle registration for* Bristol • Fenian Brotherhood • *Physics* film badge (for radiation protection) • fire brigade • fisheries (*or* fishery) board • *Railways* flat bottom • *Chemical engineering* fluidized bed • flying boat • *Med.* foreign body • Forth Bridge • Free Baptist

F-B full-bore

FBA Farm Buildings Association • (USA) Federal Bar Association • Federation of British Artists • Federation of British

Astrologers • Fellow of the British Academy • fluorescent brightening agent (used in detergents) • Freshwater Biological Association

FBAA Fellow of the British Association of Accountants and Auditors

f'ball football

f.b.c. *Insurance* fallen building clause

FBC *Chemical engineering* fluidized-bed combustion (*or* combustor)

FBCM Federation of British Carpet Manufacturers

FBCO Fellow of the British College of Ophthalmic Opticians (*or* Optometrists)

FBCP Fellow of the British College of Physiotherapists

FBCS Fellow of the British Computer Society

f.b.c.w. *Insurance* fallen building clause waiver

fbd freeboard

FBEA Fellow of the British Esperanto Association

FBEC(S) Fellow of the Business Education Council (Scotland)

FBFM Federation of British Film Makers

FBH fire brigade hydrant

FBHI Fellow of the British Horological Institute

FBHS Fellow of the British Horse Society

FBHTM Federation of British Hand Tool Manufacturers

FBI (USA) Federal Bureau of Investigation

FBIBA Fellow of the British Insurance Brokers' Association

FBID Fellow of the British Institute of Interior Design

FBIM Fellow of the British Institute of Management

FBINZ Fellow of the Bankers' Institute of New Zealand

FBIPP Fellow of the British Institute of Professional Photography

FBIS Fellow of the British Interplanetary Society

FBL flight-by-light (aircraft control system)

FBM fleet ballistic missile

FBMA Food and Beverage Managers' Association

FBO (*or* **f.b.o.**) for the benefit of

FBOA Fellow of the British Optical Association

FBOU Fellow of the British Ornithologists' Union

FBP *Chem.* final boiling point · *Biochem.* folate-binding protein

FBPS Fellow of the British Phrenological Society

FBPsS Fellow of the British Psychological Society

fbr. fibre

FBR *Nuclear engineering* fast breeder reactor · *Chemical engineering* fluidized-bed reactor

FBRAM Federation of British Rubber and Allied Manufacturers

fbro. febrero (Spanish: February)

FBS Fellow of the Botanical Society · *Med.* fetal bovine serum · *Military* forward-based system

FBSC Fellow of the British Society of Commerce

FBSE Fellow of the Botanical Society, Edinburgh

FBSM Fellow of the Birmingham School of Music

FBSS *Med.* failed back-surgery syndrome

FBT (France) Fédération des bourses du travail (Federation of Labour Exchanges; trade union) · fringe benefit tax

FBu *symbol for* Burundi franc (monetary unit)

FBU Fire Brigades Union

FBW *Aeronautics* fly-by-wire

f.b.y. future budget year

f.c. *Baseball* fielder's choice · fin courant (French: at the end of this month) · *Printing* follow copy · for cash

Fc *Immunol.* crystallizable fragment (of an immunoglobulin)

FC (Australia) Federal Cabinet · *Freemasonry* fellow craft · Fencing Club · ferrocarril (Spanish: railway) · fidei commissum (Latin: bequeathed in trust) · fieri curavit (Latin: the donor directed this to be done; on gravestones and other monuments) · fifth column · Fighter Command · fire cock · Fishmongers' Company · Football Club · Forestry Commission · *Med.* free cholesterol · Free Church · fuel cell · *Med.* full course (of therapy) · *Meteorol.* funnel cloud · *British vehicle registration for* Oxford

FCA (USA) Farm Credit Administration · Federation of Canadian Artists · Fellow of the Institute of Chartered Accountants (in England and Wales *or* in Ireland)

FCAATSI (Australia) Federal Council for the Advancement of Aborigines and Torres Strait Islanders

FCA(Aust) Fellow of the Institute of Chartered Accountants in Australia

FCAI Fellow of the New Zealand Institute of Cost Accountants

FC-AL *Computing* fibre-channel arbitrated loop

FCAnaes Fellow of the College of Anaesthetists

fcap (*or* **f/cap**, **f'cap**) foolscap

FCAR *Insurance* free of claim for accident reported

FCASI Fellow of the Canadian Aeronautics and Space Institute

FCAST *Meteorol.* forecast

FCB *Computing* file control block

FCBA Federal Communications Bar Association · Fellow of the Canadian Bankers' Association

FCBSI Fellow of the Chartered Building Societies Institute

FCC (*or* **f.c.c.**) *Crystallog.* face-centred cubic · (USA) Federal Communications Commission · Federal Council of Churches · first-class certificate · *Chemical engineering* fluid catalytic cracking

FCCA Fellow of the Chartered Association of Certified Accountants

FCCEA Fellow of the Commonwealth Council for Educational Administration

FCCEd Fellow of the College of Craft Education

FCCS Fellow of the Corporation of Secretaries (formerly Certified Secretaries)

FCCSET (USA) Federal Coordinating Council on Science, Engineering, and Technology

FCCT Fellow of the Canadian College of Teachers

FCD First Chief Directorate (of the KGB)

FCDA (USA) Federal Civil Defense Administration

FCEC Federation of Civil Engineering Contractors

FCFC Free Church Federal Council

FCFI Fellow of the Clothing and Footwear Institute

fcg facing

FCGB Forestry Committee of Great Britain

FCGI Fellow of the City and Guilds London Institute

FChS Fellow of the Society of Chiropodists

FCI Fédération cynologique internationale (French: International Federation of Kennel Clubs) • Fellow of the Institute of Commerce

FCIA Fellow of the Canadian Institute of Actuaries • Fellow of the Corporation of Insurance Agents • Foreign Credit Insurance Association

FCIArb Fellow of the Chartered Institute of Arbitrators

FCIB Fellow of the Chartered Institute of Bankers • Fellow of the Corporation of Insurance Brokers

FCIBS Fellow of the Chartered Institute of Bankers in Scotland

FCIBSE Fellow of the Chartered Institution of Building Services Engineers

FCIC (USA) Federal Crop Insurance Corporation • Fellow of the Chemical Institute of Canada

FCII Fellow of the Chartered Insurance Institute

FCILA Fellow of the Chartered Institute of Loss Adjusters

FCIM Fellow of the Chartered Institute of Marketing

FCIOB Fellow of the Chartered Institute of Building

FCIPA Fellow of the Chartered Institute of Patent Agents (now CPA, Chartered Patent Agent)

FCIPS Fellow of the Chartered Institute of Purchasing and Supply

FCIS Fellow of the Institute of Chartered Secretaries and Administrators (formerly Chartered Institute of Secretaries)

FCISA Fellow of the Chartered Institute of Secretaries and Administrators (Australia)

FCIT Fellow of the Chartered Institute of Transport • *Basketball* Four Countries International Tournament

FCMA Fellow of the Chartered Institute of Management Accountants • Fellow of the Institute of Cost and Management Accountants

FCMS Fellow of the College of Medicine and Surgery

FCMSA Fellow of the College of Medicine of South Africa

FCNA Fellow of the College of Nursing, Australia

fco *Commerce* franco (French: free of charge)

f. co. *Printing* fair copy

FCO Farmers' Central Organization • fire-control officer • Foreign and Commonwealth Office

FCOG(SA) Fellow of the South African College of Obstetrics and Gynaecology

FCollP Fellow of the College of Preceptors

FCOphth Fellow of the College of Ophthalmologists

FCOT Fellow of the College of Occupational Therapists

fcp. foolscap

FCP *Engineering* fatigue-crack propagation • Fellow of the College of Clinical Pharmacology • Fellow of the College of Preceptors • *Physics* Franck–Condon pumping (*or* principle)

FCPA Fellow of the Canadian Psychological Association • (USA) Foreign Corrupt Practices Act

FCPO Fleet Chief Petty Officer

FCPS Fellow of the College of Physicians and Surgeons

FCP(SoAf) Fellow of the College of Physicians, South Africa

FCPSO(SoAf) Fellow of the College of Physicians and Surgeons and Obstetricians, South Africa

fcs (*or* **Fcs**) francs

f.c.s. (*or* **f.c. & s.**) *Insurance* warranted free of capture, seizure, arrest, detainment, and the consequences thereof

FCS Federation of Conservative Students • Fellow of the Chemical Society (now part of the Royal Society of Chemistry; *see under* FRSC) • *Med.* fetal calf serum

FCSD Fellow of the Chartered Society of Designers

FCSP Fellow of the Chartered Society of Physiotherapy

f.c.s.r.c.c. *Insurance* warranted free of capture, seizure, arrest, detainment, and the consequences thereof, and damage caused by riots and civil commotions

FCSSA Fellow of the College of Surgeons, South Africa

FCST (USA) Federal Council for Science and Technology • Fellow of the College of Speech Therapists

FCT Fellow of the Association of Corporate Treasurers

FCTB Fellow of the College of Teachers of the Blind

FCTU Federation of Associations of Catholic Trade Unionists

fcty factory

FCU fighter control unit

fd field · fiord · ford · forward · found(ed) · fund

f.d. flight deck · focal distance · (or **f/d**) free delivery · free discharge · free dispatch

FD *British vehicle registration for* Dudley · *Physics* Fermi–Dirac (as in **FD statistics**) · Fidei Defensor (Latin: Defender of the Faith (Henry VIII); e.g. on British coins) · *Computing* finite difference · fire department · fleet duties · *British fishing port registration for* Fleetwood · *Med.* folate deficiency · (or **f/d**) free-delivered (at docks) · free delivery · Free Democrat

FDA Association of First Division Civil Servants (formerly First Division Association) · (USA) Food and Drug Administration

FD&C (USA) Food, Drug, and Color Regulations (as in **FD&C colour**)

FDC (USA) Fire Detection Center · *Philately* first-day cover · (or **f.d.c.**) *Numismatics* fleur de coin (French: mint condition)

FDDI *Computing* fibre distributed data interface (high-speed local area network system)

FDF Food and Drink Federation

fdg funding

FDG *Biochem.* fluoro-2-deoxy-D-glucose

FDHO Factory Department, Home Office

FDI Fédération dentaire internationale (French: International Dental Federation)

FDIC (USA) Federal Deposit Insurance Corporation · Food and Drink Industries Council

FDIF Fédération démocratique internationale des femmes (French: Women's International Democratic Federation)

FDM *Computing* finite-difference method · *Telecom.* frequency-division multiplexing

FDO Fleet Dental Officer · *Taxation* for declaration (purposes) only

FDP (Germany) Freie Demokratische Partei (Free Democratic Party)

fdr founder

FDR Franklin Delano Roosevelt (1882–1945, US president 1933–45)

fdry foundry

FDS Fellow in Dental Surgery · *Physics* Fermi–Dirac–Sommerfeld (in **FDS law**) · *Aeronautics* flight-director system

FDSRCPSGlas Fellow in Dental Surgery of the Royal College of Physicians and Surgeons of Glasgow

FDSRCS Fellow in Dental Surgery of the Royal College of Surgeons of England

FDSRCSE Fellow in Dental Surgery of the Royal College of Surgeons of Edinburgh

f.e. first edition · for example

Fe *Chem., symbol for* iron (Latin *ferrum*)

FE Far East · *British fishing port registration for* Folkestone · foreign editor · further education · *British vehicle registration for* Lincoln

FEA (USA) Federal Energy Administration · (USA) Federal Executive Association · Fédération internationale pour l'éducation artistique (French: International Federation for Art Education) · *Maths.* finite-element analysis

FEAF Far East Air Force

FEANI Fédération européenne d'associations nationales d'ingenieurs (French: European Federation of National Associations of Engineers)

feb. (or **febb.**) febbraio (Italian: February)

Feb February

FEB (USA) Fair Employment Board · functional electronic block

FEBS Federation of European Biochemical Societies

Feby February

fec. fecit (Latin: (he or she) made it; on works of art next to the artist's name)

FEC (USA) Federal Election Commission · First Edition Club · Fondation européenne de la culture (French: European Cultural Foundation) · Foreign Exchange Certificate (tourist currency used in China; *compare* RMB)

FECB Foreign Exchange Control Board

FECDBA Foreign Exchange and Currency Deposit Brokers' Association

FECI Fellow of the Institute of Employment Consultants

Fed. (or **fed.**) Federal(ist) · Federated · Federation

FED Federal Reserve System

FEDC Federation of Engineering Design Consultants

FedEx ('fed,eks) (USA) *Trademark* Federal Express

FEER *Banking* fundamental equilibrium exchange rate

FEF Far East Fleet

FEI Fédération équestre internationale (French: International Equestrian Federation) • (USA) Financial Executive Institute

FEIDCT Fellow of the Educational Institute of Design Craft and Technology

FEIS Fellow of the Educational Institute of Scotland

FEL free-electron laser

FELF Far East Land Forces

Fell. Fellow

FeLV *Microbiol.* feline leukaemia virus

fem. female • feminine

f.é.m. force électromotrice (French: electromotive force; emf)

FEM *Physics* field-emission microscope (or microscopy) • *Maths.* finite-element method

FEMA (USA) Federal Emergency Management Agency

FEMO *Chem., Physics* free-electron molecular orbital (in **FEMO theory**)

fenc. fencing

FEng Fellow of the Fellowship of Engineering • Fellow of the Royal Academy of Engineering

FENSA ('fɛnsə) Film Entertainments National Service Association

FEO Fleet Engineer Officer

FEOGA Fonds européen d'orientation et de garantie agriculturel (French: European Agricultural Guidance and Guarantee Fund; in the EU)

FEP fluorinated ethene propene (a plastic)

FEPA (USA) Fair Employment Practices Act

FEPC (USA) Fair Employment Practices Committee

FEPEM Federation of European Petroleum Equipment Manufacturers

Fer. Fermanagh

FERA (USA) Federal Emergency Relief Administration

FERC (USA) Federal Energy Regulatory Commission

FERDU Further Education Review and Development Unit

Ferm. Fermanagh

ferr. ferrovia (Italian: railway)

ferv. *Pharmacol.* fervens (Latin: boiling)

FES Federation of Engineering Societies • Fellow of the Entomological Society • Fellow of the Ethnological Society • *Fencing* foil, épée, and sabre • F(rederick) E(dwin) Smith, Earl of Birkenhead (1872–1930, British statesman and lawyer)

Fest. Festival

FET ('ɛf 'i: 'ti: or fɛt) (USA) federal estate tax • (USA) federal excise tax • field-effect transistor • fossil-energy technology

FEU Further Education Unit

feud. feudal(ism)

fev. fevereiro (Portuguese: February)

fév. février (French: February)

FEV *Med.* forced expiratory volume

FEX *US Navy* fleet exercise

ff (or **ff.**) fecerunt (Latin: (they) made it; on works of art next to the artists' names) • folios • following (pages, lines, etc.) • forms • (ital.) *Music* fortissimo (Italian; very loudly) • *Nautical* thick fog

f.f. fixed focus

f/f fully fitted (kitchen) (in property advertisements) • fully furnished (in property advertisements)

FF *British vehicle registration for* Bangor • Felicissimi Fratres (Latin: Most Fortunate Brothers) • Fellows • Fianna Fáil (Irish Gaelic: warriors of Ireland; Irish political party) • *Military* field force • (USA) Ford Foundation • *Military* frontier force• fully fitted (kitchen) (in property advertisements)• fully furnished (in property advertisements)

f.f.a. free foreign agency • *Commerce* free from alongside (ship)

FFA (Scotland) Fellow of the Faculty of Actuaries • Fellow of the Institute of Financial Accountants • *Biochem.* free fatty acid • Future Farmers of America

FFARACS Fellow of the Faculty of Anaesthetists of the Royal Australasian College of Surgeons

FFARCS Fellow of the Faculty of Anaesthetists of the Royal College of Surgeons of England

FFARCSI Fellow of the Faculty of Anaesthetists of the Royal College of Surgeons in Ireland

FFAS Fellow of the Faculty of Architects and Surveyors

FFA(SA) Fellow of the Faculty of Anaesthetics (South Africa)

FFB Fellow of the Faculty of Building

FFC Foreign Funds Control

FFCM Fellow of the Faculty of Community Medicine

FFCMI Fellow of the Faculty of Community Medicine of Ireland

FFD Fellow of the Faculty of Dental Surgeons

FFDRCSI Fellow of the Faculty of Dentistry of the Royal College of Surgeons in Ireland

fff (ital.) *Music* fortississimo (Italian; as loudly as possible)

FFF Free French Forces

FFHC Freedom from Hunger Campaign

FFHom Fellow of the Faculty of Homoeopathy

FFI Finance for Industry • (*or* **f.f.i.**) free from infection • French Forces of the Interior

FFJ Franciscan Familiar of Saint Joseph

FFL Forces françaises libres (Free French Forces)

ffly faithfully

FFOM Fellow of the Faculty of Occupational Medicine

FFOMI Fellow of the Faculty of Occupational Medicine of Ireland

FFPath, RCPI Fellow of the Faculty of Pathologists of the Royal College of Physicians of Ireland

FFPHM Fellow of the Faculty of Public Health Medicine

FFPM Fellow of the Faculty of Pharmaceutical Medicine

FFPS Fauna and Flora Preservation Society

FFr (*or* **Ffr**) French franc

FFRR full-frequency range recording

FFS (Algeria) Front des Forces Socialistes (French: Front of Socialist Forces)

f.f.s.s. full-frequency stereophonic sound

FFT *Computing* fast Fourier transform • *Computing* final form text

FFV (USA) First Families of Virginia

FFWM free-floating wave meter

ffy faithfully

FFY Fife and Forfar Yeomanry

fg fog

f.g. *Sport* field goal • fine grain • *Commerce* fully good

FG *airline flight code for* Ariana Afghan Airlines • *British vehicle registration for* Brighton • Federal Government • Fine Gael (Irish Gaelic: tribe of the Gaels;

Irish political party) • fire guard • *Engineering* flue gas (as in **FGD**, flue-gas desulphurization) • *Meteorol.* fog • foot guards • frais généraux (French: overheads) • full gilt

FGA Fellow of the Gemmological Association • (*or* **f.g.a.**) foreign general average

FGCH full gas central heating (in property advertisements)

FGCM Fellow of the Guild of Church Musicians • field general court martial

FGDS (France) Fédération de la gauche démocrate et socialiste (Federation of the Democratic and Socialist Left)

f.g.f. *Commerce* fully good, fair

FGI Fédération graphique internationale (French: International Graphical Federation) • Fellow of the Institute of Certified Grocers

Fgn Foreign

FGO Fleet Gunnery Officer

Fg. Off. Flying Officer

FGS Fellow of the Geological Society

FGSM Fellow of the Guildhall School of Music and Drama

fgt freight

FGT (USA) federal gift tax

FGTB Fédération générale du travail de Belgique (French: Belgian General Federation of Labour)

f.h. *Med.* fiat haustus (Latin: make a draught) • fog horn • fore (*or* forward) hatch

FH *Med.* fetal heart • *British fishing port registration for* Falmouth • *Med.* family history • *Med.* Ficoll-Hypaque (used in radiography and gradient separation) • field hospital • fire hydrant • *Rugby* fly-half • *British vehicle registration for* Gloucester

F/H freehold

FHA (USA) Farmers' Home Administration • (USA) Federal Housing Administration

FHAS Fellow of the Highland and Agricultural Society of Scotland

FHB (*or* **f.h.b.**) *Colloquial* family hold back

FHCIMA Fellow of the Hotel Catering and Institutional Management Association

FHH *Med.* fetal heart heard

FHI Fédération haltérophile internationale (French: International Weightlifting Federation)

FHLB (USA) Federal Home Loan Bank

FHLBA (USA) Federal Home Loan Bank Administration

FHLBB (USA) Federal Home Loan Bank Board

fhld (*or* **f'hold**) freehold

FHLMC (USA) Federal Home Loan Mortgage Corporation

FHM (ital.) For Him Magazine

FHNH *Med.* fetal heart not heard

fhp friction horsepower

FHR (Australia) Federal House of Representatives • *Med.* fetal heart rate

FHS Fellow of the Heraldry Society

FHSA Family Health Services Authority

FHSM Fellow of the Institute of Health Services Management

FHWA (USA) Federal Highway Administration

f.i. for instance • *Commerce* free in

FI Faeroe Islands • Falkland Islands • Fiji Islands • fire insurance • *Engineering* flow injection • *airline flight code for* Icelandair • *Irish vehicle registration for* Tipperary (N Riding)

FIA (USA) Federal Insurance Administration • Fédération internationale de l'automobile (French: International Automobile Federation) • Fellow of the Institute of Actuaries • *Chem.* flame ionization analyser • (*or* **f.i.a.**) *Commerce* full interest admitted

FIAA Fédération internationale athlétique d'amateur (French: International Amateur Athletic Federation) • Fellow of the Institute of Actuaries of Australia

FIAA&S Fellow of the Incorporated Association of Architects and Surveyors

FIAAS Fellow of the Institute of Australian Agricultural Science

FIAB Fédération internationale des associations de bibliothécaires (French: International Federation of Library Associations and Institutions) • Fellow of the International Association of Bookkeepers

FIAF Fédération internationale des archives du film (French: International Federation of Film Archives)

FIAgrE Fellow of the Institution of Agricultural Engineers

FIAI Fédération internationale des associations d'instituteurs (French: International Federation of Teachers' Associations) • Fellow of the Institute of Industrial and Commercial Accountants

FIAJ Fédération internationale des auberges de la jeunesse (French: International Youth Hostel Federation)

FIAL Fellow of the International Institute of Arts and Letters

FIAM Fellow of the Institute of Administrative Management • Fellow of the International Academy of Management

FIAP Fédération internationale de l'art photographique (French: International Federation of Photographic Art) • Fellow of the Institution of Analysts and Programmers

FIAPF Fédération internationale des associations de producteurs de films (French: International Federation of Film Producers' Associations)

FIArbA Fellow of the Institute of Arbitrators of Australia

FIASc Fellow of the Indian Academy of Sciences

Fiat (*or* **FIAT**) ('fiːət, -æt) Fabbrica Italiana Automobili Torino (Italian Motor Works in Turin)

fib. *Med.* fibula

f.i.b. free into barge • free into bond • free into bunker

FIBA Fédération internationale de basketball amateur (French: International Amateur Basketball Federation) • Fellow of the Institute of Business Administration (Australia)

FIBD Fellow of the Institute of British Decorators

FIBiol Fellow of the Institute of Biology

FIBOR Frankfurt Inter-Bank Offered Rate

FIBOT Fair Isle Bird Observatory Trust

FIBP Fellow of the Institute of British Photographers

FIBScot Fellow of the Institute of Bankers in Scotland

FIBST Fellow of the Institute of British Surgical Technicians

FIBT Fédération internationale de bobsleigh et de tobogganning (French: International Bobsleighing and Tobogganning Federation)

fic. fiction(al) • fictitious

FIC Falkland Islands Company • Fellow of Imperial College, London • frequency interference control

FICA ('fiːkə) (USA) Federal Insurance Contributions Act • Fellow of the Commonwealth Institute of Accountants • Food Industries Credit Association

FICAI Fellow of the Institute of Chartered Accountants in Ireland

FICC Fédération internationale de camping et de caravanning (French: International Federation of Camping and Caravanning) • Fédération internationale des ciné-clubs (French: International Federation of Film Societies)

FICCI Federation of Indian Chambers of Commerce and Industry

FICD Fellow of the Indian College of Dentists • Fellow of the Institute of Civil Defence

FICE Fellow of the Institution of Civil Engineers

FICeram Fellow of the Institute of Ceramics

FICFor Fellow of the Institute of Chartered Foresters

FIChemE Fellow of the Institution of Chemical Engineers

FICI Fellow of the Institute of Chemistry of Ireland

FICM Fellow of the Institute of Credit Management

FICMA Fellow of the Institute of Cost and Management Accountants

FICS Fellow of the Institution of Chartered Shipbrokers • Fellow of the International College of Surgeons

FICSA Federation of International Civil Servants' Associations

fict. fictilis (Latin: made of pottery • fiction(al) • fictitious

FICT Fellow of the Institute of Concrete Technologists

FICW Fellow of the Institute of Clerks of Works of Great Britain

fid. fidelity • fiduciary

FID Falkland Island Dependencies • Fédération internationale de documentation (French: International Federation for Documentation) • Fédération internationale du diabète (French: International Diabetes Federation) • field intelligence department

FIDA Fellow of the Institute of Directors, Australia

Fid. Def. (*or* **FID DEF**) Fidei Defensor (*see under* FD)

FIDDI *Computing* fibre-distributed data interface

FIDE Fédération internationale des échecs (French: International Chess Federation) • Fellow of the Institute of Design Engineers

FIDO ('faɪdəʊ) Film Industry Defence Organization • (*or* **Fido**) *Astronautics* (USA) Flight Dynamics Officer • Fog Investigation Dispersal Operation (in the RAF during World War II)

FIDP Fellow of the Institute of Data Processing

FIE Fédération internationale d'escrime (French: International Fencing Federation) • Fellow of the Institute of Engineers and Technicians

FIE(Aust) Fellow of the Institution of Engineers, Australia

FIED Fellow of the Institution of Engineering Designers

FIEE Fellow of the Institution of Electrical Engineers

FIEEE (USA) Fellow of the Institute of Electrical and Electronics Engineers

FIEI Fellow of the Institution of Engineers in Ireland

FIEJ Fédération internationale des éditeurs de journaux et publications (French: International Federation of Newspaper Publishers)

FIElecIE Fellow of the Institution of Electronic Incorporated Engineers

FIERE Fellow of the Institution of Electronic and Radio Engineers

FIET Fédération internationale des employés, techniciens et cadres (French: International Federation of Commercial, Clerical, and Technical Employees)

FIEx Fellow of the Institute of Export

fi. fa *Law* fieri facias (writ of execution; Latin: have it done)

FIFA ('fiːfə) Fédération internationale de football association (French: International Federation of Association Football) • Fédération internationale du film d'art (French: International Federation of Art Films) • Fellow of the International Faculty of Arts

FIFCLC Fédération internationale des femmes de carrières libérales et commerciales (French: International Federation of Business and Professional Women)

FIFDU Fédération internationale des femmes diplômées des universités (French: International Federation of University Women)

FIFF Fellow of the Institute of Freight Forwarders

FIFireE Fellow of the Institution of Fire Engineers

FIFO (*or* **fifo**) ('fiːfəʊ) *Computing, Accounting* first in first out

FIFSP Fédération internationale des fonctionnaires supérieurs de police (French: International Federation of Senior Police Officers)

FIFST Fellow of the Institute of Food Science and Technology

fig. figurative(ly) · (*or* **Fig.**) figure(s)

FIG Fédération internationale de gymnastique (French: International Gymnastic Federation)

FIGasE Fellow of the Institution of Gas Engineers

FIGC Federazione Italiana Gioco Calcio (Italian Football Association)

FIGCM Fellow of the Incorporated Guild of Church Musicians

FIGED Fédération internationale des grandes entreprises de distribution (French: International Federation of Distributors)

FIGO Fédération internationale de gynécologie et d'obstétrique (French: International Federation of Gynaecology and Obstetrics)

FIGRS Fellow of the Irish Genealogical Research Society

FIH Fédération internationale de hockey (French: International Hockey Federation) · Fédération internationale des hôpitaux (French: International Hospital Federation) · Fellow of the Institute of Housing · *Image technol.* focused-image holography

FIHE Fellow of the Institution of Health Education

FIHort Fellow of the Institute of Horticulture

FIHospE Fellow of the Institute of Hospital Engineering

FIHsg Fellow of the Institute of Housing

FIHT Fellow of the Institution of Highways and Transportation

FII franked investment income

FIIA Fellow of the Institute of Internal Auditors

FIIC Fellow of the International Institute for Conservation of Historic and Artistic Works

FIIM Fellow of the Institution of Industrial Managers

FIInfSc Fellow of the Institute of Information Scientists

FIInst Fellow of the Imperial Institute

FIISec Fellow of the International Institute of Security

FIITech Fellow of the Institute of Industrial Technicians

FIJ Fédération internationale de judo (French: International Judo Federation) · Fédération internationale des journalistes (French: International Federation of Journalists)

fil. filament · fillet · filter · filtrate

FIL Fédération internationale de laiterie (French: International Dairy Federation) · Fellow of the Institute of Linguists

FILA Fédération internationale de lutte amateur (French: International Amateur Wrestling Federation)

FILDM Fellow of the Institute of Logistics and Distribution Management

FILO first in last out

FILT Fédération internationale de lawn tennis (French: International Lawn Tennis Federation)

FIM Fédération internationale des musiciens (French: International Federation of Musicians) · Fédération internationale motocycliste (French: International Motorcycle Federation) · Fellow of the Institute of Metals · field-ion microscope (*or* microscopy)

FIMA Fellow of the Institute of Mathematics and its Applications

FIMarE Fellow of the Institute of Marine Engineers

FIMBRA ('fɪmbrə) Financial Intermediaries, Managers, and Brokers Regulatory Association

FIMC Fellow of the Institute of Management Consultants

FIMechE Fellow of the Institution of Mechanical Engineers

FIMGTechE Fellow of the Institution of Mechanical and General Technician Engineers

FIMH Fellow of the Institute of Military History

FIMI Fellow of the Institute of the Motor Industry

FIMinE Fellow of the Institution of Mining Engineers

FIMIT Fellow of the Institute of Musical Instrument Technology

FIMLS Fellow of the Institute of Medical Laboratory Sciences (formerly Technology, FIMLT)

FIMM Fellow of the Institution of Mining and Metallurgy

FIMP Fédération internationale de médecine physique (French: International Federation of Physical Medicine)

FIMS Fellow of the Institute of Mathematical Statistics · *Chem.* field-ion mass spectrometer (*or* spectrometry)

fin. ad finem (Latin: at (*or* near) the end) · final · finance · financial · financier · finis (Latin: the end) · finish

Fin. Finland · Finnish

FIN *international vehicle registration for* Finland

f.i.n.a. following items not available

FINA Fédération internationale de natation amateur (French: International Amateur Swimming Federation)

Fin. Sec. Financial Secretary

FInstAM Fellow of the Institute of Administrative Management

FInstB Fellow of the Institution of Buyers

FInstCh Fellow of the Institute of Chiropodists

FInstD Fellow of the Institute of Directors

FInstE Fellow of the Institute of Energy

FInstF Fellow of the Institute of Fuel

FInstFF Fellow of the Institute of Freight Forwarders

FInstLEx Fellow of the Institute of Legal Executives

FInstMC Fellow of the Institute of Measurement and Control

FInstO Fellow of the Institute of Ophthalmology

FInstP Fellow of the Institute of Physics

FInstPet Fellow of the Institute of Petroleum

FInstPl Fellow of the Institute of Patentees and Inventors

FInstPS Fellow of the Institute of Purchasing and Supply

FInstR Fellow of the Institute of Refrigeration

FInstSMM Fellow of the Institute of Sales and Marketing Management

FINucE Fellow of the Institution of Nuclear Engineers

f.i.o. for information only · *Commerce* free in and out

FIOA Fellow of the Institute of Acoustics

FIOP Fellow of the Institute of Plumbing · Fellow of the Institute of Printing

FIP Fédération internationale de philatélie (French: International Philatelic Federation) · Fédération internationale pharmaceutique (French: International Pharmaceutical Federation)

FIPA Fédération internationale des producteurs agricoles (French: International Federation of Agricultural Producers) · Fellow of the Institute of Practitioners in Advertising

FIPENZ Fellow of the Institution of Professional Engineers, New Zealand

FIPG Fellow of the Institute of Professional Goldsmiths

FIPM Fellow of the Institute of Personnel Management

FIPR Fellow of the Institute of Public Relations

FIProdE Fellow of the Institution of Production Engineers

FIPS (fips) *Computing* Federal Information Processing Standards

FIQ Fédération internationale de quilleurs (French: International Bowling Federation) · Fellow of the Institute of Quarrying

FIQA Fellow of the Institute of Quality Assurance

FIQS Fellow of the Institute of Quantity Surveyors

fir. firkin

f.i.r. flight information region · floating-in rate · fuel-indicator reading

FIR far-infrared radiation

FIRA Fédération internationale de rugby amateur (French: International Amateur Rugby Federation) · Furniture Industry Research Association

FIREE(Aust) Fellow of the Institute of Radio and Electronics Engineers (Australia)

FIRS Fédération internationale de patinage à roulettes (French: International Roller Skating Federation)

FIRST (fɜːst) Far Infrared and Submillimetre Telescope · *Computing* Forum of Incident and Response Teams

FIRTE Fellow of the Institute of Road Transport Engineers

FIRSE Fellow of the Institute of Railway Signalling Engineers

FIS Family Income Supplement · farm improvement scheme · Fédération internationale de sauvetage (French: International Life Saving Federation) • Fédération internationale de ski (French: International Ski Federation) • Fellow of the Institute of Statisticians • flight information service • (*or* **f.i.s.**) free into store • (fiːs) (Algeria) Front Islamique du Salut (French: Islamic Salvation Front)

FISA Fédération internationale des sociétés d'aviron (French; International Rowing Federation) • Fédération internationale du sport automobiles (French; International Motoring Federation) • Fellow of the Incorporated Secretaries Association

FISE Fédération internationale syndicale de l'enseignement (French; World Federation of Teachers' Unions) • Fellow of the Institution of Sales Engineers • Fellow of the Institution of Sanitary Engineers • Fonds international de secours à l'enfance (French; United Nations Children's Fund)

fish. fishery • fishes • fishing

FISITA Fédération internationale des sociétés d'ingénieurs des techniques de l'automobile (French: International Federation of Automobile Engineers' and Technicians' Associations)

FIST Fellow of the Institute of Science Technology

FISTC Fellow of the Institute of Scientific and Technical Communicators

FIStructE Fellow of the Institution of Structural Engineers

FISU Fédération internationale du sport universitaire (French: International University Sports Federation)

FISVA Fellow of the Incorporated Society of Valuers and Auctioneers

FISW Fellow of the Institute of Social Work

f.i.t. *Commerce* free in truck • free of income tax

FIT (USA) Federal Income Tax • Fédération internationale des traducteurs (French: International Federation of Translators)

FITA Fédération internationale de tir á l'arc (French: International Archery Federation)

FITC *Immunol.* fluorescein isothiocyanate

FITCE Fédération des ingénieurs des télécommunications de la communauté européenne (French: Federation of Telecommunications Engineers in the European Community)

FITD Fellow of the Institute of Training and Development

FITE Fellow of the Institution of Electrical and Electronics Technician Engineers

FITS *Astonomy* flexible image transport system

FITT Fédération internationale de tennis de table (French: International Table Tennis Federation)

f.i.t.w. (USA) federal income tax withholding

FIVB Fédération internationale de volleyball (French: International Volley-Ball Federation)

f.i.w. *Commerce* free in (*or* into) wagon(s)

FIWC Fiji Industrial Workers' Congress

FIWEM Fellow of the Institution of Water and Environmental Management

FIWSc Fellow of the Institute of Wood Science

fix. fixture

Fj. Fjord

FJ *airline flight code for* Air Pacific • *British vehicle registration for* Exeter

FJA Future Journalists of America

Fjd Fjord

FJI Fellow of the Institute of Journalists • *international vehicle registration for* Fiji

fk fork

f.k. flat keel

FK *British vehicle registration for* Dudley • *postcode for* Falkirk • *Astronomy* Fundamental Katalog (German: Fundamental Catalogue, as in **FK5**)

FKC Fellow of King's College (London)

FKCHMS Fellow of King's College Hospital Medical School

Fkr *symbol for* Faeroese krone (monetary unit)

fl. fleuve (French: river) • floor • *Med.* flores (Latin: flowers; powdered form of a drug) • florin • floruit (Latin: flourished; indicates the period of greatest activity of a person whose birth and death dates are not known) • flourish • fluid • *Music* flute • (Netherlands) guilder (from its former name *florin*)

f.l. falsa lectio (Latin: false reading; in a text)

Fl. Flanders • Flemish

FL Flag Lieutenant • Flight Lieutenant • *US postcode for* Florida • football league • *international vehicle registration for* Liechtenstein • *British vehicle registration for* Peterborough

Fla. Florida
FLA Fellow of the Library Association • fiat lege artis (Latin: let it be done by rules of the art) • Film Laboratory Association • Finance and Leasing Association • Future Large Aircraft
flag. *Music* flageolet
FLAI Fellow of the Library Association of Ireland
FLCD *Electronics* ferroelectric liquid-crystal display
FLCM Fellow of the London College of Music
FLCO Fellow of the London College of Osteopathy
fld field
FLD *Metallurgy* forming limit diagram
fldg folding
fl. dr. fluid dram
Flem. Flemish
flex. flexible
flg flagging • flooring • flying • following
FLHS Fellow of the London Historical Society
FLI Fellow of the Landscape Institute
FLIA Fellow of the Life Insurance Association
Flint. Flintshire
FLIP (flɪp) *US Navy* floating instrument platform
FLIR *Military* forward-looking infrared
Flli Fratelli (Italian: Brothers)
FLN (Algeria) Front de Liberation Nationale (French: National Libération Front)
FLOOD (flʌd) (USA) fleet observation of oceanographic data
flops (*or* **FLOPS**) (flɒps) *Computing* floating-point operations per second (a measure of computer power, as in **Mflops**, megaflops)
flor. floruit (*see under* fl.)
Flor. Florence • Florentine • Florida
fl oz *symbol for* fluid ounce
f.l.p. *Aeronautics* fault location panel
fl. pl. *Botany* flore pleno (Latin: with double flowers)
FLQ (Canada) Front de Liberation du Québec (French: Quebec Liberation Front)
flr. florin
FLRA (USA) Federal Labor Relations Authority
flrg flooring
fl. rt. flow rate

FLS Fellow of the Linnean Society
FLSA (USA) Fair Labor Standards Act
flst. flautist
Flt *Air force* Flight
F/Lt (*or* **F.Lt**) Flight Lieutenant
Flt Cmdr Flight Commander
fltg floating
Flt Lt Flight Lieutenant
Flt Off. Flight Officer
Flt Sgt Flight Sergeant
fluc. fluctuant • fluctuate(d) • fluctuation
FLUC *Meteorol.* fluctuating
fluor. fluorescent • fluoridation • fluoride • fluorspar
fly. *Boxing* flyweight
fm farm(er) • fathom(s) • *Physics, symbol for* femtometre(s) • from
f.m. face (*or* facial) measurement • femmes mariées (French: married women) • *Med.* fiat mistura (Latin: let a mixture be made) • fine measure(ment) • *Radio* frequency modulation
Fm *Chem., symbol for* fermium
F.m. facing matter (in advertisement placing)
FM *British vehicle registration for* Chester • *Computing* facilities management • *Metallurgy* ferritic and martensitic (describing iron) • *Electrical engineering* field magnet • Field Marshal • *Aeronautics* figure of merit • Flight Mechanic • foreign mission • Fraternitas Medicorum (Latin: Fraternity of Physicians) • fraternité mondiale (French: world brotherhood) • freemason • *Radio* frequency modulation • Friars Minor
FMA Farm Management Association • Fellow of the Museums Association • Food Machinery Association
fman foreman
FMANZ Fellow of the Medical Association of New Zealand
FMAO Farm Machinery Advisory Officer
FMAS Foreign Marriage Advisory Service
FMB (Malawi) Farmers' Marketing Board • (USA) Federal Maritime Board • Federation of Master Builders
FMC (USA) Federal Maritime Commission • Fellow of the Medical Council • Forces Motoring Club • Ford Motor Company
FMCE Federation of Manufacturers of Construction Equipment
FMCG fast-moving consumer goods

FMCP Federation of Manufacturers of Contractors' Plant

FMCS (USA) Federal Mediation and Conciliation Service

FMCW *Electronics* frequency modulated continuous wave

fmd formed

FMD foot and mouth disease

FMEA failure mode and effect analysis

FMES Fellow of the Minerals Engineering Society

FMF *Med.* fetal movements felt • Fiji Military Forces • Fleet Marine Force • Food Manufacturers' Federation

FMFPAC (USA) Fleet Marine Forces, Pacific

FMG (Nigeria) Federal Military Government • *symbol for* franc (monetary unit of Madagascar)

FMI Fellow of the Motor Industry • Filii Mariae Immaculatae (Latin: Sons of Mary Immaculate)

FMIG Food Manufacturers' Industrial Group

Fmk (*or* **FMk**) *symbol for* markka (Finnish monetary unit) • Finnmark

fml formal

FMLN (El Salvador) Farabundo Martí National Liberation Front (left-wing guerrilla movement)

fmn formation

FMN *Biochem.* flavin mononucleotide

FMO Fleet Medical Officer • Flight Medical Officer

fmr former

fmrly formerly

FMRS Foreign Member of the Royal Society

FMS (formerly) Federated Malay States • Fédération mondiale des sourds (French: World Federation of the Deaf) • Fellow of the Institute of Management Studies • Fellow of the Medical Society • *Computing* flexible manufacturing system • *Aeronautics* flight-management systems

FMSA Fellow of the Mineralogical Society of America

FMTS field maintenance test station

FMV *Computing* full-motion video

FMVSS (USA) Federal Motor Vehicle Safety Standards

f.n. footnote

FN *British vehicle registration for* Maidstone

f.n.a. for necessary action

FNA Fellow of the Indian National Science Academy • French North Africa

FNAEA Fellow of the National Association of Estate Agents

FNAL Fermi National Accelerator Laboratory

FNCB (USA) First National City Bank

FNCO Fleet Naval Constructor Officer

fnd found • foundered

fndd founded

fndr founder

fndry foundry

fne fine

FNECInst Fellow of the North East Coast Institution of Engineers and Shipbuilders

FNI Fédération naturiste internationale (French: International Naturist Federation) • Fellow of the Nautical Institute

FNIF Florence Nightingale International Foundation

FNILP Fellow of the National Institute of Licensing Practitioners

FNL Friends of the National Libraries

FNLA Frente Nacional de Libertação de Angola (Portuguese: National Front for the Liberation of Angola)

FNMA (USA) Federal National Mortgage Association

FNO Fleet Navigation Officer

FNU Forces des Nations Unies (French: United Nations Forces)

FNWC (USA) Fleet Numerical Weather Center

FNZEI Fellow of the New Zealand Educational Institute

FNZIA Fellow of the New Zealand Institute of Architects

FNZIAS Fellow of the New Zealand Institute of Agricultural Science

FNZIC Fellow of the New Zealand Institute of Chemistry

FNZIE Fellow of the New Zealand Institution of Engineers

FNZIM Fellow of the New Zealand Institute of Management

FNZPsS Fellow of the New Zealand Psychological Society

fo. folio

f.o. fast operating • *Commerce* free overside • fuel oil

f/o *Commerce* for orders • full out

Fo (ital.) *Physics, symbol for* Fourier number

FO federal official • *Army* Field Officer • *Commerce* firm offer • First Officer • *Navy*

Flag Officer · *Air force* Flying Officer · Foreign Office (now part of the FCO) · formal offer (to shareholders in business mergers) · *Military* forward observer · *Music* full organ · *British vehicle registration for* Gloucester

FOB (*or* **f.o.b.**) *Commerce* free on board

FOBFO Federation of British Fire Organizations

FOBS *Military* fractional orbital bombardment system

FOBTSU forward observer target survey unit

FoC father of the (trade-union) chapel

FOC (*or* **f.o.c.**) *Commerce* free of charge · free of claims

FOCL fibre-optic communications line

FOCOL Federation of Coin-Operated Launderettes

FOCT flag officer carrier training

FOCUS ('faʊkəs) Focus on Computing in the United States

FOD (*or* **f.o.d.**) free of damage

FODA Fellow of the Overseas Doctors' Association

FOE (USA) Fraternal Order of Eagles · (*or* **FoE**) Friends of the Earth

FOFA follow-on forces attack

FOFATUSA Federation of Free African Trade Unions of South Africa

F of F Firth of Forth

FOH *Theatre* front of house

FOI (USA) freedom of information

FOIA (USA) Freedom of Information Act

FOIC Flag Officer in charge

f.o.k. *Stock exchange* fill or kill (carry out instruction or cancel)

fol. folio · follow(ed) · following

FOL (New Zealand) Federation of Labour

folg. following

foll. followed · following

FOM *Statistics* figure of merit

FOMC (USA) Federal Open Market Committee

FONA Flag Officer, Naval Aviation

FONAC Flag Officer, Naval Air Command

FOP forward observation post

FOQ (*or* **f.o.q.**) *Commerce* free on quay

for. foreign(er) · forensic · forest(er) · forestry · *Music* forte

For *Astronomy* Fornax

FOR Fellowship of Operational Research ·

flying objects research · (*or* **f.o.r.**) free on rail

FORATOM Forum atomique européen (French: European Atomic Forum)

Ford (fɔːd) (Kenya) Forum for the Restoration of Democracy

FOREST ('fɒrɪst) Freedom Organization for the Right to Enjoy Smoking Tobacco

FOREX ('fɒrɛks) Foreign exchange

formn foreman · formation

form. wt *Chem.* formula weight

for. rts foreign rights

fort. fortification · fortified · fortify

f.o.r.t. *Commerce* full out rye terms

Fortran (*or* **FORTRAN**) ('fɔːtræn) *Computing* formula translation (a programming language)

Forts. Fortsetzung (German: continuation)

forz. *Music* forzando (Italian: with force)

f.o.s. *Finance* free of stamp · *Commerce* free on ship (*or* steamer) · *Commerce* free on station

FOS Fisheries Organization Society

FOSDIC *Computing* film optical sensing device

f.o.t. (*or* **FOT**) *Commerce* free of tax · *Commerce* free on truck(s)

Found. Foundation · Foundry

4WD (USA) four-wheel drive

FOV field of view

f.o.w. *Commerce* first open water · *Commerce* free on wagon

FOX (fɒks) *Commerce* Futures and Options Exchange (in **London FOX**)

fp (ital.) *Music* forte-piano (Italian; loud (then) soft) · freezing point

fp. fireplace · foolscap

f.p. *Med.* fiat pilula (Latin: let a pill be made) · fine paper · *Cricket* fine point · fixed price · flameproof · *Chem.* flash point · footpath · foot pound · *Sport* forward pass · freezing point · full point · fully paid (of shares)

Fp. frontispiece

FP Federal Parliament · field punishment · filter paper · *Insurance* fire policy · *Nuclear engineering* fission product · *Insurance* floating policy · *Optics* focal plane · (Australia) forensic pathologist · former pupil · fowl pest · Free Presbyterian · freezing point · fresh paragraph · fully paid (of shares) · *Computing* functional programming · *British vehicle registration for* Leicester

FPA Family Planning Association • Film Production Association of Great Britain • Fire Protection Association • *Astronomy* first point of Aries • *Computing* floating-point accelerator • Foreign Press Association • *Marine insurance* free of particular average

f.p.b. fast patrol boat

f.p.b.g. fast patrol boat with guided missiles

f.p.c. for private circulation

FPC (formerly) Family Practitioner Committee • Federation of Painting Contractors • fish protein concentrate • Flowers Publicity Council

FPEA Fellow of the Physical Education Association

FPF Federação Portuguesa de Futebol (Portuguese Football Federation)

FPGA *Computing* field-programmable gate array

FPHA (USA) Federal Public Housing Authority

FPhS Fellow of the Philosophical Society of England

FPhyS Fellow of the Physical Society

FPI *Physics* Fabry–Perot interferometer • (USA) Federal Prison Industries

FPIA Fellow of the Plastics Institute of Australia

FPLA *Electronics, Computing* field-programmable logic array

fpm (*or* **FPM**) feet per minute

FPMI Fellow of the Pensions Management Institute

FPMR (USA) Federal Property Management Regulation

FPO field post office • fire prevention officer • *US Navy* fleet post office

FPRC Flying Personnel Research Committee

FPRI Fellow of the Plastics and Rubber Institute

FPROM *Computing* fusible-link programmable (*or* field-programmable) read-only memory

fps (*or* **f.p.s.**) feet per second • *Physics* foot-pound-second (as in **fps units**) • *Photog.* frames per second

FPS Fellow of the Philharmonic Society • Fellow of the Philological Society • Fellow of the Philosophical Society • (USA) Fluid Power Society

fpsps feet per second per second

FPT *Commerce* fixed price tenders • *Nautical* forepeak tank

FPU *Computing* floating point unit

f.q. *Finance* fiscal quarter

FQ *airline flight code for* Air Aruba

FQDN *Computing* fully qualified domain name

FQS (USA) Federal Quarantine Service

fr. fragment • frame • franc(s) • free • frequent • from • front • (ital.) *Physics, symbol for* Froude number • fruit • *Botany* fruiting

f.r. folio recto (Latin: on the right-hand page)

Fr *RC Church* Brother (Latin *Frater*) • *Christianity* Father • *Chem., symbol for* francium

Fr. France • Fratelli (Italian: Brothers) • Frau (German: Mrs) • French • Friar • Friday

FR *international vehicle registration for* Faeroe Islands • Federal Republic • Federal Reserve (System) • *Chem.* fluorine rubber • Forum Romanum (Latin: Roman Forum) • *British fishing port registration for* Fraserburgh • freight release • frequency rate • *Nuclear engineering* fusion reactor • *British vehicle registration for* Preston • *airline flight code for* Ryanair

F/R folio reference

f.r.a. flame retardant additive

Fra (frɑ:) *RC Church* Brother (Italian *frate*)

FRA *Finance* forward rate agreement

FRACDS Fellow of the Royal Australian College of Dental Surgeons

FRACGP Fellow of the Royal Australian College of General Practitioners

FRACI Fellow of the Royal Australian Chemical Institute

FRACMA Fellow of the Royal Australian College of Medical Administrators

FRACO Fellow of the Royal Australian College of Ophthalmologists

FRACOG Fellow of the Royal Australian College of Obstetricians and Gynaecologists

FRACP Fellow of the Royal Australasian College of Physicians

FRACR Fellow of the Royal Australasian College of Radiologists

FRACS Fellow of the Royal Australasian College of Surgeons

FRAD Fellow of the Royal Academy of Dancing

FRAeS Fellow of the Royal Aeronautical Society

FRAgSs Fellow of the Royal Agricultural Societies

FRAHS Fellow of the Royal Australian Historical Society

FRAI Fellow of the Royal Anthropological Institute

FRAIA Fellow of the Royal Australian Institute of Architects

FRAIC Fellow of the Royal Architectural Institute of Canada

FRAIPA Fellow of the Royal Australian Institute of Public Administration

FRAM Fellow of the Royal Academy of Music • *Computing* ferroelectric random-access memory

FRAME (freɪm) Fund for the Replacement of Animals in Medical Experiments

Franc. Franciscan

f.r. & c.c. *Insurance* free of riot and civil commotions

Frank. Frankish

FRANZCP Fellow of the Royal Australian and New Zealand College of Psychiatrists

FRAP (Chile) Frente de Acción Popular (Spanish: Popular Action Front)

FRAPI Fellow of the Royal Australian Planning Institute

FRAS Fellow of the Royal Asiatic Society • Fellow of the Royal Astronomical Society

FRASE Fellow of the Royal Agricultural Society of England

FRATE (freɪt) *Railways* formulae for routes and technical equipment

fraud. fraudulent

FRB (USA) Federal Reserve Bank • (USA) Federal Reserve Board • Fisheries Research Board of Canada • Frente de la Revolución Boliviana (Spanish: Bolivian Revolutionary Front)

FRBS Fellow of the Royal Botanic Society • Fellow of the Royal Society of British Sculptors

FRC (USA) Federal Radiation Council • (USA) Federal Radio Commission • Financial Reporting Council • Flight Research Center (at NASA)

FRCA Fellow of the Royal College of Anaesthetists • Fellow of the Royal College of Art

Fr-Can. French-Canadian

FRCCO Fellow of the Royal Canadian College of Organists

FRCD *Banking* floating-rate certificate of deposit • Fellow of the Royal College of Dentists of Canada

FRCGP Fellow of the Royal College of General Practitioners

FRCM Fellow of the Royal College of Music

FRcn Fellow of the Royal College of Nursing

FRCO Fellow of the Royal College of Organists

FRCO(CHM) Fellow of the Royal College of Organists with Diploma in Choir Training

FRCOG Fellow of the Royal College of Obstetricians and Gynaecologists

FRCP Fellow of the Royal College of Physicians

FRCPA Fellow of the Royal College of Pathologists of Australasia

FRCP&S(Canada) Fellow of the Royal College of Physicians and Surgeons of Canada

FRCPath Fellow of the Royal College of Pathologists

FRCP(C) Fellow of the Royal College of Physicians of Canada

FRCPE (*or* **FRCPEd**) Fellow of the Royal College of Physicians of Edinburgh

FRCPGlas Fellow of the Royal College of Physicians and Surgeons of Glasgow

FRCPI Fellow of the Royal College of Physicians of Ireland

FRCPSGlas Fellow of the Royal College of Physicians and Surgeons of Glasgow

FRCPsych Fellow of the Royal College of Psychiatrists

FRCR Fellow of the Royal College of Radiologists

FRCS Fellow of the Royal College of Surgeons

FRCSCan Fellow of the Royal College of Surgeons of Canada

FRCSci Fellow of the Royal College of Science, Ireland

FRCSE (*or* **FRCSEd**) Fellow of the Royal College of Surgeons of Edinburgh

FRCSGlas Fellow of the Royal College of Physicians and Surgeons of Glasgow

FRCSI Fellow of the Royal College of Surgeons of Ireland

FRCSoc Fellow of the Royal Commonwealth Society

FRCVS Fellow of the Royal College of Veterinary Surgeons

frd friend

FR Dist. (USA) Federal Reserve District

fre fracture (French: invoice)

Fre. Freitag (German: Friday) • French

FREconS Fellow of the Royal Economic Society

FRED (frɛd) Fast Reactor Experiment, Dounreay • figure reading electronic device • financial reporting exposure draft

Free. (USA) freeway

FREGG (frɛg) Free Range Egg Association

FREI Fellow of the Real Estate Institute (Australia)

Frelimo (freɪ'liːməʊ) Frente de Libertação de Moçambique (Portuguese: Mozambique Liberation Front)

freq. frequency • frequent(ly) • *Grammar* frequentative

FRES Federation of Recruitment and Employment Services • Fellow of the Royal Entomological Society

FRESH (frɛʃ) foil research hydrofoil

Fr.G. French Guiana

FRG Federal Republic of Germany

FRGS Fellow of the Royal Geographical Society

FRGSA Fellow of the Royal Geographical Society of Australasia

frgt freight

FRHB Federation of Registered House Builders

FRHistS Fellow of the Royal Historical Society

Frhr Freiherr (German: Baron)

FRHS Fellow of the Royal Horticultural Society

Fri. Fribourg • Friday

FRI Fellow of the Royal Institution • Food Research Institute

FRIA Fellow of the Royal Irish Academy

FRIAI Fellow of the Royal Institute of Architects of Ireland

FRIAS Fellow of the Royal Incorporation of Architects in Scotland • Fellow of the Royal Institute for the Advancement of Science

FRIBA Fellow of the Royal Institute of British Architects

fric. (*or* **frict.**) friction(al)

FRIC Fellow of the Royal Institute of Chemistry (now part of the Royal Society of Chemistry; *see under* FRSC)

FRICS Fellow of the Royal Institution of Chartered Surveyors

FRIH (New Zealand) Fellow of the Royal Institute of Horticulture

FRIIA Fellow of the Royal Institute of International Affairs

FRIN Fellow of the Royal Institute of Navigation

FRINA Fellow of the Royal Institution of Naval Architects

FRIPA Fellow of the Royal Institute of Public Administration

FRIPHH Fellow of the Royal Institute of Public Health and Hygiene

Fris. Friesland • Frisian

Frk. Frøken (Danish *or* Norwegian: Miss) • Fröken (Swedish: Miss)

Frl. Fräulein (German: Miss)

FRL full repairing lease

frld foreland • freehold

frm from

FRMCM Fellow of the Royal Manchester College of Music

FRMCS Fellow of the Royal Medical and Chirurgical Society

FRMedSoc Fellow of the Royal Medical Society

FRMetS Fellow of the Royal Meteorological Society

FRMS Fellow of the Royal Microscopical Society

FRN *Finance* floating-rate note

FRNCM Fellow of the Royal Northern College of Music

FRNS Fellow of the Royal Numismatic Society

FRNSA Fellow of the Royal Navy School of Architects

FRO Fellow of the Register of Osteopaths • Fire Research Organization • *Insurance* fire risk only

f.r.o.f. *Insurance* fire risk on freight

Frolinat Front de Libération Nationale Tchadienne (French: Chad National Liberation Front)

front. (*or* **frontis.**) frontispiece

FRP fibre- (*or* fibreglass-)reinforced plastic • fuel-reprocessing plant

frpf fireproof

FRPharmS Fellow of the Royal Pharmaceutical Society

FRPS Fellow of the Royal Photographic Society

FRPSL Fellow of the Royal Philatelic Society, London

FRQ *Meteorol.* frequent

FRR (USA) Financial Reporting Release

FRRP Financial Reporting Review Panel

Frs. Frisian

FRS (USA) Federal Reserve System • Fellow of the Royal Society • Festiniog Railway Society • Financial Reporting Standard • fuel research station

FRSA Fellow of the Royal Society of Arts

FRSAI Fellow of the Royal Society of Antiquaries of Ireland

FRSAMD Fellow of the Royal Scottish Academy of Music and Drama

FRSC (*or* **FRSCan**) Fellow of the Royal Society of Canada • Fellow of the Royal Society of Chemistry (formerly FCS; FRIC)

FRSCM Fellow of the Royal School of Church Music

FRSE Fellow of the Royal Society of Edinburgh

FRSGS Fellow of the Royal Scottish Geographical Society

FRSH Fellow of the Royal Society of Health

FRSL Fellow of the Royal Society of Literature

FRSM Fellow of the Royal Society of Medicine

FRSNA Fellow of the Royal School of Naval Architecture

FRSNZ Fellow of the Royal Society of New Zealand

FRSSA Fellow of the Royal Scottish Society of Arts • (*or* **FRSSAF**) Fellow of the Royal Society of South Africa

FRSSI Fellow of the Royal Statistical Society of Ireland

FRSSS Fellow of the Royal Statistical Society of Scotland

FRST Fellow of the Royal Society of Teachers

FRSTM&H Fellow of the Royal Society of Tropical Medicine and Hygiene

frt (*or* **Frt**) freight

Frt fwd freight forward

FRTPI Fellow of the Royal Town Planning Institute

Frt ppd freight prepaid

FRTS Fellow of the Royal Television Society

Fru *Biochem.* fructose

frum fratrum (Latin: of the brothers)

frust. *Med.* frustillatim (Latin: in small portions)

FRVC Fellow of the Royal Veterinary College

FRVIA Fellow of the Royal Victorian Institute of Architects

frwk framework

frwy (*or* **fry.**) freeway

FRZSScot Fellow of the Royal Zoological Society of Scotland

fs. facsimile

f.s. factor of safety • faire suivre (French: please forward) • far side • film strip • fire station • flight service • flying saucer • flying status

FS *British vehicle registration for* Edinburgh • Fabian Society • Faraday Society • feasibility study • (Italy) Ferrovie dello Stato Italia (State Railways) • field security • financial secretary • financial statement • Fleet Surgeon • Flight Sergeant • Foreign Service • (USA) Forest Service • Friendly Society

f.s.a. fuel storage area

FSA (USA) Farm Security Agency • (USA) Federal Security Agency • Federation of Small Businesses • (USA) Fellow of the Society of Actuaries • Fellow of the Society of Antiquaries • Field Survey Association • Financial Services Act (1986) • Financial Services Authority • *Computing* finite-state automaton • foreign service allowance • Friendly Societies Act

FSAA Fellow of the Society of Incorporated Accountants and Auditors

FSAE Fellow of the Society of Art Education • Fellow of the Society of Automotive Engineers

FSAI Fellow of the Society of Architectural Illustrators

FSAIEE Fellow of the South African Institute of Electrical Engineers

FSAM Fellow of the Society of Art Masters

FSAScot (*or* **FSAS**) Fellow of the Society of Antiquaries of Scotland • Fellow of the Society of Arts of Scotland

FSASM Fellow of the South Australian School of Mines

FSBI Fellow of the Savings Bank Institute

FSBR Financial Statement and Budget Report

FSC (USA) Federal Supreme Court • Fellow of the Society of Chiropodists • Field Studies Council • *RC Church* Fratres Scholarum Christianorum (Latin: Brothers of the Christian Schools *or*

Christian Brothers) • Friends Service
Council

FSCA Fellow of the Society of Company
and Commercial Accountants

FSD full-scale deflection (of a measuring
instrument)

FSDC Fellow of the Society of Dyers and
Colourists

FSE Fellow of the Society of Engineers •
field support equipment

FSF Fellow of the Institute of Shipping
and Forwarding Agents • Free Software
Foundation

FSFMV *Computing* full-screen, full-motion
video

FSG Fellow of the Society of Genealogists

FSgt Flight Sergeant

FSGT Fellow of the Society of Glass
Technology

FSH *Biochem.* follicle-stimulating
hormone • full service history (in motor
advertisements)

FSHM Fellow of the Society of Housing
Managers

FSI Fédération spirite internationale
(French: International Spiritualist Federation) • Free Sons of Israel

FSK *Telecom.* frequency shift keying

FSL First Sea Lord

FSLAET Fellow of the Society of Licensed
Aircraft Engineers and Technologists

FSLIC (USA) Federal Savings and Loan
Insurance Corporation

FSLN (Nicaragua) Frente Sandinista de
Liberación Nacional (Spanish: Sandinista
National Liberation Front)

FSLTC Fellow of the Society of Leather
Technologists and Chemists

FSM Fédération syndicale mondiale
(French: World Federation of Trade
Unions) • *Electronics* flying-spot micro-
scope • (USA) Free Speech Movement

FSMC Freeman of the Spectacle-Makers'
Company

FSME Fellow of the Society of Manufacturing Engineers

FSN federal stock number

FSNA Fellow of the Society of Naval
Architects

FSO field security officer • Fleet Signals
Officer • Foreign Service Officer

FSP field security police • foreign service
pay

FSR Field Service Regulations • (ital.)
Fleet Street Patent Law Reports

FSRP Fellow of the Society for Radiological Protection

FSS Fellow of the Royal Statistical
Society • (Algeria) Front des forces
socialistes (French: Socialist Forces
Front)

FSSI Fellow of the Statistical Society of
Ireland

FSSU Federated Superannuation Scheme
for Universities

FST *Computing* flatter squarer tube

FSTD Fellow of the Society of Typographic Designers

FSU family service unit

FSUC Federal Statistics Unit Conference

FSVA Fellow of the Incorporated Society
of Valuers and Auctioneers

ft feint • *Med.* fiat (Latin: let there be
made) • *symbol for* foot (*or* feet) • fort

ft. fortification • fortify

f.t. formal training • full terms

Ft *symbol for* forint (Hungarian monetary
unit) • Fort

FT *Meteorol.* feet (unit) • Financial Times •
(*or* **F-T**) *Chem.* Fischer–Tropsch (in **FT
process**) • *Physics, Maths.* Fourier transform • *British vehicle registration for*
Newcastle upon Tyne

FTA *Med.* fluorescent treponemal
antibody • Free Trade Agreement •
Freight Transport Association • Future
Teachers of America

FTA (Index) Financial Times Actuaries
Share Index

FTAM *Computing* file transfer, access, and
method

FTASI Financial Times Actuaries All-
Share Index

FTAT Furniture, Timber, and Allied
Trades Union

FTB fleet torpedo bomber

ftbrg. footbridge

FTC (USA) Federal Trade Commission •
(USA) flight test center • flying training
command • Full Technological
Certificate (of City and Guilds Institute)

FTCD Fellow of Trinity College, Dublin

FTCL Fellow of Trinity College of Music,
London

FTD foreign technology division

FTDA Fellow of the Theatrical Designers
and Craftsmen's Association

f.t.e. full-time equivalent

FTESA Foundry Trades Equipment and
Supplies Association

FTFL *Computing* fixed-to-fixed-length (in **FTFL code**)

ftg fitting

FTG Fuji Texaco Gas

fth. (*or* **fthm**) fathom

FTI Fellow of the Textile Institute

FTII Fellow of the Institute of Taxation

FT (Index) Financial Times Ordinary Share Index

FT-IR *Chem.* Fourier-transform infrared (in **FT-IR spectroscopy**)

ft-lb foot-pound

Ft Lieut Flight Lieutenant

FTM flying training manual · *Med.* fractional test meal

ft. mist. *Med.* fiat mistura (Latin: let a mixture be made)

FT-NMR *Chem.* Fourier-transform nuclear magnetic resonance (in **FT-NMR spectroscopy**)

FTO Fleet Torpedo Officer

FT Ord Financial Times (Industrial) Ordinary Share Index

FTP Fellow of Thames Polytechnic · *Computing* file-transfer protocol

FTPA Fellow of the Town and Country Planning Association

ft. pulv. *Med.* fiat pulvis (Latin: let a powder be made)

ft/s feet (*or* foot) per second

ft/s² feet (*or* foot) per second squared

FTS Fellow of the Australian Academy of Technological Sciences and Engineering · Fellow of the Tourism Society · flying training school · *Chem.* Fourier-transform spectroscopy

FTSC Fellow of the Tonic Sol-Fa College

FTSE 100 (*or* **FT-SE 100**) Financial Times Stock Exchange 100 Index (*or* Footsie)

FTT *Med.* failure to thrive

fttr fitter

FTU (Hong Kong) Federation of Trade Unions · *Photog.* Freeman time-unit

FTVL *Computing* fixed-to-variable-length (in **FTVL code**)

FTW *Commerce* free-trade wharf

FTZ free-trade zone

FU Farmers' Union · feed unit · *Photog.* Freeman time-unit · (Germany) Freie Universität (Free University of Berlin) · *British vehicle registration for* Lincoln

FUACE Fédération universelle des associations chrétiennes d'étudiants (French: World Student Christian Federation)

FUEN Federal Union of European Nationalities

FUMIST Fellow of the University of Manchester Institute of Science and Technology

fund. fundamental

fur. furlong(s) · further

furn. furnace · furnish(ed) · furniture

fus. fuselage · fusilier

FUSE Far Ultraviolet Spectroscopic Explorer

fut. future · *Finance* futures

FUW Farmers' Union of Wales

f.v. fire vent · fishing vessel · flush valve · (*or* **fv**) folio verso (Latin: on the reverse (i.e. left-hand) page)

FV *British vehicle registration for* Preston

FVC *Med.* forced vital capacity (in respiratory function tests)

FVRDE Fighting Vehicle Research and Development Establishment

FW *Chem.* formula weight · fresh water · *airline flight code for* Isles of Scilly Skybus · *British vehicle registration for* Lincoln

FWA Factories and Workshops Act · Family Welfare Association · (USA) Federal Works Agency · Fellow of the World Academy of Arts and Sciences · Free Wales Army

FWAG Farming and Wildlife Advisory Group

f.w.b. four wheel brake (*or* braking)

FWB Free Will Baptists

FWCC Friends' World Committee for Consultation

fwd forward

f.w.d. four-wheel drive · front-wheel drive

fwdg forwarding

FWeldI Fellow of the Welding Institute

FWFM Federation of Wholesale Fish Merchants

f.w.h. flexible working hours

FWHM *Physics* full width at half maximum (in spectroscopy)

FWI Federation of West Indies · French West Indies

FWL Foundation of World Literacy

FWO Federation of Wholesale Organizations · Fleet Wireless Officer

FWPCA (USA) Federal Water Pollution Control Administration

FWS fighter weapons school · fleet work study

FWSG farm water supply grant

fwt *Boxing* featherweight

f.w.t. (*or* **FWT**) fair wear and tear

FX *British vehicle registration for* Bournemouth • (*or* **f.x.**) foreign exchange • Francis Xavier (1506–52, Spanish Jesuit missionary) • *Films, Theatre* special effects (from phonetic spelling of *effects*)

fxd fixed • foxed

fxg fixing

fxle *Nautical* forecastle

FY *postcode for* Blackpool • (*or* **f.y.**) (USA, Canada) fiscal year • *British fishing port registration for* Fowey • *British vehicle registration for* Liverpool

FYC Family and Youth Concern

FYI (*or* **f.y.i.**) for your information

FYM farmyard manure

FYP *Economics* Five-Year Plan

fz. *Music* forzando *or* forzato (Italian; to be strongly accentuated)

FZ *British vehicle registration for* Belfast • *Meteorol.* freezing • Free Zone • French Zone

FZA (USA) Fellow of the Zoological Academy

FZDZ *Meteorol.* freezing drizzle

FZFG *Meteorol.* freezing fog

FZGB Federation of Zoological Gardens of Great Britain and Ireland

FZRA *Meteorol.* freezing rain

FZS Fellow of the Zoological Society

F-Zug Fernschnellzug (German: long-distance express train)

G

g *Physics, symbol for* acceleration of free fall • (ital.) *Physics, symbol for* degeneracy • *Meteorol., symbol for* gale • gallon(s) • gas • *Chem.* gaseous (as in $H_2O(g)$) • *Physics* gerade (German: even; in spectroscopy) • *Physics, symbol for* gluon • (superscript) *Maths., symbol for* grade (as in 20^g) • *symbol for* gram(s) • *Physics, symbol for* grav • *indicating* the seventh row of vertical squares from the left on a chessboard

g. garage • gauche (French: left) • gauge • gelding • gender • general • genitive • geographical (as in **g. mile**) • gilt • goal(keeper) • gold • government • grand • great • green • grey • gros *or* grosse (French: large) • guardian • guide • guilder(s) • guinea(s) • *Navy* gunnery

G (ital.) *Physics, symbol for* conductance • *international vehicle registration for* Gabon • *Irish fishing port registration for* Galway • *Magnetism, symbol for* gauss • *Films* (Australia, USA) general exhibition (certification) • (ital.) *Thermodynamics, symbol for* Gibbs function • *symbol for* giga- (prefix indicating 10^9, as in **GHz** (gigahertz), *or* (in computing) 2^{30}) • *postcode for* Glasgow • *Biochem., symbol for* glycine • *symbol for* gourde (Haitian monetary unit) • *Slang* grand (1000 pounds or dollars) • (ital.) *Physics, symbol for* gravitational constant • *Biochem., symbol for* guanine • *Biochem., symbol for* guanosine • *symbol for* guarani (Paraguayan monetary unit) • *Botany* gynoecium (in a floral formula) • (ital.) *Physics, symbol for* shear modulus • *Meteorol., symbol for* storm • *international civil aircraft marking for* UK • *indicating* a musical note or key • *Astronomy, indicating* a spectral type

G. (*or* **G**) German(y) • good • green • Guernsey • Gulf (on maps, etc.)

G³ *Electronics* gadolinium gallium garnet

G3 Group of Three (most powerful western economies)

G5 *Finance* Group of Five (nations that agreed to exchange-rate stabilization) • *airline flight code for* Island Air

G7 Group of Seven (leading industrial nations) • *airline flight code for* Guinea Airlines

G10 *Finance* Group of Ten (nations lending money to the IMF)

G24 Group of Twenty Four (industrialized nations)

G77 Group of Seventy Seven (developing countries)

6G *airline flight code for* Las Vegas Airlines

9G *airline flight code for* Caribbean Air · *international civil aircraft marking for* Ghana

g.a. *Marine insurance* general average

Ga *Chem., symbol for* gallium

Ga. Gallic · Georgia

GA Gaelic Athletic (Club) · Gamblers Anonymous · garrison artillery · *airline flight code for* Garuda Indonesia · general agent · *Linguistics* General American · general anaesthetic · General Assembly (of the UN) · general assignment · *Marine insurance* general average · Geographical Association · Geologists' Association · *US postcode for* Georgia · *Botany, symbol for* gibberellin (as in **GA₃**) · *British vehicle registration for* Glasgow · *Sport* goal attack · government actuary · graphic arts

G/A *Marine insurance* general average · (*or* **g/a**) ground to air

GAA (Ireland) Gaelic Athletic Association

GAAP generally accepted accounting principles

GAAS generally accepted auditing standards

Gab. Gabon

GAB general arrangements to borrow (in the IMF)

GABA ('gæbə) *Biochem.* gamma-aminobutyric acid (a neurotransmitter)

GAC *Chem.* granular activated carbon

GAD *Biochem.* glutamic acid decarboxylase

Gael. (*or* **Gae.**) Gaelic

GAFTA ('gæftə) Grain and Free Trade Association

GAG *Biochem.* glycosaminoglycan

GAI (USA) guaranteed annual income · Guild of Architectural Ironmongers

gal (*or* **gal.**) gallon(s)

Gal *Geophysics, symbol for* gal · *Biochem.* galactose

Gal. *Bible* Galatians · Galicia · Galway

gall. gallery · gallon

GALP glyceraldehyde 3-phosphate

GALT *Med.* gut-associated lymphoid tissue

galv. galvanic · galvanize(d) · galvanometer

Gam. Gambia

GAM (gæm) guided aircraft missile

G&AE *Accounting* general and administrative expense

G and O *Med.* gas and oxygen (in anaesthetic)

G & S Gilbert and Sullivan

G & T (*or* **g and t**) gin and tonic

GAO (USA) General Accounting Office

GAP (gæp) general assembly programme · Great American Public

GAPAN (*or* **Gapan**) ('gæpæn) Guild of Air Pilots and Air Navigators

GAPCE General Assembly of the Presbyterian Church of England

gar. garage · garrison

GAR Grand Army of the Republic (in the American Civil War) · guided aircraft rocket

gard. garden

GARIOA government aid and relief in occupied areas

GARP (gɑːp) Global Atmospheric Research Programme

GASC German-American Securities Corporation

GASP (gɑːsp) (USA) Group Against Smokers' Pollution

gastroent. gastroenterological · gastroenterology

GATCO ('gæt,kəʊ) Guild of Air Traffic Control Officers

GATT (*or* **Gatt**) (gæt) General Agreement on Tariffs and Trade

GAUFCC General Assembly of Unitarian and Free Christian Churches

GAV *Accounting* gross annual value

GAW guaranteed annual wage

GAWF Greek Animal Welfare Fund

gaz. gazette · gazetteer

Gb *Computing, symbol for* gigabyte(s) · *Magnetism, symbol for* gilbert

GB *Med.* gall bladder · Gas Board · *Computing, symbol for* gigabyte(s) · Girls' Brigade · *British vehicle registration for* Glasgow · government and binding · *Crystallog.* grain boundary · Great Britain (abbrev. *or* IVR) · guide book · gunboat

GBA *international vehicle registration for* Alderney · Governing Bodies Association

GB and I Great Britain and Ireland

GBCW Governing Body of the Church in Wales

GBDO Guild of British Dispensing Opticians

GBE (Knight *or* Dame) Grand Cross of the Order of the British Empire

GBG *international vehicle registration for* Guernsey

GBGSA Governing Bodies of Girls' Schools Association

GBH grievous bodily harm

GBJ *international vehicle registration for* Jersey

GBM *international vehicle registration for* Isle of Man

GBN *Chem.* graphite boron nitride

g.b.o. goods in bad order

GBP *Biochem.* glutamate-binding protein • great British public

GBRE General Board of Religious Education

GBS George Bernard Shaw (1856–1950, Irish-born dramatist and critic) • *Computing* Gragg-Burlisch-Stoer (in **GBS method**)

GBSM Graduate of Birmingham and Midland Institute School of Music

GBT Green Bank Telescope

GBTA Guild of Business Travel Agents

GBZ *international vehicle registration for* Gibraltar

GC *Astronomy* galactic centre • gas chromatograph(ic) • gas chromatography • Gas Council • *Chem.* gel chromatography • *Astronomy* General Catalogue • George Cross • gliding club • Goldsmiths' College • golf club • good conduct • government chemist • Grand Chancellor • Grand Chaplain • Grand Chapter • Grand Conductor • Grand Cross • *British vehicle registration for* SW London

GCA Girls' Clubs of America • *Aeronautics* ground-controlled approach • *international vehicle registration for* Guatemala

G Capt Group Captain

GCB (Knight *or* Dame) Grand Cross of the Order of the Bath

GCBS General Council of British Shipping

GCC Gas Consumers Council • Girton College, Cambridge • Gonville and Caius College, Cambridge • Gulf Cooperation Council

GCD general and complete disarmament • (*or* **g.c.d.**) *Maths.* greatest common divisor

GCE General Certificate of Education • (USA) General College Entrance

GCF (*or* **g.c.f.**) *Maths.* greatest common factor

GCH (Knight) Grand Cross of the Hanoverian Order • Guild Certificate of Hairdressing

GCHQ Government Communications Headquarters

GCI *Aeronautics* ground-controlled interception

GCIE (Knight) Grand Commander of the Order of the Indian Empire

GC-IR gas chromatography infrared

GCIU (USA) Graphic Communications International Union

GCL Guild of Cleaners and Launderers

GCLJ Grand Cross of St Lazarus of Jerusalem

GCLH Grand Cross of the Legion of Honour

GCM *Meteorol.* general circulation model • general court martial • Good Conduct Medal • (*or* **g.c.m.**) *Statistics* greatest common measure • (*or* **g.c.m.**) *Maths.* greatest common multiple

GCMG (Knight *or* Dame) Grand Cross of the Order of St Michael and St George

GCMS gas-chromatography mass spectroscopy

GCO *Navy* Gun Control Officer

GCON Grand Cross of the Order of the Niger

GCR *Computing* grey component replacement • ground-controlled radar • *Computing* group code recording

GCRO General Council and Register of Osteopaths

GCSE General Certificate of Secondary Education

GCSG (Knight) Grand Cross of the Order of St Gregory the Great

GCSI (Knight) Grand Commander of the Order of the Star of India

GCSJ (Knight) Grand Cross of Justice of the Order of St John of Jerusalem

GCStJ (Bailiff *or* Dame) Grand Cross of the Most Venerable Order of the Hospital of St John of Jerusalem

GCVO (Knight *or* Dame) Grand Cross of the Royal Victorian Order

gd good • grand-daughter • ground

g.d. general duties • grand-daughter • gravimetric density

Gd *Chem., symbol for* gadolinium

GD *Obstet.* gestational day • *British vehicle registration for* Glasgow • *Electronics* glow discharge • *Netball* goal defence •

Graduate in Divinity · Grand Duke (*or* Duchess *or* Duchy) · Gunnery Division

GDA Glasgow Development Agency

GDBA Guide Dogs for the Blind Association

GDC General Dental Council · General Dynamics Corporation

GDI *Computing* graphical device interface

Gdk Gdansk

gdn garden · guardian

Gdns Gardens

GDP (*or* **gdp**) gross domestic product · *Biochem.* guanosine diphosphate

GDR German Democratic Republic (East Germany; now part of Germany)

gds goods

Gds Guards

Gdsm. Guardsman

g.e. *Bookbinding* gilt edges

Ge *Chem., symbol for* germanium

GE garrison engineer · general election · (USA) General Electric (Company) · *international vehicle registration for* Georgia · *British vehicle registration for* Glasgow · *British fishing port registration for* Goole · *Astronomy* greatest elongation · gross energy · *airline flight code for* TransAsia Airways

GEA Garage Equipment Association

geb. geboren (German: born) · gebunden (German: bound)

GEBCO ('gebkəʊ) general bathymetric chart of the oceans

Gebr. Gebrüder (German: brothers)

GEC (UK) General Electric Company

GED general educational development

Gedcom ('dʒed,kɒm) *Computing* genealogical data communications

gegr. gegründet (German: founded)

gel. gelatin

Gem *Astronomy* Gemini · (dʒem) *Computing* graphics environment manager

GEM genetically engineered microorganism · (dʒem) ground effect machine (air-cushion vehicle) · guidance evaluation missile · Guild of Experienced Motorists

GEMS (dʒemz) Global Environmental Monitoring System (in the UN)

gen. gender · genealogy · general(ly) · generator · generic · genetic(s) · genital · genitive · genuine · *Biology* genus

Gen. General · *Bible* Genesis · Geneva · Genoa

gen. av. *Marine insurance* general average

gend. gendarme (French: policeman)

geneal. genealogy

genit. genitive

Genl General

genn. gennaio (Italian: January)

Gen X (,dʒen 'eks) Generation X

Geo. George

GEO geostationary (*or* geosynchronous) earth orbit

geod. geodesy · geodetic

geog. geographer · geographic(al) · geography

geol. geologic(al) · geologist · geology

geom. geometric(al) · geometry

geon (*or* **GEON**) ('dʒɪɒn) gyro-erected optical navigation

geophys. geophysics

geopol. geopolitics

GEOREF ('dʒɪəʊ,ref) World Geographic Reference System

ger. gerund(ive)

Ger. German(y)

GER gross energy requirement

Gerbil ('dʒɜːbɪl) Great Education Reform Bill (1988)

Ges. Gesellschaft (German: company *or* society)

gest. gestorben (German: deceased)

Gestapo (ge'stɑːpəʊ) Geheime Staatspolizei (German: secret state police; in Nazi Germany)

GETT (USA) Grants Equal To Tax

GeV *Physics, symbol for* gigaelectronvolt(s)

gez. gezeichnet (German: signed)

g/f ground floor (in property advertisements)

GF General Foods Ltd · glass fibre · government form · gradient freezing · *Maths.* Green function · Guggenheim Foundation · *symbol for* Guinean franc · *airline flight code for* Gulf Air · *British vehicle registration for* SW London

g.f.a. *Commerce* good fair average · *Commerce* good freight agent

GFCM General Fisheries Council for the Mediterranean (in the FAO)

GFD geophysical fluid dynamics

GFG (ital.) Good Food Guide

GFH George Frederick Handel (1685–1759, German composer)

GFOFs *Stock exchange* geared futures and options and funds

GFR German Federal Republic · *Med.* glomerular filtration rate

GFS Girls' Friendly Society

GFT Glasgow Film Theatre

GFTU (USA) General Federation of Trade Unions

GFWC General Federation of Women's Clubs

g.g. gas generator

GG *Med.* gamma globulin • Girl Guides • *British vehicle registration for* Glasgow • Governor General • great gross • Grenadier Guards

GGC generalized genetic code

ggd (*or* **g.gd.**) great grand-daughter

gge garage

GGF Glass and Glazing Federation

GGG *Electronics* gadolinium gallium garnet

g.gr. great gross (144 dozen)

ggs (*or* **g.gs.**) great grandson

GGSM Graduate (in Music) of the Guildhall School of Music and Drama

GH general hospital • *international vehicle registration for* Ghana • *airline flight code for* Ghana Airways • *British fishing port registration for* Grangemouth • *Military* Green Howards • Greenwich Hospital • *Biochem.* growth hormone • *British vehicle registration for* SW London

GHA *Astronomy* Greenwich hour angle

GHB gamma hydroxybutyrate

GHCIMA Graduate of the Hotel Catering and Institutional Management Association

g.h.e. ground handling equipment

GHI Good Housekeeping Institute

GHM *Physics* generalized Hall model

GHMS Graduate in Homeopathic Medicine and Surgery

GHOST (gəʊst) global horizontal sounding technique (for collecting atmospheric data)

GHQ *Military* General Headquarters

GHRH *Biochem.* growth-hormone-releasing hormone

GHS Girls' High School

GHz *symbol for* gigahertz

gi. gill (unit of measure)

Gi *Magnetism, symbol for* gilbert

GI (*or* **g.i.**) galvanized iron • (*or* **g.i.**) gastrointestinal • generic issue • (Royal) Glasgow Institute (of the Fine Arts) • (USA) government (*or* general) issue (hence, a US serviceman) • Government of India • *Biochem.* growth inhibitor • *Irish vehicle registration for* Tipperary

GIA Garuda Indonesian Airways

Gib. Gibraltar

GIB Gulf International Bank

GIBiol Graduate of the Institute of Biology

GIC *Chem.* graphite intercalation compound • Guilde internationale des coopératrices (French: International Cooperative Women's Guild)

GIE Graduate of the Institute of Engineers and Technicians

GIEE Graduate of the Institution of Electrical Engineers

GIF (gɪf) *Computing* graphics image (*or* interchange) format

GIFT (gɪft) *Med.* gamete intrafallopian transfer (for assisting conception)

GIGO (*or* **gigo**) ('gaɪgəʊ) *Computing* garbage in, garbage out

GIMechE Graduate of the Institution of Mechanical Engineers

GINO ('dʒiːnəʊ) graphical input output (for computer graphics)

GInstAEA Graduate of the Institute of Automotive Engineer Assessors

GInstT Graduate of the Institute of Transport

GINucE Graduate of the Institution of Nuclear Engineers

gio. (*or* **giov.**) giovedì (Italian: Thursday)

GIP (*or* **g.i.p.**) glazed imitation parchment (a type of paper)

GIS *Computing* geographical (*or* geological) information system • Global Information Solutions

GISS Goddard Institute for Space Studies

giu. giugno (Italian: June)

GIUK Greenland, Iceland, United Kingdom

GIXS *Physics* grazing-incidence X-ray scattering

GJ *British vehicle registration for* SW London

GJD *Freemasonry* Grand Junior Deacon

Gk Greek

GK goalkeeper • *British fishing port registration for* Greenock • *British vehicle registration for* SW London

GKA Garter King of Arms

GKC G(ilbert) K(eith) Chesterton (1874–1936, British journalist and author)

GKS graphical kernel system (for computer graphics)

gl. gill (unit of measure) • glass • gloss

g/l grams per litre

Gl. Gloria (Latin: glory; in the liturgy)

GL *Shipping* (Germany) Germanischer Lloyd (classification society) • *postcode for* Gloucester • *Printing* gothic letter • government laboratory • *Freemasonry* Grand Lodge • Grand Luxe (of a car) • ground level • gun licence • *British vehicle registration for* Truro

4GL *Computing* fourth-generation language

glab. *Botany* glabrous

GLAB Greater London Arts Board

glam (glæm) *Colloquial* greying, leisured, affluent, married

Glam. Glamorgan

Glas. Glasgow

glau. *Botany* glaucous

GLB gay, lesbian, bisexual • (formerly) Girls' Life Brigade

GLC *Chem.* gas–liquid chromatography • Greater London Council (abolished 1986) • ground-level concentration (of radioactive material) • Guild of Lettering Craftsmen

GLCM Graduate of the London College of Music • ground-launched cruise missile

gld. guilder

GLDP Greater London Development Plan

GLM *Skiing* graduated length method

Gln (*or* **gln**) *Biochem.* glutamine

GLOBECOM ('gləʊbkɒm) *US Air Force* Global Communications System

GLOMEX ('gləʊmɛks) Global Oceanographic and Meteorological Experiment (1975–80)

GLORIA ('glɔːrɪə) Geological Long Range Asdic

Glos Gloucestershire

gloss. glossary

GLS *Freemasonry* Grand Lodge of Scotland

glt *Bookbinding* gilt

GLT greetings letter telegram

Glu (*or* **glu**) *Biochem.* glutamic acid

Gly (*or* **gly**) *Biochem.* glycine

gm gram

g.m *symbol for* gram metre

gm² grams per square metre (used in measuring weight of paper)

GM *airline flight code for* Air Slovakia • *Physics* Geiger–Müller (as in **GM counter**) • general manager • general merchandise • *Computing* General Midi • general mortgage • General Motors Corporation • Geological Museum • geometric mean • George Medal • gold

medal(list) • Grand Marshall • *Freemasonry* Grand Master • *Education* grant maintained • guided missile • *British vehicle registration for* Reading

GMAG Genetic Manipulation Advisory Group

G-man (USA) Government man (an FBI agent)

g.m.b. good merchantable brand

GMB General and Municipal Boilermakers Union • Grand Master (of the Order) of the Bath • *Archery* Grand Master Bowman

GMBE Grand Master of the Order of the British Empire

GmbH (Germany) Gesellschaft mit beschränkter Haftung (private limited company; Ltd)

Gmc Germanic

GMC general management committee • General Medical Council • *Astronomy* giant molecular cloud • Guild of Memorial Craftsmen

GMF Glass Manufacturers' Federation

GMIE Grand Master of the Order of the Indian Empire

GMKP Grand Master of the Knights of St Patrick

GMMG Grand Master of the Order of St Michael and St George

GMO genetically modified organism

GMP (USA) Glass, Molders, Pottery, Plastics, and Allied Workers International Union • Grand Master of the Order of St Patrick • gross material product • *Biochem.* guanosine monophosphate • Gurkha Military Police

g.m.q. good merchantable quality

GMR ground-mapping radar

GMRT Giant Metre-wave Radio Telescope

GMS *Education* grant-maintained status

GMSC General Medical Services Committee

GMSI Grand Master of the Order of the Star of India

GMST *Astronomy* Greenwich Mean Sidereal Time

GMT Greenwich Mean Time

GMV *Microbiol.* golden mosaic virus

GMW gram-molecular weight

gn. guinea

g.n. grandnephew • grandniece

GN *airline flight code for* Air Gabon • (USA) Graduate Nurse • *British fishing port* *registration for* Granton • *British vehicle registration for* SW London

GNAS Grand National Archery Society

GNC General Nursing Council (replaced by UKCC)

gnd ground

GND *Meteorol.* ground

GNMA (USA) Government National Mortgage Association

GNP gross national product

Gnr *Military* Gunner

GNR (Portugal) Guarda Nacional Republicana (National Republican Guard)

GnRH *Biochem.* gonadotrophin-releasing hormone

gns guineas

GNTC Girls' Nautical Training Corps

GNVQ General National Vocational Qualification

GO gas operated • General Office(r) • *Military* general order • *Music* great organ • Group Officer • *British vehicle registration for* SW London

g.o.b. good ordinary brand

GOC General Officer Commanding • General Optical Council • Greek Orthodox Church

GOC-in-C General Officer Commanding-in-Chief

GOCO government-owned, contractor-operated

GOE General Ordination Examination

GOES (USA) Geostationary Operational Environmental Satellite

GOM Grand Old Man

GONG *Astronomy* Global Oscillations Network Group

GOP (USA) Grand Old Party (the Republican Party)

Gopa ('gəʊpə) Government Oil and Pipeline Agency

GORD gastro-oesophageal reflux disease

Gos. (formerly, USSR) Gosudarstvo (Russian: state)

GOSIP ('gɒsɪp) *Computing* government open systems interconnection profile

Gosplan ('gɒs,plæn) (formerly, USSR) State Planning Commission (Russian *Gos(udarstvennaya) Plan(ovaya Comissiya)*)

Gosud. Gosudarstvo (Russian: state)

GOT *Biochem., Med.* glutamatic oxaloacetic transaminase (renamed aspartate aminotransferase, AST)

Goth. Gothic

gou. gourde (Haitian monetary unit)

Gov. (*or* **gov.**) government • governor

Gov-Gen Governor-General

Govt (*or* **govt**) government

gox (*or* **GOX**) (gɒks) *Chem.* gaseous oxygen

gp (*or* **Gp**) group

g.p. *Printing* galley proofs • *Maths.* geometrical progression • *Printing* great primer

GP Gallup Poll • gas-permeable (of contact lenses) • general paralysis • *Music* general pause • general practitioner • general purpose • general-purpose computer • Gloria Patri (Latin: glory be to the Father) • graduated pension • Graduate in Pharmacy • grand(e) passion • Grand Prix • •*British vehicle registration for* SW London

GPA General Practitioners' Association • *Education* (USA) grade point average

GPALS Global Protection Against Limited Strikes (reduced SDI programme)

GPC *Chem.* gel permeation chromatography • general purposes committee

Gp Capt Group Captain

Gp Comdr Group Commander

gpd gallons per day

GPDST Girls' Public Day School Trust

gph (*or* **GPH**) gallons per hour

GPh Graduate in Pharmacy

GPHI Guild of Public Health Inspectors

GPI *Med.* general paralysis of the insane

GPIB *Computing* general-purpose interface bus

GPKT Grand Priory of the Knights of the Temple

GPM (*or* **gpm**) gallons per minute • (USA) graduated payment mortgage • *Freemasonry* Grand Past Master

GPMU Graphical, Paper, and Media Union (formed by merger of NGA and SOGAT)

GPO General Post Office • (USA) Government Printing Office

GPR genio populi Romani (Latin: to the genius of the Roman people) • ground-penetrating radar

GPS (*or* **gps**) gallons per second • Global Positioning System (US defence satellite) • Graduated Pension Scheme • (Australia) Great Public Schools (indicating a group of mainly nonstate schools, and of sporting competitions between them)

GPT *Biochem., Med.* glutamic pyruvic transaminase (renamed alanine amino-transferase, ALT) • Guild of Professional Toastmasters

GPU General Postal Union (*see* UPU) • (USA) General Public Utilities Corporation • Gosudarstvennoye Politicheskoye Upravlenie (Russian: State Political Administration; Soviet state security system, 1922–23)

GQ *Military* general quarters

GQG Grand Quartier Général (French: General Headquarters)

gr. grade • (*or* **gr**) grain (the unit) • gram • grammar • grand • great(er) • grey • gross • ground • group

Gr (ital.) *Physics, symbol for* Grashof number • Gunner

Gr. Grecian • Greece • Greek

GR *Science* gamma ray • general reconnaissance • *Physics* general relativity • general reserve • Georgius Rex (Latin: King George) • *British fishing port registration for* Gloucester • grand recorder • *international vehicle registration for* Greece • ground rent • Gulielmus Rex (Latin: King William) • Gurkha Rifles • *British vehicle registration for* Newcastle upon Tyne

GRA Game Research Association • Greyhound Racing Association

grad (græd) *Maths.* gradient

grad. grading • gradual • graduate(d)

GradIAE Graduate of the Institution of Automobile Engineers

GradIM Graduate of the Institute of Metals

GradInstBE Graduate Member of the Institution of British Engineers

GradInstP Graduate of the Institute of Physics

GradInstR Graduate of the Institute of Refrigeration

GradIPM Graduate of the Institute of Personnel Management

GradPRI Graduate of the Plastics and Rubber Institute

GradSE Graduate of the Society of Engineers

gram. grammar(ian) • grammatical

GRAS (USA) generally regarded as safe

GRB *Astronomy* gamma-ray burst

GRBI Gardeners' Royal Benevolent Institution

Gr. Br. (*or* **Gr. Brit.**) Great Britain

GRBS Gardeners' Royal Benevolent Society

GRC General Research Corporation

Gr. Capt. Group Captain

GRCM Graduate of the Royal College of Music

GRDF Gulf Rapid Deployment Force

GRE (USA) Graduate Record Examination • grant-related expenditure • Guardian Royal Exchange Assurance plc

gr.f. *Horse racing* grey filly

GRF *Biochem.* growth-hormone releasing factor

GRI (New Zealand) guaranteed retirement income

Gr-L Graeco-Latin

grn green

GRN goods received note

gro. gross (unit of quantity)

GRO Gamma Ray Observatory • General Register Office • Greenwich Royal Observatory

GROBDM General Register Office for Births, Deaths, and Marriages

gro. t. gross tons (*or* tonnage)

grp group

GRP *Biochem.* gastrin-releasing peptide • glass- (*or* glassfibre-)reinforced plastic (*or* polyester)

grs grains • gross

GRS *Astronomy* Great Red Spot (Jupiter)

GRSC Graduate of the Royal Society of Chemistry

GRSE Guild of Radio Service Engineers

GRSM Graduate of the Royal Schools of Music

gr. t. gross ton

GRT (*or* **g.r.t.**) gross registered tonnage

Gru *Astronomy* Grus

GRU (ex-USSR) Glavnoye Razvedyvatelnoye Upravleniye (Central Intelligence Office)

gr. wt. gross weight

g.r.y. *Finance* gross redemption yield

gs. guineas

g.s. (*or* **gs**) grandson • *Aeronautics* ground-speed

Gs *Magnetism, symbol for* gauss

GS (USA) General Schedule (civil service job-classification scheme) • General Secretary • general service • *Military* General Staff • geographical survey • geological survey · *Netball* goal shooter · gold standard · grammar school · *British vehicle registration for* Luton

GSA (USA) General Services Administration · Girl Scouts of America · Girls' Schools Association · Glasgow School of Art

GS&WR (Ireland) Great Southern and Western Railway

GSB Government Savings Bank

GSC *Chem.* gas–solid chromatography · General Service Corps · (USA) General Staff Corps

GSD general supply depot · *Astronomy* Greenwich sidereal gate

GSE ground service (*or* support) equipment

GSEE Geniki Synomospondia Ergaton Hellados (Greek General Confederation of Labour)

GSI *Med.* genuine stress incontinence

GSL Geological Society of London · group scout leader · (USA) guaranteed student loan

gsm gram per square metre (unit indicating paper thickness)

g.s.m. *Commerce* good sound merchantable (quality)

GSM Garrison Sergeant-Major · general sales manager · General Service Medal · global system for mobile communications · (*or* **GSMD**) (Member of the) Guildhall School of Music and Drama

GSO General Staff Officer

GSOH good sense of humour (in personal advertisements)

GSP *Chem.* glassfibre-strengthened polyester · Good Service Pension · *Economics* gross social product

GSR galvanic skin reflex (*or* response)

GSS geostationary satellite · global surveillance system · Government Statistical Service

GST (New Zealand, Canada) goods and services tax · *Astronomy* Greenwich Sidereal Time · *Meteorol.* gust

g-st *Knitting* garter-stitch

GSVQ General Scottish Vocational Qualification

GSW gunshot wound

gt *Bookbinding* gilt · great

gt. *Med.* gutta (Latin: a drop)

g.t. gas tight · gross tonnage

GT gas turbine · *Physics* gauge theory · *airline flight code for* GB Airways · Good

Templar · *Freemasonry* Grand Tiler · Gran Turismo (Italian: grand touring; sports car) · greetings telegram · *British vehicle registration for* SW London

GTA gas–tungsten arc

Gt Brit. (*or* **Gt Br.**) Great Britain

GTC (Scotland) General Teaching Council · Girls' Training Corps · (*or* **g.t.c.**) *Commerce* good till cancelled (*or* countermanded) · Government Training Centre

GTCL Graduate of Trinity College of Music, London

gtd guaranteed

gtee guarantee

GTI Gran Turismo Injection (sports car)

g.t.m. good this month

GTO Gran Turismo Omologata (Italian: certified for grand touring; of sports cars)

GTP *Biochem.* guanosine triphosphate

gtr greater

GTR general theory of relativity

GTS gas turbine ship · (USA) General Theological Seminary

gtt. *Med.* guttae (Latin: drops) · *Med.* guttatim (Latin: drop by drop)

GTT *Med.* glucose tolerance test

g.t.w. good this week

gu. guinea · *Heraldry* gules (red)

Gu. Guinea

GU gastric ulcer · (*or* **g.u.**) genitourinary · *US postcode for* Guam · *British fishing port registration for* Guernsey · *postcode for* Guildford · *British vehicle registration for* SE London

GUALO General Union of Associations of Loom Overlookers

guar. guarantee(d)

Guat. Guatemala

gui. *Music* guitar

GUI Golfing Union of Ireland · ('guːɪ) *Computing* graphical user interface

GUIDO (*or* **Guido**) ('gaɪdəʊ) *Astronautics* (USA) Guidance Officer

guil. guilder

Guin. Guinea

Gulag ('guːlæg) Glavnoye Upravleniye Lagerei (Soviet prison and labour camp system)

GUM genito-urinary medicine · (gʊm) (Russia) Gosudarstvenni Universalni Magazin (Universal State Store)

gun. gunnery

GUR *Med.* glucose utilization rate

GUS Great Universal Stores

GUT (gʌt) *Physics* grand unified theory

guttat. *Med.* guttatim (Latin: drop by drop)

g.u.v. gerecht und vollkommen (German: correct and complete)

GUY *international vehicle registration for* Guyana

g.v. gravimetric volume • gross valuation

GV (France) grande vitesse (fast goods train) • *British vehicle registration for* Ipswich • *airline flight code for* Riga Airlines-Express

GVH *Med.* graft-versus-host

GVHD *Med.* graft-versus-host disease

gvt (*or* **Gvt**) government

GVW gross vehicle weight

GW George Washington (1732–99, US president 1789–97) • *symbol for* gigawatt(s) • *British fishing port registration for* Glasgow • guided weapons (as in **GW**

cruiser) • *British vehicle registration for* SE London

GW-Basic *Computing* gee whizz beginners' all-purpose symbolic instruction code

GWP Government White Paper • gross world product

GWR Great Western Railway

GWS *Physics* Glashow–Weinberg–Salam (in **GWS theory**)

GX *airline flight code for* Pacificair • *British vehicle registration for* SE London

Gy *Physics, symbol for* gray

GY *British fishing port registration for* Grimsby • *airline flight code for* Guyana Airways • *British vehicle registration for* SE London

gyn. (*or* **GYN, gynaecol.**) gynaecological • gynaecology

GZ *British vehicle registration for* Belfast

G-Z *Astronomy* Giacobini–Zinner (a comet)

H

h (ital.) *Physics, symbol for* heat transfer coefficient • *symbol for* hecto- (prefix indicating 100, as in **hm**, hectometre) • (ital.) *symbol for* height • *symbol for* hour • (ital.) *Physics, symbol for* Planck constant • (ital.) *Thermodynamics, symbol for* specific enthalpy • (ital.) *Maths., indicating* a small increment • *indicating* the eighth vertical row of squares from the left on a chessboard

h. harbour • hard(ness) • heat • height • high • hip • *Sport* hit • horizontal • *Music* horn • horse • hot • hour • house • hundred • husband

H (bold ital.) *Mechanics, symbol for* angular impulse • (ital.) *Physics, Med., symbol for* dose equivalent • (ital.) *Thermodynamics, symbol for* enthalpy • (ital.) *Physics, symbol for* Hamiltonian • hard (indicating the degree of hardness of lead in a pencil; also in **HB**, hard black; **HH** (*or* **2H**), double hard; etc.) • *Immunol.* heavy (in **H-chain** of an immunoglobulin

molecule) • *Physics, symbol for* henry(s) • *Slang* heroin • *Med., Pharmacol.* histamine receptor (in **H₁, H₂**; used in specifying types of antihistamines) • *Biochem.* histidine • *British fishing port registration for* Hull • *international vehicle registration for* Hungary • *Chem., symbol for* hydrogen • (ital.) *Physics, Photog., symbol for* light exposure • (bold ital.) *Physics, symbol for* magnetic field strength

H. (*or* **H**) *Advertising* halfpage • Harbour • hardness • *Cards* hearts • herbaceous • Holy • *Music* horn • hospital • hour • hydrant

H4 *airline flight code for* Hainan Airlines • *international civil aircraft marking for* Solomon Islands

5H *international civil aircraft marking for* Tanzania

9H *international civil aircraft marking for* Malta

ha *symbol for* hectare

h.a. heir apparent • *Gunnery* high angle • hoc anno (Latin: in this year)

Ha *Chem., symbol for* hahnium (element 105)

Ha. Haiti(an) • Hawaii(an)

HA *British vehicle registration for* Dudley • *Horticulture* hardy annual • *postcode for* Harrow • Hautes-Alpes (department of France) • *airline flight code for* Hawaiian Airlines • Health Authority • heavy artillery • high altitude • Highway(s) Act • *Biochem.* histamine • Historical Association • Hockey Association • *Astronomy* hour angle • *Geology* humic acid • *international civil aircraft marking for* Hungary • *Biochem.* hyaluronic acid • Hydraulic Association of Great Britain

HAA heavy anti-aircraft • *Immunol.* hepatitis-associated antigen

HAAC (USA) Harper Adams Agricultural College

HA & M (ital.) Hymns Ancient and Modern

hab. habitat • habitation

Hab. *Bible* Habakkuk

HAB high-altitude bombing

HABA *Chem.* 2-(4-hydroxyphenylazo)-benzoic acid

hab. corp. *Law* habeas corpus (writ)

habt. habeat (Latin: let him have)

HAC high-alumina cement • Honourable Artillery Company • *Chem.* hydrogenated amorphous carbon

h.a.d. hereinafter described

haem. haemoglobin • haemorrhage

HAF Hellenic Air Force

Hag. *Bible* Haggai

hagiol. hagiology

HAI hospital-acquired infection

hal. halogen

Hal *Computing* hard-array logic

Hal. Orch. Hallé Orchestra

Ham. Hamburg

Han. Hanover(ian)

H. & B. *Botany* Humboldt and Bonpland (indicating the authors of a species, etc.)

h & c (*or* **H & C**) hot and cold (water)

h & t hardened and tempered

H & W Harland and Wolff (shipbuilders)

Hants Hampshire

HAO hydrogenated anthracene oil

HAP hazardous air pollutant

h. app. heir apparent

HaPV *Microbiol.* hamster polyomavirus

har. harbour

HARCVS Honorary Associate of the Royal College of Veterinary Surgeons

harm. harmonic • harmony

HARM (hɑːm) *Military* high-speed anti-radiation missile

harp. harpsichord

HART (hɑːt) (New Zealand) Halt All Racist Tours (antiracist sports organization)

Harv. (USA) Harvard University

HAS Headmasters' Association of Scotland • Health Advisory Service • Helicopter Air Service • Hospital Advisory Service

HASAWA Health and Safety at Work Act

HAT housing action (*or* association) trust

haust. *Med.* haustus (Latin: draught)

HAV hepatitis A virus

HAWT horizontal-axis wind turbine

haz. hazard(ous)

HAZOP ('hæzɒp) *Computing* hazard and operability study

h.b. *Sport* halfback • handbook • hardback (of books) • homing beacon • human being

Hb *Biochem.* haemoglobin

HB *British vehicle registration for* Cardiff • hard-black (on pencils; indicating medium-hard lead) • *Horticulture* hardy biennial • (Spain) Herri Batasuna (Basque independentist party) • His Beatitude • *Electronics* horizontal Bridgman • House of Bishops • *international civil aircraft marking for* Switzerland and Liechtenstein

H.B. & K. *Botany* Humboldt, Bonpland, and Kunth (indicating the authors of a species, etc.)

HBC *Electrical engineering* high breaking capacity • Historic Buildings Council • Hudson's Bay Company

HBCU (USA) Historically Black Colleges and Universities

HBD had (*or* has) been drinking

Hbf. Hauptbahnhof (German: central (*or* main) station)

HBIG hepatitis B immunoglobulin

HBLV *Med.* human B-lymphotropic virus

HBM Her (*or* His) Britannic Majesty (*or* Majesty's)

H-bomb hydrogen bomb

HBP high blood pressure

hbr harbour

HBS Harvard Business School

Hbt Hobart (Australia)

HBT *Electronics* heterojunction bipolar transistor

HBV hepatitis B virus

hby hereby

h.c. habitual criminal • hand control • heating cabinet • high capacity • (*or* **hc**) honoris causa (Latin: for the sake of honour; honorary) • hot and cold (water)

h/c *Insurance* held covered

HC *British vehicle registration for* Brighton • *international civil aircraft marking for* Ecuador • Hague Convention • Headmasters' (*or* Headteachers') Conference • health certificate • Heralds' College • High Church • High Commission(er) • High Court • higher certificate • highly commended • *Computing* high-speed CMOS • Highway Code • hockey club • Holy Communion • home counties • hors concours (French: not competing) • House of Clergy • House of Commons • house of correction • housing centre • housing corporation

H/C *Insurance* held covered

HCA Hospital Caterers' Association

HCAAS Homeless Children's Aid and Adoption Society

hcap (*or* **h'cap**) handicap

HCB House of Commons Bill

HCBA Hotel and Catering Benevolent Association

h.c.c. hydraulic cement concrete

h.c.d. high current density

h.c.e. human-caused error

h.c.f. *Maths.* highest common factor • hundred cubic feet

HCF high-calorific fuel • high carbohydrate and fibre • *Maths.* highest common factor • Honorary Chaplain to the Forces

HCFC hydrochlorofluorocarbon

hCG (*or* **HCG**) *Biochem., Med.* human chorionic gonadotrophin

HCH Herbert Clark Hoover (1874–1964, US president 1929–33) • hexachlorocyclohexane (an insecticide)

HCI Hotel and Catering Institute • human–computer interface (*or* interaction)

HCIL Hague Conference on International Law

HCIMA Hotel Catering and Institutional Management Association

HCJ High Court of Justice • Holy Child Jesus

h.c.l. high cost of living

HCM High Court Master • His (*or* Her) Catholic Majesty

HCO Harvard College Observatory • Higher Clerical Officer

hcp handicap

h.c.p. *Crystallog.* hexagonal close-packed

HCP House of Commons Paper

HCPT Historic Churches Preservation Trust

hcptr helicopter

HCR High Chief Ranger • (USA) highway contract route

h.c.s. high-carbon steel

HCS Home Civil Service

HCSA Hospital Consultants and Specialists Association

HCT *Computing* high-speed CMOS with TTL inputs

hd hand • head

h.d. heavy duty • high density • *Med.* hora decubitus (Latin: at bedtime)

HD *Chemical engineering* heavy distillate • *Med.* herniated disc • high-density • *Physics* high dose (of radiation) • (51st) Highland Division • Hodgkin's disease • home defence • honourable discharge • Hoover Dam • Huddersfield (postcode *or* British vehicle registration) • *Chem.* hydrogen-deuterium

H–D *Photog.* Hurter–Driffield (in **H–D curve**)

H/D *Shipping* Havre–Dunkirk

HDA (Australia) Hawkesbury Diploma in Agriculture • high-duty alloy

HDAL *Chem.* hexadecenal

h.d.a.t.z. high-density air-traffic zone

hdbk handbook

HDC *Med.* high-dose chemotherapy • *Law* holder in due course

HDCD high-density CD-ROM (*see* CD-ROM)

HDCR Higher Diploma of the College of Radiographers

HDD *Computing* hard disk drive • *Aeronautics, Computing* head-down display • heavy-duty diesel • Higher Dental Diploma

hdg heading

HDipEd Higher Diploma in Education

hdkf handkerchief

hdl. handle

HDL *Biochem., Med.* high-density lipoprotein

HDLC *Computing* high-level data link control (a communications protocol)

hdle *Horse racing* hurdle

hdlg handling

HDM high-duty metal

hdn harden

HDODA *Chem.* 1,6-hexanediol diacrylate

HDP heavy-duty petrol • (*or* **HDPE**) *Chem.* high-density polyethylene (*or* polyethene)

hdqrs headquarters

HDR *Physics* high dose rate (of radiation) • hot dry rock (in **HDR energy**)

HDTV high-definition television

HDV heavy-duty vehicle • *Med.* hepatitis delta virus

hdw. *Computing* hardware

hdwd hardwood • headword

h.e. heat engine • hic est (Latin: this is) • hub end

He *Chem., symbol for* helium

He. Hebrew

HE *Med.* hepatic encephalopathy • high-energy • higher education • high explosive • His Eminence • His (*or* Her) Excellency • home establishment • horizontal equivalent • hydraulic engineer • *Crystallog.* hydrogen embrittlement • *British vehicle registration for* Sheffield

HEA Health Education Authority • Horticultural Education Association

HEAO High Energy Astronomy (*or* Astrophysical) Observatory

Heb. Hebrew (language) • *Bible* Hebrews

Hebr. Hebrew (language) • Hebrides

HEC (école des) hautes études commerciales (French: (college of) higher commercial studies) • Health Education Council • Higher Education Corporation • *Chem.* hydroxyethyl cellulose

HEDCOM ('hɛd,kʊm) *US Air Force* headquarters command

HEDTA *Chem.* hydroxyethylethylenediaminotriacetate

HEED (hiːd) *Physics* high-energy electron diffraction

HEF high-energy fuel

HEH His (*or* Her) Exalted Highness

h.e.i. high-explosive incendiary

HEIC Honourable East India Company

HEICS Honourable East India Company's Service

heir app. heir apparent

heir pres. heir presumptive

hel. (*or* **heli.**) helicopter

HEL high-energy laser

Hellen. Hellenic • Hellenism • Hellenistic

HELLP *Med.* haemolysis, elevated liver enzymes, low platelet count (in **HELLP syndrome**)

helo. heliport

HELP (hɛlp) helicopter electronic landing path • Help Establish Lasting Peace

HEM *Physics* hybrid electromagnetic wave

HEMM heavy earth-moving machinery

HEMS helicopter-based emergency medical services

HEMT *Electronics* high-electron-mobility transistor

HEO Higher Executive Officer • highly elliptic-inclined orbit (as in **HEO satellite**)

HEOS high-elliptic-inclined-orbit satellite

HEP *Statistics* human error probability

HEPC (USA) hydroelectric power commission

HEPCAT ('hɛp,kæt) helicopter pilot control and training

her. heraldic • heraldry • *Law* heres (Latin: heir)

Her *Astronomy* Hercules

hera high-explosive rocket-assisted

herb. herbaceous • herbalist • herbarium

hered. heredity

Herod. Herodotus (*c.* 484–*c.* 424 BC, Greek historian)

herp. (*or* **herpet.**, **herpetol.**) herpetologist • herpetology

Hert Hertford College, Oxford

Herts Hertfordshire

HERU Higher Education Research Unit

HET heavy-equipment transporter

HETP *Chem.* height equivalent per theoretical plate

HEU highly enriched uranium

HEW (USA) Department of Health, Education, and Welfare (former government department; replaced by HHS)

hex (hɛks) *Computing* hexadecimal (notation)

hex. hexachord • hexagonal

hexa. hexamethylene tetramine (type of synthetic rubber)

hf half

h.f. hold fire • *Military* horse and foot

Hf *Chem., symbol for* hafnium

HF hard firm (on pencils; indicating hard lead) • *Physics* Hartree–Fock (approximation) • *Radio, etc.* high frequency • Holy

Father • home fleet • home forces • *British vehicle registration for* Liverpool

Hfa Haifa (Israel)

HFA hydrofluoroalkane

HFARA Honorary Foreign Associate of the Royal Academy

hf. bd. *Bookbinding* half binding (*or* bound)

HFC *Electricity* high-frequency current • *Chem.* hydrofluorocarbon

HFDF high-frequency direction finder

HFO heavy fuel oil • high-frequency oscillation

HFR *Nuclear engineering* high-flux reactor

HFRA Honorary Foreign Member of the Royal Academy

HFS heated front seats (in motor advertisements)

hg *symbol for* hectogram(s)

Hg *Chem., symbol for* mercury (Latin *hydrargyrum*)

HG *postcode for* Harrogate • Haute-Garonne (department of France) • High German • high grade • His (*or* Her) Grace • Holy Ghost • Home Guard • Horse Guards • *British vehicle registration for* Preston

HGC *Computing* Hercules graphics card

HGCA Home Grown Cereals Authority

hgd hogshead

HGDH His (*or* Her) Grand Ducal Highness

HGG *Med.* hypogammaglobulinaemia

hGH (*or* **HGH**) *Biochem.* human growth hormone

HGMM Hereditary Grand Master Mason

h.g.pt. hard gloss paint

hgr hangar • hanger

hgt height

HGTAC Home Grown Timber Advisory Committee

HGV heavy goods vehicle (replaced by LGV)

HGW heat-generating waste • H(erbert) G(eorge) Wells (1866–1946, British writer)

hgy (*or* **hgwy**) highway

hh hands (height measurement for horses)

HH *British vehicle registration for* Carlisle • (*or* **2H**) double hard (on pencils; indicating very hard lead) • *international civil aircraft marking for* Haiti • *British fishing port registration for* Harwich • *Chem., Physics* heavy hydrogen • *Astronomy* Herbig–Haro (in **HH object**) • (Member of the) Hesketh Hubbard Art Society • His (*or* Her) Highness • His Holiness (title of the Pope) • His (*or* Her) Honour

HHA *Horticulture* half-hardy annual • Historic Houses Association

HHB *Horticulture* half-hardy biennial

hhd hogshead

HHD (USA) Doctor of Humanities (Latin *Humanitatum Doctor*) • hypertensive heart disease

HHDWS heavy handy deadweight scrap

HHFA (USA) Housing and Home Finance Agency

HHH (*or* **3H**) treble hard (on pencils; indicating very hard lead)

HHI Highland Home Industries

HHNK *Med.* hyperglycaemic hyperosmolar nonketoacidotic coma (in diabetes)

H-Hour Hour Hour (i.e. the specified time at which an operation is to begin)

HHP *Horticulture* half-hardy perennial

HHS (USA) Department of Health and Human Services (replaced HEW)

HHV human herpesvirus

Hi. Hindi

HI *international civil aircraft marking for* Dominican Republic • *US postcode for* Hawaii • Hawaiian Islands • hearing impaired • *Surveying* height of instrument • hic iacet (Latin: here lies; on gravestones) • high intensity • *Irish vehicle registration for* Tipperary

HIA Housing Improvement Association

hi. ac. (*or* **hiac**) high accuracy

HIAS (USA) Hebrew Immigrant Aid Society

Hib (hɪb) *Med. Haemophilus influenzae* type B (as in **Hib vaccine**)

Hib. Hibernia(n)

HIB *Physics* heavy-ion beam

Hibs. (hɪbz) (Scotland) Hibernian (football club)

hicat ('haɪ,kæt) high-altitude clear air turbulence

Hi. Com. High Command • High Commission(er)

HIDB Highlands and Islands Development Board

hier. hieroglyphics

hi-fi ('haɪ'faɪ) high fidelity

hifor ('haɪfɔː) high-level forecast

HIH His (*or* Her) Imperial Highness

HIL *Chem.* hazardous immiscible liquid

hilac (*or* **HILAC**) ('haɪlæk) *Nuclear physics* heavy-ion linear accelerator

HILAT ('haɪlæt) high-latitude (in **HILAT satellite**)

HIM His (or Her) Imperial Majesty

Hind. Hindi • Hindu • Hindustan(i)

HIP ('eɪtʃ 'aɪ 'pi: or hɪp) (USA) health-insurance plan • *Accounting* human information processing

HIPAR high-power acquisition radar

HIPDA *Chem.* hydroxyisophthalyl dihydroxamic acid

hipot ('haɪ,pɒt) high potential

Hipp. Hippocrates (c. 460–c. 375 BC, Greek physician)

Hippi high-performance parallel processor interface

HIPS (hɪpz) high-impact polystyrene (type of synthetic plastic)

hirel ('haɪ,rɛl) high reliability

hi-res ('haɪ'rɛz) *Physics* high resolution

His (or **his**) *Biochem.* histidine

HIS hic iacet sepultus or sepulta (Latin: here lies buried; on gravestones)

hist. (or **histol.**) histology • (or **histn**) historian • historic(al) • history

Hitt. Hittite

HIUS Hispanic Institute of the United States

HIV human immunodeficiency virus (the cause of AIDS)

HJ *British vehicle registration for* Chelmsford • hic jacet (Latin: here lies; on gravestones) • high jump • Hilal-e-Jurat (Pakistani honour) • Hitler Jugend (German: Hitler Youth)

HJBT *Electronics* heterojunction bipolar transistor

HJS hic jacet sepultus or sepulta (Latin: here lies buried; on gravestones)

HK *British vehicle registration for* Chelmsford • *international civil aircraft marking for* Colombia • Hong Kong (abbrev. or IVR) • *Taxation* housekeeper allowance • House of Keys (Manx Parliament)

hkf handkerchief

HKJ *international vehicle registration for* (Hashemite Kingdom of) Jordan

hl (or **hL**) *symbol for* hectolitre(s)

h.l. hoc loco (Latin: in this place)

HI. Heilige(r) (German: Saint)

HL hard labour • *British fishing port registration for* Hartlepool • Haute-Loire (department of France) • honours list • House of Laity • House of Lords • *British vehicle registration for* Sheffield •

international civil aircraft marking for South Korea

HLA *Immunol.* human lymphocyte antigen (as in **HLA system**)

HLB *Chem.* hydrophilic-lipophilic balance

HLBB (USA) Home Loan Bank Board

HLD Doctor of Humane Letters

HLE *Physics* high-level exposure (to radiation)

HLHSR *Computing* hidden-line/hidden-surface removal

HLI Highland Light Infantry

HLL *Computing* high-level language

HLNW high-level nuclear waste

hlpr helper

HLPR Howard League for Penal Reform

HLRW high-level radioactive waste

HLS Harvard Law School

HLW headlamp wipe(rs) (in motor advertisements) • high-level (radioactive) waste

HLWN highest low water of neap (tides)

HLWW headlamp wash and wipe (in motor advertisements)

hm *symbol for* hectometre(s)

h.m. hallmark • hoc mense (Latin: in this month)

HM *airline flight code for* Air Seychelles • harbour master • *Music* harmonic mean • Haute-Marne (department of France) • *Chem.* hazardous material • headmaster (or headmistress) • heavy metal • Her (or His) Majesty • home mission • *British vehicle registration for* London (central)

HMA Head Masters' Association • *Computing* high-memory area

HMAC Her (or His) Majesty's Aircraft Carrier

HMAS Her (or His) Majesty's Australian Ship

HMBDV Her (or His) Majesty's Boom Defence Vessel

HMC Headmasters' and Headmistresses' Conference • Her (or His) Majesty's Customs • Historical Manuscripts Commission • Horticultural Marketing Council • Hospital Management Committee • Household Mortgage Corporation

HMCA Hospital and Medical Care Association

HMCIC Her (or His) Majesty's Chief Inspector of Constabulary

HMCIF Her (or His) Majesty's Chief Inspector of Factories

HMCN Her (*or* His) Majesty's Canadian Navy

HMCS Her (*or* His) Majesty's Canadian Ship

HMCSC Her (*or* His) Majesty's Civil Service Commissioners

hmd humid

HMD *Computing* head-mounted display · Her (*or* His) Majesty's Destroyer

HMF *Chem.* heavy-metal fluoride · Her (*or* His) Majesty's Forces

HMFI Her (*or* His) Majesty's Factory Inspectorate

HMG heavy machine gun · Her (*or* His) Majesty's Government · *Med.* human menopausal gonadotrophin

HMHS Her (*or* His) Majesty's Hospital Ship

HMI Her (*or* His) Majesty's Inspector (of schools) · Her (*or* His) Majesty's Lieutenant · *Computing* human–machine interface

HMIED Honorary Member of the Institute of Engineering Designers

HMIP Her (*or* His) Majesty's Inspectorate of Pollution

HMIT Her (*or* His) Majesty's Inspector of Taxes

HML Her (*or* His) Majesty's Lieutenant

HMLR Her (*or* His) Majesty's Land Registry

HMML Her (*or* His) Majesty's Motor Launch

HMMS Her (*or* His) Majesty's Mine Sweeper

HMNZS Her (*or* His) Majesty's New Zealand Ship

HMO (USA) health maintenance organization

HMOCS Her (*or* His) Majesty's Overseas Civil Service

h.m.p. handmade paper

HMP Her (*or* His) Majesty's Prison · hoc monumentum posuit (Latin: he (*or* she) erected this monument)

HMPA *Chem.* hexamethylphosphoramide

HMPT *Chem.* hexamethylphosphorous triamide

HMRT Her (*or* His) Majesty's Rescue Tug

h.m.s. hours, minutes, seconds

HMS *Mining* heavy media separation · Her (*or* His) Majesty's Service · Her (*or* His) Majesty's Ship · *Biochem.* hexose monophosphate shunt

HMSO Her (*or* His) Majesty's Stationery Office

hmstd homestead

HMT Her (*or* His) Majesty's Trawler · Her (*or* His) Majesty's Treasury · Her (*or* His) Majesty's Tug

HMV His Master's Voice (music retailer)

HMW *Biochem.* high molecular weight

HMWA Hairdressing Manufacturers' and Wholesalers' Association

HMXB *Astronomy* high-mass X-ray binary

hn *Music* horn

h.n. *Med.* hac nocte (Latin: tonight)

HN *British vehicle registration for* Middlesbrough

HNC *Education* Higher National Certificate

HND *Education* Higher National Diploma

hndbk handbook

HNL *Physics* helium–neon laser

hnRNA *Biochem.* heterogeneous nuclear RNA

hnRNP *Biochem.* heterogeneous nuclear ribonucleoprotein

Hnrs Honours

ho. house

h.o. hold over

Ho *Chem., symbol for* holmium

HO *airline flight code for* Airways International · *British vehicle registration for* Bournemouth · *Law* habitual offender · head office · Home Office · Hydrographic Office

h.o.c. held on charge

HOC *Chem.* halogenated organic compound · heavy organic chemical

HoC House of Commons

HoD head of department

H of C House of Commons

H of K House of Keys

H of L House of Lords

H of R (USA) House of Representatives

Hol *Geology* Holocene

Hol. (*or* **Holl.**) Holland

HOLC (USA) Home Owners' Loan Corporation

HOLLAND hope our love lasts (*or* lives) and never dies

Holmes (hǝʊmz) Home Office Large Major Enquiry System (computer used in crime investigation)

Hom. Homer (8th century BC, Greek poet)

homeo. (*or* **homo.**) homeopath(ic) • homeopathy

HOMO *Chem.* highest occupied molecular orbital

hon. honorary • honour • honourable

Hon (*or* **Hon.**) Honorary (in titles, as in **HonFInstP**, Honorary Fellow of the Institute of Physics) • Honorary Member (in titles, as in **HonRCM**, Honorary Member of the Royal College of Music) • Honourable (title)

Hond. Honduras

Hono. Honolulu

hons honours

Hon. Sec. Honorary Secretary

HOOD (hʊd) *Computing* hierarchical object-oriented design

Hook. *Botany* (Sir William) Hooker (indicating the author of a species, etc.)

Hook. fil. *Botany* Hooker fils (Sir Joseph Hooker, son of Sir William Hooker; indicating the author of a species, etc.)

hor. horizon • horizontal • horology

Hor *Astronomy* Horologium

Hor. Horace (65–8 BC, Roman poet)

hor. decub. *Med.* hora decubitus (Latin: at bedtime)

HoReCa (International Union of National Associations of) Hotel, Restaurant, and Café Keepers

horol. horological • horology

hort. horticultural • horticulture

HORU Home Office Research Unit

Hos. *Bible* Hosea

hosp. hospital

Hotol ('həʊ,tɒl) *Astronautics* horizontal takeoff and landing (launch vehicle)

HOV *Transport* high-occupancy vehicle

how. howitzer

hp *symbol for* horsepower

h.p. half pay • heir presumptive • *Electricity* high power • hire purchase • horizontally polarized

HP *airline flight code for* America West Airlines • *British vehicle registration for* Coventry • Handley Page (aircraft) • *Horticulture* hardy perennial • Hautes-Pyrénées (department of France) • *postcode for* Hemel Hempstead • (USA) Hewlett–Packard (electronics and computing manufacturer) • high performance • *Electricity* high power • high pressure • high priest • Himachal Pradesh (India) • hire purchase • hot-pressed (paper) • house physician • Houses of Parliament • *Horticulture* hybrid perpetual (rose) • *international civil aircraft marking for* Panama

HPA Hospital Physicists' Association

HPC *Med.* history of present complaint

hpch. (*or* **hpd**) harpsichord

HPF *Computing* highest priority first

HPFS *Computing* high-performance filing system

HPGL *Computing, trademark* Hewlett-Packard graphics language

HPIB *Computing, trademark* Hewlett-Packard interface bus

HPk Hilal-e-Pakistan (Pakistani honour)

HPLC *Chem.* high-performance liquid chromatography • *Chem.* high-pressure liquid chromatography

HPP *Chem.* high-pressure polyethylene (*or* polyethene)

HPS high-pressure steam • high-protein supplement

HPT high-pressure turbine

HPTA Hire Purchase Trade Association

HPTLC *Chem.* high-performance thin-layer chromatography

HPV *Med.* human papilloma virus

h.q. headquarters • hoc quaere (Latin: look for this)

HQ headquarters

HQA Hilal-i-Quaid-i-Azam (Pakistani honour)

HQBA Headquarters Base Area

HQMC (USA) Headquarters, Marine Corps

hr *Meteorol.* hail and rain • hour

h.r. *Baseball* home run

Hr Herr (German: Mr, Sir) • Hussar

HR *international vehicle registration for* Croatia • *Med.* heart rate • *postcode for* Hereford • Highland Regiment • *Optics* high resolution • Home Rule(r) • *Baseball* home run • *international civil aircraft marking for* Honduras • (USA) House of Representatives • human resources • *British vehicle registration for* Swindon

H–R *Astronomy* Hertzsprung–Russell (in **H–R diagram**)

HRA (USA) Health Resources Administration

HRC *Chem.* high-resolution chromatography • Holy Roman Church

HRCA Honorary Royal Cambrian Academician

HRCT *Med.* high-resolution computerized (*or* computed) tomography

HRE Holy Roman Emperor (or Empire)

HREM high-resolution electron microscopy

HRGC Chem. high-resolution gas chromatography

HRGI Honorary Member of the Royal Glasgow Institute of the Fine Arts

HRH His (or Her) Royal Highness

HRHA Honorary Member of the Royal Hibernian Academy

HRI Honorary Member of the Royal Institute of Painters in Water Colours

HRIP hic requiescit in pace (Latin: here rests in peace; on gravestones)

HRM human resource management

HRMS Chem. high-resolution mass spectrometry

Hrn Herr(e)n (German: Messrs, Sirs)

HROI Honorary Member of the Royal Institute of Oil Painters

HRP Home Responsibilities Protection • Military human remains pouch

HRR Taxation higher reduced rate

hrs hours

HRSA Honorary Member of the Royal Scottish Academy

hrsg. herausgegeben (German: edited or published)

HRSW Honorary Member of the Royal Scottish Water Colour Society

HRT Med. hormone replacement therapy

HRTEM high-resolution transmission electron microscopy

hs Meteorol. hail and snow

h.s. highest score • hoc sensu (Latin: in this sense) • Med. hora somni (Latin: at bedtime)

Hs. Handschrift (German: manuscript)

HS British vehicle registration for Glasgow • Haute-Saône (department of France) • Hawker Siddeley (aircraft) • hic sepultus or sepulta (Latin: here is buried; on gravestones) • airline flight code for Highland Air • high school • Home Secretary • hospital ship • house surgeon • Numismatics, symbol for sesterce (Roman coin) • international civil aircraft marking for Thailand

HSA (USA) Health Systems Agency • Med. human serum albumin

HSAB Chem. hard and soft acids and bases

HSB Computing hue, saturation, brightness

HSC Med. haemopoietic stem cell • Health and Safety Commission • (Australia) Higher School Certificate

HSD heat-storage device • high-speed diesel

HSDU hospital sterilization and disinfection unit

hse (or **Hse**) house

HSE Health and Safety Executive • hic sepultus (or sepulta) est (Latin: here lies buried; on gravestones)

hsekpr housekeeper

hsg housing

HSH His (or Her) Serene Highness

HSI Computing human–system interface (or interaction)

HSL Computing hue, saturation, lightness • Huguenot Society of London

HSLA high-strength, low-alloy (steel)

HSM Computing hierarchical storage management • His (or Her) Serene Majesty

HSO Hamburg Symphony Orchestra

HSP Med. heat-shock protein

HSR Genetics homogeneously staining region

Hss. Handschriften (German: manuscripts)

HSS Fellow of the Historical Society (Latin Historicae Societatis Socius) • high-speed steel

HSSU hospital sterile supply unit

HST Harry S Truman (1884–1972, US president 1945–53) • Hawaii Standard Time • highest spring tide • high-speed train • Hubble Space Telescope • hypersonic transport

HSV herpes simplex virus

HSWA (USA) Hazardous and Solid Waste Act (or Amendments)

ht heat • height

h.t. Sport half time • halftone • heavy tank • hoc tempore (Latin: at this time) • hoc titulo (Latin: in (or under) this title)

HT British vehicle registration for Bristol • Sport half time • Hawaii Time • heat treatment • high temperature • Electrical engineering high tension • high tide • high treason • Horticulture hybrid tea (rose)

5-HT Biochem. 5-hydroxytryptamine

HTA Horticultural Traders' Association • Household Textile Association

h.t.b. high-tension battery

HTB Metallurgy hexagonal tungsten bronze

HTC Physics heat-transfer coefficient

htd heated

Hte (*pl.* **Htes**) Haute (French: High; used in placenames)

HTGR *Nuclear engineering* high-temperature gas-cooled reactor

HTLV *Med.* human T-cell lymphotropic virus

HTM heat-transfer medium

HTML *Computing* hypertext mark-up language

HTO *Chem.* hydrous titanium oxide (catalyst)

HTOL *Astronautics* horizontal takeoff and landing

htr heater

HTR *Nuclear engineering* high-temperature (gas-cooled) reactor

H. Trin. Holy Trinity

h.t.s. half-time survey • high-tensile steel

Hts Heights (in place names)

HTS high-temperature superconductor (*or* superconductivity)

HTT heavy tactical transport

HTTL *Computing* high-speed transistor transistor logic

HTTP *Computing* hypertext transfer (*or* transport) protocol

HTTR *Nuclear engineering* high-temperature test reactor

HTU *Chemical engineering* height of a transfer unit

HTV Harlech Television

ht wkt *Cricket* hit wicket

HU *British vehicle registration for* Bristol • Harvard University • *postcode for* Hull

HUAC ('hju:æk) House (of Representatives) Un-American Activities Committee

HUD *Aeronautics, Computing* head-up display • (USA) (Department of) Housing and Urban Development

Hugo ('hju:gəʊ) Human Genome Organization

HUJ Hebrew University of Jerusalem

HUKFORLANT *US Navy* hunter-killer forces, Atlantic

HUKS *US Navy* hunter-killer submarine

hum. human • humane • (*or* **Hum.**) humanities (classics) • humanity • humble • humorous

HUMINT (*or* **humint**) ('hju:mɪnt) *Military* human intelligence (espionage activities)

HUMRRO Human Resources Research Office

HUMV ('hʌm,vi:) *Military* human light vehicle

hund. hundred

Hung. Hungarian • Hungary

Hunts Huntingdonshire (former county)

HUP Harvard University Press • *Chem.* hydrogenuranylphosphate tetrahydrate

hur. (*or* **hurr.**) hurricane

HURCN *Meteorol.* hurricane

Husat ('hju:sæt) Human Science and Advanced Technology Research Institute

husb. husbandry

h.v. high vacuum • high velocity

HV health visitor • *Anatomy* hepatic vein • high velocity • high voltage • hoc verbum (Latin: this word) • *British vehicle registration for* London (central) • *airline flight code for* Transavia Airlines

HVA Health Visitors' Association (trade union; now a section of the MFS) • *Chem.* homovanillic acid

HVAC heating, ventilation, air conditioning • high-voltage alternating current

HVAR high-velocity aircraft rocket

HVCA Heating and Ventilating Contractors' Association

HVCert Health Visitor's Certificate

HVDC high-voltage direct current

HVEM high-voltage electron microscope

HVO Hrvatsko vijeće odbrane (Croatian Defence Council)

HVP hydrolysed vegetable protein

HVT *Physics* half-value thickness • health visitor teacher

hvy heavy

h.w. *Cricket* hit wicket

h/w herewith • husband and wife

HW *British vehicle registration for* Bristol • hazardous waste • high water • hot water

HWL Henry Wadsworth Longfellow (1807–82, US poet) • (*or* **h.w.l.**) high-water line

HWM high-water mark

HWONT high water, ordinary neap tides

HWOST high water, ordinary spring tides

HWR *Nuclear physics* heavy-water reactor

HWS hurricane warning system

HWW headlamp wash and wipe (in motor advertisements) • *Engineering* hot-and-warm worked

hwy highway

HX *postcode for* Halifax • *British vehicle registration for* London (central)

HXR hard X-ray

hy heavy • highway

Hy. Henry

HY *British vehicle registration for* Bristol •

airline flight code for Uzbekistan Airways

Hya *Astronomy* Hydra

hyb. hybrid

HYCOSY (*or* **HyCoSy**) (haɪˈkəʊzɪ) *Med.* hysterosalpingo-contrast sonography

hyd. hydrate • hydraulic • (*or* **hydrog.**) hydrographic

hydt hydrant

hyg. hygiene

Hyi *Astronomy* Hydrus

hyp. hypodermic • *Maths.* hypotenuse • (*or* **hypoth.**) hypothesis • (*or* **hypoth.**) hypothetical

Hz *Physics, symbol for* hertz

HZ *Meteorol.* dust haze • *international civil aircraft marking for* Saudi Arabia • *British vehicle registration for* Tyrone

HZD *Chem.* hydrated zirconium dioxide

HZO *Chem.* hydrated zirconium oxide

hzy *Meteorol.* hazy

I

i *symbol for* the imaginary number √−1 • (*ital.*) *Physics, symbol for* instantaneous current • (*bold ital.*) *Maths., symbol for* unit coordinate vector • (*ital.*) *Chem., symbol for* a van't Hoff factor

i. id (Latin: that) • incisor (tooth) • indicate • *Banking* interest • *Grammar* intransitive

I (*ital.*) *Physics, symbol for* electric current • *Chem., symbol for* inductive (*or* inductomeric) effect • *Biochem.* inosine • *symbol for* inti (Peruvian monetary unit) • *Chem., symbol for* iodine • (*ital.*) *Chem., symbol for* ionic strength • (*ital.*) *Chem., Physics, symbol for* ionization potential • *Biochem.* isoleucine • (*ital.*) *Physics, symbol for* isospin quantum number • *international vehicle registration or international civil aircraft marking for* Italy • (*ital.; or* I_v) *Physics, symbol for* luminous intensity • (*ital.*) *Mechanics, symbol for* moment of inertia • *Roman numeral for* one • (*ital.; or* I_e) *Physics, symbol for* radiant intensity • (*ital.*) *Maths., symbol for* unit matrix • *Logic, indicating* a particular affirmative categorial statement

I. (*or* **I**) Iesus (Latin: Jesus) • Imperator (Latin: Emperor) • Imperatrix (Latin: Empress) • Imperial • Imperium (Latin: Empire) • (single column) inch (of advertisements) • incumbent • Independence • Independent • India(n) • Infidelis (Latin: unbeliever, infidel) • Inspector • Institute • Instructor • intelligence • interceptor • International • interpreter • Ireland • Irish • Island (*or* Isle) • issue • Italian

7I *airline flight code for* Imperial Air

i.a. immediately available • in absentia (Latin: while absent) • *Aeronautics* indicated altitude • initial appearance

i.A. im Auftrage (German: by order of)

Ia. Iowa

IA *British vehicle registration for* Antrim • Incorporated Accountant • Indian Army • infected area • information anxiety • *Taxation* initial allowance • Institute of Actuaries • Inter-American • *Med.* intra-arterial • *US postcode for* Iowa • *airline flight code for* Iraqi Airways

I/A Isle of Anglesey

IA5 *Computing* International Alphabet, Number 5

IAA indoleacetic acid (plant hormone) • Institute of Industrial Administration • International Academy of Astronautics • International Actuarial Association • International Advertising Association

IAAA Irish Amateur Athletic Association • Irish Association of Advertising Agencies

IAAE (USA) Institution of Automotive and Aeronautical Engineers

IAAF International Amateur Athletic Federation

IAAP International Association of Applied Psychology

IAAS Incorporated Association of Architects and Surveyors

IAB Industrial Advisory Board • Industrial

Arbitration Board • Inter-American Bank • Internet architecture board

IABA International Association of Aircraft Brokers and Agents

IABO International Association of Biological Oceanography

i.a.c. integration, assembly, and checkout

IAC (USA) Industrial Advisory Council • Institute of Amateur Cinematographers

IACA Independent Air Carriers' Association

IACB International Advisory Committee on Bibliography (in UNESCO)

IACCP Inter-American Council of Commerce and Production

IACOMS International Advisory Committee on Marine Sciences (in the FAO)

IACP International Association for Child Psychiatry and Allied Professions • International Association of Chiefs of Police • (USA) International Association of Computer Programmers

IACS International Annealed Copper Standard

IADB Inter-American Defense Board • Inter-American Development Bank

IADR International Association for Dental Research

IAE (USA) Institute of Atomic Energy • Institute of Automobile (*or* Automotive) Engineers

IAEA International Atomic Energy Agency

IAEC Israel Atomic Energy Commission

IAECOSOC Inter-American Economic and Social Council

IAEE International Association of Energy Economists

i.a.f. *Aviation* interview after flight

IAF Indian Air Force • Indian Auxiliary Force • International Astronautical Federation

IAFD International Association on Food Distribution

IAG International Association of Geodesy • International Association of Geology • International Association of Gerontology

IAGB & I Ileostomy Association of Great Britain and Ireland

IAgrE Institution of Agricultural Engineers

IAH International Association of Hydrogeologists • International Association of Hydrology

IAHA Inter-American Hotel Association

IAHM Incorporated Association of Headmasters

IAHP International Association of Horticultural Producers

IAHR International Association for Hydraulic Research • International Association for the History of Religions

IAI International African Institute

IAL Imperial Arts League • *Computing* international algorithmic language • Irish Academy of Letters

IALA International African Law Association • International Association of Lighthouse Authorities

IALL International Association of Law Libraries

IALS International Association of Legal Science

IAM Institute of Administrative Management • Institute of Advanced Motorists • Institute of Aviation Medicine • *Anatomy* internal auditory meatus • (USA) International Association of Machinists and Aerospace Workers (trade union) • International Association of Meteorology • International Association of Microbiologists

IAMA Incorporated Advertising Managers' Association

IAMAP International Association of Meteorology and Atmospheric Physics

IAMC Indian Army Medical Corps

IAML International Association of Music Libraries

IAMS International Association of Microbiological Societies (*or* Studies)

IANC International Airline Navigators' Council

IANE Institute of Advanced Nursing Education (of the Rcn)

IANEC Inter-American Nuclear Energy Commission

IAO *Meteorol.* in and out of cloud • Incorporated Association of Organists

IAOC Indian Army Ordnance Corps

IAOS Irish Agricultural Organization Society

IAP *Physics* imaging atom probe • International Academy of Pathology • Internet access provider

IAPA Inter-American Press Association

IAPB International Association for the Prevention of Blindness

IAPC International Auditing Practices Committee

IAPG International Association of Physical Geography

IAPH International Association of Ports and Harbours

IAPO International Association of Physical Oceanography

IAPS Incorporated Association of Preparatory Schools

IAPSO International Association for the Physical Sciences of the Oceans

IAPT International Association for Plant Taxonomy

IARA Inter-Allied Reparations Agency

IARC Indian Agricultural Research Council • International Agency for Research on Cancer

IARD International Association for Rural Development

IARF International Association for Religious Freedom

IARI Indian Agricultural Research Institute

IARO Indian Army Reserve of Officers

IARU International Amateur Radio Union

IAS *Computing* immediate access store • Indian Administrative (formerly Civil) Service • (*or* **i.a.s.**) *Aeronautics* indicated air speed • *Chem.* infrared absorbed spectroscopy • (USA) Institute for Advanced Studies (as in **IAS computer**) • (USA) Institute of the Aerospace Sciences • (*or* **i.a.s.**) *Aeronautics* instrument approach system • International Accounting Standard

IASA International Air Safety Association

IASC Indian Army Service Corps • International Accounting Standards Committee

IASH International Association of Scientific Hydrology

IASI Inter-American Statistical Institute

i.a.s.o.r. ice and snow on runway

IASS International Association for Scandinavian Studies • International Association of Soil Science

i.a.t. inside air temperature

IAT International Atomic Time

IATA (aɪˈɑːtə, iːˈɑːtə) International Air Transport Association • International Amateur Theatre Association

IATUL International Association of Technological University Libraries

IAU International Association of Universities • International Astronomical Union

IAUPL International Association of University Professors and Lecturers

IAV International Association of Vulcanology

IAVG International Association for Vocational Guidance

i.a.w. in accordance with

IAW International Alliance of Women

IAWPRC International Association on Water Pollution Research and Control

ib. ibidem (*see* ibid.)

IB *British vehicle registration for* Armagh • *airline flight code for* Iberia • in bond • incendiary bomb • industrial business • information bureau • instruction book • intelligence branch • *Physics* internal bremsstrahlung • International Bank (for Reconstruction and Development) • invoice book

IBA Independent Bankers' Association • Independent Broadcasting Authority • *Horticulture* indole 3-butyric acid (rooting compound) • Industrial Bankers' Association • *Taxation* industrial buildings allowances • International Bar Association • International Bowling Association • Investment Bankers' Association • *Chem.* ion-beam analysis

IBAA Investment Bankers' Association of America

IBAE Institution of British Agricultural Engineers

IBB Institute of British Bakers • International Bowling Board • Invest in Britain Bureau

IBBISB/BF&H (USA) International Brotherhood of Boilermakers, Iron Shipbuilders, Blacksmiths, Forgers, and Helpers

IBBR *Finance* interbank bid rate

IBC International Broadcasting Corporation

IBD Incorporated Institute of British Decorators and Interior Designers • *Med.* inflammatory bowel disease • *Electronics* ion-beam deposition

IBE Institute of British Engineers • International Bureau of Education (now part of UNESCO)

IBEL *Finance* interest-bearing eligible liability

IBEW (USA) International Brotherhood of Electrical Workers

IBF Institute of British Foundrymen •

International Badminton Federation • International Boxing Federation

IBFM *Physics* interacting boson-fermion model

IBG Incorporated Brewers Guild • Institute of British Geographers • *Computing* interblock gap

IBI (*or* **i.b.i.**) *Book-keeping* invoice book inwards

ibid. ibidem (Latin: in the same place; indicating a previously cited reference to a book, etc.)

IBID ('ɪbɪd) international bibliographical description

IBiol Institute of Biology

IBK Institute of Book-Keepers

IBM *Physics* interacting boson model • intercontinental ballistic missile • International Business Machines (Corporation; computer manufacturer)

IBMBR *Finance* interbank market bid rate

IBO (*or* **i.b.o.**) *Book-keeping* invoice book outwards

i.b.p. initial boiling point

IBP Institute of British Photographers • International Biological Programme (for conservation of natural communities)

IBPAT (USA) International Brotherhood of Painters and Allied Trades

IBRC Insurance Brokers Registration Council

IBRD International Bank for Reconstruction and Development (the World Bank)

IBRO International Bank Research Organization • International Brain Research Organization

IBS (*or* **IB(Scot)**) Institute of Bankers in Scotland • *Physics* ion-beam sputtering • *Chem.* ion-beam synthesis • irritable bowel syndrome

IBST Institute of British Surgical Technicians

IBT International Brotherhood of Teamsters, Chauffeurs, Warehousemen, and Helpers of America

IBTE Institution of British Telecommunications Engineers

IBWM International Bureau of Weights and Measures

i.c. index correction • instrument correction • internal communication

i/c in charge (of) • in command

IC *Irish vehicle registration for* Carlow • identity card • Iesus Christus (Latin: Jesus Christ) • *Grammar* immediate

constituent • *Immunol.* immune complex • Imperial College (of Science and Technology, London) • *Astrology* Imum Coeli (the lowest point on the ecliptic below the horizon) • *Astronomy* Index Catalogue • *airline flight code for* Indian Airlines • industrial court • information centre • *Electronics* integrated circuit • Intelligence Corps • *Chem.* intermetallic compound • *Engineering* internal-combustion (engine) • *Physics* internal conversion • *Chem.* ion chromatography • *Physics* ionization chamber

I-C Indo-China

ICA ignition control additive (for motor vehicles) • Industrial Caterers' Association • Institute of Chartered Accountants in England and Wales • Institute of Contemporary Arts • *Med.* internal carotid artery • International Cartographic Association • International Chefs' Association • International Coffee Agreement • International Colour Authority • International Commercial Arbitration • International Commission on Acoustics • International Commodity Agreement • International Cooperation Administration • International Cooperative Alliance • International Council on Archives • International Court of Arbitration • International Cyclist Association • invalid care allowance

ICAA Invalid Children's Aid Association

ICAE International Commission on Agricultural Engineering

ICAEW Institute of Chartered Accountants in England and Wales

ICAI Institute of Chartered Accountants in Ireland • International Commission for Agricultural Industries

ICAM Institute of Corn and Agricultural Merchants

ICAN International Commission for Air Navigation

IC & CY Inns of Court and City Yeomanry

ICAO International Civil Aviation Organization

ICAP Institute of Certified Ambulance Personnel • International Congress of Applied Psychology

ICAR Indian Council of Agricultural Research

ICAS Institute of Chartered Accountants of Scotland • International Council of

Aeronautical Sciences • International Council of Aerospace Sciences

ICB Institute of Comparative Biology

ICBA International Community of Book-sellers' Associations

ICBD International Council of Ballroom Dancing

ICBHI Industrial Craft (Member) of the British Horological Institute

ICBM intercontinental ballistic missile

ICBN International Code of Botanical Nomenclature

ICBP International Council for Bird Preservation

ICC (USA) Indian Claims Commission • intercounty championship • International Chamber of Commerce • International Children's Centre • International Congregational Council • International Convention Centre (Birmingham) • International Correspondence Colleges • International (formerly Imperial) Cricket Conference • (USA) Interstate Commerce Commission

ICCA International Cocoa Agreement

ICCB Intergovernmental Consultation and Coordination Board

ICCE International Council of Commerce Employers

ICCF International Correspondence Chess Federation

ICCH International Commodities Clearing House

ICCPR International Covenant on Civil and Political Rights (of the UN)

ICCROM International Centre for Conservation at Rome

ICCS International Centre of Criminological Studies

ICD Institute of Cooperative Directors • International Classification of Diseases (WHO publication)

ICDO International Civil Defence Organization

Ice. Iceland(ic)

ICE *Med.* ice, compress, elevation (treatment for limb bruises) • *Meteorol.* icing • *Computing* in-circuit emulator • Institution of Civil Engineers • internal-combustion engine • International Cultural Exchange

ICED International Council for Educational Development

ICEF International Federation of

Chemical, Energy, and General Workers' Unions

ICEI Institution of Civil Engineers of Ireland

Icel. Iceland(ic)

IC engine internal-combustion engine

ICES *Physical chem.* internal-conversion electron spectroscopy • International Council for the Exploration of the Sea

ICETT Industrial Council for Educational and Training Technology

ICF Industrial and Commercial Finance Corporation (now part of Investors in Industry) • *Nuclear engineering* inertial-confinement fusion • International Canoe Federation

ICFTU International Confederation of Free Trade Unions

ICGS interactive computer-graphics system • International Catholic Girls' Society

ich. ichthyology

ICH *Nuclear engineering* ion cyclotron heating

ICHCA International Cargo Handling Co-ordination Association

IChemE Institution of Chemical Engineers

ICHEO Inter-University Council for Higher Education Overseas

ICHPER International Council for Health, Physical Education, and Recreation

ichth. (*or* **ichthyol.**) ichthyology

ICI Imperial Chemical Industries • International Commission on Illumination • (USA) Investment Casting Institute

ICIA International Credit Insurance Association

ICIANZ Imperial Chemical Industries, Australia and New Zealand

ICID International Commission on Irrigation and Drainage

ICIDH International Classification of Impairments, Disabilities, and Handicaps (WHO publication)

ICJ International Commission of Jurists • International Court of Justice

ICJW International Council of Jewish Women

ICL International Computers Ltd • International Confederation of Labour

ICLA International Committee on Laboratory Animals

Iclnd Iceland

ICM Institute for Complementary

Medicine • Institute of Credit Management • Intergovernmental Committee for Migrations (of the UN) • International Confederation of Midwives • *Astronomy* intracluster medium • Irish Church Missions

ICMMP International Committee of Military Medicine and Pharmacy

ICMS International Centre for Mathematical Sciences (Edinburgh)

ICN in Christi nomine (Latin: in Christ's name) • Infection Control Nurse • International Council of Nurses

ICNA Infection Control Nurses' Association

ICNB International Code of Nomenclature of Bacteria

ICNCP International Code of Nomenclature of Cultivated Plants

ICNV International Code of Nomenclature of Viruses

ICO Institute of Careers Officers • International Coffee Organization • Islamic Conference Organization

ICOM International Council of Museums

ICOMOS International Council of Monuments and Sites

icon. iconography

ICOR Intergovernmental Conference on Oceanic Research (in UNESCO)

ICorrST Institution of Corrosion Science and Technology

ICP *Physics* inductively coupled plasma • *Med.* intracranial pressure

ICPA International Commission for the Prevention of Alcoholism • International Cooperative Petroleum Association

ICPHS International Council for Philosophy and Humanistic Studies

ICPO International Criminal Police Organization (Interpol)

ICPU International Catholic Press Union

ICQ *Accounting* internal control questionnaire

ICR *Computing* intelligent character recognition • *Physics* ion cyclotron resonance

ICRC International Committee of the Red Cross

ICRF Imperial Cancer Research Fund

ICRH *Nuclear engineering* ion cyclotron resonance heating

ICRP International Commission on Radiological Protection

ICRU International Commission on Radiation Units (and Measurements)

ICS Imperial College of Science and Technology (London) • Indian Civil Service (*see under* IAS) • instalment credit selling • Institute of Chartered Shipbrokers • International Chamber of Shipping • international consultancy service • International College of Surgeons • *Finance* investors' compensation scheme

ICSA Institute of Chartered Secretaries and Administrators

ICSH *Biochem.* interstitial-cell-stimulating hormone

ICSI ('ɪksɪ) *Med.* intracytoplasmic sperm injection

ICSID International Council of Societies of Industrial Design

ICSLS International Convention for Safety of Life at Sea

ICSPE International Council of Sport and Physical Education

ICST Imperial College of Science and Technology (London)

ICSU International Council of Scientific Unions (in UNESCO)

ICTA International Council of Travel Agents

ICTP International Centre for Theoretical Physics

ICTU Irish Congress of Trade Unions

ICTV International Committee on Taxonomy of Viruses

ICU *Med.* intensive care unit • international code use (of signals)

ICVA International Council of Voluntary Agencies

i.c.w. in connection with

ICW Institute of Clerks of Works of Great Britain • International Congress of Women • *Telecom.* interrupted continuous waves • *Physics* ion-cyclotron waves

ICWA Indian Council of World Affairs • Institute of Cost and Works Accountants

ICWG International Cooperative Women's Guild

ICWU (USA) International Chemical Workers Union

ICYF International Catholic Youth Federation

ICZN International Code of Zoological Nomenclature

id. idem (Latin: the same)

i.d. inner diameter

Id. Idaho

ID *Irish vehicle registration for* Cavan • *US*

postcode for Idaho • identification • identify • infectious disease(s) • information department • (*or* **i.d.**) inside diameter • Institute of Directors • Intelligence Department • (*or* **i.d.**) *Med.* intradermal • *symbol for* Iraqi dinar (monetary unit)

IDA Industrial Diamond Association • International Development Association • Irish Dental Association • Islamic Democratic Association • *Chem.* isotope dilution analysis

IDB illicit diamond buying (*or* buyer) • Industrial Development Bank • (Northern Ireland) Industrial Development Board • Inter-American Development Bank • Internal Drainage Board

IDC industrial development certificate • insulation displacement connector

ID card identification card

IDD insulin-dependent diabetes • international direct dialling

IDDA Interior Decorators' and Designers' Association

IDDD *Telecom.* international direct distance dial(ling)

IDDM insulin-dependent diabetes mellitus

IDE *Computing* intelligent (*or* integrated) drive electronics • *Computing* interactive development environment

iden. (*or* **ident.**) identification • identify

IDF International Dairy Federation • International Democratic Fellowship • International Dental Federation • International Diabetes Federation

IDGS *Physics* isotope-dilution gamma spectrometry

IDIB Industrial Diamond Information Bureau

IDL *Biochem., Med.* intermediate-density lipoprotein • international date line

IDLH immediately dangerous to life and health

IDMS *Computing* integrated data-management system • *Chem.* isotope-dilution mass spectrometry

IDN in Dei nomine (Latin: in God's name) • *Computing* integrated data network

IDOE International Decade of Ocean Exploration (1970–80)

IDP *Biochem.* inosine diphosphate • Institute of Data Processing • *Computing* integrated data processing • International

Driving Permit • *Astronomy* interplanetary dust particle(s)

IDPM Institute of Data Processing Management

IDR *Finance* International Depository Receipt

IDRC International Development Research Centre

IDS Income Data Services • Industry Department for Scotland • Institute of Development Studies

IDSM Indian Distinguished Service Medal

IDT industrial design technology

IDV International Distillers and Vintners

i.e. id est (Latin: that is) • inside edge

IE *Irish vehicle registration for* Clare • *Med.* immunoelectrophoresis • index error • Indian Empire • Indo-European (languages) • *Chem.* ion exchange • *Chem.* ionization energy • *British fishing port registration for* Irvine • *airline flight code for* Solomon Airlines

IEA Institute of Economic Affairs • (*or* **IE(Aust)**) Institution of Engineers, Australia • International Economic Association • International Energy Agency • International Ergonomics Association

IEC industrial energy conservation • integrated environmental control • International Electrotechnical Commission • *Chem.* ion-exchange chromatography

IED improvised explosive device • Information Engineering Directorate • Institution of Engineering Designers

IEDD improvised explosive device disposal (type of bomb disposal)

IEE (USA) Institute of Energy Economics • Institution of Electrical Engineers

IEEE (USA) Institute of Electrical and Electronics Engineers

IEEIE Institution of Electrical and Electronics Incorporated Engineers

IEF Indian Expeditionary Force

IEHO Institution of Environmental Health Officers

IEI Industrial Engineering Institute • Institution of Engineers of Ireland

IEME Inspectorate of Electrical and Mechanical Engineering

IEng Incorporated Engineer

IER Institute of Environmental Research

IERE Institution of Electronic and Radio Engineers

IES Indian Educational Service • Institution of Engineers and Shipbuilders in Scotland

IET interest equalization tax

IETF Internet Engineering Task Force

IEW intelligence and electronic warfare

IExpE Institute of Explosives Engineers

i.f. information feedback • ipse fecit (Latin: he did it himself)

IF *Irish vehicle registration for* Cork • *airline flight code for* Great China Airlines • *Nuclear engineering* inertial fusion • *Baseball* infield • *Biochem.* inhibitory factor • *Genetics* initiation factor • *Sport* inside forward • *Med.* interferon • *Electronics* intermediate frequency

IFA *Med.* immunofluorescence assay • Incorporated Faculty of Arts • independent financial adviser • instrumented fuel assembly • International Federation of Actors • International Fertility Association • International Fiscal Association • Irish Football Association

IFAC International Federation of Accountants • International Federation of Automatic Control

IFAD International Fund for Agricultural Development (of the UN)

IFALPA International Federation of Air Line Pilots' Associations

IFAP International Federation of Agricultural Producers

IFATCA International Federation of Air Traffic Controllers' Associations

IFAW International Fund for Animal Welfare

IFB *Finance* invitation for bid

IFBPW International Federation of Business and Professional Women

IFC International Finance Corporation • (USA and Canada) International Fisheries Commission

IFCC International Federation of Camping and Caravanning

IFCCPTE International Federation of Commercial, Clerical, Professional, and Technical Employees

IFCO International Fisheries Cooperative Organization

IFCTU International Federation of Christian Trade Unions (now called World Confederation of Labour, WCL)

IFDM *Maths.* integrated finite-difference method

IFE (USA) Institute for Energy Technology • *Computing* intelligent front end

IFEL *Physics* induction free-electron laser

iff *Logic, Maths.* if and only if

IFF Identification, Friend or Foe (radar identification system) • *Computing* image (*or* interchange) file format • Institute of Freight Forwarders

IFFA International Federation of Film Archives

IFFPA International Federation of Film Producers' Associations

IFFS International Federation of Film Societies

IFFTU International Federation of Free Teachers' Unions

IFGA International Federation of Grocers' Associations

IFGO International Federation of Gynaecology and Obstetrics

IFIP International Federation for Information Processing

IFJ International Federation of Journalists

IFK *Maths.* integral equation of the first kind

IFL International Friendship League

IFLA International Federation of Landscape Architects • International Federation of Library Associations

IFMC International Folk Music Council

IFMSA International Federation of Medical Students' Associations

IFOR ('aɪ,fɔː) (*or* **I-For**) (NATO-led Peace) Implementation Force (in Bosnia) • International Fellowship of Reconciliation

IFORS International Federation of Operational Research Societies

Ifox ('aɪ,fɒks) Irish Futures and Options Exchange

IFP (South Africa) Inkatha Freedom Party

IFPA Industrial Film Producers' Association

IFPAAW International Federation of Plantation, Agricultural, and Allied Workers

IFPCS International Federation of Unions of Employees in Public and Civil Services

IFPI International Federation of the Phonographic Industry

IFPM International Federation of Physical Medicine

IFPW International Federation of Petroleum (and Chemical) Workers

IFR *Aeronautics* instrument flying

regulations • *Nuclear engineering* integral fast reactor

IFRB International Frequency Registration Board

IFS Indian Forest Service • International Federation of Surveyors • Irish Free State • *Computing* iterated function system

IFSPO International Federation of Senior Police Officers

IFST Institute of Food Science and Technology

IFSW International Federation of Social Workers

IFTA International Federation of Teachers' Associations • International Federation of Travel Agencies

IFTC International Film and Television Council

IFTU International Federation of Trade Unions (forerunner of WFTU)

IFUW International Federation of University Women

IFWEA International Federation of Workers' Educational Associations

IFWL International Federation of Women Lawyers

ig. ignition

Ig *Immunol.* immunoglobulin (as in **IgA**, **IgE**, **IgG**)

IG *postcode for* Ilford • Indo-Germanic (languages) • industrial group • *Astronautics* inertial guidance • Inspector General • Instructor in Gunnery • Irish Guards

IGA International Geographical Association • International Golf Association

IGADD Intergovernmental Authority on Drought and Development

IGasE Institution of Gas Engineers

IGBT *Computing* insulated-gate bipolar transistor

IGC Intergovernmental Conference

IGD illicit gold dealer

IGES *Computing* initial graphics exchange specification

IGF Inspector-General of Fortifications • *Med.* insulin-like growth factor • International Gymnastic Federation

IGFA International Game Fish Association

IGFET ('ɪgfet) *Electronics* insulated-gate field-effect transistor

IGH Incorporated Guild of Hairdressers

IGM *Chess* International Grandmaster

ign. ignite(s) • ignition • ignotus (Latin: unknown)

IGO intergovernmental organization

IGPP Institute of Geophysics and Planetary Physics

IGS Imperial General Staff • independent grammar school

IGSCC *Engineering* intergranular stress corrosion cracking

IGU International Gas Union • International Geographical Union

IGY International Geophysical Year (1.7.57 to 31.12.58)

IH *Irish vehicle registration for* Donegal • iacet hic (Latin: here lies; on gravestones) • industrial hygiene • *Computing* interrupt handler • *British fishing port registration for* Ipswich

IHA International Hotel Association

IHAB International Horticultural Advisory Bureau

IHB International Hydrographic Bureau

IHC (New Zealand) intellectually handicapped child

IHCA International Hebrew Christian Alliance

IHD International Hydrological Decade (1965–74) • ischaemic heart disease

IHE *Metallurgy* internal hydrogen embrittlement

IHEU International Humanist and Ethical Union

IHF Industrial Hygiene Foundation • International Hockey Federation • International Hospitals Federation

IHM Jesus Mundi Salvator (Latin: Jesus, Saviour of the World)

IHospE Institute of Hospital Engineering

ihp indicated horsepower

IHR Institute of Historical Research

IHS Iesus Hominum Salvator (Latin: Jesus, Saviour of Mankind) • in hoc signo (Latin: in this sign) • Jesus (Greek ΙΗΣΟΥΣ)

IHSM Institute of Health Services Management

IHT inheritance tax • Institution of Highways and Transportation

IHU Irish Hockey Union

II *Electronics* image intensifier • *Electronics* ion implantation

IIA International Institute of Agriculture

IIAC Industrial Injuries Advisory Council

IIAL International Institute of Arts and Letters

IIAS International Institute of Administrative Sciences

IIB Institut international des brevets (French: International Patent Institute)

IIBD&ID Incorporated Institute of British Decorators and Interior Designers

iid *Statistics* independent identically distributed (of random variables)

IID insulin-independent diabetes · *Electronics* ion-implantation doping

IIE Institute for International Education · International Institute of Embryology

IIEE *Physical chem.* ion-induced electron emission

IIEP International Institute of Educational Planning

IIF *Computing* Image Interchange Facility

IIHF International Ice Hockey Federation

3i Investors in Industry

III International Institute of Interpreters

IIL (*or* I²L) *Electronics* integrated injection logic

IIM Institution of Industrial Managers

IInfSc Institute of Information Scientists

IIP Institut international de la presse (French: International Press Institute) · International Ice Patrol · International Institute of Philosophy

IIR *Chem.* isobutylene-isoprene rubber

IIRS (Irish Republic) Institute for Industrial Research and Standards

IIS Institute of Information Scientists · International Institute of Sociology

IISO Institution of Industrial Safety Officers

IISS International Institute of Strategic Studies

IIT Indian Institute of Technology

i.J. im Jahre (German: in the year)

IJ *British vehicle registration for* Down

IK *Irish vehicle registration for* Dublin

IKBS *Computing* intelligent knowledge-based system

i.l. inside leg

IL *British vehicle registration for* Fermanagh · *US postcode for* Illinois · Ilyushin (aircraft) · *Sport* inside left · Institute of Linguists · *Aeronautics* instrument landing · *Immunol.* interleukin (as in **IL-1**, **IL-2**) · *Physical chem.* ionization (energy) loss (spectroscopy) · *international vehicle registration for* Israel

I/L import licence

I²L *Electronics* integrated injection logic

ILA *Physics* induction linear accelerator · Institute of Landscape Architects · *Aeronautics* instrument landing approach · International Law Association · International Longshoremen's Association

ILAA International Legal Aid Association

ILAB International League of Antiquarian Booksellers

ILBM *Computing* interleaved bit map

ILC International Law Commission (of the UN)

ILD *Med.* interstitial lung disease

Ile (*or* **ile**) isoleucine

ILEA ('ɪliə) Inner London Education Authority

ILEC Inner London Education Committee

ILF International Landworkers' Federation

ILGA Institute of Local Government Administration

ILGWU (USA) International Ladies' Garment Workers' Union

ill. illustrate(d) · illustration · illustrissimus (Latin: most distinguished)

Ill. Illinois

illegit. illegitimate

illit. illiterate

illum. illuminate(d)

illus. (*or* **illust.**) illustrate(d) · illustration · illustrator

ILN (ital.) Illustrated London News

i.l.o. in lieu of

ILO industrial liaison officer · International Labour Organization (*or* Office) (of the UN)

ILP Independent Labour Party

ILR Independent Law Reports · independent local radio

ILRM International League for the Rights of Man

ILS Incorporated Law Society · *Aeronautics* instrument landing system · *Statistics* inverse least-squares · *Chem.* ionization-loss spectroscopy

ILTF International Lawn Tennis Federation

ILU Institute of London Underwriters

ILW *Nuclear engineering* intermediate-level waste

Im. Imperial

IM *Irish vehicle registration for* Galway · Indian Marines · Institute of Metals · interceptor missile · *Chess* International Master · (*or* **i.m.**) *Med.* intramuscular

IMA Indian Military Academy · Institute of Management Accountants · Institute of Mathematics and its Applications ·

International Music Association • Irish Medical Association

imag. imaginary • imagination • imagine

IM & AWU (USA) International Molders' and Allied Workers' Union

IMarE Institute of Marine Engineers

IMARSAT (*or* **Imarsat**) ('ɪmɑː,sæt) International Maritime Satellite Organization

IMAS International Marine and Shipping Conference

IMB Institute of Marine Biology • *Computing* Intel media benchmark

IMC *Photog.* image motion compensation • Institute of Management Consultants • Institute of Measurement and Control • *Aeronautics* instrument meteorological conditions • International Maritime Committee • International Missionary Council • International Music Council

IMCO Intergovernmental Maritime Consultative Organization (of the UN)

IMEA Incorporated Municipal Electrical Association

IMechE Institution of Mechanical Engineers

IMet Institute of Metals

I. Meth. Independent Methodist

IMF International Monetary Fund • *Astrophysics* interplanetary magnetic field

IMG *Computing* image file format

IMGTechE Institution of Mechanical and General Technician Engineers

IMinE Institution of Mining Engineers

IMINT (*or* **Imint**) ('ɪmɪnt) *Military* image intelligence (gained from aerial photography)

imit. imitate • imitation • imitative

IMM Institution of Mining and Metallurgy • International Mercantile Marine • International Monetary Market

immed. immediate

IMMTS Indian Mercantile Marine Training Ship

immun. immunity • immunization • immunology

IMO International Maritime Organization • International Meteorological Organization • International Miners' Organization

imp. imperative • imperfect • imperial • impersonal • implement • import(ed) • important • importer • impression • imprimatur • imprimé (French: printed) • imprimeur (French: printer) • imprint

Imp. Imperator (Latin: Emperor) • Imperatrix (Latin: Empress) • Imperial

IMP (ɪmp) *Physics* indeterminate mass particle • *Biochem.* inosine monophosphate • (ɪmp) *Computing* interface message processor (in a network) • (ɪmp) *Bridge* International Match Point • interplanetary measurement probe

IMPA International Master Printers' Association

IMPACT ('ɪmpækt) implementation, planning, and control technique

IMPATT ('ɪmpæt) *Electronics* impact ionization avalanche transit time

impce importance

imper. imperative

imperf. imperfect • imperforate (of stamps)

impers. impersonal

impf. (*or* **impft**) imperfect

imposs. impossible

IMPR *Meteorol.* improving

impreg. impregnate(d)

improp. improper(ly)

impt important

imptr importer

impv. imperative

IMR individual medical report • infant mortality rate • Institute of Medical Research

IMRA Industrial Marketing Research Association

IMRAN ('ɪmræn) international marine radio aids to navigation

IMRO ('ɪmrəʊ) Investment Management Regulatory Organization

IMS Indian Medical Service • industrial methylated spirit • *Computing, trademark* Information Management System • Institute of Management Services • International Musicological Society

IMSM Institute of Marketing and Sales Management

IMT *Meteorol.* immediate(ly) • International Military Tribunal

IMU International Mathematical Union

IMunE Institution of Municipal Engineers (now part of the Institution of Civil Engineers)

IMVS (Australia) Institute of Medical and Veterinary Science

IMW Institute of Masters of Wine • *Biochem.* intermediate molecular weight

in *symbol for* inch(es)

in. inch(es)

In *Chem., symbol for* indium

In. India(n) • *Military* Instructor

IN *US postcode for* Indiana • Indian Navy • *Irish vehicle registration for* Kerry

INA Indian National Army • Institution of Naval Architects • International Newsreel Association

INAO (France) Institut national des appellations d'origine des vins et eaux-de-vie (body controlling wine production)

inaug. inaugurate(d)

inbd inboard (on an aircraft, boat, etc.)

Inbucon ('ɪnbjuː,kɒn) International Business Consultants

inc. include(d) • including • inclusive • income • incomplete • incorporate(d) • increase • incumbent

Inc. Incorporated (after names of business organizations; US equivalent of Ltd)

INC *Meteorol.* in cloud • Indian National Congress • in nomine Christi (Latin: in the name of Christ)

INCA International Newspaper Colour Association

incalz. *Music* incalzando (Italian; increasing speed and tone)

INCB International Narcotics Control Board (of the UN)

incho. *Law* inchoate

incid. incidental

incl. incline • include(s) • included • including • inclusive

incog. incognito

incorp. (*or* **incor.**) incorporated • incorporation

incorr. incorrect

INCPEN ('ɪŋk,pɛn) Industry Committee for Packaging and the Environment

incr. increase(d) • increasing • increment

INCR *Meteorol.* increase

incun. incunabula

in d. *Med.* in dies (Latin: daily)

ind. independence • independent • index • indicate • indication • indicative • indigo • indirect(ly) • industrial • industry

Ind *Astronomy* Indus

Ind. *Politics* Independent • India(n) • Indiana • Indies

IND *international vehicle registration for* India • in nomine Dei (Latin: in God's name) • *Pharmacol.* (USA) investigational (*or* investigative) new drug

indecl. *Grammar* indeclinable

indef. indefinite

Indeo ('ɪndɪəʊ) Intel video

indic. indicating • indicative • indicator

Ind. Imp. Indiae Imperator (Latin: Emperor of India)

indiv. (*or* **individ.**) individual

Ind. Meth. Independent Methodist

Indo-Eur. Indo-European

Indo-Ger. Indo-German(ic)

indre. indenture

induc. induction

indust. industrial • industrious • industry

ined. ineditus (Latin: unpublished)

INER Institute of Nuclear Energy Research

INET Institute of Nuclear Energy Technology

in ex. in extenso (Latin: in full)

inf *Maths.* infinum

inf. (*or* **Inf.**) infantry • inferior • infinitive • influence • information • infra (Latin: below) • *Med.* infusum (Latin: infusion)

INF intermediate-range nuclear forces (as in **INF treaty**) • International Naturist Federation • International Nuclear Force

infin. infinitive

infirm. infirmary

infl. inflammable • inflated • inflect • *Botany* inflorescence • influence(d)

infm. information

Ing. Ingenieur (German: engineer)

Ingl. Inghilterra (Italian: England)

INGO international nongovernmental organization

Inh. Inhaber (German: proprietor) • Inhalt (German: contents)

INH *Med.* isonicotinic acid hydrazide (isoniazid; antituberculosis drug)

inhab. inhabitant

INI in nomine Iesu (Latin: in the name of Jesus)

init. initial(ly) • initio (Latin: in the beginning)

inj. injection • injury

INJ in nomine Jesu (Latin: in the name of Jesus)

INLA International Nuclear Law Association • Irish National Liberation Army

in lim. in limine (Latin: at the outset)

in loc. in loco (Latin: in place of)

in loc. cit. in loco citato (Latin: in the place cited (in text))

INMARSAT (*or* **Inmarsat**) ('ɪnmɑː,sæt) International Maritime Satellite Organization

in mem. in memoriam (Latin: to the memory (of))

inn. *Cricket* innings

INO inspectorate of naval ordnance

inorg. inorganic

INP Institute of Nuclear Physics

in pr. in principio (Latin: in the beginning)

in pro. in proportion

inq. inquiry • inquisition

INR independent national radio (as in **INR licence**) • Index of Nursing Research

INRI Iesus Nazarenus Rex Iudaeorum (Latin: Jesus of Nazareth, King of the Jews)

in s. in situ

ins. inches • inscribe • inscription • inspector • insular • insulate(d) • insulation • insurance

INS Indian Naval Ship • *Physics* inelastic neutron scattering • inertial navigation system • International News Service • *British fishing port registration for* Inverness

INSA Indian National Science Academy

INSAG International Nuclear Safety Advisory Group

insce insurance

inscr. inscribe • inscription

INSEA International Society for Education through Art

INSEAD (*or* **Insead**) Institut européen d'administration des affaires (French: European Institute of Administrative Affairs)

insep. inseparable

INSET (*or* **Inset**) ('ɪnsɛt) *Education* in-service training

insol. insoluble

insolv. insolvent

insp. inspect(ed) • inspection • (*or* **Insp.**) inspector

Insp. Gen. Inspector General

inst. instance • instant (this month) • instantaneous • (*or* **Inst.**) institute • (*or* **Inst.**) institution • instruct(ion) • instructor • instrument(al)

INST in nomine Sanctae Trinitatis (Latin: in the name of the Holy Trinity)

INSTAB ('ɪnstæb) Information Service on Toxicity and Biodegradability (water pollution)

InstAct Institute of Actuaries

InstBE Institution of British Engineers

InstCE Institution of Civil Engineers

InstD Institute of Directors

InstE Institute of Energy

InstEE Institution of Electrical Engineers

instl. installation

InstMet Institute of Metals

InstMM Institution of Mining and Metallurgy

instn (*or* **Instn**) institution

InstP Institute of Physics

InstPet Institute of Petroleum

InstPI Institute of Patentees and Inventors

instr. instruction • instructor • instrument(al)

InstR Institute of Refrigeration

InstSMM Institute of Sales and Marketing Management

InstT Institute of Transport

int. (military) intelligence • intercept • interest • interim • interior • interjection • intermediate • internal • (*or* **Int.**) international • interpret(er) • interpretation • interval • intransitive • *Music* introit

INT Isaac Newton Telescope (La Palma)

int. al. inter alia (Latin: among other things)

INTAL ('ɪntæl) Institute for Latin American Integration (Buenos Aires)

Intelsat (*or* **INTELSAT**) ('ɪntɛl,sæt) International Telecommunications Satellite Consortium

intens. intensifier • intensify • intensive

inter. intermediate • interrogation mark

INTER *Meteorol.* intermittent

interj. interjection

internat. international

interp. interpreter

Interpol ('ɪntə,pɒl) International Criminal Police Organization

interrog. interrogate • interrogation • interrogative

intl international

INTO Irish National (Primary) Teachers' Organization

intr. (*or* **intrans.**) intransitive

in trans. in transit

intro. (*or* **introd.**) introduce • introduction • introductory

INTSF *Meteorol.* intensify

INTST *Meteorol.* intensity

INTUC ('ɪntʌk) Indian National Trade Union Congress

INucE Institution of Nuclear Engineers

inv. invenit (Latin: (he *or* she) designed this) • invent(ed) • invention • inventor • inversion • invert • investment • invoice

Inv. Inverness

inv. et del. invenit et delineavit (Latin: (he *or* she) designed and drew this)

invt. (*or* **invty**) inventory

IO India Office • inspecting officer • integrated optics • intelligence officer • *Irish vehicle registration for* Kildare

I/O *Computing* input/output • inspecting order

IOA Institute of Acoustics

IOB (*or* **IoB**) Institute of Bankers (renamed Chartered Institute of Bankers, CIB) • Institute of Biology • Institute of Bookkeepers • Institute of Brewing • Institute of Building (renamed Chartered Institute of Building, CIOB)

IOC Intergovernmental Oceanographic Commission • International Olympic Committee

IOCU International Organization of Consumers' Unions

IoD Institute of Directors

IODE (Canada) Imperial Order of Daughters of the Empire

IOE International Organization of Employers

IOF Independent Order of Foresters • International Oceanographic Foundation • International Orienteering Federation

I of E Institute of Export

I of M Isle of Man

IOGT International Order of Good Templars

IoJ Institute of Journalists

IOJ International Organization of Journalists

IOM Indian Order of Merit • Institute of Metals • Isle of Man

IOMTR International Office for Motor Trades and Repairs

Ion. Ionic

IONARC Indian Ocean National Association for Regional Cooperation

IOO Inspecting Ordnance Officer

IOOF Independent Order of Oddfellows

IOP *Computing* input/output processor • Institute of Painters in Oil Colours • Institute of Petroleum • Institute of Physics • Institute of Plumbing • Institute of Printing

IOPAB International Organization for Pure and Applied Biophysics

IoS (ital.) The Independent on Sunday

IOS *Computing* integrated office system

IOSCO (*or* **Iosco**) (aɪˈɒskəʊ) International Organization of Securities Commissions

IOSM Independent Order of the Sons of Malta

IoT Institute of Transport

IOU Industrial Operations Unit • I owe you

IOW Institute of Welding • (*or* **IoW**) Isle of Wight

i.p. identification point • incentive pay • indexed and paged • initial phase • *Electrical engineering* input primary

IP *Electronics* image processing • Imperial Preference • India Paper • *Baseball* innings pitched • *Biochem.* inositol phosphate • *Med.* in-patient • *Electrical engineering* input primary • instalment plan • Institute of Petroleum • Institute of Plumbing • Internet Protocol • *Chem.* ionization potential • *postcode for* Ipswich • *Irish vehicle registration for* Kilkenny

IP₃ *Biochem.* inositol triphosphate

IPA India Pale Ale • Insolvency Practitioners' Association • Institute of Park Administration • Institute of Practitioners in Advertising • International Phonetic Alphabet • International Phonetic Association • International Poetry Archives (Manchester) • International Police Academy • International Police Association • International Publishers' Association • *Biochem.* isopentenyl adenosine • *Chem.* isopropanol (*or* isopropyl alcohol)

IPAA International Petroleum Association of America • International Prisoners' Aid Association

IPARS International Programmed Airline Reservation System

IPC *Astronomy* imaging proportional counter • International Petroleum Company (Peru) • International Polar Commission • International Publishing Corporation • *Computing* interprocess communication • Iraq Petroleum Company

IPCC Intergovernmental Panel on Climatic Change (of the UN)

IPCS *Astronomy, Physics* image photon counting system • Institution of Professional Civil Servants • *Computing* intelligent process-control system

IPD individual package delivery • *Law* (Scotland) in praesentia dominorum (Latin: in the presence of the Lords (of

Session)) • *Med.* intermittent peritoneal dialysis

IPE Institution of Plant Engineers • Institution of Production Engineers • International Petroleum Exchange

IPF *Med.* interstitial pulmonary fibrosis • Irish Printing Federation

IPFA (Member or Associate of the Chartered) Institute of Public Finance and Accountancy

IPFC Indo-Pacific Fisheries Council (of the FAO)

IPG Independent Publishers' Guild • Industrial Painters' Group

IPH *Med.* idiopathic portal hypertension • industrial process heat(ing)

IPHE Institution of Public Health Engineers (*see* IWEM)

i.p.i. in partibus infidelium (Latin: in the regions of unbelievers)

IPI *Computing* Image Processing and Interchange • Institute of Patentees and Inventors • Institute of Professional Investigators • International Press Institute

IPL *Computing* initial program load

IPlantE Institution of Plant Engineers

IPLO Irish People's Liberation Organization

IPM *Freemasonry* immediate past master • (*or* **ipm**) inches per minute • Institute of Personnel Management

IPMS Institution of Professionals, Managers, and Specialists

IPO *Computing* input-process-output • Israel Philharmonic Orchestra • *Stock exchange* (USA) initial public offering (a flotation)

IPP Institute for Plasma Physics • *Med.* intermittent positive pressure (ventilation)

IPPA Independent Programme Producers' Association

IPPF International Planned Parenthood Federation

IPPR Institute for Public Policy Research

IPPS Institute of Physics and the Physical Society

IPPV *Med.* intermittent positive-pressure ventilation

IPR Institute of Public Relations

IPRA International Public Relations Association

IPRE Incorporated Practitioners in Radio and Electronics

IProdE Institution of Production Engineers

ips inches per second • *Computing* instructions per second

Ips. Ipswich

IPS inches per second • Indian Police Service • Indian Political Service • Institute of Purchasing and Supply • International Confederation for Plastic Surgery • *Astronomy* interplanetary scintillations • *Computing* interpretative programming system

IPSA International Political Science Association

IPSE ('ɪpsɪ) *Computing* integrated project support environment

i.p.s.p. (*or* **IPSP**) *Physiol.* inhibitory postsynaptic potential

IPT Institute of Petroleum Technologists

IPTO (*or* **i.p.t.o.**) *Astronautics* independent power takeoff

IPTPA International Professional Tennis Players' Association

IPTS *Physics* International Practical Temperature Scale

IPU Inter-Parliamentary Union

IPX Internet package exchange

i.q. idem quod (Latin: the same as)

IQ *airline flight code for* Augsburg Airways • Institute of Quarrying • intelligence quotient • international quota

IQA Institute of Quality Assurance

IQS Institute of Quantity Surveyors

i.r. inside radius

i.R. im Ruhestand (German: retired *or* emeritus)

Ir *Chem., symbol for* iridium

Ir. Ireland • Irish

IR *Meteorol.* ice on runway • incidence rate • index register • informal report • information retrieval • infrared (radiation) • Inland Revenue • *Sport* inside right • inspector's report • Institute of Refrigeration • instrument reading • international registration • *international vehicle registration for* Iran • *airline flight code for* Iran Air • Iranian rial (monetary unit) • *Chem.* isoprene rubber • *Irish vehicle registration for* Offaly

IRA (USA) individual retirement account • Institute of Registered Architects • Irish Republican Army

IRAD Institute for Research on Animal Diseases

IRAF *Astronomy* image reduction and analysis facility

IRAM (France) Institut de Radioastronomie Millimétrique (Institute of Millimetre Radioastronomy)

Iran. Iranian

IRAS ('aɪræs) Infrared Astronomical Satellite

IRB Irish Republican Brotherhood

IRBM intermediate-range ballistic missile

IRC Industrial Reorganization Corporation · Infantry Reserve Corps · International Red Cross · International Research Council · Internet relay chat

IRCert Industrial Relations Certificate

IRD International Research and Development Company

IRDA Industrial Research and Development Authority · Infra-Red Data Association

Ire. Ireland

IRE (USA) Institute of Radio Engineers

IREE(Aust) Institution of Radio and Electronics Engineers (Australia)

IRF International Road Federation · International Rowing Federation

IRFB International Rugby Football Board

IRFU Irish Rugby Football Union

IRI Institution of the Rubber Industry (now part of the Plastics and Rubber Institute) · (Italy) Istituto per la Ricostruzione Industriale (Industrial Reconstruction Institute)

irid. iridescent

Iris ('aɪrɪs) infrared intruder system

IRIS International Research and Information Service

IRL *international vehicle registration for* Republic of Ireland

IRLS *Navigation* interrogation recording location system

IRM *Ethology* innate releasing mechanism · Islamic Republic of Mauritania · *Geology* isothermal remanent magnetization

IRN Independent Radio News

IRO industrial relations officer · Inland Revenue Office · International Refugee Organization · International Relief Organization

IRPA International Radiation Protection Association

IRQ *Computing* interrupt request · *international vehicle registration for* Iraq

irr. *Finance* irredeemable · irregular

IRR infrared radiation · Institute of Race Relations · *Finance* internal rate of return

irreg. irregular(ly)

IRRI International Rice Research Institute

IRRV Institute of Revenues, Rating, and Valuation

IRS *Computing* information retrieval system · (USA) Internal Revenue Service

IRSF Inland Revenue Staff Federation (merged with the NUCPS to form the PTC)

IRTE Institute of Road Transport Engineers

IRTF Infrared Telescope Facility (Mauna Kea, Hawaii)

IRU industrial rehabilitation unit · International Relief Union · International Road Transport Union

IRWC International Registry of World Citizens

Is. *Bible* Isaiah · Islam(ic) · Island(s) · Isle(s) · Israel(i)

IS *international vehicle registration for* Iceland · Industrial Society · information science · information service · information system · (*or* **i.s.**) *Electrical engineering* input secondary · *Genetics* insertion sequence · International Society of Sculptors, Painters, and Gravers · Irish Society · *Physical chem.* isomer shift · *Irish vehicle registration for* Mayo

Isa ('aɪsə) individual savings account

Isa. *Bible* Isaiah

ISA Independent Schools' Association · individual savings account · *Computing* Industry Standard Architecture · International Sociological Association · *Aeronautics* International Standard Atmosphere (formerly Interim Standard Atmosphere) · International Standards in Auditing

ISAB Institute for the Study of Animal Behaviour

ISAC Industrial Safety Advisory Council

ISAM *Computing* indexed sequential access method

ISAS (Japan) Institute of Space and Astronautical Science

ISBA Incorporated Society of British Advertisers

ISBN International Standard Book Number

ISC Imperial Service College (Haileybury) · Imperial Staff College · Indian Staff Corps · *Med.* intermittent self-catheterization • International Seismological Centre • International Student Conference • International Sugar Council • *Freemasonry* International Supreme Council

ISCE International Society of Christian Endeavour

ISCh Incorporated Society of Chiropodists

ISCM International Society for Contemporary Music

ISCO Independent Schools Careers Organization • International Standard Classification of Occupations

ISD international standard depth • international subscriber dialling

ISDA International Swaps and Derivatives Association

ISDN *Computing, Telecom.* Integrated Services Digital Network

ISE Indian Service of Engineers • Institution of Structural Engineers • International Stock Exchange of the UK and the Republic of Ireland Ltd

ISEE International Sun-Earth Explorer

ISF *Computing* integrated systems factory • International Shipping Federation • International Spiritualist Federation

ISGE International Society of Gastroenterology

ISH International Society of Haematology

ISHS International Society for Horticultural Science

ISI Indian Standards Institution • International Statistical Institute • Iron and Steel Institute

ISIS ('aɪsɪs) (*or* **Isis**) Independent Schools Information Service • International Shipping Information Services

ISJC Independent Schools Joint Council

ISK *symbol for* króna (Icelandic monetary unit)

isl. (*or* **Isl.**) island • isle

ISM Iesus Salvator Mundi (Latin: Jesus, Saviour of the World) • Imperial Service Medal • Incorporated Society of Musicians • *Astronomy* interstellar medium

ISMA International Securities Market Association

ISME International Society for Musical Education

ISMRC Inter-Services Metallurgical Research Council

ISO Imperial Service Order • Infrared Space Observatory • International Standards Organization (International Organization for Standardization) • International Sugar Organization

isol. isolate(d) • isolation

ISOL *Meteorol.* isolated

ISP Institute of Sales Promotion • *Computing* instruction set processor • International Study Programme • Internet service provider

ISPA International Society for the Protection of Animals • International Sporting Press Association

ISPEMA Industrial Safety (Personal Equipment) Manufacturers' Association

ISQ *Med.* in statu quo (Latin: in the same state; unchanged)

ISR information storage and retrieval • Institute of Social Research • International Society for Radiology • *Computing* interrupt service routine • *Physics* intersecting storage ring

ISRB Inter-Services Research Bureau

ISRD International Society for Rehabilitation of the Disabled

ISRO International Securities Regulatory Organization

iss. issue

ISS Institute of Space Sciences (*or* Studies) • International Social Service • interstellar scintillations • *Physical chem.* ion-scattering spectrometry • *Physics* isotope separation system

ISSA International Social Security Association

ISSC International Social Science Council

ISSN International Standard Serial Number

ISSS International Society of Soil Science

IST Indian Standard Time • information sciences technology • Institute of Science Technology

ISTC Institute of Scientific and Technical Communicators • Iron and Steel Trades Confederation

ISTD Imperial Society of Teachers of Dancing

ISTEA Iron and Steel Trades Employers' Association

isth. (*or* **Isth.**) isthmus

IStructE Institution of Structural Engineers

ISU International Seamen's Union • International Shooting Union • International Skating Union

ISV independent software vendor • International Scientific Vocabulary

ISVA Incorporated Society of Valuers and Auctioneers

ISWG *Engineering* Imperial Standard Wire Gauge

it. italic

i.t. inspection tag · *Tools* internal thread · in transit

It. Italian · Italy

IT *airline flight code for* Air Inter · ignition temperature · income tax · (USA) Indian Territory · industrial tribunal · infantry training · information technology · *Law* Inner Temple · *Physics* International Table (in **IT calorie**) · *Irish vehicle registration for* Leitrim

i.t.a (*or* **ITA**) initial teaching alphabet

ITA Independent Television Authority (superseded by the IBA) · industrial and technical assistance · Industrial Transport Association · Information Technology Agreement

ITAI Institution of Technical Authors and Illustrators

ital. italic

Ital. Italian · Italy

ITALY I trust and love you

ITB Industry Training Board · International Time Bureau · Irish Tourist Board

i.t.c. installation time and cost

ITC Imperial Tobacco Company · Independent Television Commission · Industrial Training Council · International Tin Council · International Trade Centre · *Meteorol.* intertropical confluence · (USA) investment tax credit

ITCZ *Meteorol.* intertropical convergence zone

ITDA *Military* indirect target damage assessment

ITE Institute of Terrestrial Ecology

ITEME Institution of Technician Engineers in Mechanical Engineering

ITER international thermonuclear engineering reactor · international tokamak engineering reactor

ITF *Accounting* integrated test facility · International Tennis Federation · International Trade Federations · International Transport Workers' Federation

itin. itinerary

Itl. Italian

ITMA Institute of Trade Mark Agents ·

('ɪtmə) It's That Man Again (BBC radio series)

ITN Independent Television News

ITO *Electronics* indium–tin oxide · International Trade Organization

ITP *Biochem.* inosine triphosphate

ITS Industrial Training Service · (USA) Intermarket Trading System · International Trade Secretariat

ITT *Med.* insulin tolerance test · International Telephone and Telegraph Corporation

ITTF International Table Tennis Federation

ITU *Computing* intelligent thermal update · *Med.* intensive therapy unit · International Telecommunication Union (of the UN) · International Temperance Union · International Typographical Union

ITU-T International Telecommunications Union Technical Committee

ITV Independent Television · (USA) instructional television

ITVA International Television Association

ITWF International Transport Workers' Federation

IU immunizing unit · *Pharmacol.* international unit(s) · (Spain) Izquierda Unida (United Left; political party) · *Irish vehicle registration for* Limerick

IUA International Union Against Alcoholism · International Union of Architects

IUAA International Union of Advertisers' Associations

IUAES International Union of Anthropological and Ethnological Sciences

IUAI International Union of Aviation Insurers

IUAO International Union for Applied Ornithology

IUAPPA International Union of Air Pollution Prevention Associations

IUB International Union of Biochemistry · International Universities Bureau

IUBS International Union of Biological Sciences

IUCD *Med.* intrauterine contraceptive device

IUCN International Union for the Conservation of Nature and Natural Resources

IUCr International Union of Crystallography

IUCW International Union for Child Welfare

IUD *Med.* intrauterine death • *Med.* intrauterine (contraceptive) device

IUE *Astronomy* International Ultraviolet Explorer • (USA) International Union of Electronic, Electrical, Salaried, Machine, and Furniture Workers (trade union)

IUF International Union of Food and Allied Workers' Associations

IUFO International Union of Family Organizations

IUFRO International Union of Forest Research Organizations

IUGG International Union of Geodesy and Geophysics

IUGR *Med.* intrauterine growth retardation

IUGS International Union of Geological Sciences

IUHPS International Union of the History and Philosophy of Science

IUI *Med.* intrauterine insemination

IULA International Union of Local Authorities

IUMF (France) institut d'universitaires de formation des maîtres (advanced teacher-education institute)

IUMI International Union of Marine Insurance

IUMSWA Industrial Union of Marine and Shipbuilding Workers of America

IUNS International Union of Nutritional Sciences

IUOE (USA) International Union of Operating Engineers

IUPAB International Union of Pure and Applied Biophysics

IUPAC ('juːpæk) International Union of Pure and Applied Chemistry

IUPAP ('juːpæp) International Union of Pure and Applied Physics

IUPS International Union of Physiological Sciences

IUS *Astronautics* inertial upper stage • International Union of Students • intrauterine system (contraceptive device)

IUSP International Union of Scientific Psychology

IUSY International Union of Socialist Youth

IUT *Med.* intrauterine transfusion

IUTAM International Union of Theoretical and Applied Mechanics

i.v. increased value • *Med.* intravenous(ly) • invoice value

i.V. in Vertretung (German: by proxy)

IV *airline flight code for* Fujian Airlines • *Chem.* intermediate valency • *Med.* intravenous(ly) • *postcode for* Inverness • invoice value • *Irish vehicle registration for* Limerick

IVA individual voluntary arrangement (in bankruptcy proceedings) • invalidity allowance

IVB *Physics* intermediate vector boson • invalidity benefit

IVBF International Volley-Ball Federation

IVC *Med.* inferior vena cava

IVF *Med.* in vitro fertilization

IVP *Med.* intravenous pyelogram

IVR international vehicle registration

IVS International Voluntary Service

IVT *Astronautics* intravehicular transfer

IVU International Vegetarian Union

i.w. indirect waste • inside width

IW Inspector of Works • Isle of Wight • *British vehicle registration for* Londonderry

IWA Inland Waterways Association • Institute of World Affairs

IWC International Whaling Commission • International Wheat Council

IWD (*or* **IW&D**) Inland Waterways and Docks

IWEM Institution of Water and Environmental Management (formerly IPHE; IWPC)

IWGC Imperial War Graves Commission (now Commonwealth War Graves Commission)

IWM Imperial War Museum

IWO Institute of Welfare Officers

IWPC Institute of Water Pollution Control (*see* IWEM)

IWS International Wool Secretariat

IWW Industrial Workers of the World • International Workers of the World

IX Iesus Christus (Latin: Jesus Christ) • *Chem.* ion exchange • *Irish vehicle registration for* Longford

IY Imperial Yeomanry • *Irish vehicle registration for* Louth • *airline flight code for* Yemen Airways

IYHF International Youth Hostels Federation

IYRU International Yacht Racing Union

iyswim ('ɪz,wɪm) if you see what I mean

IZ *airline flight code for* Arkia Israeli Airlines • I Zingari (cricket club) • *Irish vehicle registration for* Mayo

IZS *Med.* insulin zinc suspension (diabetes treatment)

J

j (ital.) *Physics, symbol for* current density • *Electrical engineering, symbol for* the imaginary number √−1 • (bold ital.) *Maths., symbol for* a unit coordinate vector

j. juris (Latin: of law) • jus (Latin: law)

J *Engineering, symbol for* advance ratio • (bold ital.) *Physics, symbol for* angular momentum • (ital.) *Physics, symbol for* coupling constant • (ital.) *Physics, symbol for* current density • *British vehicle registration for* Durham • *Cards* jack • *Maths., symbol for* Jacobian determinant • *international vehicle registration for* Japan • *British fishing port registration for* Jersey • *Immunol.* joining region (of an immunoglobulin chain) • *Science, symbol for* joule(s) • *Geology* Jurassic • (bold ital.) *Physics, symbol for* magnetic polarization • (ital.) *Physics, obsolete symbol for* mechanical equivalent of heat • (ital.) *Physics, symbol for* nuclear spin quantum number • (ital.) *Chem., symbol for* rotational quantum number • (ital.) *Physics, symbol for* sound intensity • (ital.) *Physics, symbol for* total angular momentum quantum number

J. (*or* **J**) Jacobean • Jahr (German: year) • January • Jesus • jet • Jew(ish) • Journal • Judaic • Judaism • Judge • July • June • Justice

J2 *airline flight code for* Azerbaijan Hava Yollary • *international civil aircraft marking for* Djibouti

J3 *international civil aircraft marking for* Grenada

J6 *international civil aircraft marking for* St Lucia

J7 *international civil aircraft marking for* Dominica

5J *airline flight code for* Cebu Pacific Air

9J *international civil aircraft marking for* Zambia

Ja. January

JA *international vehicle registration for* Jamaica • *international civil aircraft marking for* Japan • (*or* **J/A**) *Banking* joint account • Judge Advocate • Justice of Appeal • *British vehicle registration for* Manchester

JAA Japan Aeronautic Association • Jewish Athletic Association

Jaat *Military* joint air attack team

Jac. Jacobean

JAC (USA) Joint Apprenticeship Committee • (USA) Junior Association of Commerce

Jacq. *Botany* J. F. Jacquin (1766–1839) *or* N. J. Jacquin (1727–1817) (indicating the author of a species, etc.)

JACT Joint Association of Classical Teachers

JAD Julian Astronomical Day

JADB (USA) Joint Air Defense Board

JAEC Japan Atomic Energy Commission • Joint Atomic Energy Committee (of US Congress)

JAF Judge Advocate of the Fleet

JAFC Japan Atomic Fuel Corporation

Jafo (dʒæfəʊ) *Military slang* just another fucking observer

Jag (dʒæg) Jaguar (car)

JAG Judge Advocate General

JAIEG Joint Atomic Information Exchange Group

JAL Japan Airlines • jet approach and landing chart

Jam. Jamaica • *Bible* James

JAMA ('dʒɑːmə) (ital.) Journal of the American Medical Association

jan. janitor

Jan (*or* **Jan.**) January

J. & K. Jammu and Kashmir (Indian state)

j. & w.o. *Insurance* jettisoning and washing overboard

JANET ('dʒænɪt) *Computing* Joint Academic Network

janv. janvier (French: January)

jap. japanned

Jap. Japan(ese)

JAP (dʒæp) *Colloquial* (USA) Jewish American Princess

jar. jargon

JARE Japanese Antarctic Research Expedition

Jas. *Bible* James

JAS Jamaica Agricultural Society • Junior Astronomical Society

jastop ('dʒæstɒp) *Aeronautics* jet-assisted stop

JAT Jugoslovenski Aero-Transport (Yugoslav Airlines)

JATCC Joint Aviation Telecommunications Coordination Committee

JATCRU joint air traffic control radar unit

JATO (*or* **jato**) ('dʒeɪtəʊ) *Aeronautics* jet-assisted takeoff

JATS joint air transportation service

jaund. jaundice

Jav. Java(nese) • (*or* **jav.**) *Athletics* javelin

j.b. jet bomb • joint board • junction box

Jb. Jahrbuch (German: annual *or* yearbook)

JB Bachelor of Laws (Latin *Juris Baccalaureus*) • junior beadle • *British vehicle registration for* Reading

JBAA (ital.) Journal of the British Archaeological Association

JBC Jamaica Broadcasting Corporation • Japan Broadcasting Corporation

JBCNS Joint Board of Clinical Nursing Studies

Jber. Jahresbericht (German: annual report)

JBES Jodrell Bank Experimental Station

JBL (ital.) Journal of Business Law

JBS (USA) John Birch Society

j.c. joint compound

JC *British vehicle registration for* Bangor • (*or* **J.C.**) Jesus Christ • (ital.) Jewish Chronicle • Jockey Club • (*or* **J.C.**) Julius Caesar • (USA) (member of a) Junior Chamber (of Commerce) • (USA) junior college • *Law* jurisconsult (legal adviser; jurist) • Justice Clerk • justiciary case • juvenile court

J-C Jésus-Christ (French: Jesus Christ)

JCAC (USA) Joint Civil Affairs Committee

JCAR Joint Commission on Applied Radioactivity

JCB Bachelor of Canon Law (Latin *Juris Canonici Baccalaureus*) • Bachelor of Civil Law (Latin *Juris Civilis Baccalaureus*) •

Joseph Cyril Bamford (excavating machine; named after its manufacturer)

JCC Jesus College, Cambridge • Joint Consultative Committee • Junior Chamber of Commerce

JCD Doctor of Canon Law (Latin *Juris Canonici Doctor*) • Doctor of Civil Law (Latin *Juris Civilis Doctor*)

JCI Junior Chamber International

JCL *Computing* job-control language • Licentiate in Canon Law (Latin *Juris Canonici Licentiatus*) • Licentiate in Civil Law (Latin *Juris Civilis Licentiatus*)

JCMT James Clerk Maxwell Telescope

JCNAAF Joint Canadian Navy-Army-Air Force

JC of C Junior Chamber of Commerce

JCP Japan Communist Party

JCR junior common room (in certain universities)

JCS Joint Chiefs of Staff • Joint Commonwealth Societies • (ital.) Journal of the Chemical Society

jct. (*or* **jctn**) junction

JCWI Joint Council for the Welfare of Immigrants

jd joined

JD Diploma in Journalism • Doctor of Laws *or* Jurisprudence (Latin *Jurum Doctor*) • *symbol for* Jordan dinar (monetary unit) • *Astronomy* Julian date • junior deacon • junior dean • jury duty • (USA) Justice Department • juvenile delinquent • *British vehicle registration for* (central) London

JDA Japan Defence Agency

JDB Japan Development Bank

JDipMA Joint Diploma in Management Accounting Services

JDL (USA) Jewish Defense League

j.d.s. job data sheet

JDS *Accounting* Joint Disciplinary Scheme

JE *airline flight code for* Manx Airlines • *British vehicle registration for* Peterborough

j.e.a. joint export agent

JEA Jesuit Educational Association • Joint Engineering Association

JEC Joint Economic Committee (of US Congress)

JECI Jeunesse étudiante catholique internationale (French: International Young Catholic Students)

J. Ed. (ital.) Journal of Education

JEDEC Joint Electronic Devices Engineering Council

JEIDA Japanese Electronic Industry Development Association

Jer. *Bible* Jeremiah • Jersey • Jerusalem

JERC Joint Electronic Research Committee

JERI Japan Economic Research Institute

jerob. jeroboam

Jes. Jesus

JESA Japanese Engineering Standards Association

JESSI Joint European Submicron Silicon Initiative

jet. jetsam • jettison

JET (dʒɛt) *Nuclear engineering* Joint European Torus (Culham, Oxfordshire) • Joint European Transport

JETCO ('dʒɛt,kəʊ) Jamaican Export Trading Company • Japan Export Trading Company

JETP (ital.) Journal of Experimental and Theoretical Physics

JETRO ('dʒɛtrəʊ) Japan External Trade Organization

jett. jettison

jeu. jeudi (French: Thursday)

JF *British vehicle registration for* Leicester

J/F *Book-keeping* journal folio

JFET ('dʒɛɪfɛt) *Electronics* junction field-effect transistor

JFK John Fitzgerald Kennedy (US president 1961–63)

JFM Jeunesses fédéralistes mondiales (French: Young World Federalists)

JFTC Joint Fur Trade Committee

JFU Jersey Farmers' Union

j.g. junior grade

JG *British vehicle registration for* Maidstone

JGTC Junior Girls' Training Corps

JGW *Freemasonry* Junior Grand Warden

Jh. Jahrehundert (German: century)

JH *Numismatics* Jubilee head-Victoria • *Biochem.* juvenile hormone • *British vehicle registration for* Reading

j.h.a. job hazard analysis

JHDA Junior Hospital Doctors' Association

JHMO junior hospital medical officer

JHS Jesus Hominum Salvator (Latin: Jesus Saviour of Men) • (USA) junior high school

JHU (USA) Johns Hopkins University

JHVH Jehovah

JI *British vehicle registration for* Tyrone

JIB joint intelligence bureau

jic just in case

JIC joint industrial council • (USA) joint intelligence center

JICRAR ('dʒɪkrɑː) Joint Industry Committee for Radio Audience Research

JICTAR ('dʒɪktɑː) Joint Industry Committee for Television Advertising Research

JIM Japan Institute of Metals

JINS (dʒɪnz) (USA) Juvenile(s) In Need of Supervision

JINucE Junior Member of the Institution of Nuclear Engineers

JIS Jamaica Information Service • Japan Industrial Standard • Jewish Information Society • joint intelligence staff

JIT (*or* **jit**) just-in-time (manufacturing method)

JJ *Electronics* Josephson junction • (*or* **JJ.**) Judges • (*or* **JJ.**) Justices • *British vehicle registration for* Maidstone

JK *British vehicle registration for* Brighton • *airline flight code for* Spanair

jkt jacket

j.k.t. job knowledge test

JKT Jacobus Kapteyn Telescope

Jl. (*or* **jl.**) journal • July

JL *airline flight code for* Japan Airlines • *British vehicle registration for* Lincoln

JLA Jewish Librarians' Association

JLB Jewish Lads' Brigade

JLC (USA) Jewish Labor Committee

JLP Jamaica Labour Party

JM *airline flight code for* Air Jamaica • *British vehicle registration for* Reading

JMA Japanese Meteorological Agency

JMB James Matthew Barrie (1860–1937, Scottish dramatist and novelist) • Joint Matriculation Board

JMBA (ital.) Journal of the Marine Biological Association

JMCS Junior Mountaineering Club of Scotland

JMJ *RC Church* Jesus, Mary, and Joseph

JMSAC Joint Meteorological Satellite Advisory Committee

jn join • (*or* **Jn**) junction • (*or* **Jn**) junior

JN *British vehicle registration for* Chelmsford

JNA Jordan News Agency

jnc. (*or* **Jnc.**) junction

JNC joint negotiating committee

j.n.d. just noticeable difference

JNEC Jamaican National Export Corporation

JNF Jewish National Fund

jnl (*or* **Jnl**) journal

jnlst journalist

jnr (*or* **Jnr**) junior

JNR Japanese National Railways

j.n.s. just noticeable shift

jnt joint

JNTO Japan National Tourist Organization

jnt stk joint stock

JO job order • (ital.) Journal Officiel (French: Official Gazette) • junior officer • *British vehicle registration for* Oxford

Joburg ('dʒəʊ,bɜːg) *Colloquial* Johannesburg

joc. jocose • jocular

JOC Jeunesse ouvrière chrétienne (French: Young Christian Workers) • (USA) joint operations center

j.o.d. joint occupancy date

J. of E. (ital.) Journal of Education

JOG *Yachting* junior offshore group

Johan. Johannesburg

join. joinery

Jon. *Bible* Jonah

Jos. Joseph

Josh. *Bible* Joshua

JOT joint observer team

jour. journal(ist) • journey • journeyman

JOVIAL ('dʒəʊvɪəl) *Computing* Jules' own version of international algorithmic language (named after Jules Schwarz, computer scientist)

JP *airline flight code for* Adria Airways • (*or* **j.p.**) jet propulsion (*or* jet-propelled) • Justice of the Peace • *British vehicle registration for* Liverpool

JPA Jamaica Press Association

JPC joint planning council • joint production council • Judge of the Prize Court

JPCAC Joint Production, Consultative, and Advisory Committee

JPEG ('dʒeɪpeg) *Computing* Joint Photographic Expert Group

JPL Jet Propulsion Laboratory (California) • (ital.) Journal of Planning Law

Jpn Japan

JPRS (USA) Joint Publications Research Service

JPS jet-propulsion system(s) • (USA) Jewish Publications Society • Joint Parliamentary Secretary • joint planning staff • Junior Philatelic Society

JPTO *Astronautics* jet-propelled takeoff

jr. jour (French: day)

Jr (*or* **jr**) Junior

Jr. (*or* **jr.**) journal • juror

JR *airline flight code for* Aero California • Jacobus Rex (Latin: King James) • joint resolution • Judges' Rules • Jurist Reports • *British vehicle registration for* Newcastle upon Tyne

JRA juvenile rheumatoid arthritis

JRAI (ital.) Journal of the Royal Anthropological Institute

JRC Junior Red Cross

JS *British vehicle registration for* Inverness • (USA) Japan Society • *Law* judgment summons • *Law* judicial separation

JSA Jobseeker's Allowance

JSAWC Joint Services Amphibious Warfare Centre

JSB joint-stock bank

JSC Johnson Space Center (Texas)

JSD Doctor of Juristic Science • *Computing* Jackson system development

JSDC Joint Service Defence College

JSE Johannesburg Stock Exchange

Jsey Jersey

JSLS Joint Services Liaison Staff

JSP *Computing, tradename* Jackson structured programming (named after Michael Jackson, computer scientist) • Japan Socialist Party

JSPS Japan Society for the Promotion of Science

JSS joint services standard

JSSC Joint Services Staff College • joint shop stewards' committee

J-stars ('dʒeɪ,stɑːz) joint surveillance and targeting acquisition radar system

jt joint

j.t. *Law* joint tenancy

JT *British vehicle registration for* Bournemouth

JTC Junior Training Corps

Jt Ed. Joint Editor

JTIDS Joint Tactical Information Distribution Systems

jtly jointly

JTMP Job Transfer and Manipulation Protocol

JTO jump takeoff (of an aircraft)

JTS job training standards

JTST *Meteorol.* jet stream

JTUAC Joint Trade Union Advisory Committee

j.u. joint use

Ju. June

JU *airline flight code for* Yugoslav Airlines (JAT) • *British vehicle registration for* Leicester

jud. judgment • judicial • judo

Jud. *Bible* Judah • Judaism • Judea • Judge • (*or* **Judg.**) *Bible* Judges • *Bible* Judith

JUD Doctor of Canon and Civil Law (Latin *Juris Utriusque Doctor*)

Judg. *Bible* Judges

judgt judgment

juev. jueves (Spanish: Thursday)

JUGFET ('dʒʌgfɛt) *Electronics* junction-gate field-effect transistor

juil. juillet (French: July)

jul. julho (Portuguese: July) • julio (Spanish: July)

Jul. July

Jun. June • (*or* **jun.**) junior

junc. (*or* **Junc.**) junction

jun. part. junior partner

Junr (*or* **junr**) junior

JurD Doctor of Law (Latin *Juris Doctor*)

jurisd. jurisdiction

jurisp. jurisprudence

jus. (*or* **just.**) justice

JUSMAG Joint United States Military Advisory Group

juss. *Grammar* jussive

Juss. *Botany* Jussieu (indicating the author of a species, etc.)

juv. juvenile

Juv. Juvenal (Decimus Junius Juvenalis, ?60–?140 A.D., Roman poet)

JUWTFA Joint Unconventional Warfare Task Force, Atlantic

jux. juxtaposition

Jv. Java(nese)

JV *Commerce* joint venture • (USA) junior varsity • *British vehicle registration for* Lincoln

JW *British vehicle registration for* Birmingham • Jehovah's Witness • junior warden

JWB Jewish Welfare Board • joint wages board

JWEF Joinery and Woodwork Employers' Federation

jwlr jeweller

j.w.o. *Insurance* jettisoning and washing overboard

JWS (*or* **jws**) Joint Warfare Staff

JWV Jewish War Veterans

JX *British vehicle registration for* Huddersfield • Jesus Christ

Jy *Astronomy, symbol for* jansky • July • (*or* **jy**) jury

JY *British vehicle registration for* Exeter • *airline flight code for* Jersey European Airways • *international civil aircraft marking for* Jordan

JZ *British vehicle registration for* Down

K

k (ital.) *Physics, symbol for* Boltzmann constant • *Maths., symbol for* a constant • *Maths., symbol for* curvature • *symbol for* kilo- (prefix indicating 1000, as in **km**, kilometre; *or* (in computing) 1024, as in **kbyte**, kilobyte) • (ital.) *Mechanics, symbol for* radius of gyration • (ital.) *Chem., symbol for* rate coefficient (*or* constant) • (ital.) *Physics, symbol for* thermal conductivity • (bold ital.) *Maths., symbol for* a unit coordinate vector

k. (USA) karat • keel • killed • king • knight • knot

K (ital.) *Physics, symbol for* bulk modulus • *Botany, symbol for* calyx (in a floral formula) • *international vehicle registration for* Cambodia • (ital.) *Ecology, symbol for* carrying capacity (as in **K-strategist**) • *Geology, symbol for* Cretaceous • (ital.) *Chem., symbol for* equilibrium constant • *Physics, symbol for* kaon • *Science, symbol for* kelvin(s) • (ital.) *Physics, symbol for* kerma (kinetic energy released) • *Immunol.* killer (in **K cell**) • *Computing, symbol for* kilo- (*see under* k) • *symbol for* kina (monetary unit of Papua New

Guinea) • (*ital.*) *Physics, symbol for* kinetic
energy • *Chess, symbol for* king • kip
(Laotian monetary unit; *see also under*
KN) • *Music* Kirkpatrick (preceding a
number in Ralph Kirkpatrick's catalogue
of Domenico Scarlatti's works) • *British
fishing port registration for* Kirkwall • *Music*
Köchel (preceding a number in Ludwig
von Köchel's catalogue of Mozart's
works) • *symbol for* (Zambian) kwacha
(monetary unit; *see also under* MK) •
symbol for kyat (Burmese monetary unit) •
British vehicle registration for Liverpool •
(*ital.*) *Physics, symbol for* luminous
efficacy • *Biochem., symbol for* lysine •
Chem., symbol for potassium (Latin
kalium) • *Astronomy, symbol for* solar
constant • *Baseball, symbol for* strikeout •
Astronomy, indicating a spectral type •
indicating one thousand

K. (*or* **K**) Kalt (German: cold) • King (*or*
King's) • knit

K2 *airline flight code for* Kyrgyzstan Airlines

K4 *airline flight code for* Kazakhstan Airlines

K9 *Military* canine (K9 dogs; army dogs)

8K *airline flight code for* Air Ostrava

9K *airline flight code for* Cape Air • *international civil aircraft marking for* Kuwait

KA *airline flight code for* Dragonair-Hong
Kong Dragon Airlines • *postcode for* Kil-
marnock • King of Arms • Knight of St
Andrew, Order of Barbados • *British
vehicle registration for* Liverpool

KADU ('kɑːduː) Kenya African
Democratic Union (now absorbed by
KANU)

KAK (Sweden) Kungl Automobil Klubben
(Royal Automobile Club)

Kal. Kalendae (Latin: calends; first day of
each month)

Kan. (*or* **Kans.**) Kansas

k & b kitchen and bathroom

KANTAFU Kenya African National
Traders' and Farmers' Union

KANU ('kɑːnuː) Kenya African National
Union

KANUPP ('kænʌp) (Pakistan) Karachi
Nuclear Power Plant

KAO Kuiper Airborne Observatory

Kap. *Finance* Kapital (German: capital) •
Kapitel (German: chapter)

Kar. Karachi

KAR King's African Rifles

Karel. Karelia(n)

Kash. Kashmir

KASSR (formerly) Karelian Autonomous
Soviet Socialist Republic

kb *Physics, symbol for* kilobar(s) • *Genetics,
symbol for* kilobase(s)

KB *Computing* kilobyte • King's Bench •
Chess king's bishop • Knight Bachelor •
Knitting knit into back of stitch • *Comput-
ing* knowledge base • Koninkrijk België
(Flemish: Kingdom of Belgium) • *British
vehicle registration for* Liverpool

KBASSR (formerly) Kabardino-Balkar
Autonomous Soviet Socialist Republic

KBC *Law* King's Bench Court

kbd keyboard

KBD *Law* King's Bench Division

KBE Knight Commander of the Order of
the British Empire

KBES *Computing* knowledge-based expert
system

Kbhvn København (Danish: Copenhagen)

kbp *Genetics, symbol for* kilobase pair

KBP *Chess* king's bishop's pawn

KBS Knight of the Blessed Sacrament •
Computing knowledge-based system

KBW King's Bench Walk (Temple,
London)

kbyte *Computing* kilobyte

kc *Physics* kilocycle

KC Kansas City • Kennel Club • King's
College • King's Counsel • King's Cross
(London) • *airline flight code for* Kiwi
Travel International Airlines • Knight
Commander • *RC Church* Knights of
Columbus • *British vehicle registration for*
Liverpool

KCA Keesing's Contemporary Archives

kcal *symbol for* kilocalorie(s)

KCB Knight Commander of the Order of
the Bath

KCC King's College, Cambridge •
(Knight) Commander of the Order of the
Crown, Belgium and the Congo Free
State

K cell *Immunol.* killer cell

KCH King's College Hospital (London) •
Knight Commander of the Hanoverian
Order

KCHS Knight Commander of the Order
of the Holy Sepulchre

KCIE Knight Commander of the Order of
the Indian Empire

KCL King's College, London

KCLJ Knight Commander of the Order of
St Lazarus of Jerusalem

KCMG Knight Commander of the Order of St Michael and St George

KCMS *Trademark* Kodak colour management system

kcs (*or* **kc/s**) kilocycles per second

Kčs *symbol for* koruna (Czech and Slovak monetary unit)

KCSA Knight Commander of the Military Order of the Collar of St Agatha of Paterna

KCSG Knight Commander of the Order of St Gregory the Great

KCSI Knight Commander of the Order of the Star of India

KCSJ Knight Commander of the Order of St John of Jerusalem (Knights Hospitaller)

KCSS Knight Commander of the Order of St Silvester

KCVO Knight Commander of the Royal Victorian Order

kd killed

KD kiln dried • (*or* **k.d.**) knock down (at an auction sale) • (*or* **k.d.**) *Commerce* knocked down (of goods for sale) • Kongeriget Danmark (Danish: Kingdom of Denmark) • *symbol for* Kuwaiti dinar (monetary unit) • *British vehicle registration for* Liverpool

KDC (*or* **k.d.c.**) *Computing* key distribution centre • *Commerce* knocked-down condition (of goods)

KDD *Computing* knowledge discovery in databases

KDF (*or* **KdF**) Kraft durch Freude (German: Strength through Joy; Nazi holiday scheme)

KDG King's Dragoon Guards

KDM Kongelige Danske Marine (Royal Danish Navy)

KE kinetic energy • kinetic equation • *airline flight code for* Korean Air • *British vehicle registration for* Maidstone

KEAS (*or* **k.e.a.s.**) *Aeronautics* knots equivalent airspeed

Keb Keble College, Oxford

KEH King Edward's Horse

Ken. Kensington • Kentucky • Kenya

KEO King Edward's Own

keV *Physics, symbol for* kiloelectronvolt(s)

KEY (ki:) keep extending yourself

KF *British vehicle registration for* Liverpool

KFA Kenya Farmers' Association

KFAED Kuwait Fund for Arab Economic Development

KFAT National Union of Knitwear, Footwear, and Apparel Trades

KFC *Trademark* Kentucky Fried Chicken

KFL Kenya Federation of Labour

kfm. kaufmännisch (German: commercial)

Kfm. Kaufmann (German: merchant)

Kfz. Kraftfahrzeug (German: motor vehicle)

kg keg • *symbol for* kilogram(s)

KG *British vehicle registration for* Cardiff • *Physics* Klein–Gordon (as in **KG equation**) • Knight of the Order of the Garter • Kommanditgesellschaft (German: limited partnership)

KGB Komitet Gosudarstvennoi Bezopasnosti (Russian: Committee of State Security)

KGCB Knight Grand Cross of the Bath

kgf *Physics, symbol for* kilogram-force

KGFR *Med.* kidney glomerular filtration rate

Kgl. Königlich (German: Royal)

Kgs *Bible* Kings

KGS known geological structure

KH *British vehicle registration for* Hull • *Physics* Kelvin–Helmholtz (as in **KH formula**) • kennel huntsman • *Astronautics* keyhole (as in **KH-11**, **KH-12**; imaging NASA satellites) • King's Hussars • Knight of the Hanoverian Order

KHC Honorary Chaplain to the King

KHDS Honorary Dental Surgeon to the King

KHNS Honorary Nursing Sister to the King

KHP Honorary Physician to the King

KHS Honorary Surgeon to the King • Knight of the Order of the Holy Sepulchre

kHz *symbol for* kilohertz

KI *Irish vehicle registration for* Waterford

KIA killed in action

KIAS (*or* **k.i.a.s.**) *Aeronautics* knots indicated airspeed

K-i-H Kaisar-i-Hind (Emperor of India; medal)

kil. (*or* **kild.**) *Brewing* kilderkin (size of cask)

Kild. Kildare

Kilk. Kilkenny

KIM (kɪm) *Chem.* kinetic isotope method

Kinc. Kincardine (former Scottish county)

kind. kindergarten

kingd. kingdom

Kinr. Kinross (former Scottish county)

KIO Kuwait Investment Office

Kirk. Kirkcudbright (former Scottish county)

KISS (kɪs) *Colloquial* (USA) keep it simple, stupid • *Stock exchange* (Germany) Kurs Information Service System

kJ *symbol for* kilojoule(s)

KJ *airline flight code for* British Mediterranean Airways • Med. knee jerk • Knight of St Joachim • *British vehicle registration for* Maidstone

KJV King James Version (of the Bible)

KK Kabushiki Kaisha (Japanese: joint-stock company) • *Physics* Kaluza–Klein (as in **KK theory**) • *British vehicle registration for* Maidstone

KKASSR (formerly) Kara-Kalpak Autonomous Soviet Socialist Republic

KKK Ku Klux Klan

KKt *Chess* king's knight

KKtP *Chess* king's knight's pawn

kl *symbol for* kilolitre(s)

KL *Astronomy* Kleinmann–Low (as in **KL nebula**) • *airline flight code for* KLM – Royal Dutch Airlines • Kuala Lumpur • *British vehicle registration for* Maidstone

KLH *Immunol.* keyhole-limpet haemocyanin • Knight of the Legion of Honour

KLJ Knight of the Order of St Lazarus of Jerusalem

KLM Koninklijke Luchtvaart Maatschappij (Royal Dutch Airlines)

KLSE Kuala Lumpur Stock Exchange

km *symbol for* kilometre(s)

KM *airline flight code for* Air Malta • King's Medal • Knight of Malta • *British vehicle registration for* Maidstone

KMO Kobe Marine Observatory (Japan)

KMP *Computing* Knuth-Morris-Pratt (in **KMP algorithm**)

KMT Kuomintang (Chinese Nationalist Party)

KMUL (Germany) Karl Marx Universität Leipzig

kn *Nautical, symbol for* knot • krona (Swedish monetary unit; *see also* SKr) • krone (Danish or Norwegian monetary unit; *see also* Dkr; NKr)

KN *Chess* king's knight • *Numismatics* King's Norton (mint-mark, Birmingham) • *symbol for* kip (Laotian monetary unit) • Kongeriket Norge (Norwegian:

Kingdom of Norway) • *British vehicle registration for* Maidstone

KNA Kongelig Norsk Automobil-klubb (Royal Norwegian Automobile Club)

KNM Kongelige Norske Marine (Royal Norwegian Navy)

KNP *Chess* king's knight's pawn • Kruger National Park (South Africa)

KNPC Kuwait National Petroleum Company

Knt Knight

k.o. keep off • keep out • *Football* kick off • *Colloquial* knock out (*or* knockout)

KO *airline flight code for* Alaska Central Express • *Colloquial* knock out (*or* knockout) • *British vehicle registration for* Maidstone

KOC Kuwait Oil Company

KOD *Navigation* kick-off drift

K of C *RC Church* Knights of Columbus

K of K (Lord) Kitchener of Khartoum

K of P (USA) Knights of Pythias

Komintern ('kɒmɪn,tɜːn) Communist International (Russian *Kom(munisticheski) Intern(atsionál)*)

Komp. Kompanie (German: company)

Kor. Koran • Korea(n)

KORR King's Own Royal Regiment

KOSB King's Own Scottish Borderers

KOYLI King's Own Yorkshire Light Infantry

k.p. key personnel

KP King's Parade • *Chess* king's pawn • *Military* (USA) kitchen police • Knight of the Order of St Patrick • *British vehicle registration for* Maidstone

KPD Kommunistische Partei Deutschlands (German Communist Party)

KPFSM King's Police and Fire Service Medal

kph kilometres per hour

KPM King's Police Medal

Kpmtr *Music* Kapellmeister (German: conductor)

KPNLF Khmer People's National Liberation Front

KPNO Kitt Peak National Observatory (Arizona)

KPP Keeper of the Privy Purse

kpr keeper

KPTT *Med.* kaolin partial thromboplastin time

KPU Kenya People's Union

KQ *airline flight code for* Kenya Airways

kr. krona (Swedish monetary unit; *see also* SKr) • króna (Icelandic monetary unit; *see also* ISK) • krone (Danish or Norwegian monetary unit; *see also* Dkr; NKr)

Kr *Chem.*, *symbol for* krypton

KR King's Regiment • *Military* King's Regulations • *Chess* king's rook • *British vehicle registration for* Maidstone

KRL *Computing* knowledge representation language (in artificial intelligence)

KRP *Chess* king's rook's pawn

KRR King's Royal Rifles

KRRC King's Royal Rifle Corps

KS *British vehicle registration for* Edinburgh • *US postcode for* Kansas • *Med.* Kaposi's sarcoma • King's Scholar • King's School • Kipling Society • Kitchener Scholar • Konungariket Sverige (Swedish: Kingdom of Sweden) • *international vehicle registration for* Kyrgyzstan • *airline flight code for* Peninsula Airways

KSA Kitchen Specialists' Association

KSC Kennedy Space Center (Florida) • King's School, Canterbury • Knight of St Columba • (formerly) Komunistická Strana Československá (Communist Party of Czechoslovakia)

KSG Knight of the Order of St Gregory the Great

KSh *symbol for* Kenya shilling (monetary unit)

KSI Knight of the Order of the Star of India

KSIE *Chem.* kinetic solvent isotope effect

KSJ Knight of the Order of St John of Jerusalem (Knights Hospitaller)

KSLI King's Shropshire Light Infantry

KSM Korean Service Medal • Kungliga Svenska Marinen (Royal Swedish Navy)

KSS Knight of the Order of St Silvester

KSSR Kazakh Soviet Socialist Republic (now Kazakhstan)

KSSU KLM, SAS, Swissair, UTA (international airline organization)

KStJ Knight of the Order of St John of Jerusalem (Knights Hospitaller)

KSU Kansas State University

kt (USA) karat • *symbol for* kilotonne(s) • *Nautical* knot

Kt knight

KT *postcode for* Kingston-upon-Thames • Knight Templar • Knight (of the Order) of the Thistle • *Meteorol.* knot • *British vehicle registration for* Maidstone

Kt Bach. Knight Bachelor

Kto *Banking* Konto (German: account)

Ku *Chem.*, *symbol for* kurchatovium (element 104)

KU *airline flight code for* Kuwait Airways • *British vehicle registration for* Sheffield

Kuw. Kuwait

kV *symbol for* kilovolt(s)

KV *British vehicle registration for* Coventry • *airline flight code for* Eastern Air • *Music* Köchel Verzeichnis (German: Köchel catalogue; *see under* K)

kVAr *Electrical engineering, symbol for* kilovar(s)

kVp kilovolts, peak (applied across an X-ray tube)

kW *symbol for* kilowatt(s)

KW *postcode for* Kirkwall, Orkney • *British vehicle registration for* Sheffield

kWh (*or* **kW h**) *symbol for* kilowatt hour(s)

KWIC (kwɪk) key word in context (as in **KWIC index**)

KWOC (kwɒk) key word out of context (as in **KWOC index**)

KWP Korean Workers' Party

KWT *international vehicle registration for* Kuwait

KX *airline flight code for* Cayman Airways • *British vehicle registration for* Luton

ky. kyat (Burmese monetary unit; *see also under* K)

Ky. Kentucky

KY *US postcode for* Kentucky • Kirkcaldy (postcode *or* British fishing port registration) • Kol Yisrael (Israeli broadcasting station) • *British vehicle registration for* Sheffield

kybd keyboard

Kyr. *Ecclesiast.* Kyrie eleison

kz *Meteorol., symbol for* duststorm *or* sandstorm

KZ *British vehicle registration for* Antrim • *international vehicle registration for* Kazakhstan • *Military* killing zone

L

l (ital.) *Chem.* laevorotatory (as in *l*-**tartaric acid**) • *Music* lah (in tonic sol-fa) • (ital.) *symbol for* length • (ital.) *Physics, symbol for* lepton number • *Meteorol.* lightning • *Chem.* liquid (as in $H_2O(l)$) • *symbol for* litre(s) • (ital.) *Physics, symbol for* orbital angular momentum quantum number

l. lake • lambda • land • late • lateral • law • leaf • league • leasehold • left • legitimate • length • light • line (of written matter) • link • literate • little • loch • long • lost • lough • low • *Currency* pound (Latin *libra*; symbol: £)

L (bold ital.) *Physics, symbol for* angular momentum • (ital.) *Chem., symbol for* Avogadro constant • *Roman numeral for* fifty • *Electrical engineering, symbol for* inductor • (ital.) *Physics, symbol for* Lagrangian function • *Physics, symbol for* lambert (obsolete unit of luminance) • *Linguistics* language (as in L_1, first language; L_2, second language) • (ital.) *Physics, symbol for* latent heat • learner (driver; on British motor vehicles) • *symbol for* lempira (Honduran monetary unit) • (ital.) *symbol for* length • *Biochem., symbol for* leucine • *Aeronautics* lift • *Immunol.* light (in **L-chain** of an immunoglobulin molecule) • *Irish fishing port registration for* Limerick • linear • (ital.) *Biochem., symbol for* linking number • *symbol for* litre(s) • live (on electric plugs) • *postcode for* Liverpool • (ital.) *Geography, symbol for* longitude • (ital.; *or* L_v) *Physics, symbol for* luminance • (ital.) *Astronomy, symbol for* luminosity • *international vehicle registration for* Luxembourg • (ital.) *Physics, symbol for* orbital angular momentum quantum number • (ital., *or* L_e) *Physics, symbol for* radiance • (ital.) *Electrical engineering, symbol for* self-inductance • (ital.) *Physics, symbol for* sound intensity • (small cap.) *Chem., indicating* an optically active compound having a configuration related to laevorotatory glyceraldehyde (as in **L-lactic acid**)

L. (*or* **L**) Lady • Lake • *Military* Lancers • large • Latin • law • League • left *or* (in the theatre) stage left • lethal • liber (Latin: book) • *Politics* Liberal • Licentiate (in degrees, etc.) • Lieutenant • line (of written matter) • link • *Biology* Linnaeus (indicating the author of a species, etc.) • lira *or* (pl.) lire (Italian monetary unit; *see also* Lit) • Loch • locus (Latin: place) • Lodge (fraternal) • London • Lord • *Sport* lost • Lough • low • *Currency* pound (Latin *libra*; symbol: £)

L6 *airline flight code for* Air Maldives

L9 *airline flight code for* Air Mali

4L *international civil aircraft marking for* Georgia

7L *airline flight code for* AB Shannon

9L *international civil aircraft marking for* Sierra Leone

l.a. landing account • leading article • *Med.* lege artis (Latin: as directed; in prescriptions) • lighter than air

La *Chem., symbol for* lanthanum

La. Lancastrian • Lane • Louisiana

LA *postcode for* Lancaster • *airline flight code for* Lan Chile • *Photog.* large aperture • Latin America(n) • law agent • leave allowance • *Med.* left atrium (*or* atrial) • legal adviser • Legislative Assembly • letter of authority • Library Association • Licensing Act • licensing authority • Lieutenant-at-Arms • *Physics* linear accelerator • *Computing* linear arithmetic • Literate in Arts • Liverpool Academy • *British fishing port registration for* Llanelly • Lloyd's agent • local agent • local anaesthetic • local association • local authority • *British vehicle registration for* London NW • long acting • Los Angeles • *US postcode for* Louisiana • low altitude

LAA (USA) League of Advertising Agencies • Library Association of Australia • Libyan Arab Airlines • Lieutenant-at-Arms • (USA) Life Assurance Advertisers • light anti-aircraft

LAADS (USA) Los Angeles Air Defense Sector

LAAOH Ladies' Auxiliary, Ancient Order of Hibernians

LAAR liquid-air accumulator rocket

lab. label • laboratory • labourer

Lab *Computing* lightness, red-green axis (A), yellow-blue axis (B)

Lab. (USA) Laborite • *Politics* Labour • Labrador

LAB *Chem.* linear alkyl benzene • load aboard barge • low-altitude bombing

LABA (USA) Laboratory Animal Breeders' Association

lac. lacquer • lactation

Lac *Astronomy* Lacerta

LAC Laboratory Animals Centre • leading aircraftman • Licentiate of the Apothecaries' Company • London Athletic Club

LACES London Airport cargo electronic processing scheme • (USA) Los Angeles Council of Engineering Societies

LACONIQ *Computing* laboratory computer online inquiry

LACSA Lineas Aéreas Costarricenses (Costa Rican Airlines)

LACSAB Local Authorities' Conditions of Service Advisory Board

LACW leading aircraftwoman

LAD *Linguistics* language acquisition device • light aid detachment

ladar ('leɪdɑː) laser detection and ranging

Ladp Ladyship

L. Adv. Lord Advocate

LAE *Maths.* linear algebraic equation

laev. *Med.* laevus (Latin: left)

LaF Louisiana French

LAF L'Académie Française (French Academy)

LAFC Latin-American Forestry Commission

La Font. (Jean de) La Fontaine (1621–95, French poet)

LAFTA ('læftə) Latin American Free Trade Association (*see* LAIA)

lag. lagoon

Lah. Lahore

LAH Licentiate of the Apothecaries' Hall (Dublin)

LAHS low altitude, high speed

LAI *Botany* leaf area index • Library Association of Ireland

LAIA Latin American Integration Association (formerly LAFTA)

LAK (læk) *Med.* lymphokine-activated killer (in **LAK cell**; used in cancer treatment)

LAL (France) Laboratoire de l'accélérateur linéaire (Linear Accelerator Laboratory)

lam. laminate(d)

Lam. *Botany* Lamarck (indicating the author of a species, etc.) • *Bible* Lamentations

LAM London Academy of Music • Master of the Liberal Arts (Latin *Liberalium Artium Magister*)

LAMA Locomotive and Allied Manufacturers' Association of Great Britain

LAMC Livestock Auctioneers' Market Committee of England and Wales

LAMCO Liberian-American-Swedish Mineral Corporation

LAMDA ('læmdə) London Academy of Music and Dramatic Art

LAMIDA Lancashire and Merseyside Industrial Development Association

LAMP (læmp) low-altitude manned penetration • (USA) Lunar Analysis and Mapping Program

LAMS launch acoustic measuring system

LAMSAC Local Authorities' Management Services and Computer Committee

LAN *Meteorol.* inland • (*or* **Lan**; læn) Linea Aérea Nacional (de Chile) (Chilean national airlines; Lan Chile) • local apparent noon • (læn) *Computing* local-area network

LANBY ('lænbɪ) large automatic navigation buoy

Lanc. Lancaster • *Military* Lancers

Lancs Lancashire

L & D loans and discounts • loss and damage

L&ID London and India Docks

L&NRR (USA) Louisville and Nashville Railroad

L & NWR London and North-Western Railway

L & SWR London and South-Western Railway

L & YR Lancashire and Yorkshire Railway

Lan. Fus. Lancashire Fusiliers

lang. language

Lang. Languedoc

LANICA Lineas Aéreas de Nicaragua (Nicaraguan national airlines)

LANL (USA) Los Alamos National Laboratory

LANRAC (*or* **Lanrac**) ('lænræk) Land Army Reunion Association Committee

LanR(PWV) Lancashire Regiment (Prince of Wales' Volunteers)

Lantirn ('læntən) *Military* low-altitude navigation and targeting infrared system

LAO *international vehicle registration for* Laos • Licentiate in the Art of Obstetrics

LAOAR *US Air Force* Latin American Office of Aerospace Research

Lap. Lapland • Lappish

LAP (USA) Laboratory of Aviation Psychology • Lineas Aéreas Paraguayas (Paraguayan national airlines) • *Computing* link access protocol

LAPD Los Angeles Police Department

LAPES low-altitude parachute extraction system

LAPO Los Angeles Philharmonic Orchestra

LAPT London Association for the Protection of Trade

LAR *Botany* leaf area ratio • *international vehicle registration for* Libya • *Taxation* life assurance relief • *Computing* limit address register

LARA light armed reconnaissance aircraft

larg. *Music* largamente (Italian; broadly) • largeur (French: width) • *Music* largo (Italian; very slowly)

LARO Latin American Regional Office (in the FAO)

LARSP Language Assessment, Remediation, and Screening Procedure

laryngol. laryngologist • laryngology

LAS Land Agents' Society • large astronomical satellite • League of Arab States • Legal Aid Society • London Archaeological Service • Lord Advocate of Scotland • low-altitude satellite • *Physics* low-angle scattering • lower airspace

laser ('leɪzə) light amplification by stimulated emission of radiation

LASER London and South Eastern Library Region

LASH (læʃ) *Commerce* lighter aboard ship

LASL (USA) Los Alamos Scientific Laboratory

LASMO London and Scottish Marine Oil

LASP low-altitude space platform

LASS lighter-than-air submarine simulator

lat. lateral • (*or* **lat**) *Geography* latitude • latus (Latin: wide)

Lat. Latin • Latvia(n)

LAT local apparent time • (ital.) Los Angeles Times

LATCC London air traffic control centre

LATCRS London air traffic control radar station

lat. ht latent heat

LATS (læts) *Med.* long-acting thyroid stimulator

Latv. Latvia

Latvn Latvian

LAUK Library Association of United Kingdom

laun. launched

LAUTRO (*or* **Lautro**) ('lɔːtrəʊ) Life Assurance and Unit Trust Regulatory Organization

LAV light armoured vehicle • Lineas Aéreas Venezolanas (Venezuelan Airlines) • lymphadenopathy-associated virus (original name for the Aids virus, HIV)

law. lawyer

LAW League of American Writers • light anti-tank weapon

Lawn (lɔːn) *Computing* local area wireless network

Law Rept. (*or* **Law Rpts**) Law Reports

LAWRS limited airport weather reporting system

LAWS *Meteorol.* low-altitude wind shear

lax. laxative

LAX Los Angeles international airport

lb *symbol for* binary logarithm • (*or* **lb.**) pound(s) (weight; Latin *libra*)

l.b. landing barge • *Sport* left back • *Cricket* leg bye • letter box • link belt

L.b. Lectori benevolo (Latin: to the kind reader)

LB Bachelor of Letters (*or* Literature) (Latin *Litterarum Baccalaureus*) • *international vehicle registration for* Liberia • light bomber • *airline flight code for* Lloyd Aereo Boliviano • local board • *British vehicle registration for* London NW

LBA late booking agent (euphemism for ticket tout) • linear-bounded automaton • *Computing* logical block address

LB & SCR London, Brighton, and South Coast Railway

LBB *Engineering* leak before break

LBBB *Med.* left bundle-branch block

Lbc. Lübeck

LBC Land Bank Commission • London Broadcasting Company

LBCH London Bankers' Clearing House

LBCM Licentiate of the Bandsmen's College of Music • London Board of Congregational Ministers

LBD League of British Dramatists

L/Bdr (*or* **LBdr**) Lance-Bombardier

lbf *Physics, symbol for* pound-force

lb-ft pound-foot

LBH length, breadth, height

LBI *Astronomy* long baseline interferometry

LBJ Lyndon Baines Johnson (1908–73, US president 1963–69)

LBO *Commerce* leveraged buyout

LBP length between perpendiculars

lbr labour • lumber

L.b.s. Lectori benevolo salutem (Latin: to the kind reader, greeting)

LBS Libyan Broadcasting Service • life-boat station • London Business School

LBV landing barge vehicle • Late Bottled Vintage (of port wine) • *Astronomy* luminous blue variable

lbw *Cricket* leg before wicket

LBW live body weight

l.c. label clause • law courts • lead covered • leading cases • left centre • legal currency • letter card • (*or* **lc**, **l/c**) letter of credit • loco citato (Latin: in the place cited; textual annotation) • low-calorie • low-carbon • (*or* **lc**) *Printing* lower case

LC Cross of Leo • landing craft • *Med.* Langerhans cells • Leander Club • Legislative Council • (*or* **L/C**) letter of credit • level crossing • (USA) Library of Congress • Lieutenant-Commander • *Chem.* linear chain • *Electronics* linear contact • *Military slang* line crosser (a defector) • *Chem.* liquid chromatography • *Electronics* liquid crystal • livestock commissioner • *airline flight code for* Loganair • *British vehicle registration for* London NW • Lord Chamberlain • Lord Chancellor • Lower Canada • *Electronics* lumped constant

LCA Library Club of America • Licensed Company Auditor • *Computing* logic cell array • low-cost automation

LCAD London Certificate in Art and Design

LCAO *Chem.* linear combination of atomic orbitals

LCAP *Computing* loosely coupled array of processors

l.c.b. longitudinal centre of buoyancy

LCB *Theatre* left centre back (of stage) •

(USA) Liquor Control Board • London Convention Bureau • Lord Chief Baron

LCC *Electronics* leadless chip carrier • *Accounting, Computing* life-cycle cost(ing) • *Electrical engineering* load-carrying capability • London Chamber of Commerce • London County Council (superseded by the Greater London Council, GLC)

LCCC (USA) Library of Congress Catalog Card

LCD *Electronics* liquid-crystal display • Lord Chamberlain's Department • Lord Chancellor's Department • lower court decisions • (*or* **l.c.d.**) *Maths.* lowest common denominator

LCDT London Contemporary Dance Theatre

LCE Licentiate in Civil Engineering • London Commodity Exchange

l.c.f. longitudinal centre of flotation • (*or* **LCF**) *Maths.* lowest common factor

LCFA *Chem.* long-chain fatty acid

l.c.g. longitudinal centre of gravity

LCG *Chem.* lamellar compound of graphite

LCGB Locomotive Club of Great Britain

LCh Licentiate in Surgery (Latin *Licentiatus Chirurgiae*) • Lord Chancellor

LCH London Clearing House

LCIGB Locomotive and Carriage Institution of Great Britain and Ireland

LCIOB Licentiate of the Chartered Institute of Building

LCJ Lord Chief Justice

l.c.l. lower control limit

LCL *Commerce* (USA) less-than-carload lot • *Commerce* less-than-container load • Licentiate in Canon Law

LCLS Livestock Commission Levy Scheme

LCM landing craft mechanized • *Computing* life-cycle management • London College of Music • (*or* **l.c.m.**) *Maths.* lowest (*or* least) common multiple

LC-MS *Chem.* liquid chromatography-mass spectrometry

Lcn Lincoln

LCN *Aeronautics* load classification number • local civil noon

LCO landing craft officer • launch control officer

L-Col Lieutenant-Colonel

L-Corp. Lance-Corporal

LCP last complete programme • least-cost planning • Licentiate of the College of

Preceptors • *Chem.* liquid-crystal polymer • London College of Printing • low-cost production

LCP&SA Licentiate of the College of Physicians and Surgeons of America

LCP&SO Licentiate of the College of Physicians and Surgeons of Ontario

L/Cpl Lance-Corporal

LCPS Licentiate of the College of Physicians and Surgeons

l/cr. lettre de crédit (French: letter of credit)

LCS London Cooperative Society

LCSAJ *Computing* linear code sequence and jump

LCSP London and Counties Society of Physiologists

LCST Licentiate of the College of Speech Therapists

LCT landing craft tank • local civil time

LCU *Photog.* large close-up

LCV Licentiate of the College of Violinists

ld land • *Printing* lead • load

l.d. legal dose • light difference • line of departure • line of duty

Ld Limited (company) • Lord (title)

LD Doctor of Letters *or* Literature (Latin *Litterarum Doctor*) • Lady Day • Laus Deo (Latin: Praise be to God) • *Education, Psychol.* learning-disabled • lepide dictum (Latin: wittily said) • *Pharmacol.* lethal dose (as in **LD$_{50}$**, median lethal dose) • Liberal and Democratic • Liberal Democrat • *symbol for* Libyan dinar (monetary unit) • Licentiate in Divinity • Light Dragoons • Litera Dominicalis (Latin: Dominical letter) • *postcode for* Llandrindod Wells • London Docks • *British vehicle registration for* London NW • low density • Low Dutch

L/D letter of deposit

Lda (*or* **Ld^a**) (Portugal) Sociedade de responsabilidade limitada (limited company; Ltd)

LDA Lead Development Association

l.d.b. light distribution box

l.d.c. long-distance call • lower dead centre

LDC less-developed country • local distribution company

LDDC London Docklands' Development Corporation

LDEF *Astronomy* long duration exposure facility

LDEG laus Deo et gloria (Latin: praise and glory be to God)

LDentSc Licentiate in Dental Science

Lderry Londonderry

ldg landing • leading • loading • lodging

Ldg *Navy* Leading (rank)

Ldge Lodge

L.d'H. Légion d'Honneur

LDiv Licentiate in Divinity

ldk lower deck

LDL *Biochem.* low-density lipoprotein

LDM *Physics* liquid-drop model (of the nucleus)

ldmk landmark

Ldn London

LDN less-developed nation

LDOS *Physics* local density of occupied states • Lord's Day Observance Society

Ldp Ladyship • Lordship

LDP (Japan) Liberal-Democratic Party • *Finance* London daily price • long-distance path

LDPE *Chem.* low-density polyethylene (used in packaging materials)

ldr leader • ledger • lodger

LDR low dose rate

ldry laundry

lds loads

LDS Latter-day Saints • laus Deo semper (Latin: praise be to God for ever) • Licentiate in Dental Surgery

LDSc Licentiate in Dental Science

LDV *Physics* laser Doppler velocity (*or* velocimetry) • Local Defence Volunteers (Home Guard)

LDX long-distance xerography

LDY Leicestershire and Derbyshire Yeomanry

l.e. leading edge • left eye • library edition • light equipment • limited edition • low explosive

Le *symbol for* leone (monetary unit of Sierra Leone) • (ital.) *Physics, symbol for* Lewis number

Le. Lebanese • Lebanon

LE (*or* **£E**) *symbol for* Egyptian pound (monetary unit) • *postcode for* Leicester • London Electricity • *British vehicle registration for* London NW • low energy • *Med.* lupus erythamatosus

lea. league • leather • leave

LEA Local Education Authority

LEAJ (USA) Law Enforcement and Administration of Justice

LEAP (liːp) Life Education for the Autistic Person • lift-off elevation and azimuth programmer • Loan and Educational Aid Programme

LEAR (lɪə) *Physics* low-energy antiproton ring

Leb. Lebanese • Lebanon

LEB London Electricity Board • *Astronautics* low-energy booster

LEC *Chem.* ligand-exchange chromatography • *Electronics* liquid-encapsulated Czochralski • Local Employment Committee • Local Enterprise Company

lect. lectio (Latin: lesson) • lecture(r)

lectr lecturer

led. ledger

LED *Electronics* light-emitting diode

LEDC Lighting Equipment Development Council

LEED (liːd) *Physics, Chem.* low-energy electron diffraction

LEFM linear elastic fracture mechanics

leg. legal • legate • (*or* **Leg.**) legation • *Music* legato (Italian: bound; i.e. smoothly) • (*or* **legis.**) legislation • (*or* **legis.**) legislative • (*or* **legis.**) legislature

legg. *Music* leggero (Italian: light *or* rapid)

Leics Leicestershire

Leip. Leipzig

Leit. Leitrim

LEL Laureate in English Literature • *Physics* linear energy loss

LEM (lɛm) *Astronautics* lunar excursion module

LEMA Lifting Equipment Manufacturers' Association

LEO ('liːəu) *Astronautics* low earth orbit • Lyons Electronic Office (early computer built by J. Lyons & Co.)

Lep *Astronomy* Lepus

LEP (lɛp) *Physics* Large Electron-Positron (collider; at CERN)

LEPMA (USA) Lithographic Engravers' and Plate Makers' Association

LEPORE (USA) long-term and expanded program of oceanic research and exploration

LEPRA ('lɛprə) Leprosy Relief Association

LEPT long-endurance patrolling torpedo

LES launch escape system • Liverpool Engineering Society

LèsL Licencié ès lettres (French: Bachelor of Arts)

LèsS Licencié ès sciences (French: Bachelor of Science)

LESS least-cost estimating and scheduling

LEST large earth-based solar telescope

LET *Physics* linear energy transfer

LETS Local Employment and Trade System • Local Exchange and Trading System

Leu (*or* **leu**) *Biochem.* leucine

LEU *Physics* low-enriched uranium

lev (*or* **LEV**) lunar excursion vehicle

Lev. Levant • (*or* **Levit.**) *Bible* Leviticus

lex. lexicon

LEX (lɛks) land exercise • *Computing* lexical analyser generator

lexicog. lexicographer • lexicographical • lexicography

LEY Liberal European Youth

lf leaf • *Printing* light face

l.f. ledger folio • life float

Lf limit of flocculation (in toxicology)

LF Lancashire Fusiliers • *British vehicle registration for* London NW • *Radio* low frequency • *Astronomy* luminosity function

l.f.a. local freight agent

LFB London Fire Brigade

LFBC London Federation of Boys' Clubs

l.f.c. low-frequency current

LFC Lutheran Free Church

LFCDA London Fire and Civil Defence Authority

LFD least fatal dose • low-fat diet

LFE laboratory for electronics

Lfg Lieferung (German: delivery)

LFO low-frequency oscillator

LFRD lot fraction reliability deviation

LFS laser (*or* laser-induced) fluorescent spectroscopy

lft leaflet

LFTU landing force training unit

lg *symbol for* common logarithm • long

lg. lagoon • large

LG *British vehicle registration for* Chester • Lady Companion of the Order of the Garter • landing ground • Lewis gun • Lieutenant-General • Life Guards • (David) Lloyd George (1863–1945, British statesman) • London Gazette • Low German • *airline flight code for* Luxair

LGAR (USA) Ladies of the Grand Army of the Republic

LGB Local Government Board

LGC lunar (module) guidance computer

lge large • league

LGEB Local Government Examination Board

L-Gen Lieutenant-General

LGer Low German

LGIO Local Government Information Office

LGk Late Greek

LGM Little Green Men • Lloyd's Gold Medal

LGO (USA) Lamont Geological Observatory

LGPRA Local Government Public Relations Association

LGr Late Greek

LGR leasehold ground rent • local government reports

LGSM Licentiate of the Guildhall School of Music

LGTB Local Government Training Board

lgth length

lg tn long ton

LGU Ladies' Golf Union

LGV large goods vehicle (replaced HGV)

l.h. left half • left hand(ed)

LH left hand(ed) • *British fishing port registration for* Leith • licensing hours • Licentiate in Hygiene • *Military* Light Horse • *British vehicle registration for* London NW • *airline flight code for* Lufthansa • *Biochem.* luteinizing hormone

L/H leasehold

LHA landing helicopter assault • local health authority • local hour angle • Lord High Admiral • lower hour angle

lhb *Sport* left halfback

LHC *Physics* Large Hadron Collider (at CERN) • Lord High Chancellor

LHCIMA Licentiate of the Hotel Catering and Institutional Management Association

l.h.d. left-hand drive

LHD Doctor of Humanities *or* of Literature (Latin *Litterarum Humaniorum Doctor*)

LHDC lateral homing depth charge

LHeb Late Hebrew

LHMC London Hospital Medical College

LHO livestock husbandry officer

lhr *Physics* lumen-hour

LHRC Light and Health Research Council

LH-RH *Biochem., Med.* luteinizing-hormone-releasing hormone

LHS left hand side

LHSM Licentiate of the Institute of Health Services Management

LHT Lord High Treasurer

LHWN lowest high water neap (tides)

li. link

l.i. letter of introduction • longitudinal interval

Li *Chem., symbol for* lithium

LI Leeward Islands • *airline flight code for* Liat • Liberal International • (USA) Licentiate in Instruction • Light Infantry • *Law* Lincoln's Inn • *British fishing port registration for* Littlehampton • Long Island (New York) • *Irish vehicle registration for* Westmeath

LIA (USA) Laser Industry Association • (USA) Lead Industries Association • Leather Industries of America • *Physics* linear-induction accelerator

LIAB Licentiate of the International Association of Book-Keepers

LIAT London International Arbitration Trust

lib. liber (Latin: book) • liberty • librarian • library • libretto

Lib *Astronomy* Libra

Lib. (*or* **Lib**) *Politics* Liberal • Liberia

LIB *Physical chem.* light-ion beam

LIBA Lloyd's Insurance Brokers' Association

lib. cat. library catalogue

Lib. Cong. (USA) Library of Congress

Lib Dem Liberal Democrat

LIBER Ligue des bibliothèques européennes de recherche (French: League of European Research Libraries)

LIBID ('li:bɪd) London Inter-Bank Bid Rate

LIBOR ('li:bɔ:) *Finance* London Inter-Bank Offered Rate

libst librettist

Lic. Licenciado (Spanish: Bachelor; in academic degrees) • (*or* **Lic**) Licentiate (in degrees, etc.)

LIC Lands Improvement Company • *Electronics* linear integrated circuit

LicAc Licentiate of Acupuncture

LICeram Licentiate of the Institute of Ceramics

LicMed Licentiate in Medicine

LICS *Computing* Lotus international character set

LicTheol Licentiate in Theology

LICW Licentiate of the Institute of Clerks of Works

lidar ('li:dɑ:) light detection and ranging

LIDC Lead Industries Development Council

Lieut Lieutenant

Lieut-Cdr Lieutenant-Commander

Lieut-Col Lieutenant-Colonel

Lieut-Com Lieutenant-Commander

Lieut-Gen Lieutenant-General

Lieut-Gov Lieutenant-Governor

LIF *Chem.* laser-induced fluorescence · *Med.* left iliac fossa · *Computing* low insertion force

LIFFE (laif) London International Financial Futures and Options Exchange

LIFireE Licentiate of the Institution of Fire Engineers

LIFO ('laɪfəʊ) *Accounting, Computing* last in, first out

LIFS *Physics* laser-induced fluorescence spectroscopy

Lig. Liguria · Limoges

LILO last in, last out

lim. limit

Lim. Limerick

LIM Licentiate of the Institute of Metals · linear-induction motor · *Computing* Lotus, Intel, Microsoft

LIMEAN London Inter-Bank Mean Rate

lin. lineal · linear · liniment

linac ('lɪnæk) *Physics* linear accelerator

Linc Lincoln College, Oxford

Lincs Lincolnshire

ling. linguistics

lin-log *Physics, Maths.* linear-logarithmic

Linn. (Carolus) Linnaeus (1707–78, Swedish botanist)

LInstP Licentiate of the Institute of Physics

LIOB Licentiate of the Institute of Building

LIP life insurance policy · *Med.* lymphocytic (*or* lymphoid) interstitial pneumonitis

LIPA ('lɪpə) Liverpool Institute for the Performing Arts

LIPM Lister Institute of Preventive Medicine

LIPS (*or* **lips**) (lɪps) *Computing* logical inferences per second

liq. liquid · liquor

LIRA Linen Industry Research Association

LIRMA ('lɜːmə) London International Insurance and Reinsurance Market Association

Lis. Lisbon

LIS laser isotope separation

LISA (ital.) Library and Information Science Abstracts

LISM Licentiate of the Incorporated Society of Musicians

LISP (lɪsp) *Computing* list processing (a programming language)

lit. literal(ly) · literary · literature · litre(s) · litter · little

Lit *symbol for* lira (Italian monetary unit)

LitB Bachelor of Letters *or* Literature (Latin *Litterarum Baccalaureus*)

lit. crit. (lɪt krɪt) *Colloquial* literary criticism

LitD Doctor of Letters *or* Literature (Latin *Litterarum Doctor*)

lith. lithograph(y)

Lith. Lithuania(n)

litho. (*or* **lithog.**) lithograph(ic) · lithography

lithol. lithology

Lit. Hum. Literae Humaniores (Latin: the humanities; classics course at Oxford University)

LitM Master of Letters *or* Literature (Latin *Litterarum Magister*)

Lit. Sup. (ital.) Times Literary Supplement

LittB Bachelor of Letters *or* Literature (Latin *Litterarum Baccalaureus*)

LittD Doctor of Letters *or* Literature (Latin *Litterarum Doctor*)

LittM Master of Letters *or* Literature (Latin *Litterarum Magister*)

liturg. liturgical · liturgy

LIUNA Laborers' International Union of North America

liv. *Commerce* livraison (French: delivery)

Liv. Liverpool · Livy (59 BC–17 AD, Roman historian)

liv. st. livre sterling (French: pound sterling)

Lix *Computing* legal information exchange

l.j. life jacket

LJ *British vehicle registration for* Bournemouth · *Chem.* Lennard-Jones (as in **LJ potential**) · Library Journal · *Athletics* long jump · Lord Justice (of Appeal) · *airline flight code for* Sierra National Airlines

LJC London Juvenile Courts

LJJ *Electronics* long Josephson junction · Lords Justices

LK *British fishing port registration for*

Lerwick • *British vehicle registration for* London NW

lkd locked

lkg locking

lkge *Commerce* leakage

lkr locker

ll. leaves • leges (Latin: laws) • lines (of written matter)

l.l. live load • loco laudato (Latin: in the place quoted) • lower left • lower limit

LL *Physics* Landau–Lifshitz (as in **LL splitting factor**) • Late Latin • Law Latin • (ital.) Law List • (*or* **£L, £Leb.**) *symbol for* Lebanese pound (monetary unit) • lending library • limited liability • *Physics, Maths.* linear-linear • *British fishing port registration for* Liverpool • *postcode for* Llandudno • London Library • *British vehicle registration for* London NW • Lord-Lieutenant • *Med.* lower limb • Low Latin

LL. laws • lines (of written matter) • Lords

L/L (Norway) Lutlang (limited company; Ltd)

LLA Lady Literate in Arts

LL.AA.II. Leurs Altesses Impériales (French: Their Imperial Highnesses)

LL.AA.RR. Leurs Altesses Royales (French: Their Royal Highnesses)

LLB (*or* **LIB**) Bachelor of Laws (Latin *Legum Baccalaureus*)

l.l.c. lower left centre

LLCM Licentiate of London College of Music

LLCO Licentiate of the London College of Osteopathy

LLD (*or* **LID**) Doctor of Laws (Latin *Legum Doctor*)

Llds *Insurance* Lloyd's

LLE *Chemical engineering* liquid-liquid extraction • low-level exposure (to radiation)

LL.EE. Leurs Éminences (French: Their Eminences) • Leurs Excellences (French: Their Excellencies)

LLett Licentiate of Letters

l.l.i. latitude and longitude indicator

LLI Lord-Lieutenant of Ireland

LLL Licentiate in Laws • loose leaf ledger • *Computing* low-level logic (*or* language)

LLLW liquid low-level (radioactive) waste

LLM Master of Laws (Latin *Legum Magister*)

LLMCom Master of Laws in Commercial Law

LL.MM. Leurs Majestés (French: Their Majesties)

LLN (USA) League for Less Noise

LLNL (USA) Lawrence Livermore National Laboratory

LLNW low-level nuclear waste

LLRW low-level radioactive waste

LLS *Astronautics* lunar logistics system

LLSV *Astronautics* lunar logistics system vehicle

LLU lending library unit

LLV *Astronautics* lunar logistics vehicle

LLW low-level (radioactive) waste

lm *Physics, symbol for* lumen

l.m. land mine • light metal • locus monumenti (Latin: place of the monument)

l.M. laufenden Monats (German: of the current month)

Lm *symbol for* Maltese lira (monetary unit)

LM Legion of Merit • Licentiate in Medicine • Licentiate in Midwifery • Licentiate in Music • light microscopy • liquid metal • *Electrical engineering* load management • London Museum • *British vehicle registration for* London NW • *Music, Prosody* long metre • (Scotland) Lord Marquis • Lord Mayor • *Aeronautics* lunar module

LMA *Physics* laser microprobe analyser • Linoleum Manufacturers' Association • low moisture avidity

LMBC Lady Margaret Boat Club (St John's College, Cambridge) • Liverpool Marine Biological Committee

LMBCS *Computing* Lotus multibyte character set

LMC *Astronomy* Large Magellanic Cloud • *Nuclear engineering* liquid-metal coolant • Lloyd's Machinery Certificate • Local Medical Committee

LMCC Licentiate of the Medical Council of Canada

lmd leafmould

LMD local medical doctor • *Music, Prosody* long metre double

LME London Metal Exchange

LMed Licentiate in Medicine

LMFBR *Nuclear engineering* liquid-metal-cooled fast breeder reactor

LMG light machine gun

LMH Lady Margaret Hall (Oxford University)

LMi *Astronomy* Leo Minor

LMI (USA) Logistics Management Institute

LMMS *Chem.* laser microprobe mass spectrometry

LMO lens-modulated oscillator • light machine oil

LMP *Med.* last menstrual period • lunar module pilot

LMR *Nuclear engineering* liquid-metal reactor • *Railways* London Midland Region

LMRCP Licentiate in Midwifery of the Royal College of Physicians

LMRSH Licentiate Member of the Royal Society for the Promotion of Health

LMRTPI Legal Member of the Royal Town Planning Institute

LMS *Chem.* laser mass spectrometry • Licentiate in Medicine and Surgery • *Education* local management of schools • London Mathematical Society • London Medical Schools • (*or* **LMS(R)**) (formerly) London, Midland and Scottish Railway • London Missionary Society • *Med.* loss of memory syndrome

LMSSA Licentiate in Medicine and Surgery of the Society of Apothecaries

LMT *Physics* length, mass, time • *Astronomy* local mean time

LMus Licentiate in Music

LMVD (New Zealand) Licensed Motor Vehicle Dealer

LMW *Biochem.* low molecular weight (as in **LMW protein**)

LMX London Market Excess of Loss (at Lloyd's)

LMXB *Astronomy* low-mass X-ray binary

ln *symbol for* natural logarithm

Ln Lane (in place names)

LN *British fishing port registration for* King's Lynn • *postcode for* Lincoln • liquid nitrogen • *British vehicle registration for* London NW • *Med.* lymph node • *international civil aircraft marking for* Norway

LNat Liberal National

LNB low noise blocker (on a satellite dish)

LNC League of Nations Covenant

LNER London and North Eastern Railway

LNG liquefied natural gas

Lnrk Lanark

LNS land navigation system

LNT liquid-nitrogen temperature

LNU League of Nations Union

LNWR London and North-Western Railway

l.o. lubricating oil

l/o *Commerce* leur ordre (French: their order)

Lo. Lord

LO Landsorganisationen i Sverige (General Federation of Swedish Trade Unions) • launch operator • liaison officer • *British fishing port registration for* London • *British vehicle registration for* London NW • London office • *airline flight code for* LOT Polish Airlines

l.o.a. length over all (in technical drawing)

LOA leave of absence • light observation aircraft

LOB *Baseball* left on base • line of balance • Location of Offices Bureau

LOBAL ('ləʊbæl) long-base-line buoy

LOBAR ('ləʊbɑː) long-base-line radar

loc. local • location • *Grammar* locative

l.o.c. lines of communication

LOC launch operations centre • (USA) Library of Congress • *Meteorol.* locally

LOCA *Nuclear engineering* loss-of-coolant accident

loc. cit. loco citato (*see under* l.c.)

loc. laud. loco laudato (Latin: in the place cited with approval)

locn location

loco. locomotion • locomotive

loc. primo cit. loco primo citato (Latin: in the place first cited)

LOD limit(s) of detection

LOF loss of flow (*or* fluid) • loss of function

L of C (USA) Library of Congress • lines of communication

L of N League of Nations

LOFT (lɒft) low-frequency radio telescope

log (lɒg) logarithm

log. logic(al) • logistic

log_e (,lɒg'iː) natural logarithm

LOH light observation helicopter

LOI lunar orbit insertion

LOL (Northern Ireland) Loyal Orange Lodge

LOLA ('ləʊlə) library on-line acquisition • lunar orbit landing approach

LOM (USA) Loyal Order of Moose

LOMA (USA) Life Office Management Association

Lomb. Lombard(y)

Lond. London • Londonderry

Londin. Londiniensis (Latin: (Bishop) of London)

long. (*or* **long**) longitude

Long. Longford

longl longitudinal

Lonrho ('lɒnrəʊ) *Finance* London Rhodesian

LOOM (USA) Loyal Order of Moose

l.o.p. (*or* **LOP**) *Navigation* line of position

LOPAR ('ləʊpɑː) low-power acquisition radar

loq. loquitur (Latin: he (*or* she) speaks)

LOR light output ratio • lunar orbit rendezvous

LORAC ('lɒræk) long-range accuracy

LORAD ('lɒræd) long-range active detection

LORAN ('lɒræn) long-range navigation

LORAPH ('lɒræf) long-range passive homing system

LORCS League of Red Cross and Red Crescent Societies

LORV (lɔːv) low observable re-entry vehicle

LOS Latin Old Style • Law of the Seas • line of sight • loss of signal

LOSS large-object salvage system

lot. lotion

Lot (lɒt) Polskie Linie Lotnicze (Polish Air Lines)

LOT large orbital telescope

LOTC *Finance* London over-the-counter market

Lou. Louisiana

LOX (*or* **lox**) (lɒks) liquid oxygen

loy. loyal(ty)

LOYA League of Young Adventurers

l.p. last paid • latent period • launch platform • long primer (a size of printer's type) • low pressure

Lp Ladyship • Lordship

LP Labour Party • large paper (edition of a book) • large post (a standard paper size) • last post • legal procurator • Liberal Party • life policy • Limited Partnership • *Computing* linear programming • liquid petroleum • liquid propellant • *British vehicle registration for* London NW • long-playing (of a gramophone record) • Lord Provost • low pressure • *Med.* lumbar puncture

L/P letterpress • life policy

LPA Leather Producers' Association for England • Local Productivity Association

l.p.c. low pressure chamber

LPC leaf protein concentrate (in nutrition) • Legal Practice Course • Lord President of the Council

LPE *Electronics* liquid-phase epitaxy

LPEA Licentiate of the Physical Education Association

LPed Licentiate in Pedagogy

LPF *Computing* League for Programming Freedom

LPG liquefied petroleum gas

LPGA Ladies' Professional Golf Association

LPh (*or* **LPH**) Licentiate in Philosophy

lpi lines per inch

LPLC low-pressure liquid chromatography

lpm lines per millimetre • lines per minute

LPN (USA) Licensed Practical Nurse

LPNA (USA) Lithographers' and Printers' National Association

LPO local purchasing officer • London Philharmonic Orchestra

L'pool Liverpool

LPRP Laotian People's Revolutionary Party

LPS *Biochem.* lipopolysaccharide • *Engineering* liquid-phase sintering • London Philharmonic Society • Lord Privy Seal

LPSO Lloyd's Policy Signing Office

LPU low pay unit

l.p.w. *Optics* lumens per watt

Lpz. Leipzig

l.q. lege quaeso (Latin: please read)

LQ *international civil aircraft marking for* Argentina • *Computing* letter quality (of printed output) • *Maths.* linear-quadratic

lqdr liquidator

LQR (ital.) Law Quarterly Review

LQT *Commerce* Liverpool quay terms

lr lower

l.r. landing report • log run • long range • long run

l.R. laufen de Rechnung (German: current account)

Lr Lancer • *Chem., symbol for* lawrencium • ledger

LR *British fishing port registration for* Lancaster • Land Registry • Law Report • left-right (*or* left-to-right) • liquor/liquor-to-goods ratio (in dyeing) • Lloyd's Register (of Shipping) • *British vehicle registration for* London NW • Lowland Regiment • Loyal Regiment

LRA Local Radio Association

LRAC Law Reports, Appeal Cases

LRAD Licentiate of the Royal Academy of Dancing

LRAM Licentiate of the Royal Academy of Music

LRB London Residuary Body (in local government) · London Rifle Brigade

LRC Labour Representation Committee · Langley Research Center (in NASA) · Leander Rowing Club · London Rowing Club · *Computing* longitudinal redundancy check

LRCh Law Reports, Chancery Division

LRCM Licentiate of the Royal College of Music

LRCP Licentiate of the Royal College of Physicians

LRCPE (*or* **LRCPEd**) Licentiate of the Royal College of Physicians of Edinburgh

LRCPI Licentiate of the Royal College of Physicians of Ireland

LRCPSGlas Licentiate of the Royal College of Physicians and Surgeons of Glasgow

LRCS League of Red Cross Societies · Licentiate of the Royal College of Surgeons of England

LRCSE Licentiate of the Royal College of Surgeons of Edinburgh

LRCSI Licentiate of the Royal College of Surgeons in Ireland

LRCVS Licentiate of the Royal College of Veterinary Surgeons

LREC *Med.* Local Research Ethics Committee

LRHL Law Reports, House of Lords

LRIBA Licentiate of the Royal Institute of British Architects (now Member; *see* RIBA)

LRKB Law Reports, King's Bench

LRM language reference manual

LRP Law Reports, Probate Division

LRPS Licentiate of the Royal Photographic Society

LRQB Law Reports, Queen's Bench

Lrs Lancers

LRS Land Registry Stamp · *Physical chem.* laser Raman spectroscopy · (ital.) Lloyd's Register of Shipping

LRSC Licentiate of the Royal Society of Chemistry

LRSM Licentiate of the Royal Schools of Music

LRT light rail transit · London Regional Transport · long-range transport

LRTI *Med.* lower respiratory tract infection

LRU least recently used · line replaceable unit

LRV light rail vehicle · lunar roving vehicle

LRWES long-range weapons experimental station

l.s. landing ship · left side · (*or* **L.s.**) letter signed · local sunset · *Law* locus sigilli (Latin: the place of the seal) · long sight · low speed · lump sum

LS *British vehicle registration for* Edinburgh · Law Society · Leading Seaman · *postcode for* Leeds · legal seal · *international vehicle registration for* Lesotho · letter service · Licensed Surveyor · Licentiate in Surgery · *Psychol.* liminal sensitivity · Linnean Society · locus sigilli (*see under* l.s.) · London Scottish · London Sinfonietta · *Films* long shot · loudspeaker · (*or* **£S**, **£Syr.**) *symbol for* Syrian pound (monetary unit)

LSA Land Settlement Association · leading supply assistant · Licence in Agricultural Sciences · Licentiate of the Society of Apothecaries

LSAA Linen Supply Association of America

LS&GCM Long Service and Good Conduct Medal

LSB *Computing* least significant bit · London School Board

LSByte *Computing* least significant byte

l.s.c. loco supra citato (Latin: in the place before cited)

LSC Licentiate in Sciences · liquid scintillation counting (*or* counter) · *Chem.* liquid–solid chromatography · London Salvage Corps

LScAct Licentiate in Actuarial Science

LSCF *Statistics* least squares curve fitting

LSCS *Obstetrics* lower segment Caesarean section

LSd (*or* **£Sd**) *symbol for* Sudanese pound (monetary unit)

LSD League of Safe Drivers · *Computing* least significant digit · Lightermen, Stevedores, and Dockers · lysergic acid diethylamide (hallucinogenic drug)

L.S.D. (*or* **£.s.d.**, **l.s.d.**) librae, solidi, denarii (Latin: pounds, shillings, pence)

l.s.e. limited signed edition

LSE *Chem.* liquid–solid extraction · *Chem.* liquid-solvent extraction · London School of Economics and Political Science · London Stock Exchange

L-Sgt Lance-Sergeant

LSHTM London School of Hygiene and Tropical Medicine

LSI Labour and Socialist International ·

Electronics large-scale integration (*or* integrated)

LSJ London School of Journalism

LSJM laus sit Jesu et Mariae (Latin: praise be to Jesus and Mary)

LSL landing ship logistic • low-speed logic

l.s.m. litera scripta manet (Latin: the written word remains)

LSM laser scanning microscope • *Maths., Physics* least-squares method • *Electrical engineering* linear synchronous motor

LSO London Symphony Orchestra

LSQ *Meteorol.* line squall

LSR *Astronomy* local standard of rest

LSS large-scale structure (*or* system) • Licentiate in Sacred Scripture • (USA) Lifesaving Service • life-saving station • life-support system

LSSc Licentiate in Sanitary Science

l.s.t. local standard time

LST landing ship (for) tank(s) *or* transport • Licentiate in Sacred Theology • *Astronomy* local sidereal time • local standard time

LSTTL *Computing* low-power Schottky transistor transistor logic

LSU Louisiana State University

LSZ (New Zealand) limited speed zone

lt light

l.t. landed terms • landing team • large tug • local time • locum tenens • long ton • loop test

Lt Lieutenant • *Military* Light

LT lawn tennis • *Med.* leukotriene • Licentiate in Teaching • *international vehicle registration for* Lithuania • *British vehicle registration for* London NW • London Transport • *Physics* Lorentz transformation • *British fishing port registration for* Lowestoft • low temperature • *Electrical engineering* low tension • *airline flight code for* LTU International Airways • *symbol for* Turkish lira (monetary unit)

LTA Lawn Tennis Association • *Chem.* lead tetraacetate • lighter than air (of an aircraft) • London Teachers' Association

LTAA Lawn Tennis Association of Australia

LT & SR London, Tilbury, and Southend Railway

l.t.b. low-tension battery

LTB London Tourist Board

LTBT Limited Test Ban Treaty

LTC Lawn Tennis Club • *Chemical engineering* low-temperature carbonization

Lt-Cdr Lieutenant-Commander

LTCL Licentiate of Trinity College of Music, London

Lt-Col Lieutenant-Colonel

Lt-Com Lieutenant-Commander

Ltd Limited (after the name of a private limited company)

LTDP long-term defence programme

LTE *Astronomy* local thermodynamic equilibrium

LTF Lithographic Technical Foundation

ltg lettering • lighting

ltge *Commerce* lighterage

Lt-Gen Lieutenant-General

Lt-Gov Lieutenant-Governor

LTh (*or* **LTheol**) Licentiate in Theology

LTH light training helicopter • *Biochem.* luteotrophic hormone

LTI Licentiate of the Textile Institute

LTIB Lead Technical Information Bureau

Lt Inf. Light Infantry

LTL *Commerce* (USA) less-than-truckload lot

LTM Licentiate in Tropical Medicine • London Terminal Market

LTO leading torpedo operator

LTOM London Traded Options Market

LTOS Law Times, Old Series

LTP *Biochem.* lipid-transfer protein • *Science* long-term potentiation

ltr letter • lighter

LTR *Genetics* long terminal repeat

LTRA Lands Tribunal Rating Appeals

LtRN Lieutenant, Royal Navy

LTRS Low Temperature Research Station

LTSC Licentiate of Tonic Sol-Fa College

LTTE (Sri Lanka) Liberation Tigers of Tamil Eelam

LTTL *Computing* low-power transistor transistor logic

lu. luglio (Italian: July)

l/u *Shipping* laid (*or* lying) up

Lu *Chem., symbol for* lutetium

Lu. Lucerne

LU Liberal Unionist • *British vehicle registration for* London NW • loudness unit • *postcode for* Luton

LUA Liverpool Underwriting Association

lub. (*or* **lubr.**) lubricant • lubricate • lubrication

LUC London Underwriting Centre

LUCOM ('luːkɒm) lunar communication system

lug. luggage • lugger

LUG light utility glider • *Computing* local users group

LUHF *Electronics* lowest useful high frequency

LULOP (ital.) London Union List of Periodicals

lum. lumbago • lumber • luminous

LUM lunar excursion module

LUMAS ('luːmæs) lunar mapping system

LUMO ('luːməu) *Chem.* lowest unoccupied molecular orbital

lun. lundi (French: Monday) • lunedì (Italian: Monday) • lunes (Spanish: Monday)

LUNCO Lloyd's Underwriters Non-Marine Claims Office

LUOTC London University Officers' Training Corps

Lup *Astronomy* Lupus

LUS Land Utilization Survey

LUSCS *Obstetrics* lower uterine segment Caesarean section

LUSI lunar surface inspection

lusing. *Music* lusingando (Italian: coaxing *or* caressing)

LUT *Astronautics* launch umbilical tower • *Computing* look-up table

Luth. Lutheran

lux. luxurious

Lux. Luxembourg

LuxF *symbol for* Luxembourg franc (monetary unit)

lv. leave (of absence, as from military duty) • livre (French: book)

l.v. low voltage

Lv lev (Bulgarian monetary unit)

LV *airline flight code for* Albanian Airlines • *international civil aircraft marking for* Argentina • *international vehicle registration for* Latvia • *Med.* left ventricle (*or* ventricular) • licensed victualler • *Meteorol.* light and variable • *British vehicle registration for* Liverpool • luncheon voucher

LVA Licensed Victuallers' Association

LVF *Med.* left ventricular function

LVI laus Verbo Incarnato (Latin: praise to the Incarnate Word) • low viscosity index

LVLO Local Vehicle Licensing Office

LVN (USA) licensed vocational nurse

LVO Lieutenant of the Royal Victorian Order

lvs leaves

LVS Licentiate in Veterinary Science

LVT landing vehicle, tracked

Lw (formerly) *Chem.*, *symbol for* lawrencium (*see under* Lr)

LW *airline flight code for* Air Nevada • left wing • light weight • *British vehicle registration for* London NW • *Radio, etc.* long-wave (*or* long waves) • low water

LWA London Welsh Association

l.w.b. long wheelbase

LWEST low water equinoctial spring tide

LWF Lutheran World Federation

LWL (*or* lwl) length (at) waterline (of a ship) • *Shipping* load waterline

LWM (*or* lwm) low water mark

LWONT low water, ordinary neap tides

LWOST low water, ordinary spring tides

LWR *Nuclear engineering* light-water reactor

LWRA London Waste Regulation Authority

LWT London Weekend Television

LWV (USA) League of Women Voters

lx *Physics*, *symbol for* lux

LX *British vehicle registration for* London NW • *international civil aircraft marking for* Luxembourg

Lxmbrg Luxembourg

LXX *Bible*, *symbol for* Septuagint (from its 70 translators)

l.y. *Astronomy* light year

LY *airline flight code for* El Al Israel Airlines • *international civil aircraft marking for* Lithuania • *British fishing port registration for* Londonderry • *British vehicle registration for* London NW

Lyn *Astronomy* Lynx

lyr. lyric(al)

Lyr *Astronomy* Lyra

LYR *Meteorol.* layer(ed)

Lys (*or* lys) *Biochem.* lysine

LZ *British vehicle registration for* Armagh • *airline flight code for* Balkan-Bulgarian Airlines • *international civil aircraft marking for* Bulgaria

LZT *Chem.* lead zirconate–titanate

LZW *Computing* Lempel-Ziv-Welch (in **LZW compaction**)

M

m (ital.) *Astronomy, symbol for* apparent magnitude • (bold ital.) *Physics, symbol for* magnetic moment • (ital.) *Physics, symbol for* magnetic quantum number • (ital.) *Chem., symbol for* mass • *Music* me (in tonic sol-fa) • (ital.) *Chem.* meta (as in *m*-**dichlorobenzene**) • *symbol for* metre(s) • *symbol for* milli- (prefix indicating 1/1000 (i.e. 10^{-3}), as in **mm**, millimetre) • million • (ital.) *Chem., symbol for* molality

m. *Cricket* maiden (over) • male • manipulus (Latin: handful) • mare • *Currency* mark(s) • married • masculine • medical • medicine • medium • memorandum • meridian • meridies (Latin: noon) • meridional • midday • middle • mile(s) • (USA, Canada) mill(s) (monetary unit) • mille (French: thousand) • minim (liquid measure) • minor • minute(s) • *Med.* misce (Latin: mix) • *Meteorol.* mist • mixture • moderate • mois (French: month) • molar (tooth) • month • moon • morning • mort *or* morte (French: dead) • morto (Italian: dead) • mountain

M (ital.) *Astronomy, symbol for* absolute magnitude • *Printing, symbol for* em • *symbol for* loti (monetary unit of Lesotho; pl. maloti) • (ital.; *or* M_v) *Physics, symbol for* luminous exitance • *Aeronautics* Mach (followed by a number) • (ital.) *Physics, symbol for* magnetic quantum number • (bold ital.) *Physics, symbol for* magnetization • *international vehicle registration for* Malta • *postcode for* Manchester • (ital.) *symbol for* mass, especially molar mass • *Films* (Australia) mature audience (certification) • medium (size) • *symbol for* mega- (prefix indicating one million, as in **MW**, megawatt; *or* (in computing) 2^{20}, as in **Mbyte**, megabyte) • *Chem., symbol for* mesomeric effect • *Astronomy* Messier Catalogue (followed by a number) • *Chem., symbol for* metal (in a chemical formula, as in **MOH**) • *Biochem., symbol for* methionine • *British fishing port registration for* Milford • million • minim

(liquid measure) • (bold ital.) *Physics, symbol for* moment of a force • *Economics* monetary aggregate (as in **M0**, **M1**, etc.; measures of the money supply) • motorway (as in **M1**, **M4**, etc.) • mud (on charts) • (ital.) *Electrical engineering, symbol for* mutual inductance • (ital.; *or* M_e) *Physics, symbol for* radiant exitance • *Navigation, symbol for* sea mile • *Roman numeral for* thousand • *Astronomy, indicating* a spectral type • *Logic, indicating* the middle term of a syllogism

M. (*or* **M**) Magister (Latin: Master) • magistrate • Majesty • Manitoba • March • Marquess • Marquis • martyr • Master (in titles) • May • medal • Medieval • Member (in titles) • Methodist • *Music* metronome • metropolitan • mezzo *or* mezza (Italian: half) • Middle • militia • minesweeper • Monday • Monsieur (French: Mr *or* Sir) • Monte (Italian: mount) • mother • Mountain

2M *airline flight code for* Moldavian Airlines

3M *airline flight code for* Gulfstream International Airlines

9M *international civil aircraft marking for* Malaysia

m.a. manufacturing assembly • map analysis • menstrual age

m/a *Book-keeping* my account

mA *symbol for* milliampere(s)

Ma (ital.) *Aeronautics, symbol for* Mach number

Ma. Mater (Latin: Mother)

MA *British vehicle registration for* Chester • Magistrates' Association • *airline flight code for* Malev Hungarian Airlines • (USA) Manpower Administration • (USA) Maritime Administration • *US postcode for* Massachusetts • Master of Arts • Mathematical Association • medieval archaeology • *Psychol.* mental age • Middle Ages • Military Academy • military assistant • military attaché • Missionarius Apostolicus (Latin: Apostolic

Missionary) • mobility allowance • *international vehicle registration for* Morocco • Mountaineering Association

MAA Manufacturers' Agents Association of Great Britain • master-at-arms • Mathematical Association of America • Member of the Architectural Association • *Chem.* methacrylic acid • Motor Agents' Association

MAAF Mediterranean Allied Air Forces

MAAT Member of the Association of Accounting Technicians

MABS (USA) marine air base squadron

Mac. Macao • (*or* **Macc.**) Maccabees (books of the Apocrypha)

MAc Master of Accountancy

MAC (*or* **m.a.c.**) maximum allowable concentration • *Computing* media access control (in **MAC layer**) • (mæk) *Television* multiplexed analogue components • (USA) Municipal Assistance Corporation

MACA ('mækə) Mental After Care Association

MACC military aid to the civilian community

MACE Member of the Association of Conference Executives • Member of the Australian College of Education

Maced. Macedonia(n)

mach. machine(ry) • machinist

MACHO ('mætʃəu) *Astronomy* massive astrophysical compact halo object

MACM Member of the Association of Computing Machines

macroecon. macroeconomics

MACS Member of the American Chemical Society

Mad. Madeira

MAD (mæd) magnetic anomaly detection • maintenance, assembly, and disassembly • *Psychiatry* major affective disorder • *Commerce* mean absolute deviation • *Military* mutual assured destruction

Madag. Madagascar

MADD (mæd) (USA) Mothers Against Drunk Driving

MADO Member of the Association of Dispensing Opticians

Madr. Madras • Madrid

MaE Master in Engineering

MAE Master of Aeronautical Engineering • Master of Art Education • Master of Arts in Education

MA(Econ) Master of Arts in Economics

MA(Ed) Master of Arts in Education

MAEE Marine Aircraft Experimental Establishment

maesto. *Music* maestoso (Italian: majestic, stately)

MAFA Manchester Academy of Fine Arts

MAFF (mæf) Ministry of Agriculture, Fisheries, and Food

mag. magazine • maggio (Italian: May) • magnesium • magnet(ic) • magnetism • magneto • magnitude • magnum

Mag. Magyar

MAg Master of Agriculture

Magd Magdalen College, Oxford • Magdalene College, Cambridge

MAgEc Master of Agricultural Economics

magg. maggio (Italian: May) • *Music* maggiore (Italian: major)

maglev ('mæg,lɛv) magnetic levitation

MAGPI ('mægpaɪ) *Surgery* meatal advancement and glanuloplasty (operation)

MAgr Master of Agriculture

MAgrSc Master of Agricultural Science

mah. (*or* **mahog.**) mahogany

MAI Master of Engineering (Latin *Magister in Arte Ingeniaria*) • *Forestry* mean annual increment • Member of the Anthropological Institute • *Med. Mycobacterium avium* and *intracellulare* (in **MAI complex**)

MAIAA Member of the American Institute of Aeronautics and Astronautics

MAIB Marine Accident Investigation Branch

MAICE Member of the American Institute of Consulting Engineers

MAIChE Member of the American Institute of Chemical Engineers

maint. maintenance

MAISE Member of the Association of Iron and Steel Engineers

maj. major • majority

Maj *Military* Major

Maj-Gen Major-General

Mal. *Bible* Malachi • Malay(an) • Malaysia(n) • Malta • Maréchal (French: Field Marshal)

MAL *international vehicle registration for* Malaysia

MALD Master of Arts in Law and Diplomacy

mall. malleable

MAM *Computing* multiple allocation memory

MAMBO ('mæmbəu) Mediterranean

Association for Marine Biology and Oceanography

MAMEME Member of the Association of Mining Electrical and Mechanical Engineers

man. management • manager • manual(ly) • manufacture(r)

Man. Manchester • Manila • Manitoba

MAN *British vehicle registration for* Isle of Man • (mæn) *Computing* metropolitan area network

MAnaes Master of Anaesthesiology

Manch. Manchester • Manchuria

mand. *Law* mandamus (an order from the High Court) • *Music* mandolin

M and A *Finance* mergers and acquisitions

M & B May and Baker (pharmaceutical company) • mild and bitter (beer) • Mills and Boon (publisher of romantic fiction)

M & E music and effects

Man. Dir. Managing Director

m & r maintenance and repairs

m & s maintenance and supply

M&S Marks & Spencer plc

Man. Ed. Managing Editor

manf. manufacturer

MANF May, August, November, February (months for quarterly payments)

mang. B manganese bronze

Manit. Manitoba

man. op. manually operated

man. pr. *Med.* mane primo (Latin: early in the morning)

Mans. Mansion(s)

Mansf Mansfield College, Oxford

manuf. (*or* **manufac.**) manufacture(d) • manufacturer • manufacturing

MANWEB ('mænwɛb) Merseyside and North Wales Electricity Board

MAO Master of Obstetric Art • *Biochem., Pharmacol.* monoamine oxidase

MAOI monoamine oxidase inhibitor (anti-depressant)

MAOT Member of the Association of Occupational Therapists

MAOU Member of the American Ornithologists' Union

MAP major air pollutant • *Computing* Manufacturing Automation Protocol • maximum average price • *Med.* mean arterial (blood) pressure • medical aid post • Member of the Association of Project Managers • Ministry of Aircraft Production • modified American plan (payment system in US hotels)

MAPI ('mæpɪ) messaging application program interface

MAppSc Master of Applied Science

MAPsS Member of the Australian Psychological Society

mar. *Music* marimba • marine • maritime • marriage • married • martedì (Italian: Tuesday) • marzo (Italian: March)

Mar. March

MAR *Taxation* marginal age relief • Master of Arts in Religion • *Computing* memory address register

MARAC Member of the Australasian Register of Agricultural Consultants

marc. *Music* marcato (Italian: marked; i.e. each note emphasized)

MARC (mɑːk) *Bibliog.* machine-readable cataloguing

March. Marchioness

MArch Master of Architecture

marg. margin(al)

mar. insce. marine insurance

marit. maritime

mar. lic. marriage licence

MARMAP ('mɑː,mæp) (USA) Marine Resources Monitoring Assessment and Prediction

Marq. Marquess • Marquis

MARS meteorological automatic reporting station (*or* system)

mart. martyr

MARV (mɑːv) *Military* manoeuvrable re-entry vehicle

mas. masculine

MAS Malaysian Airline System • Master of Applied Science • Military Agency for Standardization

MASAE Member of the American Society of Agricultural Engineers

masc. masculine

MASc Master of Applied Science

MASC Member of the Australian Society of Calligraphers

MASCAM ('mæskæm) *Electronics* masking-pattern adaptive sub-band coding and multiplexing

MASCE Member of the American Society of Civil Engineers

mascon ('mæskɒn) *Astronomy* mass concentration

maser ('meɪzə) microwave amplification by stimulated emission of radiation

MASH (mæʃ) (USA) mobile army surgical hospital

MASI Member of the Architects' and Surveyors' Institute

MASME Member of the American Society of Mechanical Engineers

Mass. Massachusetts

MA(SS) Master of Arts in Social Science

mat. maternity • matinée • matins • *Printing* matrix • *Finance* maturity

MAT *Insurance* marine, aviation, and transport • Master of Arts in Teaching

MATA ('mætə) multiple answering teaching aid

math. mathematical(ly) • mathematician • (USA, Canada) mathematics

MATh Master of Arts in Theology

maths. mathematics

MATIF marché à terme des instruments financiers (French: financial futures market)

matr. matrimonium (Latin: marriage)

MATS *US Air Force* Military Air Transport Service

MATSA Managerial Administrative Technical Staff Association

Matt. *Bible* Matthew

MATTS ('mætz) *Military* multiple airborne target trajectory system

MATV master antenna television

MAU *Computing* multi-station access unit

Maur. Mauritius

Mau Re (pl. **Mau Rs**) *symbol for* Mauritian rupee (monetary unit)

MAusIMM Member of the Australasian Institute of Mining and Metallurgy

MAW *Military* medium assault weapon

max. maxim • (*or* **max, MAX**) maximum

mb *Meteorol., symbol for* millibar(s)

m.b. magnetic bearing • main battery • medium bomber • *Med.* misce bene (Latin: mix well) • (*or* **m/b**) motor boat (*or* barge)

MB Bachelor of Medicine (Latin *Medicinae Baccalaureus*) • Bachelor of Music (Latin *Musicae Baccalaureus*) • *British vehicle registration for* Chester • maritime board • mark of the Beast • marketing board • maternity benefit • (Canada) Medal of Bravery • medical board • *Electronics* megabit • (*or* **Mb**) *Computing* megabyte • metropolitan borough • *Meteorol.* millibar(s) • motor boat (*or* barge) • municipal borough

Mba. Mombasa

MBA Master of Business Administration

MBAC Member of the British Association of Chemists

MBAcA Member of the British Acupuncture Association

mbar *Meteorol., symbol for* millibar(s)

MBASW Member of the British Association of Social Workers

MBBA *Chem.* N-(4-methoxybenzylidene)-4-butylaniline

m.b.c. *Med.* maximum breathing capacity

MBC *Chem.* methyl benzimidazole carbamate • metropolitan borough council • municipal borough council

MBCO Member of the British College of Ophthalmic Opticians (*or* Optometrists)

MBCPE Member of the British College of Physical Education

MBCS Member of the British Computer Society

MBD *Med.* minimal brain dysfunction

MBdgSc Master of Building Science

MBE Member of the Order of the British Empire • *Electronics* molecular-beam epitaxy

MBEng Member of the Association of Building Engineers

MBF Musicians Benevolent Fund

MBFR *Military* mutual and balanced force reduction

MBG microemulsion-based gel

mbH mit beschränkter Haftung (German: with limited liability)

MBH *Astronomy* massive black hole

MBHI Member of the British Horological Institute

MBIFD Member of the British Institute of Funeral Directors

MBK *Military* missing, believed killed

MBKSTS Member of the British Kinematograph, Sound, and Television Society

MBM Master of Business Management

MBMS *Chem.* molecular-beam mass spectrometry

MBNOA Member of the British Naturopathic and Osteopathic Association

MBNQA *Commerce* Malcolm Baldrige National Quality Award

MBO *Finance* management buyout • management by objective

MBOU Member of the British Ornithologists' Union

MBP *Med.* mean blood pressure

MBPICS Member of the British Production and Inventory Control Society

MBPsS Member of the British Psychological Society

MBPT *Physics, Astronomy* many-body perturbation theory

mbr member

MBR *Astronomy* microwave background radiation

MBS Manchester Business School

MBSc Master of Business Science

MBT *Military* main battle tank • (*or* **m.b.t.**) *Med.* mean body temperature

MBuild Master of Building

MBWA management by wandering around

Mbyte *Computing* megabyte

m.c. mois courant (French: current month) • motor cycle

m/c machine • motor cycle

Mc megacycle(s)

MC *British vehicle registration for* London NE • machinery certificate • Magistrates' Court • *Navigation* magnetic course • marginal cost • (USA) Marine Corps • (USA) Maritime Commission • marriage certificate • Master of Ceremonies • Master of Surgery (Latin *Magister Chirurgiae*) • medical certificate • medical college • (USA) Medical Corps • *Astrology* Medium Coeli (Latin: Midheaven) • (USA) Member of Congress • Member of Council • mess committee • Methodist Church • military college • Military Cross • Missionaries of Charity • *Astronomy* molecular cloud • *Chem.* molecular cluster • *international vehicle registration for* Monaco • *Politics* Monday Club • Monte Carlo • morse code • motor contact

M/C Manchester • *Finance* marginal credit

MCA Management Consultants' Association • Manufacturing Chemists' Association • Master of Commerce and Administration • Matrimonial Causes Act (1937) • *Computing, trademark* micro channel architecture • monetary compensatory amount(s) • *Electronics* multichannel (*or* multiple channel) analyser • multicriteria analysis • multiple classification analysis

MCAB (USA) Marine Corps air base

MCAM Member of the CAM Foundation

MC&G mapping, charting, and geodesy

MCAV *Computing* modified constant angular velocity

MCB (USA) Marine Corps Base • Master in Clinical Biochemistry • *Computing* memory control block • miniature circuit breaker • *Military* multiple-cratering bomblets

MCBSI Member of the Chartered Building Societies Institute

MCC Manchester Computer Centre • Marylebone Cricket Club • Maxwell Communication Corporation • Melbourne Cricket Club • member of the county council • metropolitan county council • *Navigation* mid-course correction

MCCA Minor Counties Cricket Association

MCCC Middlesex County Cricket Club

MCCD RCS Member in Clinical Community Dentistry of the Royal College of Surgeons

MCD *Chem.* magnetic circular dichroism • Master of Civic Design • *Med.* mean cell (*or* corpuscular) diameter • Movement for Christian Democracy • *Physical chem.* multiconcentration diffusion

MCE Master of Civil Engineering • *Chem.* methylene chloride extraction

MCF *Nuclear engineering* magnetic confinement fusion

MCFP (Canada) Member of the College of Family Physicians

MCG Melbourne Cricket Ground

MCGA *Computing* multicolour graphics array

MCGB Master Chef of Great Britain

McGU McGill University (Canada)

MCh (*or* **MChir**) Master of (*or* in) Surgery (Latin *Magister Chirurgiae*)

MCH *Med.* mean cell (*or* corpuscular) haemoglobin • *Biochem.* melanin-concentrating hormone

MCHC *Med.* mean cell (*or* corpuscular) haemoglobin concentration

MChD Master of Dental Surgery (Latin *Magister Chirurgiae Dentalis*)

MChE Master of Chemical Engineering

MChemA Master in Chemical Analysis

MChOrth Master of Orthopaedic Surgery (Latin *Magister Chirurgiae Orthopaedicae*)

MChS Member of the Society of Chiropodists

mcht merchant

mchy machinery

m.c.i. malleable cast iron

mCi millicurie(s)

MCIBSE Member of the Chartered Institution of Building Services Engineers

MCIM Member of the Chartered Institute of Marketing

MCIOB Member of the Chartered Institute of Building

M.CIRP Member of the International Institution for Production Engineering Research (French *Collège internationale pour recherche et production*)

MCIS Member of the Institute of Chartered Secretaries and Administrators (formerly Chartered Institute of Secretaries)

MCIT Member of the Chartered Institute of Transport

MCL Master of (*or* in) Civil Law • maximum contamination (*or* contaminant) levels

MClinPsychol Master of Clinical Psychology

MCISc Master of Clinical Science

MCLV *Computing* modified constant linear velocity

MCM *Maths.* Monte Carlo method • *Computing* multichip module • multistage conventional munitions

MCMES Member of the Civil and Mechanical Engineers' Society

Mco Morocco

MCO Managed Care Organization

mcol. musicological • (*or* **mcolst**) musicologist • musicology

MCollH Member of the College of Handicrafts

MCom Master of Commerce

MCommH Master of Community Health

MConsE Member of the Association of Consulting Engineers

MCOphth Member of the College of Ophthalmologists

MCP (*or* **m.c.p.**) male chauvinist pig • (USA) Master of City Planning • Member of Colonial Parliament • Member of the College of Preceptors • *Astronomy* microchannel plate (as in **MCP detector**)

MCPA *Chem.* 2-methyl-4-chlorophenoxyacetic acid (used in herbicides)

MCPB *Chem.* 4-(2-methyl-4-chlorophenoxy)butanoic acid (weedkiller)

MCPBA *Chem.* m-chloroperoxybenzoic acid

MCPP Member of the College of Pharmacy Practice

MCPS Mechanical Copyright Protection Society • Member of the College of Physicians and Surgeons

MCR *Engineering* maximum continuous rating • *Med.* metabolic clearance rate • middle common room (in certain universities) • mobile control room

MCS Madras Civil Service • Malayan Civil Service • Master of Commercial Science • Military College of Science • monitoring and control system • *Electronics* multichannel scaler

Mc/s megacycles per second

MCSEE Member of the Canadian Society of Electrical Engineers

MCSP Member of the Chartered Society of Physiotherapy

MCST Member of the College of Speech Therapists

MCT mainstream corporation tax • Member of the Association of Corporate Treasurers

MCU main control unit • *Photog.* medium close-up

MCV *Med.* mean cell (*or* corpuscular) volume

MCW *Telecom.* modulated continuous wave

Md *Chem., symbol for* mendelevium

Md. Maryland

M/d *Commerce* months after date

MD *airline flight code for* Air Madagascar • Doctor of Medicine (Latin *Medicinae Doctor*) • *British vehicle registration for* London NE • *Music* main droite (French: right hand) • malicious damage • managing director • *Music* mano destra (Italian: right hand) • map distance • market day • *US postcode for* Maryland • medical department • (*or* **M/D**) *Banking* memorandum of deposit • mentally deficient • mess deck • Middle Dutch • military district • mini-disc (in sound recording) • *international vehicle registration for* Moldova • molecular dynamics • Monroe Doctrine • musical director

MDA 3,4-methylenedioxyamphetamine (hallucinogenic drug; ice) • minimum detectable amount (*or* activity) • *Computing* monochrome display adaptor • Muscular Dystrophy Association

MDAM *Computing* multidimensional access memory

MD&A management discussion and analysis

MDAP (USA) Mutual Defense Assistance Program

M-day (USA) mobilization day

MdB Mitglied des Bundestages (German: Member of the Bundestag)

MDB Movimento Democrático Brasileiro (Spanish: Brazilian Democratic Movement)

MDC metropolitan district council • minimum detectable concentration • modification and design control • more developed country

MDD minimal detectable dose

Mddx Middlesex

MDEA *Chem.* *N*-methyldiethanolamine

MDentSc Master in Dental Science

MDes Master of Design

MDF medium-density fibreboard

MDG Medical Director-General

MDHB Mersey Docks and Harbour Board

MDI *Chem.* methylene diisocyanate

m. dict. *Med.* more dicto (Latin: in the manner directed; in prescriptions)

MDip Master of Diplomacy

mdise merchandise

MDiv Master of Divinity

mdl model

MDL minimum detectable level (of radioactivity)

Mdlle Mademoiselle (French: Miss)

Mdm Madam

MDMA methylenedioxymethamphetamine (hallucinogenic drug; Ecstasy)

Mdme Madame (French: Mrs)

mdn median

MDNS *Computing* managed data network service

mdnt midnight

MDP Mongolian Democratic Party

MDQ minimum detectable quantity

MDR *Computing* memory data register • (*or* **m.d.r.**) minimum daily requirement

MDRam (ˌɛmˈdiːˌræm) multibank dynamic random access memory

MDS Master of Dental Surgery • microprocessor development system

MDSc Master of Dental Science

mdse merchandise

MDT *Computing* mean downtime • (USA) Mountain Daylight Time

MDu Middle Dutch

MDU Medical Defence Union

MDV Doctor of Veterinary Medicine

MDW Military Defence Works

m.e. maximum effort • mobility equipment

Me Maine (USA) • (*or* **M^e**) Maître (French lawyer's title) • Messerschmitt (German aircraft) • *Chem., symbol for* methyl (used in formulae)

ME *British vehicle registration for* London NE • *US postcode for* Maine • managing editor • marine engineer(ing) • (USA) marriage encounter • Master of Education • Master of Engineering • mechanical engineer(ing) • (USA) Medical Examiner • *postcode for* Medway • *Med.* metabolizable energy • Methodist Episcopal • Middle East(ern) • *airline flight code for* Middle East Airlines • Middle English • military engineer • mining engineer(ing) • *British fishing port registration for* Montrose • *Physics* Mössbauer effect • Most Excellent (in titles) • mottled edges (of book pages) • *Med.* myalgic encephalomyelitis

MEA Member of the European Assembly • Middle East Airlines

MEAF Middle East Air Force

meas. measurable • measure • measurement

MEB Midlands Electricity Board

MEBA/NMU (USA) Marine Engineer Beneficial Association/National Maritime Union

MEc Master of Economics

MEC Master of Engineering Chemistry • Member of the Executive Council • Methodist Episcopal Church • Middle East Command

MECAS Middle East Centre for Arab Studies

mech. mechanic • mechanical(ly) • mechanics • mechanism

MechE Mechanical Engineer

ME(Chem) Master of Chemical Engineering

MECI Member of the Institute of Employment Consultants

MECO (ˈmiːkəʊ) *Astronautics* main engine cut off

MEcon Master of Economics

med. medallist • median • medical • medicine • medieval • medium

Med. Mediterranean

MEd Master of Education

MED maximum equivalent dose • minimum effective dose • *Chem.* molecular electronic device • (New Zealand) Municipal Electricity Department

Medico (ˈmɛdrkəʊ) Medical International

Corporation (*or* Cooperation Organization) (*or* Cooperation Organization) (US medical agency)

Medit. Mediterranean

med. jur. medical jurisprudence

MEDLARS ('mɛd,lɑːz) (USA) Medical Literature Analysis and Retrieval System

MedRC Medical Reserve Corps

MedScD Doctor of Medical Science

med. tech. medical technician • medical technology

MEE (*or* **ME(Elec)**) Master of Electrical Engineering

MEF Middle East Force

meg (mɛg) *Computing* megabyte

MEIC Member of the Engineering Institute of Canada

Mej. Mejuffrouw (Dutch: Miss)

MEK *Chem.* methyl ethyl ketone

Melan. Melanesia(n)

Melb. Melbourne

MELF Middle East Land Forces

mem. member • memento (Latin: remember) • memoir(s) • memorandum • memorial

ME(Mech) Master of Mechanical Engineering

Men *Astronomy* Mensa

MEN *Med.* multiple endocrine neoplasia

MENCAP (*or* **Mencap**) ('mɛnkæp) Royal Society for Mentally Handicapped Children and Adults

MEng Master of Engineering

MENS (mɛnz) multiple endocrine neoplasia syndromes

mensur. mensuration

mentd mentioned

MEO Marine Engineering Officer

MEP (Sri Lanka) Mahajana Eksath Peramuna (People's United Front) • Master of Engineering Physics • (*or* **m.e.p.**) mean effective pressure • Member of the European Parliament

MEPA Master of Engineering and Public Administration

mer. mercantile • merchandise • mercoledì (Italian: Wednesday) • mercredi (French: Wednesday) • mercury • meridian • meridional

Mer. Merionethshire (former Welsh county)

merc. mercantile • mercoledì (Italian: Wednesday) • mercury

MERCOSUR ('mɜːkəʊsʊə) Mercado Común del Sur (Spanish: Southern (American) Common Market)

MERLIN ('mɜːlɪn) *Astronomy* Multi-Element Radio-linked Interferometer Network (UK)

Mert Merton College, Oxford

MES *Chem.* molecular emission spectrometry • *Chem.* Mössbauer emission spectroscopy

MèsA Maître ès arts (French: Master of Arts)

MESc Master of Engineering Science

MESFET ('mɛsfɛt) *Electronics* metal-semiconductor field-effect transistor

Messrs ('mɛsəz) Messieurs (French: gentlemen *or* sirs; used in English as pl. of Mr; *see also under* MM)

met. metallurgical • metallurgist • metallurgy • metaphor • metaphysics • (*or* **met**; mɛt) meteorological (as in **met. office**) • meteorology • metronome • metropolitan

Met (*or* **met**) *Biochem.* methionine • (mɛt) Metropolitan Opera House (New York) • (mɛt) Metropolitan Police

metal. (*or* **metall.**) metallurgical • metallurgy

metaph. metaphor(ical) • metaphysical • metaphysics

METAR Meteorological Airfield Report

met. bor. metropolitan borough

MetE Metallurgical Engineer

meteorol. (*or* **meteor.**) meteorological • meteorology

Meth. Methodist

M-et-L Maine-et-Loire (department of France)

M-et-M Meurthe-et-Moselle (department of France)

m. et n. *Med.* mane et nocte (Latin: morning and night; in prescriptions)

MetR Metropolitan Railway (London)

metrol. metrological • metrology

metrop. (*or* **metropol.**) metropolis • metropolitan

metsat ('mɛtsæt) meteorological satellite

Mev. Mevrouw (Dutch: Mrs)

MeV *symbol for* megaelectronvolt(s)

MEW *Military* microwave early warning (system)

Mex. Mexican • Mexico

MEX *international vehicle registration for* Mexico

MEXE Military Engineering Experimental Establishment

Mex.Sp. Mexican Spanish

mez. *Music* mezzo *or* mezza (Italian: half *or* medium)

Mez mezzo-soprano

MEZ Mitteleuropäische Zeit (German: Central European Time)

mezzo. mezzotint

mf (ital.) *Music* mezzo forte (Italian; moderately loudly) • *Slang* (USA) motherfucker

mF *symbol for* millifarad(s)

MF *British vehicle registration for* London NE • machine finish(ed) (of paper) • magnetic field • Master of Forestry • *Telecom.* medium frequency • melamine–formaldehyde (as in **MF resin**) • *Chem.* methyl formate • Middle French • mill finish • *Slang* (USA) motherfucker • *Telecom.* multifrequency

M/F (*or* **m/f**) male or female (in advertisements)

MFA Master of Fine Arts

MFAMus Master of Fine Arts in Music

MFARCS Member of the Faculty of Anaesthetists of the Royal College of Surgeons

MFB Metropolitan Fire Brigade

MFC Mastership in Food Control • motor-fuel consumption

MFCM Member of the Faculty of Community Medicine

mfd manufactured

MFD minimum fatal dose

mfg manufacturing

MFH Master of Foxhounds • mobile field hospital

MFHom Member of the Faculty of Homeopathy

MFlem Middle Flemish

MFLOPS ('ɛmˌflɒps) *Computing* millions of floating-point operations per second

MFM *Computing* modified frequency modulation

MFN most favoured nation (in trade agreements)

MFOM Member of the Faculty of Occupational Medicine

MFP *Physics* mean free path

MFPA Mouth and Foot Painting Artists

mfr. (*or* **mfre**) manufacture • manufacturer

MFr Middle French

MFS Master of Food Science • Master of Foreign Study

m. ft. *Med.* mistura fiat (Latin: let a mixture be made; in prescriptions)

m.f.t. motor freight tariff

MFT *Computing* master file table

MFV motor fleet vessel

mg *symbol for* milligram(s)

Mg *Chem.*, *symbol for* magnesium

MG *airline flight code for* Djibouti Airlines • *British vehicle registration for* London NE • machine glazed (of paper) • machine gun • *Music* main gauche (French: left hand) • Major-General • *Building trades* make good • Morris Garages (sports car; named after its original manufacturer) • motor generator • *Med.* myasthenia gravis

MGA Major-General in charge of Administration

m.g.a.w.d. make good all works disturbed (in commercial contracts)

MGB metropolitan green belt • Ministerstvo Gosudarstvennoi Bezopasnosti (Russian: Ministry of State Security; Soviet secret police, 1946–54) • motor gunboat

MGC Machine Gun Corps • (formerly) Marriage Guidance Council (renamed Relate)

MGDS RCS Member in General Dental Surgery of the Royal College of Surgeons

mge message

MGGS Major-General, General Staff

MGI Member of the Institute of Certificated Grocers

MGk Medieval Greek

M. Glam Mid Glamorgan

MGM Metro-Goldwyn-Mayer (film studio) • mobile guided missile

MGO Master General of the Ordnance • Master of Gynaecology and Obstetrics

MGP manufactured-gas plant

Mgr Manager • Monseigneur (French: my lord) • *RC Church* Monsignor

MGr Medieval Greek

MGR *Nuclear engineering* modular gas-cooled reactor

Mgrs Managers • Monseigneurs (French: my lords) • *RC Church* Monsignors

mgt management

mH *Physics, symbol for* millihenry(s)

MH *British vehicle registration for* London NE • *Navigation* magnetic heading • *Nautical* main hatch • *airline flight code for* Malaysia Airlines • marital history • Master of Horse • Master of Horticulture • Master of Hounds • Master of Hygiene • (USA) Medal of Honor • mental health • *British fishing port registration for* Middlesbrough • military hospital • Ministry of Health

MHA (USA) Master of (*or* in) Hospital Administration • (Australia, Canada) Member of the House of Assembly • Methodist Homes for the Aged • *Chem.* methyl-3-hydroxyanthranilic acid

MHC *Immunol.* major histocompatibility complex

MHCIMA Member of the Hotel Catering and Institutional Management Association

MHD *Physics* magnetohydrodynamics

MHE Master of Home Economics

MHeb Middle Hebrew

MHF massive hydraulic fracture (*or* fracturing)

MHG Middle High German

MHK (Isle of Man) Member of the House of Keys

MHLG Ministry of Housing and Local Government

M.Hon. Most Honourable

MHortSc Master of Horticultural Science

MHR (USA, Australia) Member of the House of Representatives

MHRA Modern Humanities Research Association

MHRF Mental Health Research Fund

MHS medical history sheet • Member of the Historical Society • *Computing* message-handling system (*or* service)

MHTGR *Nuclear engineering* modular high-temperature gas-cooled reactor

MHum Master of Humanities

MHW mean high water (of tides)

MHWN mean high water neaps (tides)

MHWS mean high water springs (tides)

MHy Master of Hygiene

MHz *symbol for* megahertz

mi. mile • (USA, Canada) mill(s) (monetary unit)

Mi. minor (in names)

MI malleable iron • *US postcode for* Michigan • Military Intelligence (*see* MI5; MI6) • Ministry of Information • *Biology* mitotic index • moment of inertia • mounted infantry • *Med.* myocardial infarction • *Irish vehicle registration for* Wexford

MI5 Military Intelligence, section five (British counterintelligence agency)

MI6 Military Intelligence, section six (British intelligence and espionage agency)

MIA Master of International Affairs •

Military missing in action • (Australia) Murrumbidgee Irrigation Area

MIAA&S Member of the Incorporated Association of Architects and Surveyors

MIAeE Member of the Institute of Aeronautical Engineers

MIAgrE Member of the Institution of Agricultural Engineers

MIAM Member of the Institute of Administrative Management

MIAP Member of the Institution of Analysts and Programmers

MIAS Member of the Institute of Accounting Staff

MIB *Computing* management information base

MIBF Member of the Institute of British Foundrymen

MIBiol Member of the Institute of Biology

MIBK *Chem.* methyl isobutyl ketone

MIBritE Member of the Institution of British Engineers

MIB(Scot) Member of the Institute of Bankers in Scotland

Mic *Astronomy* Microscopium

Mic. *Bible* Micah

MICE Member of the Institution of Civil Engineers

MICEI Member of the Institution of Civil Engineers of Ireland

MICFor Member of the Institute of Chartered Foresters

Mich. Michaelmas • Michigan

MIChemE Member of the Institution of Chemical Engineers

MICorrST Member of the Institution of Corrosion Science and Technology

MICR *Computing* magnetic-ink character recognition

micRNA *Biochem.* messenger-RNA-interfering complementary RNA

Micro. Micronesia

microbiol. microbiology

micros. (*or* **micro.**) microscope • microscopist • microscopy

MICS Member of the Institute of Chartered Shipbrokers

MICV *Military* mechanized infantry combat vehicle

mid. middle • midnight

Mid. Midlands • Midshipman

MIDAS ('maɪdəs) *Computing* measurement information and data analysis system • missile defence alarm system

Middx Middlesex

MIDELEC ('mɪdɪlɛk) Midlands Electricity Board

MIDI (*or* **Midi**) ('mɪdɪ) musical instrument digital interface

Midl. Midlands • Midlothian (Scotland)

Mid. Lat. Middle Latin

MIDPM Member of the Institute of Data Processing Management

MIE(Aust) Member of the Institution of Engineers, Australia

MIED Member of the Institution of Engineering Designers

MIEE Member of the Institution of Electrical Engineers

MIEEE (USA) Member of the Institute of Electrical and Electronics Engineers

MIEI Member of the Institution of Engineering Inspection

MIE(Ind) Member of the Institution of Engineers, India

MIES Member of the Institution of Engineers and Shipbuilders, Scotland

MIEx Member of the Institute of Export

MIExpE Member of the Institute of Explosives Engineers

MIF *Computing* management information format • *Immunol.* migration inhibition factor

MIFA Member of the Institute of Field Archaeologists

MIFF Member of the Institute of Freight Forwarders

MIFG *Meteorol.* shallow fog

MIFireE Member of the Institute of Fire Engineers

MiG (mɪg) Mi(koyan and) G(urevich) (Soviet jet fighter; named after its designers)

MIG (mɪg) metal-inert gas (as in **MIG welding**)

MIGasE Member of the Institution of Gas Engineers

MIGeol Member of the Institution of Geologists

MIH Master of Industrial Health • Member of the Institute of Housing

MIHort Member of the Institute of Horticulture

MIHT Member of the Institution of Highways and Transportation

MIIE Member of the Institution of Industrial Engineers

MIIM Member of the Institute of Industrial Managers

MIInfSc Member of the Institute of Information Sciences

MIISec Member of the International Institute of Security

mil. mileage • military • militia

Mil. Milan

MIL Member of the Institute of Linguists • (USA) one million

Mil. Att. Military Attaché

MILGA Member of the Institute of Local Government Administrators

milit. military

mill. million

MILocoE Member of the Institution of Locomotive Engineers

Milw. Milwaukee

MIM Member of the Institute of Metals

MIMarE Member of the Institute of Marine Engineers

MIMC Member of the Institute of Management Consultants

MIMD *Computing* multiple instruction (stream), multiple data (stream)

MIME (maɪm) multimedia Internet mail extension • multipurpose Internet messaging extension

MIMechE Member of the Institution of Mechanical Engineers

MIMGTechE Member of the Institution of Mechanical and General Technician Engineers

MIMI Member of the Institute of the Motor Industry

MIMinE Member of the Institution of Mining Engineers

MIMM Member of the Institution of Mining and Metallurgy

MIMS (mɪmz) Monthly Index of Medical Specialities

MIMunE Member of the Institution of Municipal Engineers (now amalgamated with the Institution of Civil Engineers)

min minimum • *symbol for* minute(s)

min. mineralogical • mineralogy • minim (liquid measure) • minimum • mining • ministerial • minor • minute(s)

Min. Minister • Ministry

MIND *indicating* National Association for Mental Health

mineral. mineralogical • mineralogy

Minn. Minnesota

Min. Plen. Minister Plenipotentiary

Min. Res. Minister Resident (*or* Residentiary)

MINS (mɪnz) (USA) minor(s) in need of supervision

MInstAM Member of the Institute of Administrative Management

MInstBE Member of the Institution of British Engineers

MInstD Member of the Institute of Directors

MInstE Member of the Institute of Energy

MInstEnvSci Member of the Institute of Environmental Sciences

MInstMC Member of the Institute of Measurement and Control

MInstME Member of the Institution of Mining Engineers

MInstMM Member of the Institution of Mining and Metallurgy

MInstP Member of the Institute of Physics

MInstPet Member of the Institute of Petroleum

MInstPI Member of the Institute of Patentees and Inventors

MInstPkg Member of the Institute of Packaging

MInstPS Member of the Institute of Purchasing and Supply

MInstR Member of the Institute of Refrigeration

MInstRA Member of the Institute of Registered Architects

MInstT Member of the Institute of Transport

MInstTM Member of the Institute of Travel Managers in Industry and Commerce

MInstWM Member of the Institute of Wastes Management

MINucE Member of the Institution of Nuclear Engineers

Mio *Geology* Miocene

MIOSH Member of the Institution of Occupational Safety and Health

m.i.p. mean indicated pressure

MIP marine insurance policy · maximum investment plan · Member of the Institute of Plumbing · monthly investment plan

MIPA Member of the Institute of Practitioners in Advertising

MIPD Member of the Institute of Personnel Development

MIPM Member of the Institute of Personnel Management

MIPR Member of the Institute of Public Relations

MIProdE Member of the Institution of Production Engineers

MIPS (*or* **mips**) (mɪps) *Computing* millions of instructions per second

MIQ Member of the Institute of Quarrying

MIQA Member of the Institute of Quality Assurance

MIRA Member of the Institute of Registered Architects · Motor Industry's Research Association

MIRAS ('maɪˌræs) mortgage interest relief at source

MIRD medical internal radiation dose

MIREE(Aust) Member of the Institution of Radio and Electronics Engineers (Australia)

MIRT Member of the Institute of Reprographic Technicians

MIRTE Member of the Institute of Road Transport Engineers

MIRV (mɜːv) *Military* multiple independently targeted re-entry vehicle

Mis *Geology* Mississippian

MIS management information system · manufacturing information system · marketing information system · Member of the Institute of Statisticians · meteorological information system · Mining Institute of Scotland

misc. miscellaneous · miscellany

MISD *Computing* multiple instruction (stream), single data (stream)

MISFET ('mɪsfɛt) *Electronics* metal-insulator-semiconductor field-effect transistor

MIS(India) Member of the Institution of Surveyors (India)

Miss. Mission · Missionary · Mississippi

mist. *Med.* mistura (Latin: mixture)

mistrans. mistranslation

MIStructE Member of the Institution of Structural Engineers

Mit. Mittwoch (German: Wednesday)

MIT Massachusetts Institute of Technology

MITA Member of the Industrial Transport Association

MITD Member of the Institute of Training and Development

MITE Member of the Institution of Electrical and Electronics Technician Engineers

MITI (Japan) Ministry of International Trade and Industry

MITL magnetically insulated transmission line

MITT Member of the Institute of Travel and Tourism

Mitts (mɪts) minutes of telecommunications traffic

MIWEM Member of the Institution of Water and Environmental Management

MIX *Chem.* 1-methyl-3-isobutylxanthine

mixt. mixture

MJ *British vehicle registration for* Luton • *symbol for* megajoule(s) • Ministry of Justice

MJA Medical Journalists' Association

MJD management job description

MJI Member of the Institute of Journalists

MJQ Modern Jazz Quartet

MJS Member of the Japan Society

MJSD March, June, September, December (months for quarterly payments)

MJur Magister Juris (Latin: Master of Law)

mk *Currency* mark

Mk. (*or* **Mk**) mark (type of car) • Mark

MK *airline flight code for* Air Mauritius • *British vehicle registration for* London NE • *international vehicle registration for* Macedonia • *symbol for* Malawi kwacha (monetary unit) • *postcode for* Milton Keynes

mkd marked

MKO Mauna Kea Observatory (Hawaii)

mks *Currency* marks • (*or* **m.k.s., MKS**) metre kilogram second (as in **mks units**)

mksA (*or* **MKSA**) metre kilogram second ampere (as in **mksA system**)

mkt market

ml (*or* **mL**) *symbol for* millilitre(s)

ml. mail

ML *airline flight code for* Aero Costa Rica • Licentiate in Medicine (Latin *Medicinae Licentiatus*) • Licentiate in Midwifery • *British vehicle registration for* London NE • Master of Law(s) • Master of Letters • *Statistics* maximum likelihood • Medieval Latin • *British fishing port registration for* Methil (Scotland) • *postcode for* Motherwell • motor launch • muzzle-loading (of firearms)

MLA *Accounting* mandatory liquid assets • Master Locksmiths' Association • (*or* **MLArch**) Master of (*or* in) Landscape

Architecture • Master of the Liberal Arts • Member of the Legislative Assembly • (USA) Modern Language Association

MLC Meat and Livestock Commission • (India, Australia) Member of the Legislative Council • *Chem.* micellar liquid chromatography • *Med.* mixed lymphocyte culture (test for donor–recipient compatibility)

MLCOM Member of the London College of Osteopathic Medicine

MLCT *Chem.* metal-to-ligand charge transfer

mld. mould(ed)

MLD Master of Landscape Design • *Pharmacol.* minimal lethal dose

mldg moulding

MLF Mouvement de libération des femmes (French: Women's Liberation Movement) • *Military* multilateral (nuclear) force

MLG Middle Low German

MLib (*or* **MLibSc**) Master of Library Science

MLitt Master of Letters (Latin *Magister Litterarum*)

MLK Martin Luther King Jr. (1929–68, US Black civil-rights leader)

Mlle Mademoiselle (French: Miss)

MLNS *Med.* mucocutaneous lymph node syndrome

MLO military liaison officer • *Microbiol.* mycoplasma-like organism(s)

MLR *Banking* minimum lending rate • *Med.* mixed lymphocyte reaction (test for donor–recipient compatibility) • (ital.) Modern Language Review • *Statistics* multiple linear regression

MLRS multiple-launch rocket system

MLS main-line station • Master of Library Science • *Aeronautics* microwave landing system • mixed language system • multi-language system

MLSO Medical Laboratory Scientific Officer

MLW mean low water (of tides)

MLWN mean low water neaps (tides)

MLWS mean low water springs (tides)

mm *symbol for* millimetre(s)

m.m. (*or* **M/m**) made merchantable • mutatis mutandis (Latin: with the necessary changes)

MM *British vehicle registration for* London NE • *Music* Maelzel's metronome (indicating the tempo of a piece) • maintenance manual • Majesties • *Med.* malignant melanoma • Martyrs • *Freemasonry* Master Mason • Master Mechanic • Master of Music • Medal of Merit • mercantile marine • Messieurs (French: gentlemen *or* sirs; used in French as pl. of M (Monsieur); *compare* Messrs) • Military Medal

MMA Metropolitan Museum of Art • *Astronomy* Millimeter Array

MMath Master of Mathematics

MMB (formerly) Milk Marketing Board

MMC *Chem.* magnesium methyl carbonate • metal-matrix composite • Monopolies and Mergers Commission

MMD (Zambia) Movement for Multiparty Democracy

MMDA (USA) money market deposit account

MMDS *Radio* multipoint microwave distribution system

Mme Madame (French: Mrs)

MME Master of Mechanical Engineering • Master of Mining Engineering • Master of Music Education

MMechE Master of Mechanical Engineering

MMed Master of Medicine

MMedSci Master of Medical Science

Mmes Mesdames (French: pl. of Mme)

MMet Master of Metallurgy

MMetE Master of Metallurgical Engineering

mmf *Physics* magnetomotive force

MMG *Military* medium machine gun

MMGI Member of the Mining, Geological, and Metallurgical Institute of India

mmHg millimetre(s) of mercury (unit of pressure)

MMin Master of Ministry

MMI *Computing* man–machine interface • Municipal Mutual Insurance

MMM (Canada) Member of the Order of Military Merit

3M Minnesota Mining and Manufacturing Company

m.m.p. *Chem.* mixture melting point

MMP matrix metalloproteinase (enzyme) • Military Mounted Police

MMPI *Psychol.* Minnesota Multiphasic Personality Inventory

MMQ minimum manufacturing quantity

MMR *Med.* mass miniature radiography •

measles, mumps, rubella (in **MMR vaccine**)

MMRBM *Military* mobile medium-range ballistic missile

MMS Marine Meteorological Services • Member of the Institute of Management Services • *Chem.* methyl methanesulphonate • multimission modular spacecraft

MMSA Master of Midwifery of the Society of Apothecaries

MMSc Master of Medical Science

MMT methylcyclopentadienyl manganese tricarbonyl (antiknock petrol additive) • Multiple Mirror Telescope (Arizona)

MMTS *Chem.* methyl methylthiomethyl sulphoxide

MMU *Computing* memory management unit • *Finance* million monetary units

MMus Master of Music

MMusEd Master of Musical Education

MMX multimedia extension

mn. midnight

m.n. mutato nomine (Latin: with the name changed)

Mn *Chem., symbol for* manganese • Modern (of languages)

MN magnetic north • *British fishing port registration for* Maldon • Merchant Navy • *US postcode for* Minnesota

MNA Master of Nursing Administration • Member of the National Assembly (of Quebec) • *Chem.* molybdenum–nickel–alumina (catalyst)

MNAD Multinational Airmobile Division (of NATO)

MNAEA Member of the National Association of Estate Agents

MNAS (USA) Member of the National Academy of Sciences

MNC multinational company

MND motor neurone disease

MnE Modern English

MNE multinational enterprise

MNECInst Member of the North East Coast Institution of Engineers and Shipbuilders

mng managing

mngmt management

mngr manager

MnGr (*or* **MnGk**) Modern Greek

MNI Member of the Nautical Institute

MNIMH Member of the National Institute of Medical Herbalists

Mnl. Manila

MNM *Meteorol.* minimum

MNOS *Electronics* metal-nitride-oxide semiconductor

MNP microcomputer networking protocol

Mnr Mjnheer (Dutch: Mr *or* Sir)

MNR marine nature reserve • mean neap rise (of tides) • Mozambique National Resistance

MNT mean neap tide

MNurs Master of Nursing

mo. moment • month • mouth

m.o. mail order • modus operandi • money order

m-o months old

Mo *Computing* magneto-optical • *Chem.*, *symbol for* molybdenum

Mo. Missouri • Monday

MO mail order • manually operated • mass observation • Master of Obstetrics • Medical Officer • Medical Orderly • Meteorological Office • military operations • *US postcode for* Missouri • modus operandi • *Chem.* molecular orbital • money order • motor-operated • municipal officer • *British vehicle registration for* Reading

MoA Ministry of Aviation

MOA memorandum of agreement

MO&G Master of Obstetrics and Gynaecology

mob. mobile • (*or* **mobizn**) mobilization • mobilize

MOB (mɒb) movable object block

MOBS (mɒbz) multiple-orbit bombardment system (nuclear-weapon system)

MoC mother of the (trade-union) chapel

MOC management and operating contractor • *international vehicle registration for* Mozambique

MOCVD *Electronics* metal-organic chemical vapour deposition

mod (mɒd) *Maths., Computing* modulo

mod. moderate • *Music* moderato (Italian; at a moderate tempo) • modern • modification • modified • *Maths.* modulus

MoD Ministry of Defence

MOD mail-order department • *Meteorol.* moderate (icing, etc.)

mod. dict. *Med.* modo dicto (Latin: as prescribed)

modem ('məʊdɛm) *Computing* modulator demodulator

MODFET ('mɒd,fɛt) *Electronics* modulation-doped field-effect transistor

modif. modification

mod. praes. *Med.* modo praescripto (Latin: in the manner directed)

Mods Honour Moderations (at Oxford University)

mod^to *Music* moderato (*see under* mod.)

MODU mobile offshore drilling unit

MOEH Medical Officer for Environmental Health

MOF *Med.* multiple organ failure

MoH Ministry of Housing and Local Government

MOH Master of Otter Hounds • Medical Officer of Health

MOHLG Ministry of Housing and Local Government

MOHLL (məʊl) machine-oriented high-level language

Moho ('məʊhəʊ) *Geology* Mohorovičić discontinuity

MOI military operations and intelligence • Ministry of Information • Ministry of the Interior

mol *Chem., symbol for* mole(s)

mol. molecular • molecule

MOL *Astronautics* manned orbital laboratory • Ministry of Labour

Mold. (*or* **Moldv.**) Moldavia(n)

mol. wt. molecular weight

m.o.m. middle of month

MOM milk of magnesia

MOMA ('məʊmə) Museum of Modern Art (New York)

MOMI ('məʊmɪ) Museum of the Moving Image

MOMIMTS Military and Orchestral Instrument Makers' Trade Society

mon. monastery • monastic • monetary • monitor • monsoon

Mon *Astronomy* Monoceros

Mon. Monaco • Monaghan • Monday • Monmouthshire • Montag (German: Monday)

MON *Meteorol.* above mountains

MONEP Marché des options négotiables de Paris (French traded option market)

Mong. (*or* **Mongol.**) Mongolia(n)

monog. monograph

Mont. Montana • (*or* **Montgom.**) Montgomeryshire (former Welsh county)

Montr. Montreal

MOO (muː) *Computing* multiuser object-oriented

MOPA *Electronics* master oscillator power amplifier

MOPH (USA) Military Order of the Purple Heart

mor. *Music* morendo (Italian; dying away) · *Bookbinding* morocco

Mor. Moroccan · Morocco

MOR middle-of-the-road · *Engineering* modulus of rupture

MORB *Geology* mid-oceanic ridge basalts

MORC Medical Officers Reserve Corps

mor. dict. *Med.* more dicto (Latin: in the manner directed)

MORI (*or* **Mori**) ('mɔːrɪ) Market and Opinion Research Institute (as in **MORI poll**)

morn. morning

morph. (*or* **morphol.**) morphological · morphology

MORS *Geology* mid-oceanic ridge system

mor. sol. *Med.* more solito (Latin: in the usual manner; in prescriptions)

mort. mortal · mortality · mortar · mortgage · mortuary

mos. months

Mos. Moscow · Moselle

MOS magneto-optical system · (mɒs) *Electronics* metal–oxide–silicon (*or* –semiconductor)

MOSFET ('mɒsfɛt) *Electronics* metal–oxide–silicon (*or* –semiconductor) field-effect transistor

MOST (mɒst) *Electronics* metal–oxide–silicon (*or* –semiconductor) transistor · Molonglo Observatory Synthesis Telescope

mot. motor(ized)

MOT (*or* **MoT**) Ministry of Transport (usually referring to the road-vehicle test, as in **MOT certificate**)

MOTNE meteorological operational telecommunications network

MOUSE (*or* **mouse**) (maʊs) *Military* minimum orbital unmanned satellite of the earth (for gathering data)

mov. *Music* movimento (Italian: movement *or* speed)

MOV *Engineering* motor-operated valve · *Meteorol.* move (*or* moving)

MOVE (muːv) Men Over Violence (association for wife batterers)

movt movement

MOW (New Zealand) Ministry of Works · Movement for the Ordination of Women

MOX mixed oxide (as in **MOX fuel**)

Moz. Mozambique

mp melting point · (ital.) *Music* mezzo piano (Italian; moderately softly)

m.p. meeting point · melting point · *Cartography* mile post · months after payment · mooring post

MP *British vehicle registration for* London NE · Madhya Pradesh (Indian state) · *Metallurgy* martensitic phase · *airline flight code for* Martinair Holland · medium pressure · Member of Parliament · *Physiol.* membrane potential · *Cartography* Mercator's projection · metal pollutant · *Chem.* methyl-2-pyrrolidone · Metropolitan Police · *Cartography* mile post · Military Police(man) · mille passuum (Latin: 1000 paces; i.e. the Roman mile) · Minister Plenipotentiary · miscellaneous papers (*or* publications) · Mounted Police(man)

M/P memorandum of partnership

MPA Master of Professional Accounting · Master of Public Administration · Master Printers Association · Member of the Parliamentary Assembly of Northern Ireland · Music Publishers' Association

MPAA Motion Picture Association of America

MPAGB Modern Pentathlon Association of Great Britain

MPB (USA) Missing Persons Bureau

MPBW Ministry of Public Building and Works

MPC (*or* **m.p.c.**) mathematics, physics, chemistry · maximum permissible concentration · Metropolitan Police College · Metropolitan Police Commissioner · multimedia personal computer

MPD maximum permissible dose

MPE Master of Physical Education · maximum permissible exposure (to radiation)

MPEA Member of the Physical Education Association

MPEG ('ɛmpɛg) *Computing* Moving Picture Expert Group

MPF *Med.* maturation-promoting factor

mpg miles per gallon

MPG *Education* main professional grade (teacher's basic salary) · (Germany) Max-Planck-Gesellschaft zur Förderung der Wissenschaften (Max Planck Society for the Advancement of Science)

mph miles per hour

MPh Master of Philosophy

MPH Master of Public Health

MPhil ('em 'fɪl) Master of Philosophy

MPI Max Planck Institute

MPIA Master of Public and International Affairs

MPIR (or **MPIfR**) Max Planck Institute for Radio Astronomy

MPL maximum permissible level

MPLA Movimento Popular de Libertação de Angola (Portuguese: Popular Movement for the Liberation of Angola)

mpm metres per minute

MPO management and personnel office • Metropolitan Police Office • military post office • mobile printing (or publishing) office

m.p.p. most probable position

MPP *Computing* massively parallel processor • Member of the Provincial Parliament (of Ontario)

MPPH *Chem.* 5-(4-methylphenyl)-5-phenylhydantoin

MPR (Indonesia) Majelis Permusyawaratan Rakyat (People's Consultative Assembly) • maximum permitted radiation • Mongolian People's Republic

MPRISA Member of the Public Relations Institute of South Africa

MPRP Mongolian People's Revolutionary Party (communists)

mps metres per second

MPs Master of Psychology

MPS manufacturer's part specification • master production schedule • Medical Protection Society • Member of the Philological Society • Member of the Physical Society • *Meteorol.* metres per second • *Med.* mucopolysaccharide (as in **MPS disease**)

MPsSc Master of Psychological Science

MPsych Master of Psychology

MPsyMed Master of Psychological Medicine

m. pt. melting point

MPU *Computing* microprocessor unit

Mpy (Netherlands) Maatschappij (Company; Co.)

mq. mosque

MQ metol-quinol (photographic developer)

Mqe Martinique

MQS *Electronics* modified Q-switching

m.r. memorandum receipt

Mr Mister

MR *airline flight code for* Air Mauritanie • magnetic resonance (as in **MR scanner**) • *Computing* magneto-resistive (as in **MR heads**) • *British fishing port registration for* Manchester • map reference • *Law* Master of the Rolls • match rifle • (or **M/R**) *Commerce* mate's receipt • mental retardation • metabolic rate • Middlesex Regiment • Minister Residentiary • motivation(al) research • motorways (traffic) regulations • municipal reform • *British vehicle registration for* Swindon

MRA Maritime Royal Artillery • Moral Rearmament

MRAC Member of the Royal Agricultural College

MRACP Member of the Royal Australasian College of Physicians

MRACS Member of the Royal Australasian College of Surgeons

MRadA Member of the Radionic Association

MRAeS Member of the Royal Aeronautical Society

MRAF Marshal of the Royal Air Force

MRAIC Member of the Royal Architectural Institute of Canada

MRAO Mullard Radio Astronomy Observatory

MRAS Member of the Royal Academy of Science • Member of the Royal Asiatic Society • Member of the Royal Astronomical Society

MRB Mersey River Board

MRBM *Military* medium-range ballistic missile

MRBS Member of the Royal Botanic Society

MRC Medical Registration Council • Medical Research Council • Medical Reserve Corps • Model Railway Club

MRCA multirole combat aircraft

MRCGP Member of the Royal College of General Practioners

MRC-LMB Medical Research Council Laboratory of Molecular Biology

MRCO Member of the Royal College of Organists

MRCOG Member of the Royal College of Obstetricians and Gynaecologists

MRCP Member of the Royal College of Physicians

MRCP(UK) Member of the Royal College of Physicians (United Kingdom)

MRCPA Member of the Royal College of Pathologists of Australia

MRCPath Member of the Royal College of Pathologists

MRCPE Member of the Royal College of Physicians of Edinburgh

MRCPGlas Member of the Royal College of Physicians and Surgeons of Glasgow

MRCPI Member of the Royal College of Physicians of Ireland

MRCPsych Member of the Royal College of Psychiatrists

MRCS Member of the Royal College of Surgeons

MRCSE Member of the Royal College of Surgeons of Edinburgh

MRCSI Member of the Royal College of Surgeons of Ireland

MRCVS Member of the Royal College of Veterinary Surgeons

MRD machine-readable dictionary

MRe Mauritian rupee (monetary unit)

MRE Master of Religious Education • *Military* meal, ready to eat • Microbiological Research Establishment • Mining Research Establishment

MRes ('ɛm 'rɛz) Master of Research

MRG Minority Rights Group

MRGS Member of the Royal Geographical Society

MRH Member of the Royal Household

MRHS Member of the Royal Horticultural Society

MRI *Med.* magnetic resonance imaging (*or* image) • Member of the Royal Institution

MRIA Member of the Royal Irish Academy

MRIAI Member of the Royal Institute of the Architects of Ireland

MRIC Member of the Royal Institute of Chemistry (*see* MRSC)

MRIN Member of the Royal Institute of Navigation

MRINA Member of the Royal Institution of Naval Architects

MRIPHH Member of the Royal Institute of Public Health and Hygiene

mrkr marker

MRM mechanically recovered meat (in food processing)

MRMetS Member of the Royal Meteorological Society

MRN materials return note

mRNA *Biochem.* messenger RNA

mrng morning

MRO Member of the Register of Osteopaths

MRP manufacturers' recommended price •

Master in (*or* of) Regional Planning • *Commerce* material requirements planning

MRPharmS Member of the Royal Pharmaceutical Society

MRPhS Member of the Royal Pharmaceutical Society

Mrs ('mɪsɪz) Mistress (title of a married woman)

MRS *Physics* magnetic resonance spectroscopy • Market Research Society • *Computing* monitored retrievable storage

MRSA *Med.* methicillin-resistant *Staphylococcus aureus*

MRSC Member of the Royal Society of Chemistry (formerly MRIC)

MRSH Member of the Royal Society for the Promotion of Health

MRSL Member of the Royal Society of Literature

MRSM Member of the Royal Society of Medicine • Member of the Royal Society of Musicians of Great Britain

MRSPE Member of the Royal Society of Painter-Etchers and Engravers

MRSPP Member of the Royal Society of Portrait Painters

MRST Member of the Royal Society of Teachers

MRT *Med.* magnetic resonance tomography • mass rapid transit

MRTPI Member of the Royal Town Planning Institute

MRU manpower research unit • mobile repair unit

MRUSI Member of the Royal United Service Institution

MRV *Military* multiple re-entry vehicle

MRVA Member of the Rating and Valuation Association

ms manuscript • *symbol for* millisecond(s)

m.s. mail steamer • margin of safety • maximum stress • mild steel

m/s metre(s) per second • (*or* **M/s**) *Finance* months after sight

m/s² metre(s) per second squared

Ms (mɪz *or* məz) Miss *or* Mrs (unspecified)

MS *British vehicle registration for* Edinburgh • *airline flight code for* Egyptair • *Physics* magnetic susceptibility • (*or* **M/S**) mail steamer • *Astronomy* main sequence • *Music* mano sinistra (Italian: left hand) • manuscript • *Chem., Physics* mass spectrometer (*or* spectrometry) • (USA) Master of Science • Master of Surgery • *international vehicle registration for*

Mauritius • medical staff • *Photog.* medium shot (*or* mid-shot) • memoriae sacrum (Latin: sacred to the memory of; on gravestones) • mess sergeant • *Cartography* milestone • minesweeper • *Meteorol.* minus • *US postcode for* Mississippi • *Med.* mitral stenosis • *Physics* Mössbauer spectroscopy • (*or* **M/S**) (USA) motor ship • multiple sclerosis • municipal surveyor

MSA Malaysia-Singapore Airways • Master of Science and Arts • Master of Science in Agriculture • Media Studies Association • Member of the Society of Apothecaries • Merchant Shipping Act • *Chem.* methane sulphonic acid • metropolitan statistical area • Mineralogical Society of America • Motor Schools' Association of Great Britain • (USA) Mutual Security Agency

MSAE (USA) Master of Science in Aeronautical Engineering • (USA) Member of the Society of Automotive Engineers

MSAgr Master of Science in Agriculture

MSAICE Member of the South African Institution of Civil Engineers

MSAInstMM Member of the South African Institute of Mining and Metallurgy

MS&R Merchant Shipbuilding and Repairs

MSArch Master of Science in Architecture

MSAutE Member of the Society of Automobile Engineers

MSB Maritime Safety Board • Metropolitan Society for the Blind • *Computing* most significant bit

MSBA Master of Science in Business Administration

MSBus Master of Science in Business

MSByte *Computing* most significant byte

msc. miscellaneous

m.s.c. moved, seconded, and carried

MSc Master of Science

MSC Manchester Ship Canal • Manpower Services Commission • medical staff corps • Metropolitan Special Constabulary

MScA (*or* **MSc(Ag)**) Master of Science in Agriculture

MScApp Master of Applied Science

MSc(Arch) Master of Science in Architecture

MScD (USA) Doctor of Medical Science • Master of Dental Science

MSCE Master of Science in Civil Engineering

MSc(Econ) Master of Science in Economics

MSc(Ed) Master of Science in Education

MSChE Master of Science in Chemical Engineering

MSc(Hort) Master of Science in Horticulture

MSCI Index *Finance* Morgan Stanley Capital International World Index

MScMed Master of Medical Science

MSc(Nutr) Master of Science in Nutrition

MSCP Master of Science in Community Planning

MScTech Master of Technical Science

MSD Doctor of Medical Science • Master of Science in Dentistry • Master Surgeon Dentist • *Computing* most significant digit

MSDent Master of Science in Dentistry

MS-DOS (ˌɛmˈɛsˌdɒs) *Computing, trademark* Microsoft Disk Operating System

MSE Master of Science in Education • Master of Science in Engineering • Member of the Society of Engineers

MSEd Master of Science in Education

MSEE Master of Science in Electrical Engineering

MSEM Master of Science in Engineering Mechanics • Master of Science in Engineering of Mines

MSF Manufacturing, Science, Finance (Union) • Master of Science in Forestry • Médecins sans frontières (French: Doctors Without Frontiers; charitable organization) • mine-sweeping flotilla

MSFC (USA) Marshall Space Flight Center

msg. message

MSG monosodium glutamate (food additive)

msgr messenger

Msgr Monseigneur (French: my lord) • *RC Church* Monsignor

MSgt *US military* Master Sergeant

MSH Master of Staghounds • *Biochem.* melanocyte-stimulating hormone

MSHE (*or* **MSHEc**) Master of Science in Home Economics

Mshl Marshal

MSHyg Master of Science in Hygiene

MSI *Electronics* medium-scale integration • Member of the Chartered Surveyors' Institution • Member of the Securities

Institute • (Italy) Movimento Sociale Italiano (Italian Social Movement)

MSIAD Member of the Society of Industrial Artists and Designers

MSIE Master of Science in Industrial Engineering

MSINZ Member of the Surveyors' Institute of New Zealand

MSJ Master of Science in Journalism

MSL Master of Science in Linguistics • (or **msl**) mean sea level

MSLS Master of Science in Library Science

MSM Master of Sacred Music • Master of Science in Music • Meritorious Service Medal

MSME Master of Science in Mechanical Engineering

MSMed Master of Medical Science

MSMetE Master of Science in Metallurgical Engineering

MSMus Master of Science in Music

MSN Master of Science in Nursing • Microsoft Network

MSO Member of the Society of Osteopaths

MSocIS Membre de la société des ingénieurs et scientifiques de France (French: Member of the Society of Engineers and Scientists of France)

MSocSc Master of Social Science(s)

MSP matched sale–purchase agreement • Microsoft Paint

MSPE Master of Science in Physical Education

MSPH Master of Science in Public Health

MSPhar (or **MSPharm**) Master of Science in Pharmacy

MSPHE Master of Science in Public Health Engineering

MSQ managing service quality

MSR *Computing* magnetic stripe reader • (or **m.s.r.**) main supply route • mean spring rise (of tides) • Member of the Society of Radiographers • *Military* missile-site radar

MSRP (USA) manufacturer's suggested retail price

MSS (or **MS(S)**, **mss**) manuscripts • *Computing* mass storage system • (or **MSSc**) Master of Social Science • Master of Social Service • Member of the Royal Statistical Society • *Med.* midstream specimen (of urine) • *Physics* multichannel spectrum (or multispectral) scanner

MSSE (USA) Master of Science in Sanitary Engineering

MSt Master of Studies

MST Master of Sacred Theology • Master of Science in Teaching • mean spring tide • mean survival time • (USA) Mountain Standard Time

MStat Master of Statistics

MSTD Member of the Society of Typographic Designers

Mstr Master

MSTS *US Navy* Military Sea Transportation Service

MSU *Med.* midstream specimen of urine

MSUL Medical Schools of the University of London

MSurv Master of Surveying

MSurvSc Master of Surveying Science

MSw Middle Swedish

MSW magnetic surface wave(s) • Master in (or of) Social Work • Master of Social Welfare • medical social worker • municipal solid waste

MSX Microsoft extended Basic

MSY maximum sustainable yield (of a natural resource)

mt. *Military* megaton

m.t. metric ton(s) • missile test • (USA) Mountain Time

Mt *symbol for* metical (monetary unit of Mozambique) • Mount • Mountain

MT *British vehicle registration for* London NE • magnetic tape • mail transfer • *Med.* malignant tumour • mandated territory • *British fishing port registration for* Maryport • mass transport • mean time • mechanical transport • *Military* megaton • *Law* Middle Temple • *US postcode for* Montana • motor tanker • motor transport • (USA) Mountain Time

M/T empty (used on gas cylinders) • mail transfer

MTA *Computing* message transfer agent • minimum terms agreement • Music Teachers' Association • Music Trades' Association

MTAI Member of the Institute of Travel Agents

MTB motor torpedo boat • mountain bike

MTBE methyl *t*-butyl ether (lead-free antiknock petrol additive)

MTBF *Computing* mean time between failures

MTBI *Computing* mean time between incidents

MTC Mechanized Transport Corps • Music Teacher's Certificate

MTCA Ministry of Transport and Civil Aviation

mtd mounted

MTD maximum tolerated dose • mean temperature difference • Midwife Teacher's Diploma

mtDNA *Biochem.* mitochondrial DNA

MTech Master of Technology

MTEFL Master in the Teaching of English as a Foreign (*or* Second) Language

MTF *Image technol.* modulation transfer function

MTFA medium-term financial assistance

mtg meeting • (*or* **mtge**) mortgage • mounting

MTG (USA) methanol to gasoline

mtgee mortgagee

mtgor mortgagor

mth month

MTh Master of Theology

MTI *Radar* moving-target indicator (*or* indication)

mtl. material • monatlich (German: monthly)

MTL mean tide level

MTM methods-time measurement

mtn motion • mountain

MTN *Finance* medium-term note • multilateral trade negotiations

MTNA Music Teachers' National Association

MTO made to order • Mechanical Transport Officer

MTP Master of Town Planning

mtr. meter

MTR *US Air Force* mean time to restore • minimum time rate

Mt Rev Most Reverend (title of archbishop)

Mts Mountains • Mounts

MTS Master of Theological Studies • Merchant Taylors' School • *Computing* Michigan terminal system • motor transport service • (USA) multichannel television sound

MTT *Electronics* mean transit time

MTTR mean time to repair (*or* restore)

MTU *Computing* magnetic tape unit • *Computing* maximum transfer unit • *US Air Force* missile training unit

MTV motor torpedo vessel • (USA) music television

MTW *Meteorol.* mountain waves

m/u make up

MU *airline flight code for* China Eastern Airlines • *British vehicle registration for* London NE • maintenance unit • monetary unit • Mothers' Union • Musicians' Union

MUC Missionary Union of the Clergy

MU car *Railways* (USA) multiple-unit car

Mud (mʌd) *Computing* multi-user dungeon

MUF *Telecom.* maximum usable frequency

MUFTI ('mʌftɪ) *Military* minimum use of force tactical intervention

Mug (mʌg) *Computing* multi-user game

MUGA ('mʌɡə) *Med.* multiple-gated arteriography (as in **MUGA scan**)

Multics ('mʌltɪks) *Trademark* multiplexed information and computing service

mun. (*or* **munic.**) municipal(ity)

MUniv Master of the University

mus. museum • music(al) • musician

Mus *Astronomy* Musca

musa ('mjuːzə) multiple unit steerable aerial (*or* antenna)

MusB Bachelor of Music (Latin *Musicae Baccalaureus*)

MusD Doctor of Music (Latin *Musicae Doctor*)

MusM Master of Music (Latin *Musicae Magister*)

musn musician

mut. mutilated • mutual

MUX (*or* **mux**) (mʌks) multiplexer

mv (ital.) *Music* mezza voce (Italian: half voice; i.e. softly)

m.v. market value • mean variation • motor vessel

mV *symbol for* millivolt(s)

MV *airline flight code for* Great American Airways • *British vehicle registration for* London SE • *symbol for* megavolt(s) • merchant vessel • *Chem.* mixed valence • (*or* **M/V**) motor vessel • muzzle velocity (of firearms)

MVB Bachelor of Veterinary Medicine

MVD Doctor of Veterinary Medicine • Ministerstvo Vnutrennikh Del (Russian: Ministry of Internal Affairs; Soviet police organization, 1946–60) • *Med.* mitral valve disease

MVEE Military Vehicles and Engineering Establishment

MVetMed Master of Veterinary Medicine

MVetSc Master of Veterinary Science

MVL motor-vehicle licence

MVO Member of the Royal Victorian Order

MVP most valuable (or valued) player

MVRA Motor Vehicle Repairers' Association

MVSc (or **MVS**) Master of Veterinary Science

mvt movement

mW symbol for milliwatt(s)

MW international vehicle registration for Malawi • Master of Wine • airline flight code for Maya Airways • Radio medium wave • symbol for megawatt(s) • Middle Welsh • Nuclear engineering mixed waste • Chem. molecular weight • Most Worshipful • Most Worthy • British vehicle registration for Swindon

MWA Mystery Writers of America

MWC municipal-waste combustion

MWeldI Member of the Welding Institute

MWF Medical Women's Federation

m.w.g. music wire gauge

MWGM Freemasonry Most Worshipful (or Worthy) Grand Master

MWh (or **MW h**) symbol for megawatt hour

MWI municipal-waste incineration (or incinerator)

MWIA Medical Women's International Association

MWO Meteorological Watch Office

MWP mechanical wood pulp

MWPA Married Women's Property Act

Mx Magnetism, symbol for maxwell • Middlesex

MX British vehicle registration for London SE • airline flight code for Mexicana • missile, experimental • Meteorol. mixed (types of icing)

mxd mixed

my million years (following a number)

my. myopia

MY British vehicle registration for London SE • motor yacht

myc. (or **mycol.**) mycological • mycology

MYOB Colloquial mind your own business

MYRA Finance multiyear rescheduling agreement

myst. mystery (or mysteries)

myth. (or **mythol.**) mythological • mythology

Mz Geology Mesozoic

MZ British vehicle registration for Belfast

N

n (ital.) Chem., symbol for amount of substance • Printing, symbol for en • (ital.) Genetics, symbol for haploid chromosome number (also in **2n** (diploid), **3n** (triploid), etc.) • (ital.) Maths, symbol for indefinite number • symbol for nano- (prefix denoting 10^{-9}, as in **nm**, nanometre) • Physics, symbol for neutron • (ital.) Chem. normal (i.e. unbranched, as in **n-butane**) • Electronics n-type (semiconductor) • (ital.) Physics, Chem., symbol for number density (of atoms, particles, etc.) • (ital.) Physics, symbol for principal quantum number • (ital.) Optics, symbol for refractive index • (ital.) Physics, symbol for rotational frequency

n. nail (obsolete unit; 2¼ inches) • name • natus (Latin: born) • near • negative •

nephew • Commerce net • neuter • new • night • nominative • noon • nostro (Italian: our) • note • notre (French: our) • noun • nous (French: us or we) • number

'n' and (as in **rock 'n' roll**)

N Biochem., symbol for asparagine • Printing, symbol for en • Chess, symbol for knight • naira (Nigerian monetary unit) • Electrical engineering neutral • (ital.) Physics, symbol for neutron number • British fishing port registration for Newry • Physics, symbol for newton(s) • ngultrum (monetary unit of Bhutan; see also under Nu) • Chem., symbol for nitrogen • north(ern) • postcode for north London • international vehicle registration for Norway • nuclear (as in **N-weapon**) • Physics, symbol for nucleon •

(ital.) *Chem., Physics, symbol for* number of molecules, particles, etc • *international civil aircraft marking for* USA • (ital.) *Chem., indicating* substitution on a nitrogen atom (as in **N-phenylhydroxyl-amine**)

N. (*or* **N**) National(ist) • navigation • Navy • New • Norse • north(ern) • November • *Law* nullity • nurse (*or* nursing)

N4 *airline flight code for* National Airlines Chile

N6 *airline flight code for* Aero Continente

N7 *airline flight code for* Nordic East Airways

4N *airline flight code for* Air North

5N *airline flight code for* Aerotour Dominicano Airlines • *international civil aircraft marking for* Nigeria

7N *airline flight code for* Air Manitoba

9N *international civil aircraft marking for* Nepal • *airline flight code for* Trans States Airlines

n/a not applicable • not available

Na *Chem., symbol for* sodium (Latin *natrium*)

NA *British vehicle registration for* Manchester • (USA) Narcotics Anonymous • (USA) National Academician • National Archives • Nautical Almanac • naval architect • naval attaché • naval auxiliary • *international vehicle registration for* Netherlands Antilles • *Engineering* neutral axis • *Banking* new account • *Biochem.* nicotinic acid • *Med.* Nomina Anatomica (Latin: Anatomical Names; official anatomical terminology) • *Biochem.* noradrenaline • North America(n) • *Optics* numerical aperture • nursing auxiliary

N/A (*or* **N/a**) *Banking* new account • (*or* **N/a**) *Banking* no account • (*or* **N/a**) *Banking* no advice • *Banking, Commerce* nonacceptance • not available

n.a.a. *Shipping* not always afloat

NAA *Chem.* naphthalene acetic acid • (USA) National Aeronautic Association • National Association of Accountants • (USA) National Automobile Association • *Physics* neutron activation analysis

NAAA National Alliance of Athletic Associations

NAACP (USA) National Association for the Advancement of Colored People

NAAFA National Association to Aid Fat Americans

NAAFI (*or* **Naafi**) ('næfɪ) Navy, Army, and Air Force Institutes

NAAQS (USA) National Ambient Air Quality Standard

NAAS National Agricultural Advisory Service

NAB National Advisory Body for Public Sector Higher Education • National Alliance of Businessmen • National Assistance Board (former government department) • (USA) National Association of Broadcasters • National Australia Bank • naval air (*or* amphibious) base • New American Bible

NABC National Association of Boys' Clubs

NABS National Advertising Benevolent Society

NAC National Advisory Council • National Agriculture Centre • National Association for the Childless • *Geology* North American craton

NACA (USA) National Advisory Committee for Aeronautics

NACAB National Association of Citizens Advice Bureaux

NACCAM National Coordinating Committee for Aviation Meteorology

NACCB National Accreditation Council for Certification Bodies

NACEIC National Advisory Council on Education for Industry and Commerce

NACM National Association of Colliery Managers

NACNE National Advisory Committee on Nutrition Education

NACO National Association of Cooperative Officials

NACODS ('neɪkɒdz) National Association of Colliery Overmen, Deputies, and Shotfirers

NACOSS National Approved Council for Security Systems

NACRO (*or* **Nacro**) ('nækrəʊ) National Association for the Care and Resettlement of Offenders

NACS National Association of Chimney Sweeps

NACTST National Advisory Council on the Training and Supply of Teachers

NAD (USA) National Academy of Design • *Biochem.* nicotinamide adenine dinucleotide • *Med.* no abnormality detected • (*or* **n.a.d.**) no appreciable difference • not on active duty

NADC naval aide-de-camp

NADEC National Association of Development Education Centres

NADFAS National Association of Decorative and Fine Arts Societies

NADGE NATO Air Defence Ground Environment

NADH *Biochem.*, *indicating* a reduced form of NAD (H is the symbol for hydrogen)

NADOP (USA) North American Defense Operational Plan

NADP *Biochem.* nicotinamide adenine dinucleotide phosphate

NADW North Atlantic deep water

NADWARN ('næd,wɔːn) (USA) Natural Disaster Warning System

NAE (USA) National Academy of Engineering

NAEA National Association of Estate Agents

NAEP (USA) National Assessment of Educational Progress

NAEW NATO Airborne Early Warning

NA f. *symbol for* Netherlands Antillean guilder (monetary unit)

NAFD National Association of Funeral Directors

NAFO ('næfəʊ) National Association of Fire Officers • Northwest Atlantic Fisheries Organization

NAFTA ('næftə) New Zealand and Australia Free Trade Agreement • North American Free Trade Agreement • North Atlantic Free Trade Area

n.a.g. net annual gain

Nag. Nagasaki (Japan)

NAG *Biochem. N*-acetyl D-glucosamine • National Association of Goldsmiths

NAGC National Association for Gifted Children

NAGS National Allotments and Gardens Society

Nah. *Bible* Nahum

Nahal (nə'hɑːl) (Israel) No'ar Halutzi Lohem (Hebrew: Pioneer and Military Youth)

NAHAT National Association of Health Authorities and Trusts

NAHB (USA) National Association of Home Builders

NAHT National Association of Head Teachers

NAI nonaccidental injury

NAIR (USA) national arrangements for incidents involving radioactivity

NAIRU ('neɪruː) *Economics* nonaccelerating inflation rate of unemployment

NAITA ('neɪtə) National Association of Independent Travel Agents

NAK (*or* **nak**) (næk) *Telecom.* negative acknowledgment

NAL (USA) National Aerospace Laboratory

NALC (USA) National Association of Letter Carriers (trade union)

NALGO ('nælgəʊ) (formerly) National and Local Government Officers' Association (merged with NUPE and COHSE to form Unison)

NALHM National Association of Licensed House Managers

N. Am. North America(n)

NAM *international vehicle registration for* Namibia • (USA) National Association of Manufacturers

NAMAS National Measurement and Accreditation Service

NAMB National Association of Master Bakers

NAMCW National Association of Maternal and Child Welfare

NAMH National Association for Mental Health (now called MIND)

NAMMA NATO MRCA Management Agency

NAMS (USA) national air-monitoring sites

NAND (nænd) *Computing, Electronics* not AND (as in **NAND gate, NAND operation**)

N and Q notes and queries

NAO National Audit Office

Nap. Naples • Napoleonic

NAP (næp) National Association for the Paralysed

NAPA (USA) National Association of Performing Artists

NAPE National Association of Port Employers

NAPF National Association of Pension Funds

naph. *Chem.* naphtha

NAPLPS North American presentation level protocol syntax

NAPO ('næpəʊ) National Association of Probation Officers (trade union)

NAPT National Association for the Prevention of Tuberculosis

nar. narrow

NAR *Botany* net assimilation rate

narc. narcotic(s)

NARM (nɑːm) natural and accelerator-produced radioactive material

NAS (USA) National Academy of Sciences • National Adoption Society • National Association of Schoolmasters (now part of NAS/UWT) • naval air station • Noise Abatement Society • nursing auxiliary service

NASA (*or* **Nasa**) ('næsə) (USA) National Aeronautics and Space Administration

NASCAR ('næskɑː) (USA) National Association for Stock Car Auto Racing

NASD National Amalgamated Stevedores and Dockers • (USA) National Association of Securities Dealers

NASDA ('næsdə) (Japan) National Space Development Agency

NASDAQ ('næs,dæk) (USA) National Association of Securities Dealers Automated Quotations (system)

Nash. Nashville (USA)

Nass. Nassau (Bahamas)

NAS/UWT National Association of Schoolmasters/Union of Women Teachers

nat. national(ist) • native • natural • naturalize • naturist • natus (Latin: born)

n.a.t. normal allowed time

Nat. Natal

N. At. North Atlantic

NAT National Arbitration Tribunal

NATE National Association for the Teaching of English

NATFHE National Association of Teachers in Further and Higher Education

natl national

NATLAS ('nætlæs) National Testing Laboratory Accreditation Scheme

NATM New Austrian Tunnelling Method

NATO (*or* **Nato**) ('neɪtəʊ) North Atlantic Treaty Organization

NATS (næts) National Air Traffic Services • (USA) Naval Air Transport Service

nat. sc. (*or* **nat. sci.**) natural science(s)

NatScD Doctor of Natural Science

NATSOPA (næt'səʊpə) National Society of Operative Printers, Graphical and Media Personnel (formerly National Society of Operative Printers and Assistants; now merged with the GPMU))

N.Att. Naval Attaché

NATTS National Association of Trade and Technical Schools

natur. naturalist

Nat West (,næt 'west) National Westminster (bank)

NAU *international vehicle registration for* Nauru

naut. nautical

nav. naval • navigable • navigation • navigator

NAV *Finance* net asset value (of an organization)

NAVAIR ('næveə) (USA) Naval Air Systems Command

Nav.E. Naval Engineer

navig. navigation • navigator

NAVS National Anti-Vivisection Society

NAVSAT ('næv,sæt) navigational satellite

NAWB National Association of Workshops for the Blind

NAWC National Association of Women's Clubs

NAWO National Alliance of Women's Organizations

NAYC Youth Clubs UK (formerly National Association of Youth Clubs)

NAYT National Association of Youth Theatres

Nazi ('nɑːtsɪ) Nationalsozialisten (German: National Socialist)

nb (*or* **n.b.**) *Cricket* no ball • nota bene (Latin: note well)

Nb *Chem., symbol for* niobium

NB *British vehicle registration for* Manchester • narrow-bore • naval base • *Med.* needle biopsy • *Physics* neutral beam • New Brunswick (Canada) • North Britain (i.e. Scotland) • nota bene (Latin: note well)

NBA (USA) National Basketball Association • (USA) National Book Award • (USA) National Boxing Association • National Building Agency • Net Book Agreement • North British Academy

NBC (USA) National Basketball Committee • National Book Council (*see under* NBL) • National Boys' Club • (USA) National Broadcasting Company • National Bus Company • nuclear, biological, and chemical (of weapons or warfare)

NBCD *Computing* natural binary-coded decimal

NBD *Statistics* negative binomial distribution

NbE north by east

NBER (USA) National Bureau of Economic Research

NBG (*or* **nbg**) *Colloquial* no bloody good

NBI National Benevolent Institution • *Physics* neutral-beam injection • *Med.* no bone injury

NBK National Bank of Kuwait

NBL National Book League (formerly Council) • (*or* **nbl**) *Colloquial* not bloody likely

NBP *Chem.* normal boiling point

NBPI National Board for Prices and Incomes

NBR National Buildings Record • *Chem.* nitrile-butadiene rubber

nbre. noviembre (Spanish: November)

NBRI National Building Research Institute

NBS *Chem.* N-bromosuccinimide • (USA) National Bureau of Standards • Newcastle Business School

NBTS National Blood Transfusion Service

NBV *Accounting* net book value (of an asset)

NbW north by west

N by E north by east

N by W north by west

NC *British vehicle registration for* Manchester • National Certificate • National Congress (*or* Council) • *Education* National Curriculum • Nature Conservancy • New Caledonia • New Church • *Physics* nickel–cadmium (of electric cells) • nitrocellulose • *Chem.* nitrogen compound • *Meteorol.* no change (*or* not changing) • no charge • normally closed • North Carolina (abbrev. *or* postcode) • Northern Command • numerical control *or* numerically controlled (as in **NC machine**) • (USA) Nurse Corps

N/C (*or* **n/c**) new charter • nitrocellulose • no charge

NCA National Certificate of Agriculture • National Childminding Association • National Cricket Association • no copies available

NCAA (USA) National Collegiate Athletic Association • Northern Counties Athletic Association

NCACC National Civil Aviation Consultative Committee

NCAR National Center for Atmospheric Research (Colorado)

NCARB (USA) National Council of Architectural Registration Boards

NCB National Children's Bureau •

National Coal Board (now British Coal Corporation, BCC) • Nippon Credit Bank • *Insurance* no claim bonus

NCBA National Cattle Breeders' Association

NCBAE *Insurance* no claim bonus as earned

NCBW nuclear, chemical, and biological warfare

NCC (USA) National Climatic Center • National Computing Centre • National Consumer Council • (USA) National Council of Churches • *Education* National Curriculum Council • (formerly) Nature Conservancy Council

NCCA National Carpet Cleaners' Association

NCCI National Committee for Commonwealth Immigrants

NCCJ National Conference of Christians and Jews

NCCL National Council for Civil Liberties (now called Liberty)

NCCS national command and control system • National Council for Civic Responsibility

n.c.d. *Colloquial* no can do

NCDAD National Council for Diplomas in Art and Design

NCDL National Canine Defence League

NCERT National Council for Educational Research and Training

NCET National Council for Educational Technology

NCFT National College of Food Technology

NCH National Children's Home • National Clearing House

n.Chr. nach Christus (German: after Christ; AD)

n.c.i. no common interest

NCI (USA) National Cancer Institute • *Finance* New Community Instrument

NCIC National Cancer Institute of Canada • (USA) National Crime Information Center

NCIS National Criminal Intelligence Service

NCL National Central Library • National Chemical Laboratory • National Church League

NCLC National Council of Labour Colleges

NCNA New China News Agency

NCNC National Convention of Nigeria

and the Cameroons • National Convention of Nigerian Citizens (political party)

NCO noncommissioned officer

n.c.p. *Engineering* normal circular pitch

NCP National Car Parks Ltd • (Australia) National Country Party (now called National Party) • national cycling proficiency • *Computing* network control protocol

NCPL National Centre for Programmed Learning

NCPS noncontributory pension scheme

NCPT (USA) National Congress of Parents and Teachers

NCR National Cash Register (Company Ltd) • no carbon required (of paper)

NCRE Naval Construction Research Establishment

NCRL National Chemical Research Laboratory

NCRP (USA) National Council on Radiation Protection and Measurements

NCS (USA) National Communications System

NCSC (Australia) National Companies and Securities Commission • (USA) National Computer Security Center

NCSE National Council for Special Education

NCSS National Council of Social Service

NCT National Chamber of Trade • National Childbirth Trust • *Med.* neutron-capture therapy • *Med.* neutron computed tomography

NCTA (USA) National Community Television Association

NCTE (USA) National Council of Teachers of English

NCTU Northern Carpet Trades Union

NCU (formerly) National Communications Union (merged with the UCW to form the CWU) • National Cyclists' Union

n.c.u.p. no commission until paid

NCV (*or* **n.c.v.**) no commercial value

NCVCCO National Council of Voluntary Child Care Organizations

NCVO National Council for Voluntary Organizations

NCVQ National Council for Vocational Qualifications

NCW National Council of Women (of Great Britain)

n.d. *Photog.* neutral density • *Stock exchange* (USA) next day (delivery) • no date (*or* not dated) • no decision • no deed (*or* not deeded) • no delay • no demand • nondelivery • *Banking* not drawn • nothing doing

Nd *Chem., symbol for* neodymium

ND *airline flight code for* Airlink • *British vehicle registration for* Manchester • national debt • National Diploma • Naturopathic Diploma • *Photog.* neutral density • *Physics* neutron diffraction • no date • nondelivery • North Dakota (abbrev. *or* postcode)

N-D Notre-Dame (French: Our Lady; in church names, etc.)

NDA National Diploma in Agriculture • *Engineering* nondestructive analysis (*or* assay)

NDAC (USA) National Defense Advisory Commission

N. Dak. North Dakota

NDB *Aeronautics* nondirectional beacon

ndc National Defence College

NDC National Dairy Council • NATO Defence College • *Computing* normalized device coordinates

NDCS National Deaf Children's Society

NDD National Diploma in Dairying • National Diploma in Design

NDE (*or* **nde**) near-death experience • *Engineering* nondestructive evaluation (*or* examination)

NDF National Diploma in Forestry

NDH National Diploma in Horticulture

NDIC National Defence Industries Council

NDN National District Nurse Certificate

n.d.p. *Engineering* normal diametric pitch

NDP National Democratic Party (of various countries) • *Economics* net domestic product • (Canada) New Democratic Party • *Biochem.* nucleoside diphosphate

NDPB non-departmental public body

NDPS National Data Processing Service

NDR *Electronics* negative differential resistance (*or* resistivity) • non-domestic rates

NDRC National Defence Research Committee

NDSB Narcotic Drugs Supervisory Body (of the UN)

NDT *Engineering* nondestructive testing • *Commerce* nondistributive trade

NDTA (USA) National Defense Transportation Association

NDU Nursing Development Unit

NDV *Vet. science* Newcastle disease virus

n.e. not essential • not exceeding

n/e new edition • *Banking* no effects (i.e. no funds)

Ne *Chem., symbol for* neon

Ne. Nepal(ese) • Netherlands

NE *British vehicle registration for* Manchester • national emergency • National Executive • Naval Engineer • *US postcode for* Nebraska • Newcastle (postcode *or* British fishing port registration) • new edition • New England • news editor • *Banking* no effects (i.e. no funds) • northeast(ern) • nuclear energy • nuclear explosion

N/E new edition • *Banking* no effects (i.e. no funds) • *Accounting* not entered

NEA (USA) National Education Association • (USA) National Endowment for the Arts

NEAC New English Art Club

NEACP ('ni;kæp) (USA) National Emergency Airborne Command Post (aircraft in readiness for nuclear war)

NEAF Near East Air Force

NEAFC North-East Atlantic Fisheries Commission

NEARELF Near East Land Forces

Neb. (*or* **Nebr.**) Nebraska

NEB (USA) National Electricity Board • (USA) National Energy Board • National Enterprise Board • New English Bible

NEBOSH National Examination Board in Occupational Safety and Health

NEBSS National Examinations Board for Supervisory Studies

n.e.c. not elsewhere classified

NEC (USA) National Electric Code • National Electronics Council • National Equestrian Centre • National Executive Committee • National Exhibition Centre (Birmingham) • National Extension College (Cambridge) • *Med.* necrotizing enterocolitis • (Japan) Nippon Electric Company

NECCTA National Educational Closed Circuit Television Association

NECInst North East Coast Institution of Engineers and Shipbuilders

necr. necrosis

necrol. necrology

NED *Med.* no evidence of disease

NEDC National Economic Development Council (*or* Neddy) • North East Development Council

NEDO National Economic Development Office

NEEB North Eastern Electricity Board

NEF *Acoustics* noise exposure forecast (value)

NEFA *Chem.* nonesterified fatty acid

neg. negation • negative(ly) • negligence • negotiate

NEG negative (in transformational grammar)

Neh. *Bible* Nehemiah

NEH National Endowment for the Humanities

n.e.i. non est inventus (*or* inventa, inventum) (Latin: he (*or* she, it) has not been found) • not elsewhere indicated

NEL (USA) National Electronics Laboratory • National Engineering Laboratory

NEM *Chem.* N-ethylmaleimide

nem. con. (*or* **dis.**) nemine contradicente (*or* dissentiente) (Latin: no-one opposing; unanimously)

NE/n.d. new edition, no date given

N. Eng. New England • northern England

neol. neologism

n.e.p. new edition pending

Nep. Nepal(ese) • Neptune (planet)

NEP *international vehicle registration for* Nepal • New Economic Policy (USSR, 1921–28)

NEPA (USA) National Environmental Policy Act (1969)

NEPP (USA) National Energy Policy Plan

NEQ *Electronics* nonequivalence (as in **NEQ gate**)

NERA (USA) National Emergency Relief Administration

NERC Natural Environment Research Council

n.e.s. not elsewhere specified

NES *Physics* nuclear elastic scattering

NESC (USA) National Electric Safety Code

n.e.t. not earlier than

NET (USA) National Educational Television • *Nuclear engineering* Next European Torus

Neth. Netherlands

Neth. Ant. Netherlands Antilles

n. et m. *Med.* nocte et mane (Latin: night and morning; in prescriptions)

neurol. neurological • neurology

neut. neuter • neutral • neutralize(d)

Nev. Nevada

NEW *Economics* (USA) net economic welfare

Newf. Newfoundland

New M New Mexico

New Test. New Testament

n.f. *Engineering* near face • no fool • noun feminine

NF *airline flight code for* Air Vanuatu • *British vehicle registration for* Manchester • National Front • *Med.* neurofibromatosis • New Forest • Newfoundland • New French • (*or* **N/F**) *Banking* no funds • *Telecom.* noise factor (*or* figure) • Norman French (language) • *Pharmacol.* (USA) National Formulary

n.f.a. no further action

NFA National Farmers' Association • National Federation of Anglers • (USA) National Food Administration • (USA) National Futures Association • No Fixed Abode (*or* Address)

NFAL National Foundation of Arts and Letters

NFB National Film Board (of Canada)

NFBPM National Federation of Builders' and Plumbers' Merchants

NFBTE National Federation of Building Trades Employers

n.f.c. not favourably considered

NFC (USA) National Football Conference • National Freight Consortium

NFCO National Federation of Community Organization

Nfd Newfoundland

NFD Newfoundland • no fixed date

NFDM nonfat dry milk

NFER National Foundation for Educational Research

NFFC National Film Finance Corporation

NFFE (USA) National Federation of Federal Employees

NFFO nonfossil-fuel obligation (for electricity companies)

NFFPT National Federation of Fruit and Potato Trades

NFHA National Federation of Housing Associations

NFI National Federation of Ironmongers

NFL (USA, Canada) National Football League

Nfld Newfoundland

NFMPS National Federation of Master Printers in Scotland

NFMS National Federation of Music Societies

NFO National Freight Organization

NFPW National Federation of Professional Workers

n.f.r. no further requirements

NFRC National Federation of Roofing Contractors

NFRN National Federation of Retail Newsagents

NFS National Fire Service • National Flying Services • National Forest Service • *Computing* network filing service (*or* system) • not for sale

NFSE National Federation of Self-Employed (and Small Businesses)

NFT National Film Theatre

NFU National Farmers' Union

NFUW National Farmers' Union of Wales

NFWI National Federation of Women's Institutes

n.g. no good • not given

Ng *Geology* Neogene

Ng. Norwegian

NG *Railways* narrow gauge • National Gallery • (USA) National Guard (*or* Guardsman) • New Granada • *Med.* new growth • New Guinea • nitroglycerine • *Freemasonry* Noble Grand (*or* Guard) • no good • *British vehicle registration for* Norwich • *postcode for* Nottingham

NGA National Glider Association • (formerly) National Graphical Association (merged with SOGAT to form the GPMU)

NGC (USA) National Grid Company • *Astronomy* New General Catalogue (prefixed to a number to designate a Catalogue object)

NGF *Biochem.* nerve growth factor

NGk New Greek

NGL *Chem.* natural-gas liquid(s)

NGNP nominal gross national product

NGO (USA) National Gas Outlet • (India) nongazetted officer • nongovernmental organization

NGr New Greek

NGR *international vehicle registration for* Nigeria

NGRC National Greyhound Racing Club

NGRS Narrow Gauge Railway Society

NGS National Galleries of Scotland • National Geographic Society • nuclear generating station

NGT National Guild of Telephonists

NGTE National Gas Turbine Establishment

NGU *Med.* nongonococcal urethritis

NGV natural-gas vehicle

NH *airline flight code for* All Nippon Airways • National Hunt • naval hospital • New Hampshire (abbrev. *or* postcode) • *British vehicle registration for* Northampton • northern hemisphere • Northumberland Hussars

NHA National Horse Association of Great Britain • (USA) National Housing Agency

NHBC National Housebuilding Council

NHBRC National House-Builders' Registration Certificate

NHC National Hunt Committee

NHD Doctor of Natural History

N.Heb. (*or* **NHeb**) New Hebrew • New Hebrides

NHF National Hairdressers' Federation

NHG New High German

NHI National Health Insurance

NHK Nippon Hōsō Kyōkai (Japan Broadcasting Corporation)

NHL (USA) National Hockey League • *Med.* non-Hodgkin lymphoma

NHLBI (USA) National Heart, Lung, and Blood Institute

NHMRCA National Health and Medical Research Council of Australia

NHO Navy Hydrographic Office

NHR National Housewives Register • National Hunt Rules

NHS National Health Service

NHSTA National Health Service Training Authority

NHTPC National Housing and Town Planning Council

Ni *Chem., symbol for* nickel

NI National Insurance • Native Infantry • Naval Instructor • Naval Intelligence • new impression • *Computing* non-interfaced • Northern Ireland • North Island (New Zealand) • *Irish vehicle registration for* Wicklow

NIA (USA) National Intelligence Authority • Newspaper Institute of America

NIAAA Northern Ireland Amateur Athletic Association

NIAB National Institute of Agricultural Botany

NIACRO (*or* **Niacro**) ('naɪəkrəʊ) Northern Ireland Association for the Care and Resettlement of Offenders

NIAE National Institute of Agricultural Engineering

NIAID National Institute of Allergy and Infectious Diseases

Nibmar (*or* **NIBMAR**) ('nɪb,mɑː) no independence before majority African rule

NIBSC National Institute for Biological Standards Control

n.i.c. not in contact

Nic. Nicaragua(n)

NIC National Incomes Commission • (*or* **nic**) National Insurance contribution • network interface card • newly industrialized country • *international vehicle registration for* Nicaragua

Nica. (*or* **Nicar.**) Nicaragua(n)

NiCad (*or* **NiCd, Ni/Cd**) nickel–cadmium (battery)

NICAM ('naɪkæm) *Electronics* near-instantaneous companded audio multiplex (for digital coding of audio signals)

NICEC National Institute for Careers Education and Counselling

NICEIC National Inspection Council for Electrical Installation Contracting

NICF Northern Ireland Cycling Federation

NICG Nationalized Industries Chairmen's Group

NICRA Northern Ireland Civil Rights Association

NICS Northern Ireland Civil Service

NICU *Med.* neonatal intensive care unit

NID National Institute for the Deaf • (India) National Institute of Design • Naval Intelligence Division • Northern Ireland District

NIDC Northern Ireland Development Council

NIDD non-insulin-dependent diabetes

NIES Northern Ireland Electricity Service

NIESR National Institute of Economic and Social Research

NIF *Finance* note issuance facility

NIFES National Industrial Fuel Efficiency Service

NIFO *Accounting* next-in-first-out

NIFTP network independent file transfer protocol

Nig. Nigeria(n)

NIH (USA) National Institutes of Health • North Irish Horse (former regiment)

NIHCA Northern Ireland Hotels and Caterers Association

NIHE (Ireland) National Institute for Higher Education

NII Nuclear Installations Inspectorate

NIIP National Institute of Industrial Psychology

NILP Northern Ireland Labour Party

NIMA National Infomercial Marketing Association

nimby ('nɪmbɪ) *Colloquial* not in my back yard (indicating liberals, reformers, etc., in principle but not in practice)

NiMH nickel-metal hydride

NIMH National Institute of Medical Herbalists • (USA) National Institute of Mental Health

n.imp. new impression

NIMR National Institute for Medical Research

NIN national information network

NINO ('niːnəʊ) no inspector, no operator (system)

NIO National Institute of Oceanography

NIOSH (USA) National Institute for Occupational Safety and Health

nip. *Engineering* nipple

Nip. Nippon(ese)

ni. pri. nisi prius (Latin: unless previously)

N. Ir. Northern Ireland

NIR Nauchno-Issledovatel'skaya Rabota (Russian colour-television system) • nonionizing radiation • Northern Ireland Railways

NIRA (USA) National Industrial Recovery Act (1933)

NIRC National Industrial Relations Court

N. Ire. Northern Ireland

NIREX ('naɪreks) Nuclear Industry Radioactive Waste Executive

NIRS (USA) National Institute of Radiological Sciences

n.i.s. not in stock

NIS *symbol for* new Israeli shekel (monetary unit)

NISA National Independent Supermarkets' Association

NISC National Industrial Safety Committee

NIST (USA) National Institute of Standards and Technology

NISTRO Northern Ireland Science and Technology Regional Organization

NISW National Institute for Social Workers

NIT negative income tax

NITB Northern Ireland Tourist Board

NIV New International Version (of the Bible)

NIWAAA Northern Ireland Women's Amateur Athletic Association

n.-J. nächsten Jahres (German: next year)

NJ *British vehicle registration for* Brighton • New Jersey (abbrev. *or* postcode)

NJA National Jewellers' Association

NJAC National Joint Advisory Council

NJC National Joint Council

NJCC National Joint Consultative Committee

NJNC National Joint Negotiating Committee

NK *British vehicle registration for* Luton • *Immunol.* natural killer (as in **NK cell**) • not known

NKC *Immunol.* natural killer cell

NKGB Narodny Komissariat Gosudarstvennoi Bezopasnosti (Russian: People's Commissariat of State Security; Soviet agency, 1943–46)

NKr *symbol for* Norwegian krone (monetary unit)

NKVD Narodny Komissariat Vnutrennikh Del (Russian: People's Commissariat of Internal Affairs; Soviet agency, 1934–46)

NKz *symbol for* new kwanza (Angolan monetary unit)

n.l. (*or* **nl**) *Printing* new line • non licet (Latin: it is not permitted) • non liquet (Latin: it is not clear)

NI National

NL National Labour • (USA) National League (baseball league) • National Liberal • Navy League • Navy List • *international vehicle registration for* Netherlands • *British vehicle registration for* Newcastle-upon-Tyne • New Latin • (Australia) no liability (after the name of a public limited company; equivalent to plc) • (*or* **N.Lat.**) north latitude • *airline flight code for* Shaheen Air International

NLB National Library for the Blind

NLBD National League for the Blind and Disabled

NLC National Liberal Club • National Library of Canada

NLCS North London Collegiate School

NLD (Burma) National League for Democracy

n.l.f. nearest landing field

NLF National Labour Federation • National Liberal Federation • National Liberation Front • National Loans Fund

NLI National Library of Ireland

NLLST National Lending Library for Science and Technology

NLM National Library of Medicine

NLMC National Labour Management Council

NLN National League for Nursing

NLO naval liaison officer

NLP *Computing* natural language processing • (USA) neighborhood loan program • *Computing* neurolinguistic programming

NLQ *Computing* near letter quality

NLR *Astronomy* narrow-line region

NLRB (USA) National Labor Relations Board

NLS National Library of Scotland

n.l.t. not later than • not less than

NLW National Library of Wales

nly northerly

NLYL National League of Young Liberals

nm *symbol for* nanometre(s)

nm. nutmeg

n.m. nautical mile(s) • new moon • *Med.* nocte et mane (Latin: night and morning; in prescriptions) • nonmetallic • noun masculine

n.M. nächsten Monats (German: next month)

Nm. Nachmittag (German: afternoon)

N.m. next matter (in advertisement placing)

N/m no mark(s) (on a bill of lading)

NM *British vehicle registration for* Luton • national marketing • *Meteorol.* nautical mile(s) • New Mexico (abbrev. *or* postcode) • nuclear medicine

NMA National Management Association • National Medical Association

NMB National Maritime Board

n.m.c. no more credit

NMC National Marketing Council • (USA) National Meteorological Center

NMDA *Biochem.* N-methyl-D-aspartate (as in **NMDA receptor**)

N. Mex. New Mexico

NMFS (USA) National Marine Fisheries Service

NMI *Computing* non-maskable interrupt

n mile nautical mile

NMP *Economics* net material product

NMR *Physical chem.* nuclear magnetic resonance

NMRI *Med.* nuclear magnetic resonance imaging

NMS (USA) National Market System • National Museums of Scotland • *Stock exchange* Normal Market Size

NMSQT (USA) National Merit Scholarship Qualifying Test

NMSS (USA) National Multiple Sclerosis Society

n.m.t. not more than

NMTF National Market Traders' Federation

NMU National Maritime Union

NMW national minimum wage (as in **NMW policy**)

nn. names • notes • nouns

NN *Chem.* neutralization number • *British fishing port registration for* Newhaven • no name • *postcode for* Northampton • *British vehicle registration for* Nottingham

N/N *Banking* not to be noted

NNE north-northeast

NNEB National Nursery Examination Board

NNF Northern Nurses' Federation

NNHT Nuffield Nursing Homes Trust

NNI noise and number index (to evaluate aircraft noise) • noise nuisance index

NNMA Nigerian National Merit Award

NNOM Nigerian National Order of Merit

NNP *Economics* net national product

NNR National Nature Reserve

NNSA (USA) National Nuclear Safety Administration

NNT nuclear nonproliferation treaty

NNW north-northwest

n° numéro (French: number)

no. north(ern) • number (from Italian *numero*)

n.o. normally open • *Cricket* not out

No *Chem., symbol for* nobelium

No. north(ern) • Norway • Norwegian • number (from Italian *numero*)

NO *airline flight code for* Aus Air • *British vehicle registration for* Chelmsford • naval officer • naval operations • Navigation Officer • New Orleans • Nuffield Observatory (Jodrell Bank, Cheshire) • Nursing Officer

N/O (*or* **N/o**) *Banking* no orders

NOA National Opera Association • National Orchestral Association • not otherwise authorized

NOAA (USA) National Oceanic and Atmospheric Administration

NOAO (USA) National Optical Astronomy Observatories

nob. nobis (Latin: for (*or* on) our part) • noble

NOB naval operating base

n.o.c. notation of content • not otherwise classified

NOC National Olympic Committee • not otherwise classified

NOCD *Colloquial* not our class, dear

No. Co. northern counties

noct. *Med.* nocte (Latin: at night)

NOD Naval Ordnance Department • night observation device

NODA National Operatic and Dramatic Association

NODC non-OPEC developing country

n.o.e. notice of exception • not otherwise enumerated

NOE nuclear Overhauser effect

NOERC North of England Regional Consortium

n.o.h.p. not otherwise herein provided

n.o.i.b.n. not otherwise indexed by name

NOIC Naval Officer in Charge

NOISEP ('nɔɪsɛp) (USA) National Organization to Insure a Sound-controlled Environment

n.o.k. next of kin

NOL (USA) Naval Ordnance Laboratory

nol. con. *Law* nolo contendere (Latin: I do not wish to contend)

nol. pros. *Law* nolle prosequi (Latin: do not prosecute; procedure for ending criminal proceedings)

nom. nomenclature • nominal • nomination • nominative

nom. cap. *Finance* nominal capital

nomen. nomenclature

nomin. nominative

nomm. nomination

nom. nov. nomen novum (Latin: new name)

NOMSS (USA) National Operational Meteorological Satellite System

noncom. noncommissioned

Noncon. Nonconformist

non cul. non culpabilis (Latin: not guilty)

non obst. non obstante (Latin: notwithstanding)

non pros. *Law* non prosequitur (Latin: he does not prosecute; former judgment in favour of defendant)

non repetat. (*or* **non rep.**) *Med.* non repetatur (Latin: do not repeat; in prescriptions)

non seq. non sequitur (Latin: (a statement that) does not follow logically)

nonstand. (*or* **nonstd**) nonstandard

n.o.p. not otherwise provided (for)

NOP National Opinion Poll • not our publication

nor. normal

Nor *Astronomy* Norma

Nor. Norman • (*or* **nor.**) north • Norway • Norwegian • Norwich

NOR (nɔː) *Computing, Electronics* not OR (as in **NOR gate, NOR operation**) • *Biology* nucleolar-organizing region

NORAD ('nɔːræd) North American Air Defense Command

Norf. Norfolk

norm. normal • normalized

Norm. Norman

NORM not operationally ready maintenance

NORML (USA) National Organization for the Reform of Marijuana Laws

NORS not operationally ready supplies (*or* supply)

Northants Northamptonshire

Northd (*or* **Northumb.**) Northumberland

Norvic Norvicensis (Latin: (Bishop) of Norwich)

Norw. Norway • Norwegian

NORWEB ('nɔːwɛb) North Western Electricity Board

n^{os} numéros (French: numbers)

n.o.s. not otherwise specified

Nos. (*or* **nos.**) numbers (from Italian *numeros*)

NOS Nederlandse Omroep Stichting (Netherlands Broadcasting Corporation) • (nɒs) network operating system

NOSC *US Navy* Naval Ordnance Systems Command

NOSIG *Meteorol.* no significant change

not. notice

Not. Notary

NOT (nɒt) *Computing, Electronics* not (from its sense in logic, as in **NOT gate, NOT operation**)

NOTAR (*or* **Notar**) ('nəʊtɑː) no-tail rotor (of an aircraft)

NOTB National Ophthalmic Treatment Board

Nottm Nottingham

Notts Nottinghamshire

notwg notwithstanding

nov. novel(ist) • novembre (French *or* Italian: November) • novice • novitiate

Nov. (*or* **Nov**) November

NOW (USA) National Organization for Women • *Banking* (USA) negotiable order of withdrawal (as in **NOW account**) • New Opportunities for Women

NOX (nɒks) nitrogen oxide(s)

n.o.y. not out yet

noz. nozzle

np neap (tides) • *Printing* new paragraph • new pence

n.p. near point • *Law* net personalty • net proceeds • *Printing* new paragraph • new pattern • nickel-plated • nisi prius (Latin: unless previously) • nonparticipating • *Bibliog.* no place of publication given • no printer • no publisher • normal pitch • not paginated • nursing procedure

n/p net proceeds

Np napalm • neap (tides) • *Telecom., symbol for* neper • *Chem., symbol for* neptunium

NP national park • National (*or* Nationalist) Party • National Power plc • neuropsychiatric • neuropsychiatry • *postcode for* Newport • New Providence (Bahamas) • *Firearms* nitro proof • Nobel Prize • *Maths.* nondeterministic polynomial (in **NP-complete**) • *Chem.* nonpolar • Notary Public • noun phrase • (USA) nurse practitioner • *British vehicle registration for* Worcester

NPA National Park Authority • National Pigeon Association • (Phillipines) New People's Army • Newspaper Publishers' Association

NPACI National Production Advisory Council on Industry

NPB *Physics* neutral particle beam

NPBA National Pig Breeders' Association

NPC (China) National People's Congress • (USA) National Petroleum Council • National Ports Council • (USA) National Press Club • no-player character (in the game Dungeons and Dragons) • *Computing* normalized projection coordinates • (Nigeria) Northern People's Congress

NPCS Narrowband Personal Communications Service

NPD Nationaldemokratische Partei Deutschlands (German: National Democratic Party; neo-Nazis) • *Marketing* new

product development • *Astronomy* north polar distance

n.p.f. not provided for

NPF (Syria) National Progressive Front • Newspaper Press Fund

NPFA National Playing Fields Association

NPG National Portrait Gallery • Nuclear Planning Group (in NATO)

NPH *Chem.* normal paraffin hydrocarbon

NPK nitrogen, phosphorus, and potassium (in fertilizers; from the chemical symbols of these elements)

NPL National Physical Laboratory (Teddington, Middlesex)

npn (*or* **n-p-n**) *Electronics* n-type p-type n-type semiconductor (as in **npn transistor**)

NPN *Med.* nonprotein nitrogen

n.p.n.a. *Commerce* no protest for nonacceptance

NP/ND not published, no date given

n.p.o. *Med.* ne per oris (Latin: not by mouth)

NPO New Philharmonia Orchestra

n.p. or d. no place or date

n.p.p. no passed proof

NPP nuclear power plant

NPR (USA) National Public Radio • noise power ratio

NPRA National Petroleum Refiners Association

n.p.s. nominal pipe size • no prior service

NPS nuclear power source • nuclear power station

NPT non-proliferation treaty

n.p.u. ne plus ultra (Latin: nothing beyond; i.e. extreme or perfect state)

NPU National Pharmaceutical Union • National Postal Union • *Med.* not passed urine

NPV *Finance* net present value • no par value (of shares)

NPW nuclear-powered warship

NQ *airline flight code for* Orbi Georgian Airways

n.q.a. *Finance* net quick assets

NQOC *Colloquial* not quite our class

NQR *Physics* nuclear quadrupole resonance

nr near

n.r. net register • *Insurance* no risk

Nr Nummer (German: number)

NR *British vehicle registration for* Leicester • National Register • natural rubber • naval rating • Navy Regulations • *Insurance* no risk • (formerly) North Riding (York-shire) • *postcode for* Norwich

n.r.a. never refuse anything

NRA National Reclamation Association • (USA) National Recovery Administration (former agency) • National Recreation Area • National Rifle Association • National Rivers Authority • nuclear-reaction analysis

NRAA National Rifle Association of America

NRAO (USA) National Radio Astronomy Observatory

NRC (Ghana) National Redemption Council • National Research Council • (USA) Nuclear Regulatory Commission

NRCA National Retail Credit Association

NRCC National Research Council of Canada

NRD National Register of Designers (*or* Registered Designer)

NRDC National Research Development Corporation • (USA) Natural Resources Defense Council

NRDS *Med.* neonatal respiratory distress syndrome

NREM *Physiol.* non-rapid eye movement (as in **NREM sleep**)

NREN *Computing* national research and education network

NRF National Relief Fund

NRFL Northern Rugby Football League

NRI National Resources Institute

NRK (Norway) Norsk Rikskringkasting (state broadcasting company)

NRL National Reference Library • (USA) Naval Research Laboratory

NRM (Uganda) National Resistance Movement

NRMA (Australia) National Roads and Motorists Association

nrml normal

NRN *Physics* neutron-rich nucleus

NROTC Naval Reserve Officer Training Corps

NRP nuclear reprocessing plant

NRPB National Radiological Protection Board

NRR net reproduction rate (of popula-tions) • Northern Rhodesia Regiment

NRs *symbol for* Nepalese rupee (monetary unit)

NRS National Readership Survey • National Rose Society

NRT (*or* **n.r.t.**) net registered tonnage

NRTA (USA) National Retired Teachers Association

NRV *Finance* net realizable value • non-return valve

Nrw. Norwegian

NRZ *Computing* nonreturn to zero (method of encoding binary signals)

ns *symbol for* nanosecond(s) • (Graduate of the Royal) Naval Staff (College, Green-wich)

n.s. near side • new series • nickel steel • not satisfactory • not specified • *Banking* not sufficient (funds)

n/s nonsmoker • *Banking* not sufficient (funds)

Ns (ital.) *Meteorol., symbol for* nimbostratus

NS *British vehicle registration for* Glasgow • Nachschrift (German: postscript; PS) • *Computing* Nassi–Schneidermann (as in **NS chart**) • National Service • National Society • natural science • naval service • *Irish fishing port registration for* New Ross • new series • Newspaper Society • New Style (method of reckoning dates) • *Meteorol.* nimbostratus • (*or* **N/S**) non-smoker • not significant • Nova Scotia • nuclear science • nuclear ship • Numis-matic Society

N-S Notre-Seigneur (French: Our Lord)

NSA Abbey National Staff Association • (USA) National Security Agency • (USA) National Shipping Authority • National Skating Association • (USA) National Standards Association • (USA) National Student Association • New Society of Artists • nonsterling area

NSACS National Society for the Abolition of Cruel Sports

NSAE National Society of Art Education

NSAFA National Service Armed Forces Act

NSAID *Med.* nonsteroidal anti-inflammatory drug

NSB National Savings Bank • (USA) National Science Board

NSBA National Sheep Breeders' Associa-tion

NSC National Safety Council • National Savings Certificate(s) • (USA) National Security Council • National Sporting Club

NSCC (USA) National Securities Clearing Corporation

NSCR National Society for Cancer Relief

NSD naval supply depot

NSDAP Nationalsozialistische Deutsche Arbeiterpartei (German: National Socialist German Workers' Party; Nazi party)

NSERC (USA) Natural Sciences and Engineering Research Council

NSF (USA) National Science Foundation · (*or* **N/S/F, n.s.f.**) *Banking* not sufficient funds · Nuclear Structure Facility (Daresbury, Cheshire)

NSFGB National Ski Federation of Great Britain

NSG *Education* nonstatutory guidelines (concerning the National Curriculum)

NSGT non-self-governing territory (*or* territories)

NSHEB North of Scotland Hydroelectric Board

NSI (USA) National Security Information

NSL National Sporting League

NSLR *Physics* nuclear spin-lattice relaxation

NSM new smoking material · nonstipendiary minister

NSO National Solar Observatory · Naval Staff Officer

NSPCC National Society for the Prevention of Cruelty to Children

NSPE (USA) National Society of Professional Engineers

n.s.p.f. not specially provided for

NSRA National Small-bore Rifle Association

NSS national sample survey · New Shakespeare Society · *Med.* normal saline solution

NSSA National School Sailing Association

NSSU National Sunday School Union

NST Newfoundland Standard Time

NSTC Nova Scotia Technical College

NSTP Nuffield Science Teaching Project

NSU *Med.* nonspecific urethritis

NSW New South Wales

NSY New Scotland Yard

n.t. net terms · net tonnage

NT (Ireland) National Teacher · National Theatre · National Trust · neap tide(s) · *British fishing port registration for* Newport · new technology · New Testament · New Translation · Northern Territory (Australia) · *Bridge* no-trump(s) · not titled · Nurse Teacher · *airline flight code for* Pacific Eagle Airlines · *British vehicle registration for* Shrewsbury

NTA (USA) National Technical Association · net tangible assets

NTB *Economics* nontariff barrier

NTC *Physics* negative temperature coefficient

NTD *Colloquial* not top drawer

NTDA National Trade Development Association

NTEU (USA) National Treasury Employees Union

NTFS new technology filing system

ntfy notify

NTG North Thames Gas

NTGB North Thames Gas Board

NTGk New Testament Greek

Nth North

nthn northern

NTI noise transmission impairment

NTIA (USA) National Telecommunications and Information Administration

n.t.l. no time lost

NTL *Electronics* nonthreshold logic

NTM *Economics* nontariff measure

n.t.o. not taken out

NTO Naval Transport Officer

n.t.p. no title page

NTP *Physics* normal temperature and pressure · *Biochem.* nucleoside triphosphate

NTS National Trust for Scotland · Nevada Test Site · not to scale

NTSB (USA) National Transportation Safety Board

NTSC (USA) National Television System Committee

NTT New Technology Telescope

NTUC (Singapore) National Trades Union Congress

NTV Nippon Television

NTVLRO National Television Licence Records Office

nt. wt. (*or* **nt wt**) net weight

n.u. name unknown · number unobtainable

Nu *symbol for* ngultrum (monetary unit of Bhutan) · (ital.) *Physics, symbol for* Nusselt number

NU National Union · Nations Unies (French: United Nations; UN) · natural uranium · Northern Union · *British vehicle registration for* Nottingham · number unobtainable

NUAAW National Union of Agricultural and Allied Workers

nuc. (*or* **nucl.**) nuclear

NUCPS (formerly) National Union of Civil and Public Servants (merged with the IRSF to form the PTC)

NUCUA National Union of Conservative and Unionists Associations

nud. nudism • nudist

NUDAGO National Union of Domestic Appliances and General Operatives

NUDETS ('nju:dɛtz) nuclear detection system

NUFLAT National Union of Footwear, Leather, and Allied Trades (now amalgamated with the NUHKW to form the National Union of Knitwear, Footwear, and Apparel Trades)

NUHKW National Union of Hosiery and Knitwear Workers (*see* NUFLAT)

NUI National University of Ireland

NUIW National Union of Insurance Workers

NUJ National Union of Journalists

NUJMB Northern Universities Joint Matriculation Board

NUL (USA) National Urban League

NULMW National Union of Lock and Metal Workers

num. number • numeral(s) • numerical • numerology

Num. *Bible* Numbers

NUM National Union of Mineworkers • New Ulster Movement

NUMAST ('nju:mæst) National Union of Marine, Aviation, and Shipping Transport Officers

numis. (*or* **numism.**) numismatic(s)

NUOS naval underwater ordnance station

NUPE ('nju:pɪ) (formerly) National Union of Public Employees (merged with COHSE and NALGO to form Unison)

NUR National Union of Railwaymen (*see* RMT)

NURBS (nɜːbz) *Computing* nonuniform rational B-splines

NUS National Union of Seamen (*see* RMT) • National Union of Students

NUT National Union of Teachers

N-u-T Newcastle-upon-Tyne

NUTG National Union of Townswomen's Guilds

NUTGW (*or* **NUT&GW**) National Union of Tailors and Garment Workers (now amalgamated with the GMB)

NUTN National Union of Trained Nurses

nutr. nutrition

NUU New University of Ulster

NV (Netherlands) Naamloze Vennootschap (after the name of a public limited company; equivalent to plc) • (*or* **n.v.**) needle valve • *US postcode for* Nevada • New Version • nonvintage (of wine, etc.) • (*or* **n.v.**) *Finance* nonvoting (shares) • *Shipping* (Norway) Norske Veritas (classification society) • *British vehicle registration for* Northampton • *airline flight code for* Northwest Territorial Airways

N/V nonvintage (of wine etc.) • *Banking* no value

NVB National Volunteer Brigade

NVG *Military* night-vision goggles

NVGA (USA) National Vocational Guidance Association

NVI nonvalue indicator (type of postage stamp)

NVM Nativity of the Virgin Mary • *Chem.* nonvolatile matter

NVQ National Vocational Qualification

NVRam non-volatile random access memory

NVRS National Vegetable Research Station

n.w. net weight • no wind

NW *British vehicle registration for* Leeds • net worth • North Wales • northwest(ern) • *airline flight code for* Northwest Airlines • *postcode for* northwest London

NWC (USA) National War College

NWEB North Western Electricity Board

Nwfld Newfoundland

NWFP North-West Frontier Province (Pakistan)

NWGA National Wool Growers' Association

NWIDA North West Industrial Development Association

NWP North-Western Province (India)

NWS (USA) National Weather Service • normal water surface

n. wt. net weight

n.w.t. nonwatertight

NWT Northwest Territories (Canada)

NWTV North West Television

NX *airline flight code for* Air Macau • *British vehicle registration for* Dudley

NY *British vehicle registration for* Cardiff • New Year • New York (abbrev. *or* state postcode) • *airline flight code for* Norlandair (Iceland)

NYA (USA) National Youth Administration

NYC New York City

NYCSCE New York Coffee, Sugar, and Cocoa Exchange

NYD *Med.* not yet diagnosed

NYFE New York Futures Exchange

nyl. nylon

NYMEX ('naɪmɛks) New York Mercantile Exchange

NYMT National Youth Music Theatre

NYO National Youth Orchestra

NYOS National Youth Orchestra of Scotland

NYP not yet published

NYPD New York Police Department

NYS New York State

NYSE New York Stock Exchange

NYT National Youth Theatre • (ital.) New York Times

NZ *airline flight code for* Air New Zealand • *British vehicle registration for* Londonderry • neutral(ity) zone • New Zealand (abbrev. *or* IVR)

NZAP (formerly) New Zealand Associated Press

NZBC New Zealand Broadcasting Corporation

NZDSIR New Zealand Department of Scientific and Industrial Research

N. Zeal. New Zealand

NZEF New Zealand Employers' Federation • New Zealand Expeditionary Force (in World Wars I and II)

NZEFIP New Zealand Expeditionary Force in the Pacific (in World War II)

NZEI New Zealand Educational Institute

NZFL New Zealand Federation of Labor

NZIA New Zealand Institute of Architects

NZLR New Zealand Law Reports

NZMA New Zealand Medical Association

NZPA New Zealand Press Association

NZRFU New Zealand Rugby Football Union

NZRN New Zealand Registered Nurse

O

o *Computing, Maths., symbol for* order (*see under* O) • (ital.) *Chem.* ortho (as in **o-cresol**)

o. occasional • *Printing* octavo • off • old • only • optimus (Latin: best) • order • organ • *Baseball* out(s) • over • *Meteorol.* overcast • overseer

O *Genetics, symbol for* operator • *Slang* opium • *Computing, Maths., symbol for* order (followed in brackets by limiting value of a function) • *Geology* Ordovician • *Chem., symbol for* oxygen • *Pharmacol.* pint (Latin *octarius*) • *Med.*, indicating a blood group (*see also under* ABO) • *Logic*, indicating a particular negative categorial proposition • (ital.) *Chem.*, indicating substitution on an oxygen atom (in names of compounds)

O. (*or* **O**) observer • occupation • Ocean • *Printing* octavo • October • *Med.* oculus (Latin: eye) • Oddfellows • oeste (Spanish

or Portuguese: west) • Office • officer • Ohio • old • operation • orange • order (of knighthood, etc.) • ordinary • Orient • Osten (German: east) • *Music* ottava (Italian: octave) • ouest (French: west) • *Cricket* over • ovest (Italian: west) • owner

6O *international civil aircraft marking for* Somalia

7O *international civil aircraft marking for* Yemen

8O *airline flight code for* West Coast Air

o/a on account (of) • on or about

OA *British vehicle registration for* Birmingham • objective analysis • office address • (*or* **o.a.**) office automation • Officers' Association • Officier d'Académie (French: Officer of the Academy; award for services in education) • *Banking* old account • *Chem.* oleic acid • *airline flight code for* Olympic Airways • operational analysis • osteoarthritis • (*or* **o.a.**) overall

OAA Organisation pour l'alimentation et l'agriculture (French: Food and Agriculture Organization; FAO) • Outdoor Advertising Association of Great Britain

OAC *Chem.* organo-aluminium compound(s)

OACI Organisation de l'aviation civile internationale (French: International Civil Aviation Organization; ICAO)

o.a.d. overall depth

OAG (USA) Official Airline Guide

o.a.h. overall height

o. alt. hor. *Med.* omnibus alternis horis (Latin: every other hour)

OAM Medal of the Order of Australia

OAMDV omnia ad majorem Dei gloriam (Latin: all to the greater glory of God)

OANA Organization of Asian News Agencies

O & A October and April (on bills)

O & C Oxford and Cambridge (Schools Examination Board)

O & E (USA) Operations and Engineers

O & M organization and method(s)

O & O Oriental and Occidental Steamship Company • owned and operated

o.a.o. off and on

OAO one and only • Orbiting Astronomical Observatory

OAP old age pension(er)

OAPC (USA) Office of Alien Property Custodian

OAPEC (əʊ'eɪpɛk) Organization of Arab Petroleum-Exporting Countries

OAr Old Arabic

OAR *RC Church* (of the) Order of Augustinian Recollects

OAS *Military* offensive air support • on active service • *Physics* optical absorption spectroscopy • Organisation de l'armée secrète (French: Secret Army Organization; opposed Algerian independence) • Organization of American States

OASIS (əʊ'eɪsɪs) optimal aircraft sequencing using intelligent systems

OAT outside air temperature

OATC Oceanic Air Traffic Control

OATUU Organization of African Trade Union Unity

OAU Organization of African Unity

ob. obiit (Latin: (he *or* she) died) • obiter (Latin: incidentally) • obligation • oboe • observation • obsolete • obstetric(s)

o.b. ordinary building (grade; of timber)

o/b on or before (preceding a date)

Ob. *Bible* Obadiah

OB *British vehicle registration for* Birmingham • *British fishing port registration for* Oban • observed bearing • (USA) obstetric(s) • (USA) obstetrician • off Broadway • official board • Old Bailey • old bonded (whisky) • old boy • Order of Barbados • order of battle • ordinary business (in life-assurance policies) • ordnance board • outside broadcast • *international civil aircraft marking for* Peru

OBA optical bleaching agent (in detergents)

Obad. *Bible* Obadiah

obb. (*or* **obbl.**) *Music* obbligato (Italian: obligatory)

ÖBB Österreichische Bundesbahnen (German: Federal Railways of Austria)

OBC old boys' club • on-board computer (in motor advertisements)

ob. dk observation deck

obdt obedient

OBE Officer of the Order of the British Empire • out-of-the-body experience

OBEV (ital.) Oxford Book of English Verse

ob-gyn (*or* **ob-gyne**) obstetrics–gynaecology

OBI Order of British India

obit. obituary

obj. *Grammar* object(ive) • (*or* **objn**) objection

obl. obligation • oblige • oblique • oblong

OBLI Oxford and Buckinghamshire Light Infantry

OBM *Surveying* Ordnance benchmark

obo or best offer

OBO ore-bulk-oil (ship)

Obogs ('əʊbɒgz) *Military* on-board oxygen-generating system

ob. ph. *Surveying* oblique photograph(y)

obre. octobre (French: October)

obs. obscure • observation • observatory • observe(d) • observer • obsolete • obstetric(s)

Obs. Observatory • (ital.) The Observer

OBS *Meteorol.* observe(d) • *Med.* organic brain syndrome

obsc. obscure

OBSC *Meteorol.* obscure(d)

obscd obscured

OBSF *Accounting* off balance sheet finance

obsol. obsolescent • obsolete

o.b.s.p. obiit sine prole (Latin: died without issue)

obst. oboist

obstet. obstetric(s) • obstetrician

obstn obstruction

obt obedient

obtd obtained

OBU offshore banking unit • One Big Union

obv. obverse

OBV ocean boarding vessel

o.c. office copy • official classification • *Architect.* on centre • (*or* **oc**) only child • *Shipping* open charter • open cover • opere citato (*see* op. cit.)

o/c officer commanding • overcharge

Oc. Ocean

OC *British vehicle registration for* Birmingham • Observer Corps • Officer Commanding • *Electrical engineering* operating characteristic • operations centre • oral contraceptive • (Officer of the) Order of Canada • Orienteering Club • *Philately* original cover • Oslo Convention • Ottawa Convention • overseas command • overseas country

oca. *Music* ocarina

OCA Old Comrades Association

OCAM (*or* **OCAMM**) Organisation commune africaine et malgache (*or* africaine, malgache, et mauritienne) (French: African and Malagasy (*or* African, Malagasy, and Mauritian) Common Organization)

O.Carm. (of the) Order of Carmelites

O.Cart. (of the) Order of Carthusians

OCAS Organization of Central American States

OCAW (USA) Oil, Chemical, and Atomic Workers International Union

Oc.B/L *Shipping* ocean bill of lading

occ. (*or* **occn**) occasion • (*or* **occas.**) occasional(ly) • occident(al) • occupation • occurrence

OCD *Med.* obsessive compulsive disorder • (USA) Office of Civil Defense • *Computing* on-line communications driver • Ordo (*or* Ordinis) Carmelitarum Discalceatorum (Latin: (of the) Order of Discalced Carmelites; Discalced Carmelite Fathers)

OCDE Organisation de coopération et de développement economiques (French: Organization for Economic Cooperation and Development; OECD)

OCDM (USA) Office of Civil Defense Mobilization (1958–61)

OCDS (*or* **ocds Can**) (Canada) Overseas College of Defence Studies

O/Cdt Officer-Cadet

oceanog. oceanography

OCelt Old Celtic

OCF Officiating Chaplain to the Forces

och. ochre

OCH oil central heating (in property advertisements)

OCNL *Meteorol.* occasional(ly)

OCorn Old Cornish

OCP *Chem.* organic conducting polymer

OCPSF organic chemicals, plastics, and synthetic fibers

OCR *Computing* optical character recognition (*or* reader) • Ordo (*or* Ordinis) Cisterciensium Reformatorum (Latin: (of the) Order of Reformed Cistercians; Trappists)

OCS (USA) Officer Candidate School • Old Church Slavonic • outer continental shelf

OCSO (of the) Order of Cistercians of the Strict Observance (Trappists)

ocst. *Meteorol.* overcast

oct. *Music* octave • *Printing* octavo

Oct *Astronomy* Octans • (*or* **Oct.**) October

OCTU ('ɒktuː) Officer Cadet Training Unit

OCTV open-circuit television

OCU Operational Conversion Unit

OCUC Oxford and Cambridge University Club

OCV open-circuit voltage

od. oder (German: or)

o.d. *Med.* oculus dexter (Latin: right eye) • *Military* olive drab • outer (*or* outside) diameter

OD *Med.* oculus dexter (Latin: right eye) • Doctor of Optometry • *British vehicle registration for* Exeter • *international civil aircraft marking for* Lebanon • officer of the day • Old Dutch (language) • *Military* olive drab • operations division • (Jamaica) (Officer of the) Order of Distinction • ordinary seaman • Ordnance datum (standard sea level of the Ordnance Survey) • ordnance department • ordnance depot • organization development • other denominations • outer (*or* outside) diameter • (*or* **o/d**) overdose

O/D (*or* **O/d, o/d**) on deck • *Banking* on demand • *Banking* overdraft (*or* overdrawn)

ODA *Chem.* 4,4′-oxydianiline • *Computing* open (formerly office) document architecture • *Med.* Operating Department Assistant • Overseas Development Administration (former government department; replaced by the Department of International Development)

ODan Old Danish

ODAS Ocean Data Station

ODC (of the) Order of Discalced Carmelites

ODCh Chaplain for Other Denominations

ODE *Maths.* ordinary differential equation

ODECA Organización de estados centroamericanos (Spanish: Organization of Central American States)

ODESSA (əu'dɛsə) Organisation der SS-Angehörigen (German: Organization of SS members)

ODI *Computing* open datalink interface • Overseas Development Institute

ODO outdoor officer (at customs)

ODP official (*or* overall) development planning • *Computing* open distributed processing • open-door policy

ODS *Metallurgy* oxide dispersion strengthening

ODV *Colloquial* eau de vie (French: cognac)

o.e. omissions excepted • open end

Oe *Physics, symbol for* oersted

OE *international civil aircraft marking for* Austria • *British vehicle registration for* Birmingham • (USA) Office of Education • Old English (language) • Old Etonian • *Med.* on examination • (Guyana) Order of Excellence • original error • outboard engine

OEA Overseas Education Association • oxygen-enriched air

OEC oxygen-enriched combustion

OECD Organization for Economic Cooperation and Development

OECS Organization of Eastern Caribbean States

OED (ital.) Oxford English Dictionary

OEEC Organization for European Economic Cooperation (superseded by OECD)

OEF Organization of Employers' Federations and Employers in Developing Countries

OEL occupational exposure limit (of radiation)

OEM *Computing* original equipment manufacturer

OEO (USA) Office of Economic Opportunity

OEP (USA) Office of Economic Preparedness

OER Officers' Emergency Reserve • Organization for European Research • *Chem.* oxygen enhancement ratio

OES ocean (wave and tidal) energy system(s) • (USA) Office of Economic Stabilization • *Physics* optical emission spectroscopy • Order of the Eastern Star • Organization of European States

o.f. optional form • outside face

OF *British vehicle registration for* Birmingham • Oddfellows • oil-filled • oil-fired • *Printing* old-face (type) • Old French (language) • operational (*or* operating) forces • Order of the Founder (of the Salvation Army) • oxidizing flame • *airline flight code for* Travaux Aériens de Madagascar

OFA *Med.* oncofetal antigen

O factor *Psychol.* oscillation factor

OFB *Computing* output feedback

OFEMA Office français d'exportation de matériel aéronautique (French: Office for the Export of Aeronautical Material)

off. offer(ed) • office • (*or* offr) officer • (*or* offcl) official • officinal

OFFER (*or* Offer) ('ɒfə) Office of Electricity Regulation

offic. official(ly)

OFGAS (*or* Ofgas) ('ɒf,gæs) Office of Gas Supply

OFHC *Engineering* oxygen-free high-conductivity copper

OFlem Old Flemish

OFLOT (*or* Oflot) ('ɒf,lɒt) Office of the National Lottery

OFM (of the) Order of Friars Minor (Franciscans)

OFMCap. (of the) Order of Friars Minor Capuchin (Franciscan order)

OFMConv. (of the) Order of Friars Minor Conventual (Franciscan order)

OFr Old French

OFR Order of the Federal Republic of Nigeria

OFris Old Frisian

OFS Orange Free State

OFSTED (*or* Ofsted) ('ɒf,stɛd) Office for Standards in Education

OFT Office of Fair Trading

OFTEL (*or* **Oftel**) ('ɒf,tɛl) Office of Telecommunications

OFWAT (*or* **Ofwat**) ('ɒf,wɒt) Office of Water Services

o.g. *Philately* original gum · *Sport* own goal

OG *British vehicle registration for* Birmingham · *Astronomy* object glass · Officer of the Guard · *Architect.* ogee · Olympic Games · (USA) original gangster (used in the criminal world for a respected elder) · *Brewing* original gravity · *Philately* original gum · outside guard

OGael Old Gaelic

ÖGB Österreichischer Gewerkschaftsbund (German: Austrian Federation of Trade Unions)

OGCM *Meteorol.* ocean general circulation model

OGD *Med.* oesophagogastroduodenoscopy

OGL *Commerce* open general licence

OGM ordinary general meeting

OGO Orbiting Geophysical Observatory

Ogpu (*or* **OGPU**) ('ɒgpuː) Obyedinyonnoye Gosudarstvennoye Politicheskoye Upravleniye (Russian: United State Political Administration; Soviet state security system, 1923–34)

OGS Oratory of the Good Shepherd · *Electronics* oxide glassy (*or* glasslike) semiconductor

o.h. observation helicopter · office hours · *Med.* omni hora (Latin: hourly) · on hand

OH *British vehicle registration for* Birmingham · *international civil aircraft marking for* Finland · *US postcode for* Ohio · *Physics* ohmic heating · *Numismatics* old head (of Victoria)

OHBMS On Her (*or* His) Britannic Majesty's Service

OHC (*or* **o.h.c.**) overhead cam (*or* camshaft)

OHDETS ('əʊdɛtz) over-horizon detection system

OHG *Commerce* Offene Handelsgesellschaft (German: partnership) · Old High German

OHMS On Her (*or* His) Majesty's Service

OHN occupational health nurse

OHNC occupational health nursing certificate

OHP overhead projector

OHS occupational health service · *Metallurgy* open-hearth steel

OHV (*or* **o.h.v.**) overhead valve

OI *British vehicle registration for* Belfast ·

office instruction · Old Irish (language) · operating instructions · *Med.* osteogenesis imperfecta

OIC (*or* **O i/c**) officer in charge · Organisation internationale du commerce (French: International Trade Organization; ITO) · Organization of Islamic Conference

OIcel Old Icelandic

OIEO offers in excess of

OIG organisation intergouvernementale (French: intergovernmental organization)

OII *Physics* optical imaging instrument

OIPC Organisation internationale de police criminelle (French: International Criminal Police Organization; Interpol)

OIr Old Irish

OIRO offers in the region of

OIS organizer industrial safety · overnight indexed swap

OIt Old Italian

OIT Organisation internationale du travail (French: International Labour Organization; ILO)

o.j. open joist (*or* joint)

OJ *British vehicle registration for* Birmingham · Official Journal (of the EU) · (USA, Australia) orange juice · Order of Jamaica

OJAJ October, January, April, July (months for quarterly payments)

OJCS Office of the Joint Chiefs of Staff

OJR old Jamaica rum

OJT on-the-job training

OK all correct (from *orl korrect,* phonetic spelling) · *British vehicle registration for* Birmingham · *airline flight code for* Czech Airlines · *international civil aircraft marking for* the Czech Republic · *US postcode for* Oklahoma · *Chem.* oxidation kinetics

o.k.a. otherwise known as

OKE *Physical chem.* optical Kerr effect

OKH Oberkommando der Heeres (German Army High Command, World War II)

Okla. Oklahoma

Okt. Oktober (German: October)

OKTA *Meteorol., indicating* one-eighth of the sky area

ol. *Med.* oil (Latin *oleum*) · olive

Ol. Olympiad · Olympic

OL *British vehicle registration for* Birmingham · (*or* **o.l.**) *Med.* oculus laevus (Latin: left eye) · *postcode for* Oldham · Old

Latin • *airline flight code for* OLT-Ostfriesische Lufttransport • *Computing* on-line • operating licence • (Officer of the) Order of Leopold • Ordnance Lieutenant • *Sport* outside left • overflow level • overhead line

Old Test. Old Testament

OLE *Computing* object linking and embedding

O level *Education* Ordinary level (replaced by GCSE)

OLG Old Low German

Oli *Geology* Oligocene

OLQ officer-like qualities

OLR off-line reader

OLRT *Computing* on-line real time

OLS *Statistics* ordinary least squares

o.m. old measurement • *Med.* omni mane (Latin: every morning)

Om. Oman

OM *British vehicle registration for* Birmingham • *airline flight code for* MIAT-Mongolian Airlines • old man • optical microscopy • Optimus Maximus (Latin: greatest and best; title given by Romans to Jupiter) • Order of Merit • ordnance map • organic matter • *international civil aircraft marking for* Slovakia

OMA *Physics* optical multichannel analyser • ('əumə) *Finance* (USA) orderly marketing agreement

OMB (USA) Office of Management and Budget

OMC operation and maintenance costs • *Chem.* organometallic compound

OMC-1 *Astronomy* Orion molecular cloud

OMCS Office of the Minister for the Civil Service

OMI Oblate(s) of Mary Immaculate

OMM (Canada) Officer of the Order of Military Merit

omn. hor. *Med.* omni hora (Latin: every hour)

omn. noct. *Med.* omni nocte (Latin: every night)

OMO one-man operator *or* operation (of buses)

OMPA *Chem.* octamethyl pyrophosphoramide

OMR *Computing* optical mark reading

o.m.s. output per man shift

OMS *Computing* object management system • *Astronautics* orbital manoeuvring system • Organisation mondiale de la santé (French: World Health Organization; WHO)

OMT *Computing* object modelling technique

OMV open-market value

o.n. *Med.* omni nocte (Latin: every night)

ON *airline flight code for* Air Nauru • *British vehicle registration for* Birmingham • octane number • Old Norse • Ontario • (Jamaica) Order of the Nation • orthopaedic nurse

ONC Ordinary National Certificate • Orthopaedic Nursing Certificate

OND Ophthalmic Nursing Diploma • Ordinary National Diploma

ONERA Office national d'études et de recherches aerospatiales (French: National Office of Aerospace Study and Research)

ONF Old Norman (*or* Northern) French

ONG organisation non-gouvernementale (French: nongovernmental organization)

ONGC (USA) Oil and Natural Gas Commission

ONI Office of Naval Intelligence

o.n.o. or near(est) offer

onomat. onomatopoeia • onomatopoeic

ONorw Old Norwegian

ONR (USA) Office of Naval Research

ONS Office for National Statistics

Ont. Ontario

ONZ Order of New Zealand

o/o (*or* **o.o.**) on order • order of

O/o offers over

OO *international civil aircraft marking for* Belgium • *British vehicle registration for* Chelmsford • Observation Officer • *Colloquial* once-over (i.e. a preliminary inspection) • operation order • Orderly Officer

OOD *Computing* object-oriented design • officer of the day • officer of the deck

OODBMS *Computing* object-oriented database management system

OOG officer of the guard

OOL *Computing* object-oriented language

OON Officer of the Order of the Niger

OOP *Computing* object-oriented programming

OOT out of town

OOW officer of the watch

op. opaque • opera (Latin: works) • operation • operator • opinion • opposite • optical • optimus (Latin: excellent) • opus (Latin: a work; *see also* Op.)

o.p. *Theatre* opposite prompt side (i.e. actor's right) • out of print • overproof (of alcohol)

Op. *Music* Opus (preceding a number; indicating a piece by a particular composer)

OP *British vehicle registration for* Birmingham • *Military* observation post • Old Persian (language) • old prices • *Insurance* open policy • *Theatre* opposite prompt side (i.e. actor's right) • Ordo (*or* Ordinis) Praedicatorum (Latin: (of the) Order of Preachers; Dominicans) • organophosphate • osmotic pressure • *Colloquial* other people (*or* people's; *see also under* OPM; OP's) • out of print • outpatient • overproof (of alcohol) • *Chem.* oxide precipitate

OPA (USA) Office of Price Administration (during World War II) • *Meteorol.* opaque (rime ice)

op-amp *Electronics* operational amplifier

op art optical art

OPAS Occupational Pensions Advisory Service

OPB Occupational Pensions Board

OPC *Computing* optical photoconductor • (*or* **opc**) ordinary Portland cement • outpatients' clinic • Overseas Press Club of America

op. cit. opere citato (Latin: in the work cited; used in textual annotations)

OPCON ('ɒp,kɒn) operational control

OPCS Office of Population Censuses and Surveys

OPD outpatients' department

OPDAR ('ɒp,dɑ:) optical detection and ranging

OPE *Chem.* organic phosphate ester

OPEC ('əʊ,pɛk) Organization of Petroleum-Exporting Countries

Op-Ed ('ɒp,ɛd) (USA) opposite editorial (page; of a newspaper)

OPEIU (USA) Office and Professional Employees International Union

OPEP Organisation des pays exportateurs de pétrole (French: Organization of Petroleum Exporting Countries; OPEC)

OPers Old Persian

OPEX ('əʊ,pɛks) operational, executive, and administrative personnel (in the UN)

OPg Old Portuguese

OPG *Dentistry* orthopantomogram

Oph *Astronomy* Ophiuchus

ophthal. (*or* **oph.**) ophthalmic • (*or* **ophthalmol.**) ophthalmologist • (*or* **ophthalmol.**) ophthalmology

OPIC (USA) Overseas Private Investment Corporation

opl operational

OPM (USA) Office of Personnel Management • operations per minute • *Colloquial* (USA) other people's money • output per man

OPMA Overseas Press and Media Association

opn operation • opinion • option

o.p.n. ora pro nobis (Latin: pray for us)

OPO one-person operator *or* operation (of buses)

OPol Old Polish

opp. (*or* **Opp.**) opuses (*or* opera; *see also under* Op.) • opportunity • opposed • opposite • opposition

OPP oriented polypropene (in **OPP film**, used for packaging) • out of print at present

oppy opportunity

OPQ occupational personality questionnaire

opr. operate • operator

OPr (*or* **OProv**) Old Provençal

OPRAF ('ɒpræf) Office of Passenger Rail Franchising

OPruss Old Prussian

ops. operations

OP's *Colloquial* other people's (as in **OP's cigarettes**)

OPS Office of Public Service

opt. *Grammar* optative • optical • optician • optics • optimal • optimum • optional

OPT optimized production technology • *Chem.* orientation phase transition

OptD Doctor of Optometry

OQ Officer of the National Order of Quebec • *airline flight code for* Zambian Express Airways

or. orange • oratorio • orient(al) • original • other

o.r. operationally risky • operational (*or* operations) requirement • operations room • out of range • overhaul and repair • *Insurance* owner's risk

Or Oriel College, Oxford

OR *Statistics* odds ratio • official receiver • official referee • Old Roman • *Med.* (USA) operating room • operational (*or* operations) requirement • operational (*or*

operations) research • *Electronics* or (from its sense in logic, as in **OR gate**, **OR operation**) • orderly room • *US postcode for* Oregon • *Military* other ranks • *Sport* outside right • *Insurance* owner's risk • *British vehicle registration for* Portsmouth

ÖR Österreichischer Rundfunk (Austrian broadcasting service)

Oracle ('ɒrək°l) optional reception of announce!ments by coded line electronics (teletext service of Independent Television)

orat. oration • orator • oratorical(ly) • oratorio • oratory

o.r.b. *Insurance* owner's risk of breakage

ORB oceanographic research buoy • omnidirectional radio beacon

ORBIS ('ɔːbɪs) orbiting radio beacon ionospheric satellite

ORC (USA) Officers' Reserve Corps • Overseas Research Council

ORCA Ocean Resources Conservation Association

orch. orchestra(l) • (*or* **orchd**) orchestrated by • orchestration

ord. ordain(ed) • order • ordinal • ordinance • ordinary • ordnance

o.r.d. *Insurance* owner's risk of damage

ORD *Chem.* once-run distillate • *Chem.* optical rotary dispersion

ordn. ordnance

Ore. Oregon

ORE occupational radiation exposure

Oreg. Oregon

ORELA (USA) Oak Ridge Electron Linear Accelerator

ORESCO (ɒ'rɛskəʊ) Overseas Research Council

o.r.f. *Insurance* owner's risk of fire

ORF *Genetics* open reading frame

org. organ • organic • organism • organist • organization • organized

ORGALIME Organisme de liaison des industries métalliques européennes (French: Liaison Group for the European Metal Industries)

organ. organic • (*or* **orgzn**) organization

orgst. organist

Ori *Astronomy* Orion

orient. oriental • orientalist

orig. origin • original(ly) • originate(d)

ORIT Organización regional interamericana de trabajadores (Spanish: Inter-American Regional Organization of Workers)

Ork. (*or* **Orkn.**) Orkney Islands

o.r.l. *Insurance* owner's risk of leakage

ORL *Med.* otorhinolaryngology (ear, nose, and throat specialty; ENT)

orn. ornament • ornithology

ornithol. (*or* **ornith.**) ornithological • ornithology

ORNL (USA) Oak Ridge National Laboratory

OROM *Computing* optical read-only memory

orph. orphan(age)

o.r.r. *Insurance* owner's risk rates

ORR Office of the Rail Regulator

ors. others

ORS *Geology* Old Red Sandstone • Operational Research Society

ORSL Order of the Republic of Sierra Leone

ORSORT Oak Ridge School of Reactor Technology

ORT Ooty Radio Telescope • *Med.* oral rehydration therapy • (USA) Organization for Rehabilitation by Training

ORTF (France) Office de Radiodiffusion-Télévision Française (former state radio and television service)

orth. orthography • orthopaedic(s)

Orth. Orthodox (religion)

ORuss Old Russian

ORV (USA) off-road vehicle

o.s. ocean station • *Med.* oculus sinister (Latin: left eye) • oil switch • old series • (*or* **os**) only son • on station • outside (measurement)

o/s on sale • out of service • (*or* **O/s**) out of stock • outsize (of clothing) • (*or* **O/s**) *Banking* outstanding

Os *Chem., symbol for* osmium

OS *airline flight code for* Austrian Airlines • *British vehicle registration for* Glasgow • *Med.* oculus sinister (Latin: left eye) • Old Saxon (language) • Old School • old series • Old Side • Old Style (method of reckoning dates) • *Computing* operating system (*see also* OS/2) • Ordinary Seaman • Ordnance Survey • out of stock • outsize (of clothing)

OS/2 *Computing* Operating System/2 (produced by IBM and Microsoft)

OSA (USA) Office of the Secretary of the Army • Official Secrets Act • *Typography* old style antique • (of the) Order of St Augustine (Augustinians) • Overseas Sterling Area

OSAF (USA) Office of the Secretary of the Air Force

OS&W *Building trades* oak, sunk and weathered

OSax Old Saxon

OSB (of the) Order of St Benedict (Benedictines)

osc *Military* (graduate of) overseas staff college

osc. oscillator

OSC (of the) Order of St Clare (Poor Clares) • *Chem.* organic sulphur compound

OScan (*or* **OScand**) Old Scandinavian

OSCAR ('ɒskə) Orbital Satellites Carrying Amateur Radio • Organization for Sickle Cell Anaemia Research

OSD (USA) Office of the Secretary of Defense • (of the) Order of St Dominic (Dominicans)

OSE operational support equipment

OSerb Old Serbian

OSF *Computing* Open Software Foundation • (of the) Order of St Francis (Franciscans)

OSFC (of the) Order of St Francis, Capuchin

OSHA ('əʊʃə) (USA) Occupational Safety and Health Administration

OSI (USA) Office of Scientific Integrity • on-site inspection • *Computing* open systems interconnection

O/Sig Ordinary Signalman

Osl. Oslo

OSl (*or* **OSlav**) Old Slavonic

OSL Old Style Latin

OSM (of the) Order of the Servants of Mary (Servites)

OSN (USA) Office of the Secretary of the Navy

OSNC Orient Steam Navigation Company

OSO orbiting solar observatory

osp off-street parking (in property advertisements)

o.s.p. obiit sine prole (Latin: died without issue)

OSp Old Spanish

OSR (USA) Office of Science and Research

OSRB Overseas Service Resettlement Bureau

OSRD (USA) Office of Scientific Research and Development

OSRO Office for Special Relief Operations (in the FAO)

OSS Office for the Supervision of Solicitors • (USA) Office of Space Sciences (in NASA) • (USA) Office of Strategic Services (during World War II)

O.SS.S Ordo (*or* Ordinis) Sanctissimi Salvatoris (Latin: (of the) Order of the Most Holy Saviour; Bridgettines)

O.SS.T Ordo (*or* Ordinis) Sanctissimae Trinitatis Redemptionis Captivorum (Latin: (of the) Order of the Most Holy Trinity for the Redemption of Captives; Trinitarians)

OST (USA) Office of Science and Technology

osteo. osteopath(ic) • osteopathy

OSTI (USA) Office of Scientific and Technical Information • (USA) Organization for Social and Technological Innovation

OStJ Officer of the Order of St John of Jerusalem

OSU (of the) Order of St Ursula (Ursulines)

OSUK Ophthalmological Society of the United Kingdom

OSw Old Swedish

OT *airline flight code for* Evergreen Alaska • occupational therapy (*or* therapist) • off time • Old Testament • *Med.* operating theatre • (Australia) Overland Telegraph (from Adelaide to Darwin) • overseas trade • overtime • *British vehicle registration for* Portsmouth

OTA (USA) Office of Technology Assessment • *Chem.* organic trace analysis

OTAN Organisation du traité de l'Atlantique nord (French: North Atlantic Treaty Organization; NATO)

OTASE Organisation du traité de défense collective pour l'Asie du sud-est (French: South-East Asia Treaty Organization; SEATO)

OTB (USA) off-track betting • oxide titanium bronze

otbd outboard

OTC (USA) officer in tactical command • Officers' Training Corps • Officers' Transit Camp • one-stop- inclusive tour charter • Organization for Trade Cooperation • over the counter (as in **OTC market** (in securities), **OTC medicines**) • oxytetracycline (an antibiotic)

OTE on-target earnings (for a salesman)

OTEC ('əʊtɛk) ocean thermal-energy conversion

OTeut Old Teutonic

OTF *Photog.* off the film

OTG outside temperature gauge (in motor advertisements)

OTH *Telecom.* over the horizon (as in **OTH radar**)

otol. otological · otology

OTP *Meteorol.* on top

OTP EPRom *Computing* one-time programmable erasable programmable read-only memory

OTR *Physics* optical transition radiation

OTS Office of Technical Services · Officers' Training School · *Advertising* opportunities to see

ott. *Music* ottava (Italian: octave) · ottobre (Italian: October)

Ott. Ottawa

OTT *Med.* oesophageal transit test · *Colloquial* over the top (i.e. excessive)

OTU *Biology* operational taxonomic unit · operational training unit

OTurk Old Turkish

OU *British vehicle registration for* Bristol · *airline flight code for* Croatia Airlines · official use · Open University · Oxford University

OUAC Oxford University Appointments Committee · Oxford University Athletic Club

OUAFC Oxford University Association Football Club

OUBC Oxford University Boat Club

OUCC Oxford University Cricket Club

OUDS ('əʊ 'juː 'diː 'es *or* aʊdz) Oxford University Dramatic Society

OUP (Northern Ireland) Official Unionist Party · Oxford University Press

OURC Oxford University Rifle Club

OURFC Oxford University Rugby Football Club

OURT Order of the United Republic of Tanzania

out. outlet

outbd outboard

ov. ovary · over · overture

Ov. Ovid (43 BC–AD 17, Roman poet)

OV *British vehicle registration for* Birmingham · *airline flight code for* Estonian Air

ovbd overboard

ovc. (*or* **OVC**) *Meteorol.* overcast

o.v.c. other valuable consideration (in contract law)

ovfl. overflow

OVH overhead projector

ovhd overhead

ovld overload

o.v.n.o. or very near offer

ÖVP (Austria) Österreichische Volkspartei (People's Party)

ovpd overpaid

Ovra Opera di vigilanza e di repressione dell'anti-fascismo (Italian secret police of the Fascist regime)

ovrd. override

OVV *Astronomy* optically violently variable (in **OVV quasars**)

o.w. one way · out of wedlock

OW Office of Works · Old Welsh (language) · *British vehicle registration for* Portsmouth

O/W oil in water (emulsion)

OWC Ordnance Weapons Command

OWF optimum working frequency

OWI (USA) Office of War Information (during World War II) · (USA) operating (a motor vehicle) while intoxicated

OWRS (USA) Office of Water Regulations and Standards

OWS ocean weather service · ocean weather ship (*or* station)

ox *Chem.* oxalate ion (used in formulae)

Ox. Oxford

OX *British vehicle registration for* Birmingham · *postcode for* Oxford

Oxbridge ('ɒks,brɪdʒ) Oxford and Cambridge (Universities; regarded collectively)

Oxf. Oxford(shire)

OXFAM (*or* **Oxfam**) ('ɒksfæm) Oxford Committee for Famine Relief

Oxon Oxfordshire (Latin *Oxonia*) · Oxoniensis (Latin: of Oxford)

OY *international civil aircraft marking for* Denmark · *British vehicle registration for* London NW

oz (*or* **oz.**) ounce(s) (Italian *onza*)

OZ *airline flight code for* Asiana Airlines · *British vehicle registration for* Belfast

oz ap apothecaries' ounce

oz av (*or* **oz avdp**) avoirdupois ounce

oz T troy ounce

P

p (bold ital.) *Physics, symbol for* electric dipole moment • (ital.) *Physics, symbol for* momentum (bold ital. in vector equations) • (ital.) *Chem.* para (as in **p-cresol**) • *symbol for* penny (*or* pence) • (bold ital.) *Chem., symbol for* permanent dipole moment of a molecule • *Biochem., symbol for* (terminal) phosphate (in a polynucleotide) • (ital.) *Music* piano (Italian; softly, quietly) • *symbol for* pico- (prefix indicating 10^{-9} as in **ps**, picosecond) • (ital.) *Physics, symbol for* pressure • *Physics, symbol for* proton • *Electronics* p-type (semiconductor) • (ital.) *Biochem., symbol for* pyranose • *Meteorol., symbol for* shower • *Physics, Chem., indicating* the electron state $l=1$ (where l is orbital angular momentum quantum number)

p. page • pamphlet • paragraph • part • participle • particle • partim (Latin: in part) • pass(ed) • *Nautical* passing showers • past • peak • *Ichthyol.* pectoral • per (Latin: by, for) • perch (solid measure for stone) • *Music* percussion • *Grammar* person • *Currency* piastre • pied (French: foot) • pint • pipe • pius (Latin: holy) • polar • pole • pondere (Latin: by weight) • population • port • post (Latin: after) • pouce (French: inch) • pour (French: for *or* per) • primus (Latin: first) • pro (Latin: for, in favour of) • professional • *Knitting* purl

P (bold ital.) *Physics, symbol for* dielectric polarization • *medieval Roman numeral for* four hundred • *international civil aircraft marking for* North Korea • *Genetics, symbol for* parental generation • (ital.) *Physics, symbol for* parity • *symbol for* parking (on road signs) • *Education* (USA) passing (performance rating) • *Chess, symbol for* pawn • pedestrian crossing • *Botany* perianth (in a floral formula) • *Geology* Permian • *symbol for* peta- (prefix indicating 10^{-15} as in **PJ**, petajoule) • pharmacy (on medicines obtained without a prescription from a pharmacy) • *Biochem., symbol for* phosphate • *Chem., symbol for* phosphorus • *Botany, symbol for* phytochrome • *Physics, symbol for* poise • *Computing* polynomial (referring to a class of formal languages recognizable in polynomial time) • *British fishing port registration for* Portsmouth • *international vehicle registration for* Portugal • Post Office (on maps) • (ital.) *Physics, symbol for* power • (ital.) *Physics, symbol for* pressure • *Biochem., symbol for* proline • (ital.) *Genetics, symbol for* promoter • proprietary (product) • *symbol for* pula (monetary unit of Botswana)

P. (*or* **P**) *RC Church* Papa (Latin: Pope) • parson • Pastor • pater (Latin: father) • *Music* pedal • *Ecclesiast.* Père (French: Father) • *Horticulture* perennial • period • personnel • pitch • populus (Latin: people) • positive • post • postage • *Med.* posterior • Presbyterian • President • Priest • Prince • probate • pro-consul • Progressive (party, movement, etc.) • *Theatre* prompt side (i.e. the actor's left) • Protestant • public • pupil

P2 *international civil aircraft marking for* Papua New Guinea • Propaganda due (Italian masonic lodge)

P4 *international civil aircraft marking for* Aruba

P5 *airline flight code for* AeroRepublica

P6 *airline flight code for* Trans Air

P45 *indicating* a form relating to unemployment benefit issued by the DSS

7P *international civil aircraft marking for* Lesotho

8P *international civil aircraft marking for* Barbados • *airline flight code for* Pacific Coastal Airlines

pa. past

p.a. participial adjective • per annum • permanent address • personal appearance • press agent

p/a *Book-keeping* personal account

p.A. per Adresse (German: care of)

Pa *Physics, symbol for* pascal • *Chem., symbol for* protactinium

Pa. Pennsylvania

PA *British vehicle registration for* Guildford • *postcode for* Paisley • Pakistan Army • *airline flight code for* Pan Air Lineas Aereas • *international vehicle registration for* Panama • Parents' Association • (*or* **P/A**) *Insurance* particular average • *US postcode for* Pennsylvania • performance assessment • *Med.* pernicious anaemia • *Insurance* personal accident • *Accounting* personal account • *Taxation* personal allowance • personal appearance • personal assistant • *Chem.* phosphoric acid • (USA) physician assistant • Pierre Allain (climbing boot; named after its inventor) • Piper Aircraft (as in **PA-28**) • *Engineering* pitch angle • *Med.* plasminogen activator • political agent • por autorización (Spanish: by authority of) • *Astronomy* position angle • *Military* Post Adjutant • power amplifier • (*or* **P/A**) power of attorney • *RC Church* Prefect Apostolic • press agent • Press Association • press attaché • (*or* **P/A**) *Banking, Book-keeping* private account • product analysis • programme assistant • *Med.* psoriatic arthritis • (USA) public accountant • public address (system) • publicity agent • Publishers Association • *Engineering* pulsed annealing • purchasing agent • (USA) Prosecuting Attorney

PAA *Chem.* peracetic acid • *Physics* photon activation analysis • *Chem.* polyacrylic acid • *Chem.* primary aromatic amine

PAADC Principal Air Aide-de-camp

PAB *Chem.* *p*-aminobenzamidine

PABA ('pɑːbə) *Biochem., Pharmacol.* *p*-aminobenzoic acid

PABLA ('pæblə) problem analysis by logical approach

PABP *Biochem.* poly(A)-binding protein

PABX *Telephony* private automatic branch exchange

pac passed (final examination of the) advanced class (of the Military College of Science)

Pac. Pacific

PAC *US Air Force* Pacific Air Command • Pan-African(ist) Congress • Pan-American Congress • *Building trades* passive air cycle • (pæk) (USA) political action committee • *Chem.* polycyclic aromatic compound • polymer–asphalt composite • *Chem.* powdered activated carbon • Public Accounts Committee • Public Assistance Committee • *Stock exchange* put-and-call (option)

P-A-C *Psychol.* parent, adult, child (in transactional analysis)

PACAF *US Air Force* Pacific Air Forces

PACE (peɪs) performance and cost evaluation • Police and Criminal Evidence Act (1984) • precision analogue computing equipment • Protestant and Catholic Encounter

Pacif. Pacific

PACOM ('pækɒm) (USA) Pacific Command

PACS (pæks) Pacific area communications system • *Med.* picture archiving and communication system

PACT (*or* **Pact**) (pækt) Producers' Alliance for Cinema and Television

PaD Pennsylvania Dutch

PAD (pæd) *Computing* packet assembler/disassembler • (pæd) passive air defence • payable after death

PADAR ('peɪdɑː) *Military* passive detection and ranging

PADLOC ('pædlɒk) *Military* passive detection and location of countermeasures

p. Adr. per Adresse (German: care of; c/o)

p. ae. *Pharmacol.* partes aequales (Latin: equal parts)

p.a.f. puissance au frein (French: brake horsepower)

PAF *Computing* peripheral address field • *Med.* platelet-activating factor • (Northern Ireland) Protestant Action Force

PAFC phosphoric acid fuel cell

PaG Pennsylvania German

PAg Professional Agronomist

PAGB Proprietary Association of Great Britain

PAGE *Biochem.* polyacrylamide gel electrophoresis

PAH *Chem.* polycyclic aromatic hydrocarbon

PAHO Pan American Health Organization

PAI *Med.* platelet accumulation index

PAICV (Cape Verde) Partido Africano da Independencia de Cabo Verde (Portuguese: African Party for the Independence of Cape Verde)

PAIS (USA) Public Affairs Information Service

Pak. Pakistan(i)

Pak Re Pakistan rupee (monetary unit)

pal. palaeography • palaeontology

Pal *Geology* Palaeocene

Pal. Palace • Palestine

PAL (pæl) (USA) Parcel Air Lift • *Computing* peripheral availability list • *Television* phase alternation (*or* alternating) line • *Botany* phenylalanine ammonia lyase • Philippine Airlines • (USA) Police Athletic League • present atmospheric level • *Computing, Electronics* programmable array logic

palaeob. (*or* **palaeobot.**) palaeobotanical • palaeobotany

palaeog. palaeographical • palaeography

palaeontol. palaeontology

PALS (pælz) *Computing* permissive action link systems

pam. (*or* **pamph.**) pamphlet

PAM (pæm) *Telecom.* pulse-amplitude modulation

PAMA Pan-American Medical Association • Press Advertisement Managers' Association

pan. panchromatic • panoramic • pantomime • pantry

Pan. Panama

PAN (Mexico) Partido Acción Nacional (Spanish: National Action Party) • peroxyacetyl nitrate (atmospheric pollutant) • polyacrylonitrile (polymer)

PANAFTEL (pæn'æftɛl) Pan-African Telecommunications Network

Pan Am Pan American World Airways

Pan. Can. Panama Canal

P&E plant and equipment

P & F chart point-and-figure chart

P & G Procter and Gamble (pharmaceutical company)

P & L profit and loss (as in **P & L account**)

P & O Peninsular and Oriental (Steamship Company)

P&OSNCo Peninsular and Oriental Steam Navigation Company

p & p postage and packing

P&RT physical and recreational training

P&S *Stock exchange* (USA) purchase and sales

PANH *Chem.* polycyclic aromatic nitrogen heterocyclic (compound)

PANS procedures for air navigation services

PAO *Med.* peak acid output • Prince Albert's Own (regiment) • public affairs officer

Pap. Papua(n)

PAP (Singapore) People's Action Party • Polska Agencja Prasowa (Polish Press Agency) • *Med.* pulmonary arterial pressure

PAPS (pæps) periodic armaments planning system

par. paragraph • parallax • parallel • paraphrase • parenthesis • parish • parochial

p.a.r. planed all round (in woodworking)

Par. Paraguay

PAR (pɑː) *Military* perimeter acquisition radar • phased-array radar • *Botany* photosynthetically active radiation • *Computing* positive acknowledgment and retransmission • *Aeronautics* precision approach radar • programme analysis review • *Electronics* pulse acquisition radar

para. paragraph

parab. parabola

par. aff. *Pharmacol.* pars affecta (Latin: (to the) part affected)

parch. parchment

Par. Ch. parish church

paren. parenthesis

parens. parentheses

Parl. Parliament • (*or* **parl.**) parliamentary

parl. agt. parliamentary agent

parl. proc. parliamentary procedure

Parly Sec. Parliamentary Secretary

PARM programme analysis for resource management

part. partial • participate • participial • participle • particle • particular • partition • partner(ship)

part. aeq. *Pharmacol.* partes aequales (Latin: in equal parts)

pas. *Grammar* passive

p.a.s. power-assisted steering

PAS *Med.* p-aminosalicylic acid (tuberculosis treatment) • *Chem.* periodic acid–Schiff (as in **PAS reaction**) • *Physics* plasma activated source • *Physics* positron annihilation spectroscopy • public-address system

PASCAL (*or* **Pascal**) ('pæs,kæl) *Computing, indicating* a programming language (named after Blaise Pascal (1623–62), French philosopher, mathematician, and physicist)

PASH (pæʃ) *Chem.* polynuclear aromatic sulphur heterocyclic (compound)

Pasok Panhellenic Socialist Movement

pass. passage • passenger • passim (Latin:

here and there throughout) • *Grammar* passive

Pass. Passover

PASSIM ('pæsɪm) (USA) Presidential Advisory Staff on Scientific Management

pat. patent(ed) • pattern

p.a.t. *Chem.* poids atomique (French: atomic weight)

PAT *Football* (USA) point(s) after touchdown • (pæt) *Physics* positron annihilation technique • *Banking* preauthorized automatic transfer • Professional Association of Teachers

Pata. Patagonia

patd patented

path. (*or* **pathol.**) pathological • pathology

Pat. Off. Patent Office

pat. pend. patent pending

PAU Pan American Union • programmes analysis unit

pav. pavilion

Pav *Astronomy* Pavo

PAW plasma-arc welding • powered all the way

PAWA Pan American World Airways

PAWR (USA) Public Authority for Water Resources

PAX *Telephony* private automatic exchange

PAYE pay as you earn (income tax) • pay as you enter

paymr paymaster

payt payment

PAYV pay as you view

p.b. *Baseball* passed ball(s)

Pb *Chem., symbol for* lead (Latin *plumbum*) • *symbol for* petabyte(s)

PB *airline flight code for* Air Burundi • Bachelor of Philosophy (Latin *Philosophiae Baccalaureus*) • *British vehicle registration for* Guildford • pass book • permanent base • *Athletics* personal best • Pharmacopoeia Britannica (Latin: British Pharmacopoeia) • *Pharmacol.* phenobarbitone • plastic-bonded • Plymouth Brethren • power brakes • Prayer Book • premium bond • Primitive Baptists • (USA) Publications Board • *Knitting* purl into back of stitch

PBA *Colloquial* poor bloody assistant • (USA) Professional Bowlers Association • (USA) Public Buildings Administration

PBAB please bring a bottle

PBB *Chem.* polybrominated biphenyl (toxic constituent of plastics, etc.)

PBC powerboat club

PBI *Colloquial* poor bloody infantry(man) • *Med.* protein-bound iodine

pbk paperback (book)

PBM *Surveying* permanent benchmark • play by mail (of games)

PBR payment by results

PBS *Med.* phosphate-buffered saline • Public Broadcasting Service

p.b.t. *Finance* profit before tax

PBT *Accounting* pay-back time

PBX *Telephony* private branch exchange

pc *Astronomy, symbol for* parsec

pc. percentage • *Printing* pica • piece • price

p.c. per cent • postcard • *Med.* post cibum (Latin: after meals; in prescriptions)

PC *airline flight code for* Air Fiji • *British vehicle registration for* Guildford • Panama Canal • *Chem.* paper chromatography • *Chem.* paraffin concentration • Parish Council(lor) • Parliamentary Commissioner (ombudsman) • (France) Partie Communiste (Communist Party) • Past Commander • Paymaster Captain • (Ireland) Peace Commissioner • (USA) Peace Corps • perpetual curate • personal computer • *Physics* phase conjugation • *Military* pioneer corps • *Machinery* pitch circle • *Politics* Plaid Cymru • Police Constable • (USA) politically correct (*or* political correctness) • *Chem.* polycarbonate • polymer concrete • *Chem.* polypropylene (*or* polypropene) carbonate (plastic) • Portland cement • Post Commander • potentially correct • preparatory commission • Press Council • prestressed concrete • Prince Consort • *Electronics* printed circuit • Prison Commission • Privy Council • Privy Counsellor • process control • (USA) professional corporation • (Canada) Progressive Conservative • propositional calculus • *Chem.* propylene (*or* propene) carbonate • public convenience (on maps) • pulverized coal • *Microbiol.* pure culture

P/C (*or* **p/c**) petty cash • price(s) current

PCA Parliamentary Commissioner for Administration (ombudsman) • *Med.* patient-controlled analgesia • Permanent Court of Arbitration • *Statistics* principal component analysis

PCAS (formerly) Polytechnics Central Admissions System (merged with UCCA to form UCAS)

PC-AT personal computer, advanced technologies

PCB petty cash book · *Chem.* polychlorinated biphenyl (toxic constituent of plastics, etc.) · (*or* **pcb**) *Electronics* printed-circuit board · *Insurance* private car benefits

pcc *Chem.* precipitated calcium carbonate

PCC parochial church council · Partido Comunista de Cuba (Spanish: Communist Party of Cuba) · political consultative committee · Press Complaints Commission · Privy Council cases

PCD *Electronics* photo compact disc

PC-DOS (ˌpiːˈsiːˌdɒs) *Computing, trademark* Personal Computer Disk Operating System

PCE Partido Comunista de España (Communist Party of Spain) · Postgraduate Certificate of Education · *Engineering* pyrometric cone equivalent

P-Celtic *Linguistics, indicating* one of two main groups of languages that developed from Common Celtic. *See also* Q-Celtic

pcf pounds per cubic foot

PCF Parti Communiste Français (French Communist Party) · pistol, centre fire (calibre of pistol)

PCFC Polytechnics and Colleges Funding Council

PCGC pulverized coal gasification and combustion

PCGN Permanent Committee on Geographical Names

P. Ch. parish church

pci pounds per cubic inch

PCI Partito Comunista Italiano (Italian Communist Party; now renamed Partito Democratico della Sinistra, Democratic Party of the Left) · *Computing* peripheral component interconnect · personal computer interface

PCIUG personal computer independent user group

pcl parcel

PCL *Computing* printer control language

pcm per calendar month

PCM *Physics* phase-change material · phase-contrast microscope · photochemical machinery · (USA) plug-compatible manufacturer · polyatomic ceramic material · protein-calorie malnutrition · *Telecom.* pulse-code modulation

PCMCIA Personal Computer Memory Card International Association

PCMI photochromic microimage

PCMO Principal Colonial Medical Officer

PCN (El Salvador) Partido de Conciliación Nacional (Spanish: National Conciliation Party) · *Computing* personal communications network

PCNB *Chem.* pentachloronitrobenzene

PCOB Permanent Central Opium Board (in the UN)

PCOD polycystic ovary disease

p-code *Computing, indicating* an intermediate language designed as the target language for UCSD Pascal

PCP Past Chief Patriarch · *Chem.* pentachlorophenol (wood preservative) · *Finance* permissible capital payment · phencyclohexylpiperidine (the drug phencyclidine; angel dust) · *Med. Pneumocystis carinii* pneumonia (complication of Aids) · polychloroprene (rubber) · prime commercial paper

PCPA *Pharmacol.* p-chlorophenylalinine

PCR Pedestrian Crossings Regulations · *Med.* plasma clearance rate · *Biochem.* polymerase chain reaction · *Physics* primary cosmic rays

PCRS Poor Clergy Relief Society

pcs. pieces · prices

PCS *Physics* photon-correlation spectroscopy · *Electronics* plasma-current switch · (Scotland) Principal Clerk of Session

PCSP Permanent Commission for the South Pacific

pct per cent

PCT *Med.* positron computed tomography · product consistency test

PCTCT *Accounting* profit chargeable to corporation tax

PCTE *Computing* portable common tool environment

PCTFE polychlorotrifluoroeth(yl)ene (plastic coating)

p.c.u. passenger car unit

PCU plant control unit · power-control (*or* -conversion) unit · pressurization-control unit

PCV *Med.* packed cell volume · passenger-carrying vehicle · passenger-controlled vehicle · (USA) Peace Corps Volunteers · positive crankcase ventilation (in a car) · pressure containment vessel

PCX *Computing* picture exchange format

PC-XT personal computer extended

PCZ Panama Canal Zone

PCZST Panama Canal Zone Standard Time

pd paid • passed

p.d. per diem (Latin: daily) • *Engineering* pitch diameter • poop deck • postage due • (*or* **p/d**) post-dated • *Physics* potential difference • preliminary design • printer's devil

Pd *Chem., symbol for* palladium

PD Doctor of Pharmacy (Latin *Pharmaciae Doctor*) • *British vehicle registration for* Guildford • per diem (Latin: daily) • *British fishing port registration for* Peterhead • Pharmacopoeia Dublinensis (Latin: Dublin Pharmacopoeia) • polar distance • (USA) Police Department • port dues • posdata (Spanish: postscript; PS) • postal district • preventive detention (*or* detainee) • (Germany) Privatdozent (unsalaried university teacher) • probability of detection • production department • (Ireland) Progressive Democrat(s) • progressive disease • *Insurance* property damage • (USA) Public Defender • *Computing* public domain (software) • *Chem.* pyrimidine dimer

P/D price–dividend (in **P/D ratio**)

PDA *Med.* patent ductus arteriosus • *Computing* personal digital assistant • pour dire adieu (French: to say goodbye) • *Navigation* predicted drift angle

PDAD Probate, Divorce, and Admiralty Division (former division of the High Court)

PdB (USA) Bachelor of Pedagogy

PDC (El Salvador) Partido Demócrata Cristiano (Spanish: Christian Democratic Party) • personnel dispatch (*or* dispersal) centre • *Engineering* pulse-discharge cleaning • *Chem.* pyridinium dichromate

PDCI Parti Démocratique de la Côte d'Ivoire (French: Democratic Party of Côte d'Ivoire)

PdD (USA) Doctor of Pedagogy

PDD precise delay detonator

PDE *Maths.* partial differential equation • Projectile Development Establishment

P-de-C Pas-de-Calais (department of France)

P-de-D Puy-de-Dôme (department of France)

PDF *Statistics* probability density function

PDFLP Popular Democratic Front for the Liberation of Palestine

PDG Parti Démocratique de Guinée (French: Democratic Party of Guinea) • Paymaster Director-General • président directeur général (French: chairman and managing director)

PDGF *Med., Biochem.* platelet-derived growth factor

PDH *Computing* plesiochronous digital hierarchy

p.d.i. predelivery inspection

PDI *Astronautics* powered descent initiation

pdl *Physics, symbol for* poundal

PDL *Computing* page description (*or* definition) language • poverty datum line • *Computing* program design language

PdM (USA) Master of Pedagogy

PDM physical distribution management • *Telecom.* pulse-duration modulation

pdn production

PDN public data network

PDOS *Physics* partial density of states

PDP *Computing* parallel distributed processing • program development plan (in NASA) • programmed data processor

PDPA People's Democratic Party of Afghanistan

pdq (*or* **PDQ**) *Colloquial* pretty damn quick

PDR People's Democratic Republic • *Astronomy* photodissociation region • *Engineering* pitch-to-diameter ratio • price–dividend ratio

PDRA postdoctoral research assistant

P/D ratio price–dividend ratio

PDS Parkinson's Disease Society • (Germany) Party of Democratic Socialism (formerly the Communist Party) • *Physics* phase differential scattering • previously digested sludge • *Computing* programming documentation standards

PDSA People's Dispensary for Sick Animals

PDSR Principal Director of Scientific Research

PDT Pacific Daylight Time • personal development technology • *Med.* photodynamic therapy

PDTC Professional Dancer's Training Course Diploma

PDU process-development unit

p.e. *Law* personal estate • printer's error

Pᵉ *RC Church* (Spain) Padre

PE *British vehicle registration for* Guildford • peat extract • *Meteorol.* (ice) pellets • permissible error • *Insurance* personal effects • *international vehicle registration for* Peru • *postcode for* Peterborough • (USA)

Petroleum Engineer • Pharmacopoeia Edinburgensis (Latin: Edinburgh Pharmacopoeia) • *Computing* phase-encoded (of tape format) • physical education • *airline flight code for* Pine State Airlines • plastic explosive • pocket edition • *British fishing port registration for* Poole • polyeth(yl)ene • Port Elizabeth (South Africa) • *Physics* potential energy • Presiding Elder • printer's error • *Statistics* probable error • *Computing* processing element • procurement executive • (USA) Professional Engineer • Protestant Episcopal • *Med.* pulmonary embolism

P/E part exchange (in property or motor advertisements) • port of embarkation • (*or* **p/e**) price–earnings (in **P/E ratio**)

PEA Physical Education Association of Great Britain and Northern Ireland

PEC (*or* **p.e.c.**) photoelectric cell • photoelectrochemical cell • Protestant Episcopal Church

PECD *Electronics* photoelectric conversion device

ped. pedal • pedestal • pedestrian

PedD (USA) Doctor of Pedagogy

pediat. pediatrics

PEDir Director of Physical Education

Peeb. Peebles (former Scottish county)

PEEP (piːp) pilot's electronic eye-level presentation

PEF *Insurance* (USA) personal effects floater • *Electronics* pulsed electric field

PEFCO (ˈpɛfkəu) Private Export Funding Corporation

Peg *Astronomy* Pegasus

PEG *Chem.* polyethylene glycol

PEI (Canada) Prince Edward Island

p.ej. por ejemplo (Spanish: for example; e.g.)

Pek. Peking

PEL permissible exposure level (*or* limit)

Pemb Pembroke College (Oxford *or* Cambridge)

Pembs Pembrokeshire

Pemex (ˈpɛmɛks) Petróleos Mexicanos (Mexican oil company)

pen. penal • penetration • peninsula(r)

Pen *Geology* Pennsylvanian

Pen. Peninsula • Penitentiary

PEN (pɛn) International Association of Poets, Playwrights, Editors, Essayists, and Novelists

PEng Member of the Society of Professional Engineers • (Canada) Registered Professional Engineer

Penn. Pennsylvania

penol. penology

pent. pentagon

Pent. Pentateuch • Pentecost

PEO (USA) Philanthropic Educational Organization • *Chem.* polyeth(yl)ene oxide

PEP *Radio* peak envelope power • (*or* **Pep**; pɛp) personal equity plan • *Biochem.* phosphoenolpyruvate • political and economic planning

PEPP (USA) Professional Engineers in Private Practice

per. percentile • period • person

Per *Astronomy* Perseus

Per. Persia(n)

PER price–earnings ratio • Professional and Executive Recruitment • Professional Employment Register

PERA Production Engineering Research Association of Great Britain

per an. (*or* **per ann.**) per annum (Latin: yearly)

P/E ratio price–earnings ratio

perc. *Music* percussion

per con *Book-keeping* per contra (Latin: on the other side)

perd. *Music* perdendosi (Italian: dying away)

perf. perfect • perfection • perforated • perforation • performed (by) • performance

peri. (*or* **perig.**) perigee

PERK (pɜːk) perchloreth(yl)ene (solvent)

PERL (pɜːl) *Computing* practical extraction and report language

Perm (pɜːm) pre-embossed rigid magnetic technology

PERME Propellants, Explosives, and Rocket Motor Establishment

perp. perpendicular • perpetual

per pro. per procurationem (*see under* pp)

pers. person • personal(ly) • (*or* **persp.**) perspective

Pers. Persia(n)

pert. pertaining

PERT (pɜːt) *Computing, Management* project (*or* program, performance) evaluation and review technique

Peruv. Peruvian

PES *Chem.* photoelectron spectroscopy • *Chem.* potential-energy surface • *Computing* programmable electronic system

PESA *Physics* particle elastic scattering analysis • *Med.* percutaneous epididymal sperm aspiration

PESC Public Expenditure Survey Committee

Pesh. Peshawar (Pakistan)

PESM photoelectron spectromicroscopy

PEST (pɛst) Political, Environmental, Social, and Technological (framework for analysing these aspects of a business environment) • Pressure for Economic and Social Toryism (left-wing Conservative group)

pet. petroleum • petrological • petrologist • petrology

Pet. *Bible* Peter • Peterhouse (Cambridge college) • (Gaius) Petronius (1st-century Roman satirist)

PET *Chem.* polyeth(yl)ene terephthalate (plastic used in food packaging) • (pet) *Med.* positron-emission tomography (as in **PET scan**) • *Taxation* potentially exempt transfer

PETA *Chem.* pentaerythritol triacrylate • People for Ethical Treatment of Animals

Pet.E (USA) Petroleum Engineer

petn petition

PETN pentaerythritol tetranitrate (explosive)

PETP polyethylene (*or* polyethene) terephthalate (polyester)

petr. petrification • petrify • petrology

PETRAS ('pɛtrəs) Polytechnic Educational Technology Resources Advisory Service

Petriburg. Petriburgensis (Latin: (Bishop) of Peterborough)

petro. petrochemical

petrog. petrography

petrol. petrology

PETS *Computing* posting and enquiry terminal system

p.ex. par exemple (French: for example; e.g.)

P/Ex part exchange (in property or motor advertisements)

PEX *indicating* a discounted airline fare (probably a back formation from APEX, advance-purchase excursion)

pf pfennig (German currency) • *Music* piano (instrument; from *pianoforte*)

pf. perfect • *Finance* preferred (stock) • proof

p.f. (ital.) *Music* piano e forte (Italian; soft and then loud) • (ital.) *Music* più forte

(Italian; louder) • pneumatic float • pro forma (invoice)

pF *Physics, symbol for* picofarad(s)

PF *British vehicle registration for* Guildford • panchromatic film • Patriotic Front • phenol–formaldehyde (as in **PF resin**) • *Building trades* plain face • *Physics* poloidal field • *Physical chem.* polycrystalline film • power factor • Procurator-Fiscal • *Finance* public funding • pulverized fuel

PFA (USA) Private Fliers' Association • Professional Footballers' Association • pulverized fuel ash

P factor *Psychol.* preservation factor

PFB *Building trades* preformed beams • *Chemical engineering* pressurized fluidized bed

PFBC *Chemical engineering* pressurized fluidized-bed combustion (*or* combustor)

PFBG *Chemical engineering* pressurized fluidized-bed gasification (*or* gasifier)

PFBR *Nuclear engineering* prototype fast breeder reactor

pfc passed flying college (in the RAF)

PFC *Chem.* perfluorocarbon • polychlorinated fluorocarbon (synthetic resin) • *Colloquial* (USA) poor foolish (*or* forlorn) civilian • (*or* **Pfc**) *Military* (USA) Private first class

pfce performance

pfd *Finance* preferred

Pfd Pfund (German: pound)

PFD personal flotation device • position-fixing device

pfd sp. preferred spelling

PFF pathfinder force

pfg pfennig (German currency)

PFGE *Biochem.* pulsed-field gel electrophoresis

PFLO Popular Front for the Liberation of Oman

PFLP Popular Front for the Liberation of Palestine

PFM *Telecom.* pulse-frequency modulation

PFP (NATO) Partnership for Peace • personal financial planning • (South Africa) Progressive Federal Party

PFR *Nuclear engineering* prototype fast reactor

PFRT preliminary flight rating test

PFSA pour faire ses adieux (French: to say goodbye)

PFT *Med.* pulmonary function test

pft acct *Music* piano(forte) accompaniment

PFV pour faire visite (French: to make a call)

pfx prefix

pg. page

p.g. pay group • paying guest • persona grata (Latin: acceptable person) • proof gallon (of alcohol) • proving ground

Pg *Geology* Palaeogene

Pg. Portugal • Portuguese

PG *airline flight code for* Bangkok Airways • *British vehicle registration for* Guildford • *Films* parental guidance (certification) • *Freemasonry* Past Grand • paying guest • postgraduate • Preacher General • *Colloquial* (USA) pregnant • prisonnier de guerre (French: prisoner of war) • Procurator-General • *Pharmacol.* prostaglandin (as in **PGE₁**, etc.) • *Chem.* pyrolytic graphite

PGA *Biochem., Botany* phosphoglyceric acid • *Electronics* pin grid array • Professional Golfers' Association • *Computing* programmable gate array • *Biochem.* pteroylglutamic acid (folic acid)

PGAL *Biochem.* phosphoglyceraldehyde

p.g.c. *Navigation* per gyrocompass

PGCE Postgraduate Certificate of Education

PG Cert Postgraduate Certificate

PGD *Freemasonry* Past Grand Deacon

PG Dip Postgraduate Diploma

PgDn page down (on a keyboard)

PGDRS psychogeriatric dependency rating scale

PGF *Med.* polypeptide growth factor

PGJD *Freemasonry* Past Grand Junior Deacon

PGL *Med.* persistent generalized lymphadenopathy (a stage of Aids) • *Freemasonry* Provincial Grand Lodge

PGM *Freemasonry* Past Grand Master • precision-guided munition

PGP *Computing* pretty good privacy

PGR *Films* (Australia) parental guidance recommended (certification) • population growth rate • *Psychol.* psychogalvanic response

PGSD *Freemasonry* Past Grand Senior Deacon

p.g.t. per gross ton

PgUp page up (on a keyboard)

ph *Optics, symbol for* phot(s)

ph. phase

p.h. *Engineering* precipitation hardening

pH *Chem.* potential of hydrogen ions (measure of acidity or alkalinity)

Ph *Geology* Phanerozoic • *Chem., symbol for* phenyl group (in formulae) • (*or* **Ph.**) Philosophy (in degrees)

PH *British vehicle registration for* Guildford • *international civil aircraft marking for* Netherlands • *postcode for* Perth • petroleum hydrocarbon • *British fishing port registration for* Plymouth • *airline flight code for* Polynesian Airlines • previous (medical) history • public health • public house (on maps) • *Med.* pulmonary hypertension • (USA) Purple Heart (military decoration)

PHA *Immunol.* phytohaemagglutinin • Public Health Act • (USA) Public Housing Administration • (USA) public housing authority • *Physics* pulse-height analyser (*or* analysis)

PHAB (fæb) Physically Handicapped and Able-Bodied (a charity)

phal. phalanx

phar. pharmaceutical • pharmacist • (*or* **Phar.**) pharmacopoeia • pharmacy

PharB (*or* **PharmB**) Bachelor of Pharmacy (Latin *Pharmaciae Baccalaureus*)

PharD (*or* **PharmD**) Doctor of Pharmacy (Latin *Pharmaciae Doctor*)

pharm. pharmaceutical • pharmacist • pharmacology • (*or* **Pharm.**) pharmacopoeia • pharmacy

PharM (*or* **PharmM**) Master of Pharmacy (Latin *Pharmaciae Magister*)

pharmacol. pharmacology

pharm. chem. pharmaceutical chemistry

PhB Bachelor of Philosophy (Latin *Philosophiae Baccalaureus*)

ph. brz. phosphor bronze

PHC Pharmaceutical Chemist • primary health care

PhD Doctor of Philosophy (Latin *Philosophiae Doctor*)

PHD Doctor of Public Health

PhDEd Doctor of Philosophy in Education

Phe (*or* **phe**) *Biochem.* phenylalanine • *Astronomy* Phoenix

PHE Public Health Engineer

phen *Chem.* 1,10-phenanthroline (used in formulae)

PhG (USA) Graduate in Pharmacy

PHI permanent health insurance • Public Health Inspector

PHIBLANT ('fɪblænt) *US Navy* Amphibious Forces, Atlantic

PHIBPAC ('fɪb,pæk) *US Navy* Amphibious Forces, Pacific

PHIGS (fɪgs) *Computing* programmers' hierarchical interactive graphics standard

phil. philological • philology • philosopher • philosophical • philosophy

Phil. (*or* **Phila.**) Philadelphia • Philharmonic • *Bible* Philippians • Philippines

Philem. *Bible* Philemon

Phil. I. (*or* **Phil. Is.**) Philippine Islands

philol. philological • philology

philos. philosopher • philosophical • philosophy

Phil. Soc. Philharmonic Society

Phil. Trans. (ital.) Philosophical Transactions of the Royal Society of London

PhL Licentiate in (*or* of) Philosophy

PHLS Public Health Laboratory Service

PhM Master of Philosophy (Latin *Philosophiae Magister*)

PhmB Bachelor of Pharmacy

PHN Public Health Nurse

Phoen. Phoenician

phon. (*or* **phonet.**) phonetics • (*or* **phonol.**) phonology

phot. (*or* **photog.**) photograph(ic) • photographer • photography

photom. photometrical • photometry

p.h.p. pounds per horsepower • pump horsepower

phr. phrase • phraseology

PHR *Engineering* power-to-heat ratio

phren. (*or* **phrenol.**) phrenological • phrenology

PHS (USA) Public Health Service

PHTS (USA) Psychiatric Home Treatment Service

PHWR *Nuclear engineering* pressurized heavy-water reactor

phys. physical(ly) • physician • physicist • physics • physiological • physiology

phys. ed. physical education

physiog. physiography

physiol. physiological • physiologist • physiology

phys. sc. physical science

p.i. *Insurance* professional indemnity (policy)

PI *Irish vehicle registration for* Cork • *Med.* parainfluenza virus • Pasteur Institute • *Maths.* path integral • *Law* (USA) personal injury • petrol-injected •

Pharmacopoeia Internationalis (Latin: International Pharmacopoeia) • Philippine Islands • photographic interpretation (*or* interpreter) • *Colloquial* (USA) pimp • (USA) principal investigator • private investigator • *Accounting* profitability index • *Computing* programmed instruction

PIA Pakistan International Airlines Corporation • *Computing* peripheral interface adaptor • Personal Investment Authority

piang. *Music* piangendo (Italian: plaintive)

pianiss. *Music* pianissimo (Italian: very soft)

PIARC Permanent International Association of Road Congresses

PIAT *Military* projector infantry antitank (portable weapon)

PIB Petroleum Information Bureau • *Chem.* polyisobutylene • Prices and Incomes Board (superseded by NBPI)

PIBOR ('piːbɔː) *Finance* Paris InterBank Offered Rate

PIBS permanent interest-bearing share

pic. *Music* piccolo • (*or* **pict.**) pictorial

Pic *Astronomy* Pictor

PIC *Computing* picture format • *Computing* problem isolation code • product of incomplete combustion • *Computing* programmable interrupt controller

PICS platform for Internet content selection

PID pelvic inflammatory disease • *Computing* personal identification device • prolapsed intervertebral disc (slipped disc)

PIDE (Portugal) Polícia Internacional e de Defesa do Estado (International Police for the Defence of the State; former state security system)

PIDS (pɪdz) *Med.* primary immune deficiency syndrome

PIE *Physics* positive-ion emission • Proto-Indo-European (language)

PIF (pɪf) *Computing* program information file

PIGE *Physics* proton-induced gamma-ray emission

pigmt pigment

pigmtn pigmentation

PIH pregnancy-induced hypertension

PIK payment in kind

pil. pilula (Latin: pill; in prescriptions)

PIL *Computing* paper interchange language • payment in lieu

PILL (pɪl) *Computing* programmed instruction language learning

PILOT ('paɪlət) *Computing* programmed inquiry, learning, or teaching

PIM (*or* **Pim**) *Computing* personal information manager • *Telecom.* pulse-interval modulation

PIMS profit impact of market strategy

p-i-n *Electronics* p-type, intrinsic, n-type (semiconductor)

PIN (pɪn) personal identification number (used, with cash or credit card, to access computer-based bank accounts, etc.)

P-in-C Priest-in-Charge

PINC property income certificate

Pind. Pindar (518–438 BC, Greek lyric poet)

Ping (pɪŋ) *Computing* Packet Internet groper

PINS (pɪnz) (USA) person in need of supervision

pinx. pinxit (Latin: (he *or* she) painted it)

PIO *Computing* parallel input/output • photographic interpretation officer • *Military* (USA) public information office (*or* officer)

PIP (Peru) Policía de Investigiones del Peru Federales (Spanish: Peruvian Federal Investigation Police; Peruvian equivalent of the FBI)

PIPO *Computing* parallel in, parallel out

PIPPY (*or* **Pippy**) ('pɪpi:) *Colloquial* person inheriting parent's property

PIR passive infrared

PIRA Paper Industries Research Association • Provisional Irish Republican Army

Pis. *Astrology* Pisces

PISO *Computing* parallel in, serial out

PITCOM ('pɪtkɒm) Parliamentary Information Technology Committee

PIX *Computing* picture exchange format

PIXE ('pɪksɪ) *Physics* particle-induced X-ray emission

pixel ('pɪksəl) *Computing* picture element

pizz. *Music* pizzicato (Italian: pinched *or* plucked)

p.j. physical jerks • (*or* **p-j**) *Colloquial* pyjama

PJ *British vehicle registration for* Guildford • *international civil aircraft marking for* Netherlands Antilles • *Physics, symbol for* petajoule(s) • Presiding Judge • Probate Judge • *Colloquial* pyjama

pk pack • package • park • peak • peck (unit)

pK *Chem.* potential of *K* (symbol for the dissociation constant; a measure of the strength of acids)

PK *British vehicle registration for* Guildford • *international civil aircraft marking for* Indonesia and West Irian • *international vehicle registration for* Pakistan • *airline flight code for* Pakistan International • personal knowledge • *Immunol.* Prausnitz-Kustner (as in **PK test**) • psychokinesis

PKD polycystic kidney disease

pkg. (*or* **pkge**) package • packing

PKI Partai Komunis Indonésia (Indonesian Communist Party)

PKP Polskie Koleje Panstwowe (Polish State Railways)

pkt packet • pocket

PKU *Med.* phenylketonuria

pkwy (USA) parkway

pl. place • plate • platoon • plural • pole (measure)

Pl. Place (in street names) • Plate(s) (in books, etc.) • (Germany) Platz (in street names)

PL *airline flight code for* Aeroperu • *British vehicle registration for* Guildford • (*or* **P/L**) *Marine insurance* partial loss • *Insurance* passenger liability • patrol leader (in the Scouting movement) • Paymaster Lieutenant • *British fishing port registration for* Peel • Pharmacopoeia Londiniensis (Latin: Pharmacopoeia of London) • Plimsoll line (on a ship) • *postcode for* Plymouth • Poet Laureate • *international vehicle registration for* Poland • position line • Primrose League • (*or* **P/L**) *Law* product liability • product licence (on labels of medicinal products) • programmed learning • *Computing* programming language (as in **PL/I, PL/M**, etc.) • public law • public library

P–L *Astronomy* period–luminosity (as in **P–L relation**)

PL/I (*or* **PL/1**) *Computing* Programming Language I

p.l.a. passengers' luggage in advance

Pla. Plaza

PLA (China) People's Liberation Army • Port of London Authority • *Computing, Electronics* programmed (*or* programmable) logic array

plan. planet • planetarium

plas. plaster • plastic

plat. plateau • (or **platf.**) platform • platinum • platonic • platoon

PLATO (or **Plato**) ('pleɪtəʊ) Computing programmed logic for automatic teaching operation

Plaut. (Titus Maccius) Plautus (c. 254–184 BC, Roman dramatist)

plc public limited company (following the name of a company)

PLC Poor Law Commissioners • Marketing product life cycle • Computing programmable logic controller • public limited company (following the name of a company)

PLCC Computing plastic leadless chip carrier

PLCWTWU Power Loom Carpet Weavers' and Textile Workers' Union

plcy policy

pld payload

PLD potentially lethal damage • Computing programmable logic device

Ple Geology Pleistocene

Plen. Plenipotentiary

plf (or **plff**) plaintiff

PLF Med. (serum) placental fementin

PLG private/light goods (vehicle)

Pli Geology Pliocene

PLI Partito Liberale Italiano (Italian Liberal Party) • President of the Landscape Institute

PLL Computing phase-locked loop

PLM Paris–Lyons–Mediterranean (Railway) • Telecom. pulse-length modulation

PL/M Computing Programming Language for Microcomputers

plmb. plumber • plumbing

plng planning

PLO Palestine Liberation Organization

PLP Parliamentary Labour Party • (Bermuda) Progressive Labour Party • (Bahamas) Progressive Liberal Party

PLR public lending right

Pls Plates (in books, etc.)

PLS Statistics partial least-squares • Computing programmable logic sequencer

Pl Sgt Platoon Sergeant

PLSS (plss) Astronautics personal (or portable) life-support system

plstr plasterer

plt pilot

PLT Princeton Large Torus (experimental fusion reactor)

pltc. political

pltf plaintiff

plu. plural

PLU Colloquial people like us

Pluna (or **PLUNA**) ('pluːnə) Primeras Líneas Uruguayas de Navegación Aérea (Uruguayan airline)

plup. (or **plupf.**) pluperfect

plur. plural • plurality

Pluto (or **PLUTO**) ('pluːtəʊ) pipe line under the ocean (conveying fuel to Allied forces, World War II)

Ply. Plymouth

plywd plywood

PL/Z Computing Programming Language Zilog

pm. premium

p.m. permanent magnet • post meridiem (Latin: after noon) • postmortem (examination) • (or **pm**) premolar (tooth)

Pm Chem., symbol for promethium

PM British vehicle registration for Guildford • Pacific mail • parachute mine • Music particular (or peculiar, proper) metre • Past Master (of a fraternity) • Paymaster • Physics perturbation method • Telecom. phase modulation • piae memoriae (Latin: of pious memory) • Pipe Major • Police Magistrate • polícia militar (Portuguese: military police) • polizia militare (Italian: military police) • Pope and Martyr • Postmaster • post meridiem (see under p.m.) • postmortem (examination) • powder metallurgy • preventive (or predictive) maintenance (esp. of computers) • Prime Minister • product manager • Military Provost Marshal • airline flight code for Tropic Air

PMA Pakistan Medical Association • Dentistry papillary, marginal, attached (gingivitis; in **PMA index**) • paramethoxyamphetamine (hallucinogenic drug) • personal military assistant • Chem. phenylmercuric acetate • Chem. polymethyl acrylate (synthetic polymer) • (USA) Purchasing Management Association

PMAF (USA) Pharmaceutical Manufacturers' Association Foundation

PM & ATA Paint Manufacturers' and Allied Trades Association

PM&R physical medicine and rehabilitation

PMB Potato Marketing Board

PMBX *Telephony* private manual branch exchange

PMC Personnel Management Centre · *Building trades* plaster-moulded cornice

PMD (USA) Program for Management Development

PME *Electronics* protective multiple earthing

PM-ECM *Electrical engineering* permanent-magnet electronically commutated motor

PMF probable maximum flood · *Physics* pulsating magnetic field

PMG (ital.) Pall Mall Gazette · Paymaster General · *Electrical engineering* permanent-magnet generator · Postmaster General · *Military* Provost Marshal General

p.m.h. per man-hour

PMH previous medical history

pmk postmark

PML Prime Minister's list · *Insurance* probable maximum loss

PMLA (ital.) (USA) Publications of the Modern Language Association

PMM platinum-metal minerals · *Telecom.* pulse-mode multiplex

PMMA polymethylmethacrylate (synthetic resin)

PMN *Med.* polymorphonuclear (leucocyte or neutrophil)

PMO Principal Medical Officer

pmr paymaster

PMR Pacific missile range · *Med.* polymyalgia rheumatica

PMRAFNS Princess Mary's Royal Air Force Nursing Service

PMS *Printing* Pantone Matching System · *Med.* pregnant mare's serum · *Med.* premenstrual syndrome · President of the Miniature Society · *Computing* processor-memory-switch (notation) · project management system

PMSF *Chem.* phenylmethylsulphonyl fluoride

PMSG *Med.* pregnant mare's serum gonadotrophin

pmt payment

PMT *Photog.* photomechanical transfer · (USA) *Finance* post-market trading · *Med.* premenstrual tension · project management team

PMTS *Commerce* predetermined motion-time standards

PMV predicted mean vote

PMX *Telephony* private manual exchange

pn *Chem.* propylenediamine (used in formulae) · (or **p-n**) *Electronics* p-type n-type semiconductor (as in **pn junction**)

p.n. percussion (or percussive) note · please note · *Commerce* promissory note · *Physics* proton number

PN *British vehicle registration for* Brighton · Pakistan Navy · *Med.* parenteral nutrition · *Engineering* performance number · *Astronomy* planetary nebula · *Med.* postnatal · *Physics* post-Newtonian · *British fishing port registration for* Preston · *Commerce* promissory note · *Computing* pseudonoise (as in **PN sequence**) · psychoneurotic

P/N part number · *Commerce* promissory note

Pna Panama

PNA Pakistan National Alliance · paranitroaniline (dye) · *Chem.* polynuclear aromatic (compound) · Psychiatric Nurses Association

PNAA *Physics* prompt-neutron activation analysis

PNB Philippine National Bank

PNC Palestinian National Council · *Physics* parity nonconserving · (Guyana) People's National Congress

PND *Med.* postnatal depression

PNdB *symbol for* perceived noise decibel(s)

pndg pending

pneu. (or **pneum.**) pneumatic

PNEU Parents' National Educational Union

p.n.g. persona non grata

PNG Papua New Guinea (abbrev. *or* IVR)

PNI *Med.* psychoneuroimmunology

pnl panel

PNL Pacific Northwest Laboratory · *Med.* polymorphonuclear leucocyte

PNM (Trinidad and Tobago) People's National Movement

PNO Principal Nursing Officer · *Computing* public network operator

pnp (or **p-n-p**) *Electronics* p-type n-type p-type semiconductor (as in **pnp transistor**)

PnP *Computing* plug 'n' play

PNP (Puerto Rico) Partido Nuevo Progresista (New Progressive Party) · (Jamaica) People's National Party

pnr pioneer

p.n.r. prior notice required

PNR *Physics* prompt nuclear reaction

PNS *Med.* parasympathetic nervous system

PNSB *Microbiol.* purple nonsulphur bacteria

Pnt. (USA) Pentagon

PNTO Principal Naval Transport Officer

pntr painter

PNV (Spain) Partido Nacional Vasco (Basque Nationalist Party)

pnxt pinxit (Latin: (he *or* she) painted it)

PNYA Port of New York Authority

Pnz. Penzance

p.o. part of • *Med.* per os (Latin: by mouth; in prescriptions) • postal order • power oscillator • previous order(s) • *Baseball* putout(s)

Po *Chem., symbol for* polonium

PO parcels office • parole officer • par ordre (French: by order) • Passport Office • Patent Office • personnel officer • petty officer • Philharmonic Orchestra • pilot officer • Portsmouth (postcode *or* British vehicle registration) • postal order • Post Office • power-operated • *Chem.* propylene oxide • Province of Ontario • public office(r) • Pyrénées-Orientales (department of France)

POA *Engineering* post-oxidation annealing • price on application (in property or motor advertisements) • primary optical area (in graphic design) • Prison Officers' Association (trade union)

POAC Post Office Advisory Council

POB Post Office Box

POC *Chem.* particulate organic carbon • (*or* **p.o.c.**) port of call • *Chem.* product of combustion

Pod *Computing* power-on diagnostics

POD pay on death • pay(ment) on delivery • (ital.) Pocket Oxford Dictionary • port of debarkation • (USA) Post Office Department

POE port of embarkation • port of entry

POED Post Office Engineering Department

poet. poetic(al) • poetry

POETS day ('pəʊɪtz) *Colloquial, indicating* Friday (piss off early tomorrow's Saturday)

POEU Post Office Engineers Union

P. of W. Prince of Wales

POG passion fruit, orange, and guava juice

POGO ('pəʊgəʊ) Polar Orbiting Geophysical Observatory

poi. poison(ous)

pol. polar • polarize • police • political • politician • politics

Pol. Poland • Polish

POL Patent Office Library • petroleum, oil, and lubricants • *Computing* problem-oriented language

pol. ad. political adviser

pol. econ. political economy

pol. ind. pollen index

Polis ('pɒlɪs) Parliamentary On-Line Information Service

polit. political • politics

poll. pollution

pol. sci. political science

Poly. (*or* **Polyn.**) Polynesia(n)

Polyb. Polybius (*c.* 200–*c.* 120 BC, Greek historian)

POM particulate organic matter • prescription-only medicine (*or* medication)

POMEF Political Office Middle East Force

PON *Computing* passive optical network

PONI ('pəʊnɪ) product of Northern Ireland

Ponsi ('pɒnsɪ) *Military slang* person of no strategical importance

pont. br. pontoon bridge

Ponti ('pɒntɪ) *Military slang* person of no tactical importance

POO Post Office order

POOF *Computing* peripheral on-line oriented function

pop. popular(ly) • population

POP plaster of Paris • (*or* **p.o.p.**) point of purchase • point of presence • Post Office preferred (size of envelopes, etc.) • (pɒp) *Computing* post office protocol • *Photog.* printing-out paper • *Med.* progestogen-only pill • proof of purchase

POPA Property Owners Protection Association

POPL ('pɒpᵊl) *Computing* Principles of Programming Languages

por. porosity • porous • portion

p.o.r. pay(able) on receipt • pay on return • port of refuge

port. portable • portrait(ure)

Port. Portugal • Portuguese

pos. position • positive

POS point of sale • *Med.* polycystic ovary syndrome • Port of Spain (Trinidad) • *Computing* product of sums (in **POS expression**)

posn position

pos. pro. possessive pronoun

poss. possession • *Grammar* possessive • possible • possibly

POSS Palomar Observatory Sky Survey • passive optical suveillance system • prototype optical surveillance system

POSSLQ ('pɒsəl,kjuː) *Colloquial* (USA) person of the opposite sex sharing living quarters

Possum ('pɒsəm) patient-operated selector mechanism (phonetic spelling of POSM)

POST (pəʊst) Parliamentary Office of Science and Technology • point-of-sales terminal (in supermarkets, etc.) • *Computing* power-on self test

posth. (*or* **posthum.**) posthumous(ly)

pot. potash • potassium • potential • potentiometer

poul. poultry

POUM (Spain) Partido Obrero de Unificación Marxista (Workers' Party of Marxist Unity)

POUNC Post Office Users' National Council

p.o.v. privately owned vehicle

POV *Films* point of view

POW please oblige with • Prince of Wales • prisoner of war

powd. powder

POY *Textiles* partially oriented yarn

pp past participle • per procurationem (Latin: by authority of; in correspondence, used by signatory on behalf of someone else) • (ital.) *Music* pianissimo (Italian; very quietly)

pp. pages

p.p. parcel post • per person • per procurationem (*see under* pp) • play or pay • post-paid • *Med.* post prandium (Latin: after a meal; in prescriptions) • prepaid • present position • privately printed

PP *international civil aircraft marking for* Brazil • *British vehicle registration for* Luton • *Geophysics* Pacific plate • parcel post • parish priest • parliamentary papers • (Spain) Partido Popular (Popular Party) • Pastor Pastorum (Latin: Shepherd of the Shepherds) • Past President • Pater Patriae (Latin: Father of his Country) • Patres (Latin: Fathers) • *Med.* pellagra-preventive (in **PP factor**, former name for the vitamin nicotinic acid) • permanent pass • petrol point • *Chem.* phenolphthalein (test paper for alkalinity) • pilot plant • *Chem.* polyprop(yl)ene • prepositional phrase

p.p.a. *Chem.* polyphosphoric acid

PPA Pakistan Press Association • Parti Populaire Algérian (French: Popular Party of Algeria) • Periodical Publishers' Association • Pre-School Playgroups Association

PPARC Particle Physics and Astronomy Research Council

ppb parts per billion

PPB paper, printing, and binding • party political broadcast • planning-programming-budgeting (system) • private posting box

PPBAS planning-programming-budgeting-accounting system

PPBS planning-programming-budgeting system

PPC Patres Conscripti (Latin: Conscript Fathers, members of the Roman Senate) • (*or* **p.p.c.**) pour prendre congé (French: to take leave) • Professional Purposes Committee • *Med.* progressive patient care • (USA) Public Power Corporation

PPCLI Princess Patricia's Canadian Light Infantry

PPCS (New Zealand) Primary Producers' Cooperation Society

ppd post-paid • prepaid

PPD (Puerto Rico) Partido Popular Democrático (Popular Democratic Party) • *Med.* purified protein derivative (of tuberculin)

PPE personal protective equipment • philosophy, politics, and economics (course at Oxford University)

PPF *Insurance* (USA) personal property floater

PPFA Planned Parenthood Federation of America

PPG Pacific proving grounds

pph. pamphlet

PPH (USA) paid personal holidays • *Med.* post-partum haemorrhage

p.p.i. parcel post insured

PP$_i$ *Biochem.*, *symbol for* pyrophosphate

PPI (USA) patient package insert (instructions included with a prescription) • *Radar* plan-position indicator • *Insurance* policy proof of interest • *Economics* producer price index

PPInstHE Past President of the Institute of Highway Engineers

PPIStructE Past President of the Institution of Structural Engineers

PPITB Printing and Publishing Industry Training Board

PPK Polizei Pistole Kriminal (German: police criminal pistol)

PPL Phonographic Performance Limited • private pilot's licence

pple participle

PPLO *Microbiol.* pleuropneumonia-like organism(s)

ppm pages per minute • parts per million • pulse per minute

PPM pages per minute • *Electronics* peak programme meter • *Telecom.* pulse-position modulation

PPMA Produce Packaging and Marketing Association

PPN *Computing* public packet network

PPO (USA) *Med.* preferred-provider organization

ppp (ital.) *Music* pianississimo (Italian; as quietly as possible)

PPP Pakistan's People's Party • *Biochem.* pentose phosphate pathway • (Gambia, Guyana) People's Progressive Party • personal pension plan • *Computing* point-to-point protocol • private patients plan • psychology, philosophy, and physiology (course at Oxford University) • *Economics* purchasing power parity

PPPS (or **ppps**) post post postscriptum (Latin: third postscript)

ppr paper • (or **p.pr.**) present participle • proper

PPR printed paper rate (of postage)

PPRA Past President of the Royal Academy

PPRBA Past President of the Royal Society of British Artists

PPRBS Past President of the Royal Society of British Sculptors

PPRE Past President of the Royal Society of Painter-Etchers and Engravers

p.pro. per procurationem (*see under* pp)

PPROI Past President of the Royal Institute of Oil Painters

PPRTPI Past President of the Royal Town Planning Institute

PPS Parliamentary Private Secretary • *Med.* pelvic pain syndrome • *Biochem.* pentose phosphate shunt • (or **pps**) post postscriptum (Latin: further postscript) • (Australia) prescribed payments system • Principal Private Secretary • (ital.)

Proceedings of the Prehistoric Society • purchasing power standard

PPSIAD Past President of the Society of Industrial Artists and Designers

ppt. *Chem.* precipitate

PPT Parti Progressiste Tchadien (Chad Progressive Party) • *Electrical engineering* peak power transfer

PPTA (New Zealand) Post-primary Teachers Association

pptd *Chem.* precipitated

pptg *Chem.* precipitating

pptn *Chem.* precipitation

ppty property

PPU Peace Pledge Union • Primary Producers' Union

pq previous (*or* preceding) question

PQ parliamentary question • *Politics* (Canada) Parti Québecois (French: Quebec Party) • personality quotient • Province of Quebec

pr painter • pair • paper • per • power

pr. prayer • *Finance* (USA) preferred (stock) • present • pressure • price • print(ed) • printer • printing • pronoun • proper • prove • provincial

p.r. parcel receipt • *Med.* per rectum (Latin: by the rectum)

Pr (ital.) *Physics, symbol for* Prandtl number • *Chem., symbol for* praseodymium • *Chem., symbol for* propyl group (in formulae)

Pr. Praça (Portuguese: square; in place names) • *Finance* (USA) preferred (stock) • Priest • Prince • Protestant • Provençal

PR *British vehicle registration for* Bournemouth • parliamentary report • *Med.* partial remission • partial response • (Chile) Partido Radical (Radical Party) • (France) Parti Républicain (Republican Party) • *Computing* pattern recognition • (or **P/R**) payroll • percentile rank • performance ratio • *Law* personal representative • *airline flight code for* Philippine Airlines • photographic reconnaissance • Pipe Rolls (public records, 1130–1832) • plotting and radar • Populus Romanus (Latin: the Roman people) • postal regulations • preliminary report • Pre-Raphaelite • press release • press representative • *postcode for* Preston • *Boxing* prize ring • production rate • profit rate • progress report • project report • proportional representation • public relations • Puerto Rican • Puerto Rico • purchase request

PRA *Med.* plasma renin activity • President of the Royal Academy • probabilistic risk analysis (*or* assessment) • (USA) Public Roads Administration

prag. pragmatic • pragmatism

PRB People's Republic of Bulgaria • Pre-Raphaelite Brotherhood

PRBS President of the Royal Society of British Sculptors

PRC People's Republic of China • *Med.* plasma renin concentration • (USA) Postal Rate Commission • post Romam conditam (Latin: after the foundation of Rome) • (USA) Price Regulation Committee

PRCA President of the Royal Cambrian Academy • Public Relations Consultants' Association

prchst parachutist

prcht parachute

PRCP President of the Royal College of Physicians

prcs. process

PRCS President of the Royal College of Surgeons

PRE Petroleum Refining Engineer • President of the Royal Society of Painter-Etchers and Engravers

Preb. Prebend(ary)

prec. preceding • precision

Prec. Precentor

PRECIS ('preɪsɪ) preserved context index system

pred. predicate • (*or* **predic.**) predicative

pref. preface • prefatory • preferably • preference • preferred • prefix

Pref. Prefect

prehist. prehistoric(al) • prehistory

prej. prejudice

prelim. preliminary

prem. premium

pre-mRNA *Biochem.* precursor messenger RNA

PrEng. (USA) Professional Engineer

prep. preparation • preparatory • preposition

prepd prepared

prepg preparing

prepn preparation

PREPP *Med.* Post-Registration Education and Practice Project (for establishing standards of nursing practice)

pres. present (time) • presentation •

presidency • presidential • presumed • presumptive

Pres. (*or* **Presb.**) Presbyter(ian) • President

press. pressure

PRESTO ('prestəʊ) (USA) program reporting and evaluation system for total operation

presv. preservation • preserve

pret. *Grammar* preterite

prev. previous(ly)

prf proof

PRF Petroleum Research Fund • *Electronics* pulse repetition (*or* recurrence) frequency

prfnl professional

prfr proofreader

PRHA President of the Royal Hibernian Academy

pri. primate • primer • priority • private

PRI (Mexico) Partido Revolucionario Institucional (Institutional Revolutionary Party) • Partito Repubblicano Italiano (Italian Republican Party) • Plastics and Rubber Institute • President of the Royal Institute of Painters in Water Colours • *Computing* primary-rate ISDN

PRIA President of the Royal Irish Academy

PRIAS President of the Royal Incorporation of Architects in Scotland

PRIBA President of the Royal Institute of British Architects

PRII Public Relations Institute of Ireland

prim. primary • primate • primer • primitive

primip. *Med.* primipara (woman who has borne one child)

prin. principal • principle

Prin. Principal • Principality

print. printing

PRISA Public Relations Institute of South Africa

prism. prismatic

PRISM ('prɪzəm) (USA) program reliability information system for management

priv. private • privative

prm premium

PRM personal radiation monitor

PRML *Computing* partial-response maximum-likelihood

p.r.n. *Med.* pro re nata (Latin: as the situation demands; in prescriptions)

pro. procedure • proceed • procure • profession(al)

Pro (*or* **pro**) *Biochem.* proline

Pro. Provost

Pr.O Press Officer

PRO Public Record Office • public relations officer

pro-am ('prəu'æm) *Sport* professional–amateur

prob. probability • probable • probably • probate • problem

prob. off. probation officer

proc. procedure • proceedings • process

Proc. Proceedings • Proctor

Proc. Roy. Soc. (ital.) Proceedings of the Royal Society

prod. produce(d) • producer • product • production

prof. profession(al)

Prof. Professor

Prof. Eng. (USA) Professional Engineer

prog. prognosis • *Computing* program • programme • progress • progressive

Prog. Progressive (party, movement, etc.)

PROI President of the Royal Institute of Oil Painters

proj. project • projectile • projection • projector

prol. prologue

PROLOG (*or* **Prolog**) ('prəulɒg) *Computing* programming in logic (a programming language)

prom. promontory • promote(r) • promotion

PROM (prɒm) *Computing* programmable read-only memory

pron. pronominal • pronoun • pronounceable • pronounce(d) • pronouncement • pronouncing • pronunciation

pron. a. pronominal adjective

PRONED ('prəu‚nɛd) Promotion of Non-Executive Directors

pro. note *Finance* promissory note

pronunc. pronunciation

prop. proper(ly) • property • proposition • proprietary • proprietor

PROP Preservation of the Rights of Prisoners

propl proportional

propn proportion

propr proprietor

PRORM Pay and Records Office, Royal Marines

pros. prosodical • prosody • *Advertising* prospectus rate

PROS preventive maintenance, repair, and operational services

Pros. Atty Prosecuting Attorney

prost. *Med.* prostate • prostitution

Prot. Protectorate • Protestant

pro tem. pro tempore (Latin: for the time being)

prov. proverb • (*or* **provb.**) proverbial(ly) • province • provincial • provisional

Prov. Provençal • Provence • *Bible* Proverbs • Province • Provost

Prov. GM *Freemasonry* Provincial Grand Master

Provo. ('prəuvəu) Provisional (a member of PIRA)

prox. proximo (Latin: in (*or* of) the next (month))

prox. acc. proxime accessit (Latin: (he *or* she) came nearest; next in order of merit to the winner)

prox. luc. proxima luce (Latin: the day before)

pr.p. present participle

PrP *Biochem.* prion protein (constituent of scrapie agent)

PRP petrol refilling point • profit- (*or* performance-)related pay

pr. pr. praeter propter (Latin: about, nearly)

PRR *Electronics* pulse repetition rate

prs pairs

PRs *symbol for* Pakistan rupee (monetary unit)

PRS Performing Right Society Ltd • President of the Royal Society

PRSA President of the Royal Scottish Academy • Public Relations Society of America

prsd pressed

PRSE President of the Royal Society of Edinburgh

PRSH President of the Royal Society for the Promotion of Health

Pr.ST (USA) Prairie Standard Time

PRSW President of the Royal Scottish Water Colour Society

PRT (USA) personal rapid transit • petroleum revenue tax

prtg printing

PRU photographic reconnaissance unit(s) (in the RAF)

PRUAA President of the Royal Ulster Academy of Arts

Prus. Prussia(n)

p.r.v. pour rendre visite (French: to return a call)

PRV pressure-reducing valve

PRWA President of the Royal West of England Academy

PRWS President of the Royal Society of Painters in Water Colours

ps *Military* passed school of instruction (of officers) • *Physics, symbol for* picosecond(s) • postscript

ps. pieces • pseudonym

p.s. particle size • pull switch

Ps *Physics, symbol for* positronium

Ps. *Bible* (Book of) Psalms • Psalm

PS *British vehicle registration for* Aberdeen • paddle steamer • Parliamentary Secretary • (Portugal) Partido Socialista (Socialist Party) • (France) Parti Socialiste (Socialist Party) • passenger steamer • Pastel Society • penal servitude • Permanent Secretary • *Biochem.* phosphatidyl serine • *Botany* photosystem (in **PSI, PSII**) • *Linguistics* phrase structure • *Electronics* plasma switch • *Meteorol.* plus • Police Sergeant • *Chem.* polystyrene • postscript • *Computing* PostScript • power steering • press secretary • private secretary • Privy Seal • *Theatre* prompt side (i.e. the actor's left) • *Biochem.* protein synthesis • *Physics* proton synchrotron • Provost Sergeant • (Italy) Pubblica Sicurezza (police force) • (USA) public school • (Australia) public service (equivalent to the civil service) • *airline flight code for* Ukraine International Airlines

PS/2 *Computing* Personal System/2

psa *indicating* Graduate of RAF Staff College

Psa. *Bible* (Book of) Psalms • Psalm

PsA *Astronomy* Piscis Austrinus

PSA Passenger Shipping Association • Petty Sessions Area • *Physics* phase-shift analysis • Photographic Society of America • pleasant Sunday afternoon • Political Studies Association of the United Kingdom • President of the Society of Antiquaries • (Australia) Prices Surveillance Authority • probabilistic safety analysis (*or* assessment) • Property Services Agency • *Med.* prostatic specific antigen • (New Zealand) Public Service Association

PSAB Public Schools Appointments Bureau

PSAC (USA) President's Science Advisory Committee

p's and q's *Colloquial, indicating* manners

(phonetic spelling of p(lea)se and (than)k you's)

PSAT (USA) Preliminary Scholastic Aptitude Test

PSB pistol, small-bore • (Japan) Postal Savings Bureau

PSBA Public School Bursars' Association

PSBR public sector borrowing requirement

psc *Military* passed staff college

Psc *Astronomy* Pisces

PSC (Belgium) Parti Social Chrétien (Francophone Christian Social Party) • *Meteorol.* polar stratospheric clouds • Professional Services Committee • (USA) Public Service Commission

PSCD patrol service central depot

PSCM *Meteorol.* progression, southerly, cyclonicity, and meridionality (in **PSCM index**)

PSD (Portugal) Partido Social Democrata (Social Democratic Party) • (Madagascar) Parti Social Démocrate (Social Democratic Party) • pay supply depot • *Law* Petty Sessional Divison • *Physics* position-sensitive detector • (USA) prevention of significant deterioration • *Physics* proportional scintillation detector

PSDI Partito Socialista Democratico Italiano (Italian Democratic Socialist Party)

PSDR public sector debt requirement

PSE Pacific Stock Exchange • pale soft exudate (in meat processing) • Pidgin Sign English • *Computing* programming (*or* project) support environment • psychological stress evaluator (lie detector)

pseud. pseudonym

psf (*or* **p.s.f.**) pounds per square foot

PSF *Image technol.* point spread function

PSG *Linguistics* phrase-structure grammar

PSHFA Public Servants Housing Finance Association

psi (*or* **p.s.i.**) pounds (*or* pound-force) per square inch

PSI Partito Socialista Italiano (Italian Socialist Party) • *Education* personalized system of instruction • Pharmaceutical Society of Ireland • Policy Studies Institute

psia pounds per square inch, absolute

PSIAD President of the Society of Industrial Artists and Designers

psid pounds per square inch, differential

psig pounds per square inch, gauge

PSIS Permanent Secretaries Committee on the Intelligence Services

PSIUP Partito Socialista Italiano di Unità Proletaria (Italian Socialist Party of Proletarian Unity)

PSK *Telecom.* phase shift keying

PSL Paymaster Sublieutenant • (USA) Primary Standards Laboratory • *Economics* private-sector liquidity • public-sector loan(s)

PSL/PSA *Computing* problem statement language/problem statement analyser

psm passed school of music (Certificate of the Royal Military School of Music)

PSM product sales manager

PSMA President of the Society of Marine Artists

PSN *Computing* packet-switching network • *Meteorol.* position • *Computing* public switched network

PSNC Pacific Steam Navigation Company

PSO personal staff officer • Principal Scientific Officer

PSOE (Spain) Partido Socialista Obrero Español (Spanish Workers Socialist Party)

PSP (Netherlands) Pacifistich Socialistische Partij (Pacifist Socialist Party) • *Med.* phenolsulphonphthalein (in **PSP test** for kidney function)

PSPACE *Computing* polynomial space

PSR *Commerce* profit-sharing ratio • *Physics* proton storage ring

PSRAM (ˌpiːˈɛsˌræm) pseudo static RAM

PSRO (USA) Professional Standards Review Organization

Pss Psalms

PSS *Computing* packet switching service • *Computing* Packet SwitchStream (PPN of British Telecom) • Palomar Sky Survey • *Med.* physiological saline solution • (*or* **pss.**) postscripts • power-system stabilizer • Printing and Stationery Service • professional services section

PSSC Personal Social Services Council

psso *Knitting* pass slipped stitch over

PST (USA, Canada) Pacific Standard Time

pstl postal

PSTN public switched telephone network

PSU (Italy) Partito Socialista Unitario (Unitary Socialist Party) • police support unit • *Computing* power supply unit • process support unit

p. surg. plastic surgery

PSV public service vehicle

PSW *Computing* processor (*or* program) status word • psychiatric social worker

psych. psychic(al) • psychological • psychologist • psychology

psychoanal. psychoanalysis

psychol. psychological • psychologist • psychology

pt part • patient • payment • pint (abbrev. *or* symbol) • point (abbrev. *or* (in printing) symbol) • port

pt. *Grammar* preterite

p.t. part time • past tense • point of turn(ing) • primary target • pro tempore (Latin: for the time being)

Pt *Chem., symbol for* platinum • Point (in place names) • Port (in place names)

PT *international civil aircraft marking for* Brazil • *British vehicle registration for* Newcastle-upon-Tyne • (USA) Pacific Time • (USA) patrol torpedo (in **PT boat**) • *Insurance* perte totale (French: total loss) • *Physics* perturbation theory • *Physics* phase transition • physical therapy • physical training • physiotherapist • *British fishing port registration for* Port Talbot • postal telegraph • post town • preferential treatment • *Med.* previously treated • Public Trustee • pupil teacher • purchase tax • *airline flight code for* West Air Sweden

Pta *symbol for* peseta (Spanish monetary unit) • Pretoria

PTA Parent–Teacher Association • Passenger Transport Authority • *Med.* percutaneous transluminal angioplasty • Pet Traders' Association • *Med.* plasma thromboplastin antecedent • *Med.* post-traumatic amnesia • preferential trade area • Printing Trades Alliance • *Chem.* p-terephthalic acid

ptbl. portable • potable

PT boat (USA) patrol torpedo boat

PTBT partial test-ban treaty

PTC *Med.* percutaneous transhepatic cholangiography • *Diving* personnel transfer capsule • *Genetics* phenylthiocarbamide (referring to the inherited ability to taste it) • photographic type composition • *Med.* plasma thromboplastin component • primary training centre • Public Services, Tax, and Commerce Union (formed from merger of IRSF and NUCPS)

PTCA *Med.* percutaneous transluminal coronary angioplasty

ptd painted • printed

Pte Plate (in books, etc.) • *Military* Private • (India, etc.) private limited company (after company name; equivalent to Ltd)

PTE Passenger Transport Executive • post-test examination

pt ex. (*or* **pt exch.**) part exchange

PTFCE *Chem.* polytrifluorochloroeth(yl)ene

PTFE polytetrafluoroeth(yl)ene

ptg printing

Ptg *Geology* Pleistogene

Ptg. Portugal • Portuguese

PTH *Biochem.* parathyroid hormone • (USA) public teaching hospital

PTI physical training instructor • Press Trust of India • *Computing* public tool interface

PTIME *Computing* polynomial time

PTM *Telecom.* pulse-time modulation

PTMA phosphotungstomolybdic acid (used in manufacture of certain pigments)

ptn partition • portion

PTN *Colloquial* pay through the nose • public telephone network (of British Telecom) • public transportation network

ptnr partner

PTO (USA) Patent and Trademark Office • (*or* **pto**) please turn over • *Astronautics* power takeoff • public telecommunications operator • Public Trustee Office

ptp past participle

pt/pt point-to-point

ptr printer

pts parts • payments • pints • points • ports

Pts. Portsmouth

PTS Philatelic Traders' Society • pressurized thermal shock • printing technical school

ptsc *Military* passed technical staff college

PTSD *Med.* post-traumatic stress disorder

pts/hr parts per hour

Ptsmth Portsmouth

PTT *Med.* partial thromboplastin time • Postal, Telegraph, and Telephone Administration

pt-tm. part-time

PTV (USA) public television (noncommercial television)

p.t.w. per thousand words

pty party

Pty (Australia, South Africa, etc.) Proprietary (after a company name; equivalent to Ltd)

p.u. paid up

Pu *Chem., symbol for* plutonium

PU *British vehicle registration for* Chelmsford • *Med.* passed urine • pick-up • *airline flight code for* Pluna • *Chem.* polyurethane • *Computing* processing unit • public utility

pub. public • publican • publication • publish(ed) • publisher • publishing

Pub. *Advertising* publishers' announcement

pubd published

pub. doc. public document

publ. public • publican • publication • publicity • published • publisher • publishing

pubn publication

pubr publisher

pub. wks public works

PUC papers under consideration • pick-up car • (USA) Public Utilities Commission

PUD pick-up and delivery • (USA) planned unit development (a large condominium)

pug. pugilist

PUHCA (USA) Public Utility Holding Company Act

pulv. *Pharmacol.* pulvis (Latin: powder)

p.u.m.s. permanently unfit for military service

pun. punish(ment)

punc. punctuation

Punj. Punjab

PUO *Med.* pyrexia (fever) of unknown origin

Pup *Astronomy* Puppis

PUP People's United Party • (USA) Princeton University Press • (Northern Ireland) Progressive Unionist Party

pur. (*or* **purch.**) purchase(r) • purification • purify • (*or* **purp.**) purple • pursuit

PURV powered underwater research vehicle

p.u.s. permanently unfit for service

PUS Parliamentary Undersecretary • Permanent Undersecretary

PUVA ('puːvə) *Med.* psoralen ultraviolet A (treatment for psoriasis)

PUWP Polish United Workers' Party

pv. *Microbiol.* pathovar

p.v. *Med.* per vaginam (Latin: by the vagina)

PV *British vehicle registration for* Ipswich • *Physics* parity violation • patrol vessel • petite vitesse (French: goods *or* slow train) • *Sport* pole vault • positive vetting • power–voltage • pressure vessel • (*or* **P–V**) pressure–volume • profit–volume (as in **PV chart, PV ratio**)

PVA polyvinyl acetate (synthetic resin) • *Chem.* polyvinyl alcohol

PVB *Chem.* polyvinyl butyral

p.v.c. pigment volume concentration (in paint technology)

PVC *Computing* permanent virtual circuit • polyvinyl chloride (synthetic resin)

PVCH *Chem.* polyvinylcyclohexane

PVD *Med.* peripheral vascular disease • physical vapour deposition

PVDA (*or* **PvdA**) (Netherlands) Partij van de Arbeid (Labour Party)

PVDC *Chem.* polyvinylidenechloride

PVF polyvinyl fluoride (synthetic resin)

PVG *Chem.* polyvinylene glycol

PVO Principal Veterinary Officer

PVP polyvinyl pyrrolidone (synthetic resin)

PVS *Med.* persistent vegetative state • post-Vietnam syndrome • *Med.* postviral (fatigue) syndrome (myalgic encephalomyelitis)

PVSM (India) Param Vishishc Seva Medal

p.v.t. par voie télégraphique (French: by telegraph)

Pvt. *Military* Private

PVT *Chem.* polyvinyltoluene • pressure, volume, temperature

PVTCA *Chem.* polyvinyltrichloro-acetate

PVX *Microbiol.* potato virus X

PVY *Microbiol.* potato virus Y

p.w. per week

PW *British vehicle registration for* Norwich • *British fishing port registration for* Padstow • *Physics* plane wave • policewoman • power windows (in vehicles) •

prisoner of war • public works • *Electronics* pulse width

PWA *Med.* person with Aids • (USA) Public Works Administration (1933–43)

PWC (USA) personal watercraft • post-war credits (after World War II)

pwd powered

PWD Public Works Department

PWE Political Welfare Executive

PWLB Public Works Loan Board

PWM *Telecom.* pulse-width modulation

PWO *Military* Prince of Wales's Own

p.w.p. price when perfect

pwr power

PWR *Nuclear engineering* pressurized-water reactor

pwr sup. power supply

pwt pennyweight

PWV (formerly) Pretoria-Witwatersrand-Vereeniging (now Gauteng province)

px Pedro Ximénez (grape; referring to sweet wines and sherries)

PX part exchange (in property or motor advertisements) • physical examination • please exchange • *British vehicle registration for* Portsmouth • (USA) Post Exchange (army or navy retail store) • *Telephony* private exchange

pxt pinxit (Latin: (he *or* she) painted it)

py *Chem.* pyridine (used in formulae)

PY *British vehicle registration for* Middlesbrough • *international vehicle registration for* Paraguay • *airline flight code for* Surinam Airways

PYB *Accounting* preceding-year basis

pyo (*or* **PYO**) pick your own (fruit, etc.)

pyro. (*or* **pyrotech.**) pyrotechnics

Pyx *Astronomy* Pyxis

Pz *Geology* Palaeozoic

PZ *British vehicle registration for* Belfast • *British fishing port registration for* Penzance • *international civil aircraft marking for* Surinam

PZI *Med.* protamine zinc insulin (diabetes treatment)

PZS President of the Zoological Society

PZT *Astronomy* photographic zenith tube

Q

q (ital.) *Physics, symbol for* density of heat flow rate • (ital.) *Physics, symbol for* electric charge • (bold ital.; *or* **q**ᵢ) *Maths., symbol for* a generalized coordinate • (ital.) *Chem., symbol for* partition function (individual entity) • *Physics, symbol for* quark • (ital.) *symbol for* quintal (100 kg) • (ital.) *Meteorol., symbol for* specific humidity • *Meteorol., symbol for* squall

q. quaere (Latin: inquire) • quaque (Latin: every) • quart • quarter • quarterly • quarto • quasi (Latin: almost) • quench • query • question • quick • quire

Q *Biochem., symbol for* glutamine • (ital.) *Chem., symbol for* partition function (whole system) • *international vehicle registration for* Qatar • *Electronics* quality (in **Q factor**) • (ital.) *Electrical engineering, symbol for* quality factor • (ital.) *Physics, symbol for* quantity of electricity (i.e. electric charge) • (ital.) *Physics, symbol for* quantity of heat • (ital.) *Physics, symbol for* quantity of light • *Geology* Quaternary • *Chess, symbol for* queen • query (hence camouflaged or disguised, as in **Q-boat** (*or* **Q-ship**), **Q car**) • *symbol for* quetzal (Guatemalan monetary unit) • (ital.) *Electrical engineering, symbol for* reactive power • (ital.) *Chemical engineering, symbol for* throughput • *indicating* the highest level of security according to the Nuclear Regulatory Commission (as in **Q clearance**)

Q. (*or* **Q**) quantity • quarterly • Quartermaster • quarter-page (in advertisement placing) • Quarto (Shakespeare manuscript) • Quebec • Queen (*or* Queen's) • Queensland • question • *pseudonym of* (Sir Arthur Thomas) Quiller-Couch (1863–1944, British writer)

Q4 *airline flight code for* Mustique Airways

Q7 *airline flight code for* Qatar Airways

Q9 *airline flight code for* Interbrasil Star

3Q *airline flight code for* Yunnan Airlines

8Q *international civil aircraft marking for* the Maldives

9Q *international civil aircraft marking for* Zaïre (now Democratic Republic of Congo)

q.a. quick assembly

QA (*or* **Q/A**) qualification approval • quality assurance • quarters allowance

QAB *Church of England* Queen Anne's Bounty

QADS quality-assurance data system

QAIMNS Queen Alexandra's Imperial Military Nursing Service

QAM *Telecom.* quadrature amplitude modulation

Q & A question and answer

Qantas ('kwɒntəs) Queensland and Northern Territory Aerial Service (Australian national airline)

QARANC Queen Alexandra's Royal Army Nursing Corps

QARNNS Queen Alexandra's Royal Naval Nursing Service

QAS *Chem.* quaternary ammonium sulphanilamide

QB (*or* **q.b.**) *Football* (USA) quarterback • *Law* Queen's Bench • *Chess* queen's bishop

Qbc Quebec

QBD Queen's Bench Division

QBE *Computing* query by example

QBI *Slang* quite bloody impossible

QBO *Meteorol.* quasi-biennial oscillation

QBP *Chess* queen's bishop's pawn

QC quality control • Quartermaster Corps • Queen's College • Queen's Counsel • *Law* quit claim

QCD *Physics* quantum chromodynamics

QCE quality-control engineering

Q-Celtic *Linguistics, indicating* one of two main groups of languages developed from Common Celtic. *See also* P-Celtic

QC Is Queen Charlotte Islands (Canada)

QCM *Physics* quark cluster (*or* confinement) model

QCR quality-control reliability

QCT quality-control technology · *Med.* quantitative computed tomography

QCVSA Queen's Commendation for Valuable Service in the Air

q.d. *Med.* quaque die (Latin: every day) · quasi dicat (Latin: as if one should say) · quasi dictum (Latin: as if said)

QDE *Physics* quantum detection efficiency

QDRI (USA) qualitative development requirement information

q.d.s. *Med.* quater in die sumendus (Latin: to be taken four times a day)

q.e. quod est (Latin: which is)

QE *Physics* quantum efficiency · quantum electronics

QE2 (ital.) Queen Elizabeth II (passenger liner)

QED *Physics* quantum electrodynamics · quod erat demonstrandum (Latin: which was to be proved)

QEF quod erat faciendum (Latin: which was to be done)

QEH Queen Elizabeth Hall (London)

QEI quod erat inveniendum (Latin: which was to be discovered *or* found out)

QEO Queen Elizabeth's Own

QER (ital.) Quarterly Economic Review

QF *airline flight code for* Qantas Airways · *Electronics* quality factor · quick-firing

QFA *Computing* quick file access

QFD *Commerce* quality function deployment · *Physics* quantum flavourdynamics

QFSM Queen's Fire Service Medal for Distinguished Service

QFT *Physics* quantum field theory

QG Quartermaster-General · quartiere generale (Italian: headquarters) · quartier-général (French: headquarters)

QGM Queen's Gallantry Medal

QGP *Physics* quark-gluon plasma

q.h. *Med.* quaque hora (Latin: every hour)

QHC Queen's Honorary Chaplain

QHDS Queen's Honorary Dental Surgeon

QHM Queen's Harbour Master

QHNS Queen's Honorary Nursing Sister

QHP Queen's Honorary Physician

QHS Queen's Honorary Surgeon

QI quartz–iodine (in **QI lamp**)

QIC *Computing* quarter-inch committee (*or* cartridge)

q.i.d. *Med.* quater in die (Latin: four times a day)

QIP quiescat in pace (Latin: may (he *or* she) rest in peace)

QISAM *Computing* queued indexed sequential access method

qk quick

QKt *Chess* queen's knight

QKtP *Chess* queen's knight's pawn

ql quintal (100 kg)

q.l. *Med.* quantum libet (Latin: as much as you please; in prescriptions)

QL *airline flight code for* Lesotho Airways · Queen's Lancers · *Computing* query language

Qld (*or* **QLD**) Queensland

QLF *Commerce* quality loss function

qlty quality

qly quarterly

qm. quomodo (Latin: by what means)

q.m. *Med.* quaque mane (Latin: every morning; in prescriptions)

QM *airline flight code for* Air Malawi · *Physics* quantum mechanics · *Physics* quark model · Quartermaster · Queen's Messenger

QMAAC Queen Mary's Army Auxiliary Corps

QMC Quartermaster Corps · Queen Mary College, London (now part of QMW)

QMD *Physical chem.* quantum molecular dynamics

Q. Mess. Queen's Messenger

QMG Quartermaster-General

QMGF Quartermaster-General to the Forces

Qmr Quartermaster

QMR qualitative material requirement

QMS *Physics* quadrupole mass spectrometry (*or* spectrometer) · Quartermaster Sergeant

QMW Queen Mary and Westfield College, London

qn question · quotation

q.n. quaque nocte (Latin: every night; in prescriptions)

Qn Queen

QN *Chess* queen's knight

QNI Queen's Nursing Institute

QNP *Chess* queen's knight's pawn

QNS quantity not sufficient

qnt. quintet

qnty quantity

QO *Navy* qualified in ordnance · qualified officer

QOCH Queen's Own Cameron Highlanders

QOOH Queen's Own Oxfordshire Hussars

Q(ops) Quartering (operations)

QOR qualitative operational requirement

q.p. *Med.* quantum placet (*see* q.pl.)

QP *airline flight code for* Airkenya Aviation • qualification pay • *Chess* queen's pawn • *Computing* query processing

QPC Qatar Petroleum Company

QPFC Queen's Park Football Club

q.pl. *Med.* quantum placet (Latin: as much as seems good; in prescriptions)

QPM Queen's Police Medal

QPO *Astronomy* quasi-periodic oscillation

QPR Queen's Park Rangers (football club)

qq. questions

Qq. Quartos (Shakespeare manuscripts)

QQ *airline flight code for* Reno Air

qq. hor. *Med.* quaque hora (Latin: every hour; in prescriptions)

qq.v. quae vide (Latin: which (words, items, etc.) see; textual cross reference)

qr. quarter • quarterly • quire

QR Quarterly Review • *Chess* queen's rook • *Marketing* quick response • *symbol for* riyal (monetary unit of Qatar)

QRA quick reaction alert (of military aircraft)

QRIH Queen's Royal Irish Hussars

QRP *Chess* queen's rook's pawn

QRR Queen's Royal Rifles

qrs quarters

QRV Qualified Valuer, Real Estate Institute of New South Wales

q.s. *Med.* quantum sufficit (Latin: as much as will suffice; in prescriptions) • quarter section (of land)

QS quadraphonic-stereophonic (of audio equipment) • quantity surveyor • quarantine station • quarter sessions • Queen's Scholar • *Building trades* quick sweep • *airline flight code for* Tatra Air

QSAR quantitative structure-activity relationship

QSM *Physics* quantum-statistical model • (New Zealand) Queen's Service Medal

QSMR quantitative structure-metabolism relationship

QSO *Astronomy* quasistellar object • (New Zealand) Queen's Service Order

QSS *Astronomy* quasistellar source

QSTOL ('kjuː,stɒl) *Aeronautics* quiet short takeoff and landing

QSTS quadruple screw turbine ship

qt *symbol for* quart

qt. quantity • quartet

q.t. (*or* **QT**) *Colloquial* quiet (as in **on the q.t.**)

qtly quarterly

qto quarto

QTOL ('kjuː,tɒl) *Aeronautics* quiet takeoff and landing

qtr quarter

qty quantity

qu. quart • quarter • quarterly • queen • query • question

QU *airline flight code for* Uganda Airlines

quad. quadrant • quadraphonic • quadrilateral • quadruple • (*or* **quadr., quadrupl.**) quadruplicate

qual. qualification • qualitative • quality

qualgo ('kwælgəʊ) quasi-autonomous local government organization

quango ('kwæŋgəʊ) quasi-autonomous nongovernmental (*or* national government) organization

quant. quantitative • quantity

quant. suff. *Med.* quantum sufficit (*see under* q.s)

quar. (*or* **quart.**) quarter • quarterly

quasar ('kweɪzɑː) *Astronomy* quasistellar object

quat. *Med.* four (Latin *quattuor*; in prescriptions)

QUB Queen's University, Belfast

Que. Quebec

ques. question

questn. questionnaire

quint. quintuplicate

QUIP (kwɪp) *Computing* query interactive processor

quor. quorum

quot. quotation • quote

quotid. *Med.* quotidie (Latin: daily)

q.v. *Med.* quantum vis (Latin: as much as you wish) • quod vide (Latin: which (word, item, etc.) see; textual cross reference)

QV *airline flight code for* Lao Aviation

QVR Queen Victoria Rifles

QW *airline flight code for* Turks and Caicos Airways

QWERTY ('kwɜːtɪ) *indicating* a standard keyboard in English-speaking countries (from the first six of the upper row of keys on a typewriter)

QWL quality of work(ing) life

qy quay • query

7QY *international civil aircraft marking for* Malawi

qz quartz

R

r (ital.) *Electricity, symbol for* internal resistance • (ital.) *Maths., symbol for* a polar coordinate • (bold ital.) *Maths., Physics, symbol for* position vector • (ital.) *symbol for* radius • (bold ital.) *Maths., symbol for* radius vector • (ital.) *Ecology, symbol for* rate of increase (as in **r-strategist**) • *Music* ray (in tonic sol-fa) • *Biochem., symbol for* ribonucleoside (preceding the nucleoside symbol; as in **rA**, ribosyladenine)

r. radius • (USA) railroad • railway • rain • range • rare • ratio • *Commerce* received • recipe • *Printing* recto • red • replacing • reply • reserve • residence • resides • response • retired • right • rises • river • road • rod (unit of length) • rouble • *Cards* rubber • ruled • *Cricket, Baseball, etc.* run(s) • rupee

R *Biochem., symbol for* arginine • *Medieval Roman numeral for* eighty • (ital.) *Chem., symbol for* molar gas constant • *Chem., symbol for* a radical (in formulae, as in **ROH**) • *symbol for* radius • *British fishing port registration for* Ramsgate • *symbol for* rand (South African monetary unit) • *Maths., symbol for* ratio • *Physics, symbol for* Réaumur (temperature scale) • (ital.) *Chem.* rectus (as in **(R)-butan-2-ol**) • *Electrical engineering, symbol for* resistance • *Chem., symbol for* resonance effect • *Films* (Australia, USA) restricted (certification) • retarder (French: slow down; on a clock or watch regulator) • return (ticket *or* fare) • reverse (on selector mechanism of vehicle with automatic transmission) • *Physics, symbol for* roentgen • *Chess, symbol for* rook • *Microbiol.* rough (in **R-form** of a bacterial colony)

R. (*or* **R**) rabbi • *Politics* Radical • radius • (USA) railroad • railway • Rapido (Italian: express train) • Recht (German: law) • *Med.* recipe (Latin: take; in prescriptions) • recommendation • rector • red • redactor • Regiment • Regina (Latin: Queen) • registered • Regius •

regular (clothing size) • *Navigation* relative • report • Republic • Republican • reserve • response (in Christian liturgy) • Respublica (Latin: Republic) • reward • Rex (Latin: King) • *Military* Rifles • right *or* (in the theatre) stage right • River • road • Roman • Romania(n) • Rome • rosary • rouble • route • Royal • Rue (French: street) • *Nautical* run (i.e. deserted)

R3 *airline flight code for* Armenian Airlines

3R *airline flight code for* Air Moldova International

3 Rs (*or* **three Rs**) reading, (w)riting, and (a)rithmetic

4R *international civil aircraft marking for* Sri Lanka

5R *international civil aircraft marking for* Madagascar

6R *airline flight code for* Air Affaires Afriques

8R *airline flight code for* Avia Air • *international civil aircraft marking for* Guyana

ra. radio

Ra *Chem., symbol for* radium • (ital.) *Physics, symbol for* Rayleigh number

RA *international vehicle registration for* Argentina • *British vehicle registration for* Nottingham • *Meteorol.* rain • Ramblers' Association • Rear-Admiral • reduction of area • *Soccer* Referees' Association • (USA) Regular Army • República Argentina (Spanish: Argentine Republic) • (USA) Resettlement Administration • *Med.* rheumatoid arthritis • *Astronomy* right ascension • Road Association • Royal Academy (*or* Academician) • Royal Artillery • *airline flight code for* Royal Nepal Airlines • *international civil aircraft marking for* Russia

R/A *Finance* refer to acceptor (on a bill of exchange) • return to author

RAA Rabbinical Alliance of America • Regional Arts Association • Royal Academy of Arts • Royal Artillery Association • Royal Australian Artillery

RAAF Royal Australian Air Force • Royal Auxiliary Air Force

RAAFNS Royal Australian Air Force Nursing Service

RAAMC Royal Australian Army Medical Corps

RAANC Royal Australian Army Nursing Corps

Rab. Rabat (Morocco)

Rabb. Rabbinate • Rabbinic(al)

RABDF Royal Association of British Dairy Farmers

RABI Royal Agricultural Benevolent Institution

RAC Railway Association of Canada • Regional Advisory Committee (of the TUC) • Regional Advisory Council • Royal Agricultural College • Royal Armoured Corps • Royal Automobile Club

RACA Royal Automobile Club of Australia

RACE (reɪs) rapid automatic checkout equipment • Research and Development in Advanced Communication Technologies for Europe

RACGP Royal Australian College of General Practitioners

RAChD Royal Army Chaplains' Department

RACI Royal Australian Chemical Institute

RACO Royal Australian College of Ophthalmologists

RACOG Royal Australian College of Obstetricians and Gynaecologists

RACON (or **racon**) ('reɪkɒn) radar beacon

RACP Royal Australasian College of Physicians

RACS Royal Arsenal Cooperative Society • Royal Australasian College of Surgeons

rad Maths., symbol for radian

rad. radar • radiator • radical • radio • radiologist • radiology • radiotherapist • radiotherapy • radius • Maths., Anatomy radix

r.a.d. radiation absorbed dose (of radiation) • rapid automatic drill

Rad. Politics Radical • Radnorshire (former Welsh county)

RAD Med. reflex anal dilation • Royal Academy of Dancing • Royal Albert Docks

RADA ('rɑːdə) Royal Academy of Dramatic Art

radar ('reɪdɑː) radio detection and ranging

RADAR Royal Association for Disability and Rehabilitation

RADC Royal Army Dental Corps

raddol. Music raddolcendo (Italian: becoming calmer)

RADIUS ('reɪdɪəs) Religious Drama Society of Great Britain

RAdm (or **RADM**) Rear-Admiral

radmon ('ræd,mɒn) radiological monitor(ing)

radn radiation

RAE Royal Aerospace (formerly Aircraft) Establishment • Royal Australian Engineers

RAEC Royal Army Educational Corps

RAeroC Royal Aero Club of the United Kingdom

RAeS Royal Aeronautical Society

RAF Rote Armee Faktion (German: Red Army Faction; terrorist group) • Royal Aircraft Factory • Royal Air Force

RAFA Royal Air Forces Association • Royal Australian Field Artillery

RAFBF Royal Air Force Benevolent Fund

RAFES Royal Air Force Educational Service

RAFG Royal Air Force Germany

RAFMS Royal Air Force Medical Services

RAFR Royal Air Force Regiment

RAFRO Royal Air Force Reserve of Officers

RAFSAA Royal Air Force Small Arms Association

RAFSC Royal Air Force Staff College

RAFT Banking revolving acceptance facility by tender

RAFTC Royal Air Force Transport Command

RAFVR Royal Air Force Volunteer Reserve

RAGA Royal Australian Garrison Artillery

RAH Royal Albert Hall

RAHS Royal Australian Historical Society

RAI Radiotelevisione Italiana (Italian broadcasting corporation; originally Radio Audizioni Italiane) • Royal Anthropological Institute

RAIA Royal Australian Institute of Architects

RAIC Royal Architectural Institute of Canada

RAID (reɪd) Computing redundant array of independent (or inexpensive) disks

Raj. Rajasthan (India)

RAL Rutherford Appleton Laboratory (Harwell, Oxfordshire)

rall. *Music* rallentando (Italian: becoming slow)

RALS *Aeronautics* remote augmented lift system

r.a.m. relative atomic mass

RAM *Aeronautics* radar absorbing material • (ræm) *Computing* random-access memory • (ræm) (USA) reverse annuity mortgage • rocket-assisted motor • (Member of the) Royal Academy of Music • Royal Air Maroc (Moroccan airline) • *Freemasonry* Royal Arch Masons

RAMAC Radio Marine Associated Companies

ramb. rambler (rose)

RAMC Royal Army Medical Corps

RAN request for authority to negotiate • Royal Australian Navy

RANC Royal Australian Naval College

R & A Royal and Ancient (Golf Club, St Andrews)

R & B (*or* **r & b**) rhythm and blues • ring and ball (game)

r. & c.c. riot and civil commotion

R & D research and development

R & E research and engineering

R & I Regina et Imperatrix (Latin: Queen and Empress) • Rex et Imperator (Latin: King and Emperor)

r & m reports and memoranda

R & M reliability and marketing

R & R (*or* **R and R**) *Med.* rescue and resuscitation • rest and recreation • rock and roll

R & T research and technology

R&VA Rating and Valuation Association

RANN (USA) Research Applied to National Needs

RANR Royal Australian Naval Reserve

RANVR Royal Australian Naval Volunteer Reserve

RAOB Royal Antediluvian Order of Buffaloes

RAOC Royal Army Ordnance Corps

RAOU Royal Australian Ornithologists' Union

rap. rapid

RAP ready-assembled price • Regimental Aid Post • remedial action programme (*or* plan) • *Med.* right atrial pressure

RAPC Royal Army Pay Corps

RAPID ('ræpɪd) Register for the Ascertainment and Prevention of Inherited Diseases

RAPRA Rubber and Plastics Research Association of Great Britain

RAR Royal Australian Regiment

RARDE Royal Armament Research and Development Establishment

RARE *Computing* réseaux associés pour la recherche européenne (French: Associate Networks for European Research)

RARO Regular Army Reserve of Officers

RAS *Aeronautics* rectified air speed • Royal Agricultural Society • Royal Asiatic Society • Royal Astronomical Society

RASC Royal Army Service Corps (now called Royal Corps of Transport, RCT)

RASE Royal Agricultural Society of England

raser ('reɪzə) radio-frequency amplification by stimulated emission of radiation

RASH *Meteorol.* rain shower

RASN *Meteorol.* rain and snow

RAST *Immunol.* radio-allergosorbent test

rat. rateable • rating • ration

RAT (ræt) rocket-assisted torpedo

RATAN ('reɪtæn) radar and television aid to navigation

RATO ('reɪtəʊ) rocket-assisted takeoff

RATP (France) Régie autonome des transports parisiens (Paris transport authority)

RAuxAF Royal Auxiliary Air Force

RAVC Royal Army Veterinary Corps

RAX *Telephony* rural automatic exchange

r.b. *Sport* right back • rubber band

Rb *Chem., symbol for* rubidium

RB *British vehicle registration for* Nottingham • *Astronomy* radiation belt • radiation burn • reconnaissance bomber (aircraft; as in **RB-57**) • representative body • República Boliviana (Spanish: Republic of Bolivia) • *international vehicle registration for* Republic of Botswana • review body • Rifle Brigade • Royal Ballet • *airline flight code for* Syrian Arab Airlines

RBA (Member of the) Royal Society of British Artists

RBAF Royal Belgian Air Force

RBC *Med.* red blood cell • *Med.* red blood (cell) count • Royal British Colonial Society of Artists

RBE relative biological effectiveness (of radiation)

RBerks Royal Berkshire Regiment

RBG Royal Botanic Gardens (Kew)

r.b.i. require better information • *Baseball* run(s) batted in

RBI resource-based industry • *Skating* right back inside (of skate) • ('rıbı) *Baseball* run(s) batted in

RBK&C Royal Borough of Kensington and Chelsea

rbl. rouble

RBL Royal British Legion

RBNA Royal British Nurses' Association

RBO *Skating* right back outside (of skate)

RBP *Biochem.* retinol-binding protein

RBS *Med.* radionuclide bone scintigraphy • Royal Society of British Sculptors • *Physics* Rutherford backscattering spectroscopy (*or* spectrometry)

RBSA (Member of the) Royal Birmingham Society of Artists

rbt roundabout

RBT random breath testing • *Computing* remote batch terminal

RBY Royal Bucks Yeomanry

r.c. radio code (*or* coding) • reinforced concrete • release clause • reverse course • right centre • rotary combustion • rubber-cushioned

RC *airline flight code for* Atlantic Airways Faeroe Islands • *British vehicle registration for* Nottingham • *Cycling* racing club • *Med.* red (blood) cell • Red Cross • Reformed Church • reinforced concrete • reproductive capacity • (USA) Republican Convention • research centre • Reserve Corps • *Photog.* resin-coated (as in **RC paper**) • *Electronics* resistor-capacitor (*or* resistance-capacitance) • rifle club • *Cycling* road club • Roman Catholic • *Building trades* rough cutting • Royal Commission • *international vehicle registration for* Taiwan (Republic of China)

R/C recredited

RCA *international vehicle registration for* Central African Republic • Rabbinical Council of America • Racecourse Association • Radio Corporation of America • *Med.* right coronary artery • (Member of the) Royal Cambrian Academy • (Member of the) Royal Canadian Academy of Arts • Royal College of Art • Royal Company of Archers • Rural Crafts' Association

RCAC Royal Canadian Armoured Corps

RCAF Royal Canadian Air Force

RCamA (Member of the) Royal Cambrian Academy

RCAMC Royal Canadian Army Medical Corps

RCASC Royal Canadian Army Service Corps

RCB *international vehicle registration for* Republic of the Congo • *Theatre* right centre back (of stage)

RCC recovery control centre • *Med.* renal cell carcinoma • rescue coordination centre • Roman Catholic Chaplain • (*or* **RCCh**) Roman Catholic Church • Rural Community Council

rcd received

RCD (Tunisia) Rassemblement Constitutionnel Démocratique (pro-government political party) • Regional Cooperation for Development (association of Asian countries) • *Electronics* residual current device

RCDC Royal Canadian Dental Corps

RCDS Royal College of Defence Studies

RCE *Engineering* rotary combustion engine

RCFCA Royal Canadian Flying Clubs' Association

RCGA Royal Canadian Golf Association

RCGP Royal College of General Practitioners

RCGS Royal Canadian Geographical Society

RCH railway clearing house • *international vehicle registration for* Republic of Chile

RCHA Royal Canadian Horse Artillery

RCHM Royal Commission on Historical Monuments

r.c.i. radar coverage indicator

RCI Royal Canadian Institute

RCJ Royal Courts of Justice

RCL Royal Canadian Legion • ruling case law

RCM radar (*or* radio) countermeasures • *Med.* radiological contrast medium • regimental court martial • Royal College of Midwives • (Member of the) Royal College of Music

RCMP Royal Canadian Mounted Police (formerly RNWMP)

RCN Royal Canadian Navy • (*or* **Rcn**) Royal College of Nursing

RCNC Royal Corps of Naval Constructors

RCNR Royal Canadian Naval Reserve

RCNT Registered Clinical Nurse Teacher

RCNVR Royal Canadian Naval Volunteer Reserve

RCO Royal College of Organists

RCOG Royal College of Obstetricians and Gynaecologists

RCP Royal College of Physicians

RCPath Royal College of Pathologists

RCPE (*or* **RCPEd**) Royal College of Physicians, Edinburgh

RCPI Royal College of Physicians of Ireland

RCPSG Royal College of Physicians and Surgeons of Glasgow

RCPsych Royal College of Psychiatrists

rcpt receipt

RCR Royal College of Radiologists

RCS reaction control system (in a spacecraft) • *Nuclear engineering* reactor coolant (*or* cooling) system • remote control system • Royal Choral Society • Royal College of Science • Royal College of Surgeons of England • Royal Commonwealth Society • Royal Corps of Signals • Royal Counties Show

RCSB Royal Commonwealth Society for the Blind

RCSE (*or* **RCSEd**) Royal College of Surgeons of Edinburgh

RCSI Royal College of Surgeons in Ireland

RCSLT Royal College of Speech and Language Therapists

RCSS random communications-satellite system

rct receipt • recruit

RCT regimental combat team • remote control transmitter (for TV operation, etc.) • Royal Corps of Transport

RCU remote-control unit • road-construction unit • rocket-countermeasure unit

rcvr receiver

RCVS Royal College of Veterinary Surgeons

rd rendered • road • round • *Physics, symbol for* rutherford

r.d. *Physics* relative density • rive droite (French: right bank) • *Shipping* running days

Rd Road

RD radiation dose • *Metallurgy* radiation-induced defect • *British vehicle registration for* Reading • récemment dégorgé (French: recently disgorged; referring to wines having a longer ageing period on their first cork) • (*or* **R/D**) refer to drawer (on a cheque) • (USA) registered dietician • República Dominicana (Spanish: Dominican Republic) • research department • Royal Dragoons • Royal Naval and Royal Marine Forces Reserve Decoration • Rural Dean • (New Zealand) Rural Delivery

RDA Rassemblement Démocratique Africain (French: African Democratic Rally; political party) • recommended dietary (*or* daily) allowance (of nutrients) • Retail Distributors' Association • Royal Defence Academy • Royal Docks Association

RD&D research, development, and demonstration

RD&E research, development, and engineering

RDAT (*or* **R-DAT**) rotary-head digital audio tape

RDB *Military* Research and Development Board • Royal Danish Ballet • Rural Development Board

RDBMS *Computing* relational database management system

RDC Royal Defence Corps • *Insurance* running-down clause • Rural District Council

RDCA Rural District Councils' Association

r.d.d. required delivery date

RDD *Marketing* (USA) random digital dialing

RDE Research and Development Establishment • *Chem.* rotating-disc electrode

RDF radio direction finder (*or* finding) • *Military* (USA) Rapid Deployment Force • Royal Dublin Fusiliers

RDI Royal Designer for Industry (of the Royal Society of Arts)

rDNA *Biochem.* ribosomal DNA

RDP (South Africa) Reconstruction and Development Programme • *Botany* ribulose diphosphate (*see* RuBP)

RDPL *international civil aircraft marking for* Laos

rdr radar

rds. *Shipping* roadstead

RDS radio data system (for automatic tuning of receivers) • *Med.* respiratory distress syndrome • Royal Drawing Society • Royal Dublin Society

RDT&E research, development, testing, and engineering

RDV (*or* **rdv**) rendezvous

RDX Research Department Explosive (cyclonite)

rdy ready

RDy (*or* **RDY**) Royal Dockyard

RDZ radiation danger zone

r.e. *Football* (USA) right end

Re (ital.) *Physics, symbol for* Reynolds number • *Chem., symbol for* rhenium • *symbol for* rupee (Indian monetary unit; *see also* Rs)

RE *Chem.* rare earth • (*or* **R/E**) (USA) real estate • Reformed Episcopal • religious education • renewable energy • revised edition • Right Excellent • *Med.* right eye • Royal Engineers • Royal Exchange • Royal Society of Painter-Etchers and Engravers • *British vehicle registration for* Stoke-on-Trent

REA Radar and Electronics Association • request for engineer's authorization • Rubber Export Association • (USA) Rural Electrification Administration

reac. reactor

REAC Regional Education Advisory Committee (of the TUC)

REACH (riːtʃ) Retired Executives' Action Clearing House

react (rɪˈækt) research education and aid for children with potentially terminal illness

Rear-Adm. Rear-Admiral

reasm. reassemble

REB regional examining body • *Physics* relativistic electron beam

rec. receipt • receive • *Med.* recens (Latin: fresh; in prescriptions) • recent • reception • recipe • record(ed) • recorder • recording • recreation

REC Railway Executive Committee • (rɛk) regional electricity company

recd received

recep. reception

RECHAR (ˈriːtʃɑː) Reconversion de Bassins Charbonniers (French: Reconversion of Coal Fields; an EU funding programme for the redevelopment of depressed mining areas)

recip. reciprocal • reciprocity

recirc. recirculate

recit. recitation • *Music* recitative

reclam. reclamation

recm. recommend

RECMF Radio and Electronic Component Manufacturers' Federation

recog. recognition • recognize

recom. recommend

recon. reconciliation • (*or* **recond.**)

recondition • reconnaissance • reconnoitre • reconsign(ment) • (*or* **reconst.**) reconstruct(ion)

REconS Royal Economic Society

recpt receipt

recryst. *Chem.* recrystallized

rec. sec. recording secretary

rect. receipt • rectangle • rectangular • *Med.* rectificatus (Latin: rectified; in prescriptions) • rectify

Rect. Rector(y)

rec-v (*or* **recvee**) (ˈrɛkˈviː) (USA) recreational vehicle (e.g. a motorized caravan)

red. *Finance* redeemable • reduce(d) • reduction

redisc. rediscount

redox (ˈrɛdɒks) *Chem.* reduction–oxidation

Red R Register of Engineers for Disaster Relief

redupl. reduplicate • reduplication

REE *Chem.* rare-earth element

ref. refer(red) • referee • reference • refining • reform(ed) • reformation • reformer • refrigerated ship • refund(ing) • refuse

refash. refashioned

Ref. Ch. Reformed Church

refd referred • refund

refl. reflect(ion) • reflective • reflex • reflexive

Ref. Pres. Reformed Presbyterian

refrig. refrigerate • refrigeration • refrigerator

Ref. Sp. Reformed Spelling

reg. regiment • region • register(ed) • registrar • registration • registry • regular(ly) • regulation • regulator

Reg. Regent • (*or* **Reg**) Regent's Park College, Oxford • Regina (Latin: Queen)

REGAL (ˈriːg°l) *Aeronautics* range and elevation guidance for approach and landing

regd registered

Reg-Gen. Registrar-General

Reg. Prof. Regius Professor

regr. registrar

Regt Regent • Regiment

regtl regimental

Reg. TM Registered Trade Mark

REHAB (ˈriːhæb) *Med., Psychiatry* Rehabilitation Evaluation of Hall and Baker

reinf. reinforce

reinfmt reinforcement

reit. *Printing* reiteration

REIT (USA) real-estate investment trust

rej. reject

rel. relate • relating • relation • relative(ly) • release(d) • relic • *Bibliog.* relié (French: bound) • religion • religious • reliquiae (Latin: relics)

REL recommended exposure limit (for radiation)

rel. pron. relative pronoun

rem (rɛm) roentgen equivalent man (former unit of radioactivity)

rem. remark(s) • remit • remittance

REM (rɛm *or* 'ɑː 'iː 'ɛm) *Physiol.* rapid eye movement (as in **REM sleep**) • *Chem.* rare-earth metal

REME ('riːmiː) Royal Electrical and Mechanical Engineers

remitt. remittance

Ren. Renaissance

Renf. Renfrewshire (former Scottish county)

Renfe (*or* **RENFE**) Red Nacional de Ferrocarriles Españoles (Spanish state railways)

REngDes Registered Engineering Designer

renv. renovate • renovation

REO regional education officer

rep (rɛp) repetition

rep. repair • repeat • *Med.* repetatur (Latin: let it be repeated; in prescriptions) • report(ed) • reporter • represent(ing) • (rɛp) representative • reprint

Rep. (USA) Representative • Republic • (USA) Republican

REPC Regional Economic Planning Council

repl. replace(d) • replacement

repo. repossess

repr. (*or* **repres.**) represent(ative) • represented • representing • reprint(ed)

repro. reproduced • reproduction

rept receipt • report

repub. republish

Repub. Republic • (USA) Republican

req. request • require(d) • requisition

reqd required

reqn requisition

reqs requires

RER renewable energy resource(s) • *Cytology* rough endoplasmic reticulum

RERO Royal Engineers Reserve of Officers

res. rescue • research(er) • reservation • reserve(d) • residence • resident • reside(s) • resigned • resolution

RES renewable energy source (*or* system) • *Physiol.* reticuloendothelial system • Royal Entomological Society of London

resgnd resigned

resig. resignation

resp. respective(ly) • respiration • respondent • responsibility

res. phys. resident physician

res. sec. resident secretary

rest. (*or* **restr.**) restaurant • restoration • restrict(ion)

Rest. *History* Restoration

ret. retain • retire(d) • return(ed)

Ret *Astronomy* Reticulum

RET *Computing* resolution enhancement technology

retd retained • retired • returned

R. et I. Regina et Imperatrix (Latin: Queen and Empress) • Rex et Imperator (Latin: King and Emperor)

retnr retainer

RETRA ('rɛtrə) Radio, Electrical, and Television Retailers' Association

RETRO (*or* **Retro**) ('rɛtrəʊ) *Astronautics* (USA) Retrofire Officer

rev. revenue • reverse(d) • review(ed) • revise(d) • revision • revolution • revolve • revolver • revolving

Rev. *Bible* Revelation • (*or* **Rev**) Reverend • Review

REV *Astronautics* re-entry vehicle

rev. a/c revenue account(s)

Revd Reverend

Rev. Stat. Revised Statutes

Rev. Ver. Revised Version (of the Bible)

rew. reward

REXX *Computing* restructured extended executor

rf reef

rf. *Music* rinforzando (Italian: reinforcing)

r.f. radio frequency • range finder • rapid fire • *Telecom.* reception fair • relative flow • *Baseball* right field(er) • rough finish (of paper)

Rf *symbol for* rufiyaa (monetary unit of Maldives) • *Chem., symbol for* rutherfordium

RF radio frequency • reconnaissance fighter (aircraft; as in **RF-4E**) • *Military* regular forces • *Genetics* release factor • *Cartography* representative fraction • République française (French Republic) •

research foundation • *Military* Reserve Force • *Chem.* retention factor (in chromatography) • Rockefeller Foundation • Royal Fusiliers • rugby football • *symbol for* Rwanda franc (monetary unit) • *British vehicle registration for* Stoke-on-Trent

RFA Royal Field Artillery • Royal Fleet Auxiliary

RFAC Royal Fine Art Commission

R factor *Genetics* resistance factor (*see* R plasmid)

r.f.b. (*or* **RFB**) *Football* (USA) right fullback

RFC (USA) Reconstruction Finance Corporation • *Computing* request for comments • Royal Flying Corps • Rugby Football Club

RFD radio-frequency device • reporting for duty • (USA) rural free delivery (postal service)

RFDS (Australia) Royal Flying Doctor Service

RFE Radio Free Europe

RFH Royal Festival Hall (London)

RFI radio-frequency interference • request for information • *Skating* right forward inside (of skate)

RFL Rugby Football League

RFLP *Genetics* restriction fragment length polymorphism

Rfn Rifleman

RFN Registered Fever Nurse

RFO *Skating* right forward outside (of skate)

r.f.p. retired on full pay

RFP *Nuclear engineering* reversed-field pinch

RFPC *Electronics* radio-frequency pulse compression

RFQ *Physics* radio-frequency quadrupole • *Commerce* request for quotation

RFR Royal Fleet Reserve

rfrd referred

RFS Registry of Friendly Societies • Royal Forestry Society

RFSU Rugby Football Schools' Union

RFT *Computing* revisable form text • *Astronomy* richest-field telescope

RFU Rugby Football Union

rfz. *Music* rinforzando (Italian: reinforcing)

r.g. *Football* (USA) right guard • rive gauche (French: left bank)

RG *international vehicle registration for* Guinea • *British vehicle registration for* Newcastle upon Tyne • *postcode for*

Reading • reserve guard • *airline flight code for* Varig

RGA remote geological analysis • residual-gas analyser • Royal Garrison Artillery • Royal Guernsey Artillery

RGB *Electrical engineering, Image technol.* red, green, blue (as in **RGB signals**)

RGBI red, green, blue intensity

rgd registered • reigned

rge range

R-Genl Registrar-General

RGG Royal Grenadier Guards

RGH Royal Gloucestershire Hussars

RGI Royal Glasgow Institute of the Fine Arts

RGJ Royal Green Jackets

rgn region

Rgn Rangoon

RGN Registered General Nurse (formerly SRN)

RGNP *Economics* real gross national product

RGO Royal Greenwich Observatory (now in Cambridge)

RGS Royal Geographical Society

RGSA Royal Geographical Society of Australasia

Rgt Regiment

RGT *Physics* relativistic gravitational theory

rgtl regimental

r.h. right half • right hand(ed)

Rh *Med.* rhesus (factor) (as in **Rh positive**) • *Chem., symbol for* rhodium

RH *British vehicle registration for* Hull • *postcode for* Redhill • *Meteorol.* relative humidity • remote handling (*or* handled) • *international vehicle registration for* Republic of Haiti • right hand(ed) • *British fishing port registration for* Rochester • Royal Highness • Royal Hospital

RHA Regional Health Authority • Road Haulage Association • Royal Hibernian Academy • Royal Horse Artillery

RHamps Royal Hampshire Regiment

rhap. rhapsody

RHAS Royal Highland and Agricultural Society of Scotland

RHB Regional Hospital Board • (*or* **r.h.b.**) *Football* (USA) right halfback

RHBNC Royal Holloway and Bedford New College, London

r.h.d. right-hand drive (of a motor vehicle)

RHEED *Physics* reflection high-energy electron diffraction

rheo. rheostat

rheol. rheological • rheology

rhet. rhetoric(al)

RHF Royal Highland Fusiliers

RHG Royal Horse Guards

RHHI Royal Hospital and Home for Incurables (Putney)

RHIC *Physics* relativistic heavy-ion collider (*or* collisions)

RHistS Royal Historical Society

RHM Ranks Hovis McDougall

RHMS Royal Hibernian Military School

rhom. (*or* **rhomb.**) rhombic • rhomboid • rhombus

r.h.p. rated horsepower

RHQ regimental headquarters

r/h/r rearhead restraints (in motor advertisements)

RHR rearhead restraints (in motor advertisements) • Royal Highland Regiment (Black Watch)

r.h.s. right-hand side • round-headed screw

RHS Royal Highland Show • Royal Historical Society • Royal Horticultural Society • Royal Humane Society

RHSI Royal Horticultural Society of Ireland

RHV Registered Health Visitor

r.i. reflective insulation • rubber insulation

Ri (ital.) *Meteorol., symbol for* Richardson number

RI *Irish vehicle registration for* Dublin • radio interference • radioisotope • Railway Inspectorate • refractive index • Regimental Institute • Regina et Imperatrix (Latin: Queen and Empress) • *Shipping* (Italy) Registro Italiano Navale (classification society) • (*or* **R/I**) reinsurance • religious instruction • report of investigation • *international vehicle registration for* Republic of Indonesia • Rex et Imperator (Latin: King and Emperor) • Rhode Island (abbrev. *or* postcode) • Rockwell International Corporation • Rotary International • (Member of the) Royal Institute of Painters in Water Colours • Royal Institution

RIA *Med.* radioimmunoassay • Royal Irish Academy

RIAA Recording Industry Association of America

RIAC Royal Irish Automobile Club

RIAF Royal Indian Air Force

RIAI Royal Institute of the Architects of Ireland

RIAM Royal Irish Academy of Music

RIAS Royal Incorporation of Architects in Scotland

RIASC Royal Indian Army Service Corps

Rib *Biochem.* ribose

RIB Racing Information Bureau • rigid-hull inflatable boat • Rural Industries Bureau

RIBA Member of the Royal Institute of British Architects • (*or* **Riba**) Royal Institute of British Architects

RIBI Rotary International in Great Britain and Ireland

RIC *Chem.* radiation-induced change • *Physics* radiation-induced conductivity • Radio Industry Council • Royal Institute of Chemistry (now called Royal Society of Chemistry, RSC) • Royal Irish Constabulary

RICA Research Institute for Consumer Affairs

RICE (raɪs) rest, ice, compression, elevation (for treating sports injuries)

RICO (USA) Racketeer Influenced and Corrupt Organizations Act

RICS Royal Institution of Chartered Surveyors

RID *Physics* radiation-induced defect • *Physics* radiation-induced diffusion

RIE *Commerce* recognized investment exchange • Royal Indian Engineering (College)

RIF *Military* reduction in force • Royal Inniskilling Fusiliers

Rif. Brig. Rifle Brigade

RIFF (rɪf) *Computing* raster image file format

RIIA Royal Institute of International Affairs

RIM *international vehicle registration for* Islamic Republic of Mauritania • Royal Indian Marines

RIMB (South Africa) Research Institute for Medical Biophysics

RIN *Aeronautics* reference indicator number • Royal Indian Navy

RINA Royal Institution of Naval Architects

rinf. *Music* rinforzando (Italian: reinforcing)

RINVR Royal Indian Naval Volunteer Reserve

RIO reporting in and out

RIOP Royal Institute of Oil Painters

rip. *Music* ripieno (Italian; all instrumentalists)

RIP *Computing* raster input processor • requiescat (*or* requiescant) in pace (Latin: may (he, she, *or* they) rest in peace)

RIPA Royal Institute of Public Administration

RIPH&H (*or* **RIPHH**) Royal Institute of Public Health and Hygiene

RIR Royal Irish Regiment

RIrF Royal Irish Fusiliers

RIS Research Information Service • resonance ionization spectroscopy

RISC (rɪsk) reduced-instruction-set computer

rit. *Music* ritardando (Italian: holding back) • *Music* ritenuto (Italian: held back)

RIT *Psychiatry* Rorschach inkblot test

riten. *Music* ritenuto (Italian: held back)

riv. (*or* **Riv.**) river

RIY *Physics* relative ion yield

RJ *British vehicle registration for* Manchester • ramjet • *Military* road junction • *airline flight code for* Royal Jordanian

RJA Royal Jersey Artillery

RJE *Computing* remote job entry

RJLI Royal Jersey Light Infantry

RJM Royal Jersey Militia

RK *airline flight code for* Air Afrique • *British vehicle registration for* London NW • *Med.* radical keratotomy (used in treatment of shortsightedness) • religious knowledge

RKO (USA) Radio-Keith-Orpheum (film studio, now broadcasting company)

rky rocky

RL reference library • *international vehicle registration for* Republic of Lebanon • research laboratory • rocket launcher • Rugby League • *British vehicle registration for* Truro

RLD (USA) retail liquor dealer

RLF Royal Literary Fund

R Lincolns Royal Lincolnshire Regiment

RLL *Computing* run length limited

RLO railway liaison officer • returned letter office (formerly dead letter office, DLO)

RLPO Royal Liverpool Philharmonic Orchestra

RLPS Royal Liverpool Philharmonic Society

Rls *symbol for* rial (monetary unit of Iran)

RLS Robert Louis Stevenson (1850–94, Scottish writer)

RLSS Royal Life Saving Society

rly (*or* **Rly**) railway • relay

rm ream • room

RM *British vehicle registration for* Carlisle • radiation monitoring • radio monitoring • *Computing* Reed-Muller (in **RM code**) • Registered Midwife • *symbol for* Reichsmark (former German currency) • remote monitoring • *international vehicle registration for* Republic of Madagascar • Resident Magistrate • riding master • *postcode for* Romford • Royal Mail • Royal Marines

RMA Royal Marine Artillery • Royal Marines Association • Royal Military Academy (Sandhurst; formerly Woolwich) • Royal Musical Association

RMB (*or* **Rmb**) (China) renminbi (Chinese: people's money (the currency used by the indigenous population); *compare* FEC, Foreign Exchange Certificate)

RMC regional meteorological centre • (formerly) Royal Military College (Sandhurst; *see under* RMA)

RMCC Royal Military College of Canada

RMCM (Member of the) Royal Manchester College of Music

RMCS Royal Military College of Science

r.m.d. ready money down

RMedSoc Royal Medical Society, Edinburgh

RMetS Royal Meteorological Society

RMFVR Royal Marine Forces Volunteer Reserves

RMH Royal Marsden Hospital (London)

RMI Resource Management Initiative (for the NHS)

RMIT Royal Melbourne Institute of Technology

RMLI Royal Marine Light Infantry

r.m.m. *Chem.* relative molecular mass

RMM *international vehicle registration for* Republic of Mali

RMN Registered Mental Nurse

RMO Regimental Medical Officer • Regional Medical Officer • resident medical officer • Royal Marine Office

RMP Royal Marine Police • Royal Military Police

RMPA Royal Medico-Psychological Association

RMRA Royal Marines Rifle Association

rms *Maths.* root mean square

RMS radiation-monitoring system • (USA) Railway Mail Service • remote monitoring system • *Maths.* root mean square • Royal Mail Service • Royal Mail Ship (*or* Steamer) • Royal Microscopical Society • Royal Society of Miniature Painters

RMSchMus Royal Marines School of Music

RMSM Royal Military School of Music

RMT National Union of Rail, Maritime, and Transport Workers (formed by amalgamation of the NUR with the NUS)

r.n. *Telecom.* reception nil

Rn *Chem., symbol for* radon

RN *British vehicle registration for* Preston • Registered Nurse • *international vehicle registration for* Republic of Niger • Royal Navy (*or* Naval) • *British fishing port registration for* Runcorn

RNA *Biochem.* ribonucleic acid • Royal Naval Association

RNAS Royal Naval Air Service • Royal Naval Air Station

RNase (*or* **RNAase**) *Biochem.* ribonuclease

RNAW Royal Naval Aircraft Workshop

RNAY Royal Naval Aircraft Yard

R 'n' B rhythm and blues

RNB Royal Naval Barracks

RNBT Royal Naval Benevolent Trust

RNC (USA) Republican National Committee • Royal Naval College

RNCM (Member of the) Royal Northern College of Music

rnd round

RND Royal Naval Division

RNEC Royal Naval Engineering College

Rnf. Renfrewshire (former Scottish county)

RNF Royal Northumberland Fusiliers

RNIB Royal National Institute for the Blind

RNID Royal National Institute for the Deaf

RNLAF Royal Netherlands Air Force

RNLI Royal National Lifeboat Institution

RNLO Royal Naval Liaison Officer

RNMDSF Royal National Mission to Deep Sea Fishermen

RNMH Registered Nurse for the Mentally Handicapped

RNMS Royal Naval Medical School

RNoN Royal Norwegian Navy

RNP *Biochem.* ribonucleoprotein

RNPFN Royal National Pension Fund for Nurses

R 'n' R rock and roll

RNR Royal Naval Reserve

RNRA Royal Naval Rifle Association

rns runs

RNS *Stock Exchange* Regulatory News Service • *Maths.* residue number system • Royal Numismatic Society

RNSA Royal Naval Sailing Association

RNSC Royal Naval Staff College

RNSR Royal Naval Special Reserve

RNSS Royal Naval Scientific Service

RNT Registered Nurse Tutor • Royal National Theatre

RNTE Royal Naval Training Establishment

RNTNEH Royal National Throat, Nose, and Ear Hospital

RNTU Royal Naval Training Unit

RNVR Royal Naval Volunteer Reserve

RNVSR Royal Naval Volunteer Supplementary Reserve

RNWMP (formerly, in Canada) Royal Northwest Mounted Police (*see* RCMP)

rnwy runway

RNXS Royal Naval Auxiliary Service

RNZAC Royal New Zealand Armoured Corps

RNZAF Royal New Zealand Air Force

RNZIR Royal New Zealand Infantry Regiment

RNZN Royal New Zealand Navy

RNZNVR Royal New Zealand Naval Volunteer Reserve

ro. (*or* **r°**) *Printing* recto • roan

r.o. *Rowing* rowed over • *Cricket* run out

RO *British vehicle registration for* Luton • radar operator (*or* observer) • radio operator • Radio Orchestra • *Psychol.* reality orientation • receiving office(r) • receiving order • record(s) office • recruiting officer • regimental order • registered office • relieving officer • reserved occupation • returning officer • *Chem.* reverse osmosis • *symbol for* rial Omani (monetary unit of Oman) • *international vehicle registration for* Romania • *British fishing port registration for* Rothesay • *airline flight code for* Tarom-Romanian Air Transport

ROA (*or* **RoA**) *Education* record of achievement • Reserve Officers' Association • *Finance* return on assets

ROAM *Finance* return on assets managed

ROAR right of admission reserved

ROB remaining on board

ROC *Finance* return on capital • Royal Observer Corps

ROCE *Finance* return on capital employed

ROE *Finance* return on equity • Royal Observatory, Edinburgh

ROF Royal Ordnance Factory

Roffen. Roffensis (Latin: (Bishop) of Rochester)

R of O Reserve of Officers

ROG (*or* **r.o.g.**) receipt of goods

ROI region of interest • *Finance* return on investment • (Member of the) Royal Institute of Oil Painters

ROK *international vehicle registration for* Republic of Korea

Rolls (rəulz) Rolls-Royce (car)

rom. roman (type)

Rom. Roman • Romance *or* Romanic (languages) • Romania(n) • *Bible* Romans

ROM (rɒm) *Computing* read-only memory

Rom. Cath. Roman Catholic

RONA *Finance* return on net assets

ROP *Advertising* run of paper (as in **ROP printing**)

RORC Royal Ocean Racing Club

ro-ro (*or* **RORO**) ('rəurəu) roll-on/roll-off (ferry)

Ros. Roscommon

ROS (rɒs) *Computing* remote operations service

Rosa ('rəuzə) *Computing* recognition of open systems achievement

ROSAT *Astronomy* Röntgenstrahlen Satellit (German: X-ray satellite)

ROSE *Computing* Research Open Systems in Europe

ROSLA ('rɒzlə) raising of school-leaving age

RoSPA (*or* **Rospa**) ('rɒspə) Royal Society for the Prevention of Accidents

rot (rɒt) *Maths., symbol for* curl (of a function)

rot. rotary • rotating • rotation • rotor

Rot. Rotterdam

ROT remedial occupational therapy • rule of thumb

ROTC (USA) Reserve Officers' Training Corps

ROU *international vehicle registration for* Republic of Uruguay

rout. routine

ROV remotely operated vehicle

ROW right of way

ROWPU reverse osmosis water-purification unit

Rox. Roxburghshire (former Scottish county)

Roy. Royal

r.p. *Telecom.* reception poor • regimental policeman • reply paid

Rp *symbol for* rupiah (Indonesian monetary unit)

RP *British vehicle registration for* Northampton • radiation protection • *Chem.* reaction product • (*or* **rp**) Received Pronunciation • recommended practice • recommended price • recovery phase • *Chem.* redox potential • redundancy payment • Reformed Presbyterian • Regimental Police(man) • registered plumber • Regius Professor • reinforced plastic • reply paid • (*or* **R/P**) reprint(ing) • República Portuguesa (Republic of Portugal) • *international vehicle registration or international civil aircraft marking for* Republic of the Philippines • *Finance* (USA) repurchase agreement • research paper • *Med.* retinitis pigmentosa • *Insurance* return (of) premium • Révérend Père (French: Reverend Father) • rocket projectile • (Member of the) Royal Society of Portrait Painters • rules of procedure

RPA radiation protection adviser • *Education* record of personal achievement • Registered Plumbers' Association

RPB *Finance* recognized professional body

RPC rapid Portland cement • *Computing* remote procedure call • (USA) Republican Party Conference • request the pleasure of your company • Royal Pioneer Corps

RPD Doctor of Political Science (Latin *Rerum Politicarum Doctor*) • regional port director

RPE radio production executive • Reformed Protestant Episcopal • *Med.* retinal pigment epithelium (of the eye)

RPF *Med.* renal plasma flow • *Med.* retroperitoneal fibrosis

r.p.g. rounds per gun

RPG *Computing* report program generator • *Military* rocket-propelled grenade • role-playing game

rph revolutions per hour

RPhilS Royal Philharmonic Society

RPI retail price index

R plasmid *Genetics* resistance plasmid (formerly factor)

RPLC *Chem.* reverse-phase liquid chromatography

rpm revolutions per minute

RPM reliability performance measure • resale price maintenance

RPMS Royal Postgraduate Medical School

RPN Registered Psychiatric Nurse • *Computing, Maths.* reverse Polish notation

RP/ND reprinting, no date

RPO railway post office • regional personnel officer • Royal Philharmonic Orchestra

RPQ request for price quotation

RPR (France) Rassemblement pour la République (French: Rally for the Republic; Gaullists)

r-protein *Biochem.* ribosomal protein

rps revolutions per second

RPS radiological protection service • rapid processing system • Royal Philharmonic Society • Royal Photographic Society

RPSGB Royal Pharmaceutical Society of Great Britain

rpt repeat • report • reprint

RPV *Nuclear engineering* reactor pressure vessel • *Military* remotely piloted vehicle

RQ (*or* **r.q.**) regraded (*or* remoulded) quality (of tyres) • (*or* **R/Q**) *Commerce* request for quotation • *Med.* respiratory quotient

RQL reference quality level

RQMS regimental quartermaster sergeant

rqmt requirement

rqr. require(ment)

rr. rare

r.r. ready reckoner

RR *British vehicle registration for* Nottingham • radiation resistance • (USA) railroad • Remington Rand (US company) • research report • return rate • Right Reverend • *Cycling* road race • *British fishing port registration for* Rochester • (*or* **R-R**) Rolls-Royce • (USA) rural route

RRA Royal Regiment of Artillery

RRB Race Relations Board

RRC *Cycling* Road Racing Club • (Lady of the) Royal Red Cross

RRE Royal Radar Establishment (*see* RSRE)

RRF Royal Regiment of Fusiliers

RRI Rowett Research Institute (Aberdeen)

RRK *Chem.* Rice-Ramsperger-Kassel (in **RRK theory**)

RRKM *Chem.* Rice-Ramsperger-Kassel-Marcus (in **RRKM theory**)

RRL Registered Record Librarian • Road Research Laboratory

RRM (USA) renegotiable-rate mortgage

rRNA *Biochem.* ribosomal RNA

RRP recommended retail price

RR. PP. Révérends Pères (French: Reverend Fathers)

RRR (USA) return receipt requested (of registered mail)

3 Rs (*or* **three Rs**) reading, (w)riting, and (a)rithmetic

RRS Royal Research Ship

RRT rail rapid transit

rs *Meteorol., symbol for* sleet (rain and snow)

r.s. right side

Rs *symbol for* rupees (*see also under* Re)

RS *British vehicle registration for* Aberdeen • *Physics* Raman scattering • Received Standard (English) • reconnaissance squadron • reconnaissance strike • recording secretary • recruiting service • *Computing* Reed–Solomon (as in **RS code**) • Reformed Spelling • remote sensing • research station • *Med.* respiratory system • *Law* Revised Statutes • *Military* Royal Scots • Royal Society

R/S rejection slip

RSA Republic of South Africa • (New Zealand) Returned Services Association • *Computing* Rivest, Shamir, and Adelman (denoting a method of encryption, as in **RSA cipher, RSA system**) • Road Safety Act • Royal Scottish Academy (*or* Academician) • Royal Society for the Encouragement of Arts, Manufactures, and Commerce • Royal Society of Arts • Royal Society of Australia

RSAA Royal Society for Asian Affairs

RSAD Royal Surgical Aid Society

RSAF Royal Small Arms Factory

RSAI Royal Society of Antiquaries of Ireland

RSAMD Royal Scottish Academy of Music and Drama

RSAS Royal Surgical Aid Society

r.s.b. range safety beacon

RSC Royal Shakespeare Company • Royal Society of Canada • Royal Society of Chemistry • Rules of the Supreme Court

RSCDS Royal Scottish Country Dance Society

rsch research

RSCJ Religiosae Sacratissimi Cordis Jesus (Latin: Nuns of the Most Sacred Heart of Jesus; Sacred Heart Society)

RSCM Royal School of Church Music

RSCN Registered Sick Children's Nurse (formerly SRCN)

RSD recovery, salvage, and disposal • Royal Society of Dublin

RSE Received Standard English • Royal Society of Edinburgh

RSF *Building trades* rough sunk face • Royal Scots Fusiliers

RSFS Royal Scottish Forestry Society

RSFSR (formerly) Russian Soviet Federative Socialist Republic

RSG rate-support grant • recirculating steam generator • regional seat of government (in civil defence) • *Military* Royal Scots Greys

RSGB Radio Society of Great Britain (amateur radio operators)

RSGS Royal Scottish Geographical Society

RSH Royal Society for the Promotion of Health

RSHA Reichssicherheitshauptamt (Reich Security Central Office; in Nazi Germany)

RSI *Med.* repetitive strain (*or* stress) injury

R. Signals Royal Corps of Signals

RSJ *Building trades* rolled-steel joist

RSL (Australia) Returned Services League • Royal Society of Literature

RSLA raising of school-leaving age

RSM regimental sergeant major • regional sales manager • *international vehicle registration for* Republic of San Marino • Royal School of Mines • Royal Society of Medicine • Royal Society of Musicians of Great Britain

RSMA Royal Society of Marine Artists

RSME Royal School of Military Engineering

rsn reason

RSNA Radiological Society of North America

RSNC Royal Society for Nature Conservation

RSNZ Royal Society of New Zealand

RSO radiological safety officer • Radio Symphony Orchestra • railway sorting office • railway suboffice • *Military* range

safety officer • recruiting staff officer • resident surgical officer • Royal Scottish Orchestra • rural suboffice

RSocMed Royal Society of Medicine

r.s.p. rain stopped play

RSPB Royal Society for the Protection of Birds

RSPCA Royal Society for the Prevention of Cruelty to Animals

RSPP Royal Society of Portrait Painters

rsq. rescue

RSRE Royal Signals and Radar Establishment (formerly RRE)

RSS Fellow of the Royal Society (Latin *Regiae Societatis Socius*) • Royal Statistical Society

RSSA Royal Scottish Society of Arts

RSSPCC Royal Scottish Society for the Prevention of Cruelty to Children

RSTM&H Royal Society of Tropical Medicine and Hygiene

rstr. restricted

RSU road safety unit

RSUA Royal Society of Ulster Architects

RSV *Med.* respiratory syncytial virus • Revised Standard Version (of the Bible) • *Microbiol.* Rous sarcoma virus

RSVP répondez s'il vous plaît (French: please reply)

rsvr reservoir

RSW (Member of the) Royal Scottish Society of Painters in Water Colours

RSWC right side up with care

RSwN Royal Swedish Navy

rt right

r.t. *Football* (USA) right tackle

RT *British vehicle registration for* Ipswich • radiation therapy • (*or* **R/T**) radio telegraph (*or* telegraphy) • radio telephone (*or* telephony) • reaction time • reading test • received text • return ticket • room temperature • round table • round trip

RTA reciprocal trade agreement(s) • *Med.* road traffic accident • Road Traffic Act

RTB *Military* return to base

RTBA rate to be agreed

RTBF (Belgium) Radio-Télévision Belge de la Communauté Française (French broadcasting company)

rtc. ratchet

RTC *Computing* real-time clock • (India) Road Transport Corporation • Round Table Conference

rtd returned

rtd ht *Cricket* retired hurt

RTDS *Computing* real-time data system

rte route

RTE Radio Telefís Éireann (Gaelic: Irish Radio and Television) • *Computing* real-time execution

RTECS ('ɑːtɛks) Registry of Toxic Effects of Chemical Substances

RTF Radiodiffusion-Télévision Française (French television network) • *Genetics* resistance transfer factor • *Computing* rich text format

RTFM *Computing* read the flippin' manual

rtg rating

RTG *Astronomy* radioisotope thermoelectric generator • *Physics* relativistic theory of gravitation

Rt Hon. Right Honourable

RTI Round Table International

RTITB Road Transport Industry Training Board

RTK right to know

RTL *Computing* real-time language • *Electronics* resistor-transistor logic

RTM *Computing* read the manual

rtn retain • return

rtng returning

RTO (USA) railroad transportation officer • railway transport officer

RTOL ('ɑːˌtɒl) *Aeronautics* reduced takeoff and landing

RTP *Engineering* rated thermal power • *Physics* room temperature and pressure

RTPI Royal Town Planning Institute

RTR Royal Tank Regiment

RTRA Road Traffic Regulation Act

Rt Rev. Right Reverend

RTS Religious Tract Society • *Computing* request to send • reserve tug service • Royal Television Society • Royal Toxophilite Society

RTSA Retail Trading Standards Association

RTT (*or* **RTTY**) radioteletype

RTTC *Cycling* Road Time Trials Council

RTU *Military* returned to unit

RTV real-time video

Rt W Right Worshipful

RTW (*or* **rtw**) ready to wear

RTYC Royal Thames Yacht Club

RTZ Rio Tinto Zinc Corporation Limited

Ru *Chem., symbol for* ruthenium

Ru. Russia(n)

RU *British vehicle registration for* Bournemouth • Readers' Union • *Computing* registered user • reprocessed uranium • *international vehicle registration for* Republic of Burundi • *Med., Pharmacol.* Roussel-Uclaf (French pharmaceutical company; in **RU-486**, abortion pill) • Rugby Union

RUA Royal Ulster Academy of Painting, Sculpture, and Architecture

RUAS Royal Ulster Agricultural Society

RUBISCO (ruːˈbɪskəu) *Botany* RuBP carboxylase

RuBP (*or* **RUBP**) *Botany* ribulose 1,5-bisphosphate (formerly diphosphate, RuDP *or* RDP)

RUBSSO Rossendale Union of Boot, Shoe, and Slipper Operatives

RUC Royal Ulster Constabulary

RUCR Royal Ulster Constabulary Reserve

rud. rudder

RuDP *Botany* ribulose diphosphate (*see* RuBP)

RUF *Banking* revolving underwriting facility

RUG *Computing* restricted users group

RUI Royal University of Ireland

RUKBA (*or* **Rukba**) Royal United Kingdom Beneficent Association

rumpie ('rʌmpɪ) *Colloquial* rural upwardly mobile professional

RUPP road used as public path

RUR Royal Ulster Regiment

RURAL ('ruərəl) Society for the Responsible Use of Resources in Agriculture and on the Land

Rus. Russia(n)

RUS *international vehicle registration for* Russian Federation

RUSI Royal United Services Institute for Defence Studies (formerly Royal United Service Institution)

RUSM Royal United Service Museum

Russ. Russia(n)

Rut. (*or* **Rutd**) Rutland

r.v. *Statistics* random variable • rendezvous

RV *British vehicle registration for* Portsmouth • rateable value • (USA) recreational vehicle (e.g. a motorized caravan) • *Astronautics* re-entry vehicle • *airline flight code for* Reeve Aleutian Airways • rendezvous • research vessel • *Med.* residual volume • Revised Version (of the Bible) • Rifle Volunteers • *Med.* right ventricle (*or* ventricular)

RVB *Chem.* resonating valence bond

RVC Rifle Volunteer Corps • Royal Veterinary College

RVCI Royal Veterinary College of Ireland

RVLR Road Vehicles Lighting Regulations

RVM Royal Victorian Medal

RVO Royal Victorian Order

RVR runway visual range

RVSVP répondez vite, s'il vous plaît (French: please reply quickly)

RVU research vessel unit

Rw. Rwanda

RW *British vehicle registration for* Coventry • rainwater • *airline flight code for* Rheinland Air Service • (*or* **R/W**) right of way • Right Worshipful • Right Worthy • Royal Warrant • runway

RWA Race Walking Association • (Member of the) Royal West of England Academy • *international vehicle registration for* Rwanda

RWAFF Royal West African Frontier Force

R War. R Royal Warwickshire Regiment

RWAS Royal Welsh Agricultural Society

r.w.d. rear-wheel drive

RWD radioactive-waste disposal

RWEA (Member of the) Royal West of England Academy

RWF Radio Wholesalers' Federation • Royal Welch Fusiliers

RwFr Rwanda franc

RWGM *Freemasonry* Right Worshipful Grand Master

RWGR *Freemasonry* Right Worthy Grand Representative

RWGS *Freemasonry* Right Worthy Grand Secretary

RWGT *Freemasonry* Right Worthy Grand Templar • *Freemasonry* Right Worthy Grand Treasurer

RWGW *Freemasonry* Right Worthy Grand Warden

R Wilts Yeo. Royal Wiltshire Yeomanry

RWK Queen's Own Royal West Kent Regiment

RWM radioactive waste management • *Computing* read-write memory

RWP rainwater pipe

RWS (Member of the) Royal Society of Painters in Water Colours

rwy (*or* **Rwy**) railway

Rx *Broadcasting* recording

RX *British vehicle registration for* Reading • *British fishing port registration for* Rye

ry (*or* **Ry**) railway

RY *airline flight code for* Air Rwanda • *British vehicle registration for* Leicester • *British fishing port registration for* Ramsey

RYA Royal Yachting Association

RYC in reply to your cable

RYS Royal Yacht Squadron

r.z. return to zero

RZ *British vehicle registration for* Antrim

RZSI Royal Zoological Society of Ireland

RZSScot (*or* **RZSS**) Royal Zoological Society of Scotland

S

s *Physics, symbol for* electron state $l=0$ (where l is orbital angular momentum quantum number) • (ital.) *Chem., Physics, symbol for* path length • *symbol for* second(s) • (ital.) *Chem.* secondary (isomer; as in **s-butyl alcohol**) • (ital.) *Chem., symbol for* sedimentation coefficient • *Spectroscopy, symbol for* singlet • *Music* soh (in tonic sol-fa) • *Chem.* solid (as in **NaCl(s)**) • *Physics, symbol for* specific entropy • (ital.) *Physics, symbol for* spin quantum

number • *Physics* strange (a quark flavour) • *Spectroscopy, symbol for* strong absorption • (ital.) *Chem.* symmetric (as in **s-dichloroethane**)

s. school • sea • seaman • section • see • semi- • series • sermon • sets • shilling • siècle (French: century) • siehe (German: see) • sign(ed) • sine (Latin: without) • *Grammar* singular • sinister (Latin: left) • sinistra (Italian: left) • sire • sister • small • snow • society • solidus (Latin: shilling) • solo • son • *Music* soprano (instrument) • spherical • steamer • steel • stem • stock • stratus (cloud) • *Grammar* substantive • succeeded • suit • sun(ny)

s/ sur (French: on; in place names)

S (ital.) *Electrical engineering, symbol for* apparent power • (ital.) *Physics, symbol for* entropy • (bold ital.) *Physics, symbol for* Poynting vector • *symbol for* Schilling (Austrian monetary unit) • *Biochem., symbol for* serine • *Medieval Roman numeral for* seven(ty) • *postcode for* Sheffield • *Electrical engineering, symbol for* siemens • *Med.* signā (*see under* Sig.) • *Geology* Silurian • (ital.) *Chem.* sinister (as in **(S)-butan-2-ol**) • *Irish fishing port registration for* Skibbereen • slow (on a clock or watch regulator) • small (size) • *Microbiol.* smooth (in **S-form** of a bacterial colony) • *Astronomy, symbol for* solar mass • south(ern) • (ital.) *Physics, symbol for* spin quantum number (of a system) • *Astronomy, symbol for* spiral galaxy (followed by letter(s)) • (ital.) *Physics, symbol for* strangeness quantum number • *Chem., symbol for* substitution reaction • sucre (Ecuadorian monetary unit) • *Chem., symbol for* sulphur • *Shipping* summer loading (on load line) • *Biochem., symbol for* Svedberg unit (as in **S value, 70S ribosome**) • *international vehicle registration for* Sweden • (ital.) *Chem., indicating* substitution on a sulphur atom (in names of compounds)

S. (*or* **S**) Sabbath • *Heraldry* sable • *Anatomy* sacral (of vertebrae) • Saint • San (Italian: Saint) • Sankt (German: Saint) • Santo *or* Santa (Italian: Saint) • São (Portuguese: Saint) • satisfactory • Saturday • Saxon • School • Scotland • Scottish • Sea • secondary • secretary • (ital.) *Music* segno (Italian: sign) • Seite (German: page) • Senate • Señor (Spanish: Mr) • sentence • September • sepultus (Latin: buried) • series • ship • signaller • signature • Signor (Italian: Mr) • Signora (Italian: Mrs) • Socialist • Society • Socius (Latin: Fellow; in titles) • solar • *Music* soprano (voice) • south(ern) • *Cards* spades • staff • *Law* statute • *Microbiol.* strain (followed by a number) • submarine • summer • sun • Sunday • Sweden

S2 *international civil aircraft marking for* Bangladesh

S5 *international civil aircraft marking for* Slovenia

S7 *international civil aircraft marking for* Seychelles • *airline flight code for* Siberia Airlines

S8 *airline flight code for* Estonian Aviation Company

2S *airline flight code for* Island Express

4S *airline flight code for* East West Airlines

7S *airline flight code for* Ryan Air

sa. *Heraldry* sable

s.a. safe arrival • *Med.* secundum artem (Latin: by skill) • see also • *Horticulture* semiannual • sex appeal • siehe auch (German: see also) • sine anno (Latin: without date; undated) • *Chem.* soluble in alkali • special agent • storage area • subject to approval • subsistence allowance

Sa. Saturday

SA *British vehicle registration for* Aberdeen • Salvation Army • *Meteorol.* sandstorm • Saudi Arabia (abbrev. *or* IVR) • Saudi Arabian • seaman apprentice • (USA) Secretary of the Army • *Horticulture* semiannual • sex appeal • *Microbiol.* simian agent (as in **SA12**) • *Med.* sinoatrial (as in **SA node**) • small arms • (Spain, Portugal) sociedad anónima *or* (in Portugal) sociedade anónima (public limited company; plc) • (France, Belgium, Luxembourg, Switzerland) société anonyme (public limited company; plc) • Society of Antiquaries • Society of Arts • Society of Authors • Soil Association • Son Altesse (French: His *or* Her Highness) • South Africa(n) • *airline flight code for* South African Airways • South America(n) • South Australia(n) • *Astronomy* spherical aberration • *Computing* structured systems analysis • Sturmabteilung (German: storm troopers; Nazi terrorist militia) • surface area • *Military* surface-to-air (missile; as in **SA-9**) • Swansea (postcode *or* British fishing port registration)

S/A subject to acceptance • *Banking* (USA) survivorship agreement

SAA small arms ammunition • South African Airways • Speech Association of America • Standards Association of Australia • *Chem.* surface-active agent • *Computing* systems application architecture

SAAA Scottish Amateur Athletic Association

SAAAU South African Amateur Athletic Union

SAAB (*or* **Saab**) (sɑːb) Svensk Aeroplan Aktiebolag (Swedish aircraft and car company)

SAAD small-arms ammunition depot

SAAF South African Air Force

SAAFA Special Arab Assistance Fund for Africa

SAAO South African Astronomical Observatory

SAARC South Asian Association for Regional Cooperation

SAAU South African Agricultural Union

sab. sábado (Portuguese: Saturday) · sabato (Italian: Saturday)

Sab. Sabbath

SAB Science (*or* Scientific) Advisory Board · Society of American Bacteriologists · *Music* soprano, alto, bass · South Atlantic Bight

SABA Scottish Amateur Boxing Association

Sabat. Sabbatical

SABC Scottish Association of Boys' Clubs · South African Broadcasting Corporation

Sabena (*or* **SABENA**) (səˈbiːnə) Société anonyme belge d'exploitation de la navigation aérienne (Belgian World Airlines)

SABMIS (ˈsæb‚mɪs) seaborne antiballistic missile intercept system

sabo. sabotage

SABRA (ˈsæbrə) South African Bureau of Racial Affairs

SABS South African Bureau of Standards

sac *Military, indicating* qualified at a small-arms technical long course

SAC Scientific Advisory Committee · Scottish Arts Council · Scottish Automobile Club · Senior Aircraftman · small-arms club · South Atlantic coast · (USA) State Athletic Commission · (USA) Strategic Air Command

SACEUR (*or* **Saceur**) (ˈsæk‚jʊə) Supreme Allied Commander Europe

SACL South African Confederation of Labour

SACLANT (ˈsæk‚lænt) Supreme Allied Commander Atlantic

SACO Sveriges Akademikers Centralorganisation (Swedish Confederation of Professional Associations; trade union)

SACP South African Communist Party

Sacr. Sacramento (USA) · Sacrist

SACSEA (ˈsæk‚siː) Supreme Allied Command, SE Asia

SACSIR South African Council for Scientific and Industrial Research

SACU single-application computer user · Southern African Customs Union

SACW Senior Aircraftwoman

Sad systems analysis and design

SAD *Psychiatry* seasonal affective disorder

SADC South African Development Community

SADCC Southern Africa Development Coordinating Conference (*or* Committee)

SADF South African Defence Force

SADG Société des architectes diplômés par le gouvernement (French: Society of Government-Certified Architects)

SADT *Computing, trademark* structured analysis and design technique

SAE (*or* **s.a.e.**) self-addressed envelope · (USA) Society of Automotive Engineers (designating a viscosity scale for motor oils) · (*or* **s.a.e.**) stamped addressed envelope

SAEF Stock Exchange Automatic Execution Facility

SAF (USA) Secretary of the Air Force · Society of American Foresters · (USA) Strategic Air Force

SAFA South Africa Freedom Association

s.a.f.e. stamped addressed foolscap envelope

S. Afr. South Africa(n)

S.Afr.D. South African Dutch

SAFU Scottish Amateur Fencing Union

SAG (USA) Screen Actors' Guild

SAGA Society of American Graphic Artists

SAGB Spiritualist Association of Great Britain

SAGE (seɪdʒ) *Military* (USA, Canada) semiautomatic ground environment

SAH *Med.* subarachnoid haemorrhage · Supreme Allied Headquarters

Sai. Saigon

SAIDS (seɪdz) simian acquired immune deficiency syndrome

SAIF South African Industrial Federation

SAIMR South African Institute of Medical Research

SAIRR South African Institute of Race Relations

sal. salary

SAL South Arabian League · surface airlifted mail

SALA South African Library Association

SALC symmetry-adapted linear combinations

Salop ('sæləp) Shropshire

SALP South African Labour Party

SALR *Meteorol.* saturated adiabatic lapse rate · South African Law Reports

SALT (sɔːlt) Strategic Arms Limitation Talks (*or* Treaty)

salv. salvage

Salv. Salvador

sam. samedi (French: Saturday)

Sam. Samaria · Samaritan · Samoa · Samstag (German: Saturday) · *Bible* Samuel

S. Am. South America(n)

SAM (sæm) *Physics* scanning Auger microprobe (*or* microscopy) · (USA) shared-appreciation mortgage · (USA) Space Available Mail · surface-to-air missile

SAMA Saudi Arabian Monetary Agency

Samar. Samaritan

SAMC South African Medical Corps

S. Amer. South America(n)

SAMH Scottish Association for Mental Health

san. sanitary

SAN *Chem.* styrene–acrylonitrile (polymer)

sanat. sanatorium

SANCAD ('sænkæd) Scottish Association for National Certificates and Diplomas

s & d song and dance

sand. sandwich

S&F *Colloquial* shopping and fucking (type of popular fiction) · *Insurance* stock and fixtures

S&FA shipping and forwarding agents

S&H (*or* **S and H**) shipping and handling (charges)

S&L (USA) savings and loan association

s & m sausages and mash

S & M sadism and masochism · *Insurance* stock and machinery

S&P 500 (USA) Standard and Poors 500 Stock Index

s & s *Colloquial* sex and shopping (type of popular fiction)

SANDS (*or* **Sands**) (sændz) Stillbirth and Neonatal Death Society

s. & s.c. sized and supercalendered (of paper)

S & T Salmon & Trout Association · supply and transport

Sane (seɪn) Schizophrenia – A National Emergency

SANE (seɪn) (USA) Committee for a Sane Nuclear Policy

sanit. sanitary · sanitation

s.a.n.r. subject to approval, no risk

SANROC ('sæn,rɒk) South African Non-Racial Olympics Committee

sans. *Printing* sanserif

Sans. (*or* **Sansk.**) Sanskrit

SANS *Physics* small-angle neutron scattering

SANZ Standards Association of New Zealand

SAO Scottish Association of Opticians

SAOS Scottish Agricultural Organization Society

s.a.p. soon as possible

SAP South African Police

SAPA (sæpə) South African Press Association · South African Publishers' Association

sapfu ('sæpfuː) *Slang* surpassing all previous foul-ups (*or* fuck-ups)

s.a.p.l. *Shipping* sailed as per list (i.e. Lloyd's List)

sap. no. *Chem.* saponification number

Sar. Sarawak · Sardinia(n)

SAR search and rescue · Son Altesse Royale (French: His (*or* Her) Royal Highness) · Sons of the American Revolution · South African Republic · (China) Special Administrative Region · *Physics* specific absorption rate · synthetic-aperture radar

SARAH ('seərə) search and rescue homing (radar system) · surgery assistant robot acting on the head (in brain surgery)

SARBE ('sɑːbɪ) search and rescue beacon equipment (of SARAH)

Sarl. (*or* **Sàrl.**) (France, Belgium, Luxembourg, Switzerland) société à responsabilité limitée (private limited company; Ltd)

SARs Substantial Acquisition Rules (for takeovers and mergers)

SARSAT ('sɑːsæt) search and rescue satellite-aided tracking

SA/RT *Computing* structured systems analysis for real time

Sarum. Sarumensis (Latin: (Bishop) of Salisbury)

Sas. (Italy) società in accomandita semplice (limited partnership)

SAS Fellow of the Society of Antiquaries

(Latin *Societatis Antiquariorum Socius*) • Scandinavian Airline System • *Physics* small-angle scattering • Small Astronomical Satellite • Son Altesse Sérénissime (French: His (*or* Her) Most Serene Highness • *Military* Special Air Service • Statement of Auditing Standards

SASC Small Arms School Corps

SASE (*or* **sase**) (USA) self-addressed stamped envelope

Sasi *Trademark* Shugart Associates system interface

Sask. Saskatchewan

SASO Senior Air Staff Officer • South African Students' Organization

SASR Special Air Service Regiment

sat. satellite • saturate(d)

Sat. Saturday • *Astronomy* Saturn

S. At. South Atlantic

SAT (USA) scholastic aptitude test • Senior Member of the Association of Accounting Technicians • ship's apparent time • South Australian Time • (sæt) *Education* standard assessment task

SATB *Music* soprano, alto, tenor, bass

SATCO ('sætkəu) signal automatic air-traffic control system

SATEX ('seɪtɛks) semiautomatic telegraph exchange

satn saturation

SATRO Science and Technology Regional Organization

SATS South African Transport Services

S. Aus. (*or* **S. Austral.**) South Australia(n)

s.a.v. sale at valuation • stock at valuation

SAV *Microbiol.* simian adenovirus

SAVAK (*or* **Savak**) ('sævæk *or* 'sɑːvæk) (Iran) Sāzmān-i-Attalāt Va Amnīyat-i-Keshvar (National Security and Intelligence Organization, 1957–79)

SAVS Scottish Anti-Vivisection Society

SAW space at will (of advertisements) • submerged arc welding • *Telecom.* surface acoustic wave

SAWS synoptic automatic weather station

sax. saxophone

Sax. Saxon • Saxony

SAXS *Physics* small-angle X-ray scattering

SAYE save as you earn (savings scheme)

sb *Physics, symbol for* stilb

sb. *Grammar* substantive

s.b. single-breasted • small bore (rifle) • smooth bore • *Baseball* stolen base(s)

Sb *Chem., symbol for* antimony (Latin *stibium*)

SB *airline flight code for* Air Calédonie International • (USA) Bachelor of Science (Latin *Scientiae Baccalaureus*) • *Astronomy, symbol for* barred spiral galaxy (as in **SBa**, **SBb**) • *British vehicle registration for* Glasgow • sales book • Sam Browne (military officer's belt and strap) • savings bank • selection board • Serving Brother (of an order) • *Commerce* short bill (of exchange) • sick bay • Signal Boatswain • signal book • simultaneous broadcast(ing) • *Chemical engineering* slurry bed • *Commerce* small business • *Chem.* sodium borate • South Britain (i.e. England and Wales) • Special Branch (of police) • Statute Book • *Med.* stillborn • stretcher bearer • sub-branch

SBA School of Business Administration • sick-bay (*or* -berth) attendant • (USA) Small Business Administration • *Aeronautics* standard beam approach

SBAA Sovereign Base Areas Administration

SBAC Society of British Aerospace Companies (formerly Aircraft Constructors)

SBB Schweizerische Bundesbahnen (Swiss Federal Railways)

SBBNF Ship and Boat Builders' National Federation

SBC School Broadcasting Council • single-board computer • *Electrical engineering* small bayonet cap

SBD Soviet Block Division (of the CIA) • *Electronics* surface-barrier (*or* Schottky-barrier) diode

SbE south by east

SBE Southern British English (dialect)

SBGI Society of British Gas Industries

SBH Scottish Board of Health

SBIC (USA) small business investment company

SBL *Electronics* surface boundary layer

SBLI (USA) savings bank life insurance

SBM single buoy mooring

SBN Standard Book Number (replaced by ISBN)

SBNO Senior British Naval Officer

s'board starboard

SBOT Sacred Books of the Old Testament

SBP *Med.* systolic blood pressure

SBR styrene–butadiene rubber

sbre. septiembre (Spanish: September)

SBS *Med.* sick-building syndrome • *Military* Special Boat Service

SBStJ Serving Brother of the Order of St John of Jerusalem

SBT *Shipping* segregated ballast tanks

SBU strategic business unit

SBV seabed vehicle

SbW south by west

S by E south by east

S by W south by west

sc *Printing* small capitals • *Military* (student at the) staff college

sc. scale • scene • science • scientific • scilicet (namely *or* that is; from Latin *scire licet*) • screw • scruple (unit of weight) • sculpsit (Latin: (he *or* she) carved (*or* engraved) this)

s.c. salvage charges • self-contained • *Printing* single column • *Printing* small capitals • steel casting • supercalendered (of paper)

s/c self-contained • *Commerce* son compte (French: (on) his account)

Sc *Chem.*, *symbol for* scandium • (ital.) *Physics, symbol for* Schmidt number • (ital.) *Meteorol., symbol for* stratocumulus

Sc. Scandinavia(n) • Scotch • Scotland • Scots • Scottish

SC *British vehicle registration for* Edinburgh • safe custody • sailing club • Salvage Corps • *Law* same case • *Military* (USA) Sanitary Corps • (Australia, New Zealand) School Certificate • Schools Council • *British fishing port registration for* Scilly • Security Council (of the UN) • self-contained • Senatus Consultum (Latin: decree of the Senate) • Senior Counsel • service certificate • *Law* Sessions Cases • *airline flight code for* Shandong Airlines • shooting club • short course • Signal Corps • *Printing* single column • *Chem.* single crystal • skating club • skiing club • small craft • social club • solar cell • South Carolina (abbrev. *or* postcode) • *Military* Southern Command • special case • Special Constable (*or* Constabulary) • sports club • Staff Captain • staff college • Staff Corps • standing committee • standing conference • (Canada) Star of Courage • statutory committee • *Meteorol.* stratocumulus • *Engineering* stress corrosion • structural change (in transformational grammar) • Suffolk and Cambridgeshire (regiment) • supercalendered (paper) •

Supreme Court • surface contamination • swimming club

SCA sickle-cell anaemia • *Computing* synchronous concurrent algorithm

SCAAA Southern Counties Amateur Athletic Association

Scada ('skɑːdə) system control and data acquisition

SCAHT Scottish Churches Architectural Heritage Trust

Scand. (*or* **Scan.**) Scandinavia(n)

SCAO Senior Civil Affairs Officer

SCAP (skæp) Supreme Command (*or* Commander) Allied Powers

SCAPA Society for Checking the Abuses of Public Advertising

s. caps *Printing* small capitals

SCAR (skɑː) Scientific (*or* Special) Committee on Antarctic Research

SCARA ('skɛərə) selective compliance assembly robot arm

Scarab ('skærəb) submerged craft assisting repair and burial

SCARF *Accounting* Systems Control and Review File

Scart (*or* **SCART**) (skɑːt) Syndicate des Constructeurs des Appareils Radio Récepteurs et Téléviseurs (French designers of a plug and socket system used in video equipment, as in **Scart socket**)

SCAT *Med.* sheep-cell agglutination test

ScB Bachelor of Science (Latin *Scientiae Baccalaureus*)

SCB Solicitors Complaints Bureau • Speedway Control Board (in motorcycling)

ScBC Bachelor of Science in Chemistry

ScBE Bachelor of Science in Engineering

SCBU *Med.* special-care baby unit

s.c.c. *Printing* single column centimetre • *Electrical engineering* single cotton-covered (wire)

SCC Sea Cadet Corps • Society of Church Craftsmen • *Med.* squamous-cell carcinoma • *Engineering* stress-corrosion cracking

SCCA Sports Car Club of America

SCCAPE Scottish Council for Commercial, Administrative, and Professional Education

SCCL Scottish Council for Civil Liberties

scd scheduled

ScD Doctor of Science (Latin *Scientiae Doctor*)

SCD sickle-cell disease

SCDA Scottish Community Drama Association

SCDC Schools Curriculum Development Committee

ScDHyg Doctor of Science in Hygiene

ScDMed Doctor of Science in Medicine

sce. scenario

SCE schedule compliance evaluation • Scottish Certificate of Education • *Genetics* sister-chromatid exchange

SCEC Scottish Community Education Council

SCET (skɛt) Scottish Council for Educational Technology

scf standard cubic feet

SCF Save the Children Fund • *Physics* self-consistent field • Senior Chaplain to the Forces

scfh standard cubic feet per hour

scfm standard cubic feet per minute

scg scoring

SCG Sydney Cricket Ground

Sc. Gael. Scottish Gaelic

SCGB Ski Club of Great Britain

sch. scholar • scholarship • scholastic • scholium (Latin: note) • school • schooner

Sch. *Taxation* schedule (followed by a letter) • Schilling (Austrian monetary unit) • School

sched. schedule

schem. schematic

scherz. *Music* scherzando (Italian: playful, humorous)

SchMusB Bachelor of School Music

schol. scholar • scholarship • scholastic • scholium (Latin: note)

schr schooner

sci. science • scientific

s.c.i. *Printing* single-column inch

SCI Scottish Central Institutions • Society of the Chemical Industry

SCID (*or* **skid**) (skɪd) *Med.* severe combined immune deficiency

sci. fa. scire facias (Latin: that you cause to know)

sci-fi ('saɪ'faɪ) *Colloquial* science fiction

scil. scilicet (*see under* sc.)

SCIT Special Commissioners of Income Tax

SCK Servants of Christ the King

Scl *Astronomy* Sculptor

SCL Scottish Central Library • Student in (*or* of) Civil Law

SCLC *Med.* small-cell lung cancer (*or* carcinoma) • (USA) Southern Christian Leadership Conference

SCLI Somerset and Cornwall Light Infantry

ScM Master of Science (Latin *Scientiae Magister*)

SCM (Ireland) *Stock exchange* smaller companies market • State Certified Midwife (replaced by RM, Registered Midwife) • Student Christian Movement • summary court-martial

SCMA Society of Cinema Managers of Great Britain and Ireland (Amalgamated)

SCMES Society of Consulting Marine Engineers and Ship Surveyors

ScMHyg Master of Science in Hygiene

SCNE Select Committee on National Expenditure

SCNO Senior Canadian Naval Officer

Sco *Astronomy* Scorpius

SCOBEC ('skəubɛk) Scottish Business Education Council

SCODL *Computing* scan conversion object description language

S. Con. Res. (USA) Senate concurrent resolution

SCONUL Standing Conference of National and University Libraries

SCOR Scientific Committee on Oceanic Research • Standing Committee on Refugees

Scot. Scotch (whisky) • Scotland • Scottish

ScotBIC ('skɒt,bɪk) Scottish Business in the Community

SCOTEC ('skəutɛk) Scottish Technical Education Council

SCOTUS ('skəutəs) Supreme Court of the United States

SCOTVEC ('skɒt,vɛk) (formerly) Scottish Vocational Education Council (merged with SEB to form SQA)

SCOUT (skaut) *Commerce* Shared Currency Option Under Tender

SCP single-cell protein (in food technology) • (Canada) Social Credit Party

SCPC *Telecom.* single channel per carrier

SCPR Scottish Council of Physical Recreation

SCPS Society of Civil and Public Servants

scr. *Finance* scrip • scruple (unit of weight)

SCR *Chemical engineering* selective catalytic

reactor (*or* reaction) • senior common room (in a university) • *Electronics* silicon-controlled rectifier • Society for Cultural Relations with the USSR

SCRE Scottish Council for Research in Education

Script. Scriptural • Scripture(s)

scRNA *Biochem.* small cytoplasmic RNA

scRNP *Biochem.* small cytoplasmic ribonucleoprotein

SCS (USA) Soil Conservation Service • space communications system

SCSA signal computing system architecture • Soil Conservation Society of America

SCSI ('skʌzɪ) small computer systems interface

SCSS Scottish Council of Social Service

Sct *Astronomy* Scutum

SCTR Standing Conference on Telecommunications Research

SCU Scottish Cricket Union • Scottish Cycling (*or* Cyclists') Union

SCUA Suez Canal Users' Association

scuba ('skju:bə) self-contained underwater breathing apparatus

sculp. (*or* **sculps.**, **sculpt.**) sculpsit (*see under* sc.) • sculptor (*or* sculptress) • sculptural • sculpture

SCV Stato della Città del Vaticano (Italian: Vatican City State)

SCWS Scottish Co-operative Wholesale Society

SCY *British vehicle registration for* Truro (Isles of Scilly)

SCYA Scottish Christian Youth Assembly

sd said • sailed • sewed (of books) • signed • sound

s.d. safe deposit • same date • sans date (French: no date) • semidetached • *Philosophy* sense datum • several dates • *Commerce* short delivery • siehe dies (German: see this) • sine die (Latin: without a day (being fixed)) • *Statistics* standard deviation

SD Diploma in Statistics • Doctor of Science (Latin *Scientiae Doctor*) • *British vehicle registration for* Glasgow • salutem dicit (Latin: (he *or* she) sends greeting) • sea-damaged • (USA) Secretary of Defense • semi-detached • Senatus Decreto (Latin: by decree of the Senate) • send direct • *Med.* senile dementia • Senior Deacon • sequence date • *Military* service dress • *Commerce* short delivery •

Sicherheitsdienst (German: Security Service; in Nazi Germany) • *Finance* sight draft • Signal Department • Signal Division • *Computing* single density • South Dakota (abbrev. *or* postcode) • special delivery • special duty • staff duties • stage door • *Statistics* standard deviation • (USA) State Department • structural description (in transformational grammar) • submarine detector • *airline flight code for* Sudan Airways • *British fishing port registration for* Sunderland • supply depot • *international vehicle registration for* Swaziland

S/D (USA) school district • *Finance* sight draft

SDA (formerly) Scottish Development Agency • Scottish Dinghy Association • Scottish Diploma in Agriculture • Seventh Day Adventists • Social Democratic Alliance • spray-drying absorption (*or* absorber)

S. Dak. South Dakota

SD&T staff duties and training

SDAT *Med.* senile dementia of the Alzheimer type

S-DAT stationary digital audio tape

SDC Society of Dyers and Colourists • submersible decompression chamber

SDD Scottish Development Department • *Telephony* subscriber direct dialling

SDECE (France) Service de documentation étrangère et de contre-espionage (counterintelligence agency)

SDF Social Democratic Federation

SDG Soli Deo Gloria (Latin: Glory to God Alone)

SDH *Computing* synchronous digital hierarchy

SDHE spacecraft data-handling equipment

SDI selective dissemination of information • Strategic Defense Initiative (US Star Wars programme)

SDIO (USA) Strategic Defense Initiative Office

sdl. saddle

SDL special duties list

SDLC *Computing* synchronous data link control

SDLP (Northern Ireland) Social Democratic and Labour Party

SDMJ September, December, March, June (dates for quarterly payments)

SDO Senior Dental Officer • senior duty

officer • station duty officer • subdivisional officer

S. Doc. (USA) Senate document

SDP (formerly) Social Democratic Party (merged with Liberal Party to form SLD) • *Insurance* social, domestic, and pleasure

SDPM software development process model

SDR special despatch rider • *Finance* special drawing right(s)

SDRAM (ˌɛsˈdiːˌræm) synchronous dynamic random access memory

SDS scientific data system • Sisters of the Divine Saviour • sodium dodecyl sulphate (detergent) • (Germany) Sozialistischer Deutscher Studentenbund (Federation of Socialist Students) • (USA) strategic defense system • (USA) Students for a Democratic Society

SDS-PAGE *Biochem.* sodium dodecyl sulphate–polyacrylamide gel electrophoresis

SDT Society of Dairy Technology

s.e. single end(ed) • single engine • *Bookkeeping* single entry • special equipment • *Statistics* standard error • straight edge

Se *Chem., symbol for* selenium

SE *British vehicle registration for* Aberdeen • *British fishing port registration for* Salcombe • sanitary engineering • Society of Engineers • *Computing* software engineering • Son Eminence (French: His Eminence) • Son Excellence (French: His (*or* Her) Excellency) • southeast(ern) • *postcode for* southeast London • Staff Engineer • Standard English • Stirling engine • (*or* **S/E**) stock exchange • *Building trades* stopped end • *international civil aircraft marking for* Sweden

SEA South-East Asia

SEAAC South-East Asia Air Command

SEAC (ˈsiːæk) School Examination and Assessment Council • South-East Asia Command • Standard Eastern Automatic Computer

SEAF (ˈsiːæf) Stock Exchange Automated Exchange Facility

SEAL (siːl) *US Navy* sea-air-land

SEALF South-East Asia Land Forces

SE & CR South Eastern and Chatham Railway

SEAQ (ˈsiːæk) Stock Exchange Automated Quotations System

SEATO (ˈsiːtəʊ) Southeast Asia Treaty Organization (1954–77)

SEATS (ˈsiːæts) Stock Exchange Alternative Trading Service

SEB (formerly) Scottish Education Board (merged with SCOTVEC to form SQA) • Southern Electricity Board

sec (sɛk) *Maths.* secant

sec. second (of time or an angle) • secondary • seconded • (*or* **Sec.**) secretary • section • sector • secundum (Latin: according to) • security

SEC (USA) Securities and Exchange Commission • *Chem.* size-exclusion chromatography • *Military* Southeastern Command

SECAM (ˈsiːˌkæm) séquentiel couleur à mémoire (colour-television broadcasting system developed in France)

SECC Scottish Exhibition and Conference Centre

Sec. Gen. (*or* **Sec-Gen**) Secretary General

sech (sɛtʃ) *Maths.* hyperbolic secant

sec. leg. secundum legem (Latin: according to law)

Sec. Leg. Secretary of the Legation

sec. nat. secundum naturam (Latin: according to nature, naturally)

sec. reg. secundum regulam (Latin: according to rule)

sect. section

secy (*or* **Secy**) secretary

sed. sedative • sediment

SED Scottish Education Department • shipper's export declaration • (formerly, in East Germany) Sozialistische Einheitspartei Deutschlands (Socialist Unity (i.e. Communist) Party)

SEDAR (ˈsiːdɑː) *Navigation* submerged electrode detection and ranging

sedt sediment

sedtn sedimentation

SEE *Physics* secondary electron emission • Senior Electrical Engineer • Society of Environment Engineers

SEEB (*or* **Seeboard**) Southeastern Electricity Board

SEF Shipbuilding Employers' Federation

seg. segment • segregate • *Music* segue (Italian: follows, comes after)

SEG *Taxation* Self Employment Group • socioeconomic grade

SEH St Edmund Hall (Oxford University)

SEIF (Portugal) Secretaria de Estado da

Informação e Turismo (State Information and Tourist Board)

SEIS submarine escape immersion suit

seismol. seismological • seismology

SEIU (USA) Service Employees International Union

sel. select(ed) • selection • selig (German: deceased)

Sel (*or* **Selw**) Selwyn College, Cambridge

Selk. Selkirk

SELNEC ('sɛl,nɛk) South-East Lancashire, North-East Cheshire

sem. semester • semicolon • seminary

Sem. Seminary • Semitic

s.e.(m.) (*or* **SE(M)**) *Statistics* standard error (of the mean)

SEM scanning electron microscope (*or* microscopy)

semp. *Music* sempre (Italian: always)

sen. senior • *Music* senza (Italian: without)

Sen. Senate • Senator • (Marcus Annaeus) Seneca (*c.* 55 BC–AD 41, Roman writer) • Senior

SEN special educational needs • (formerly) State Enrolled Nurse (*see* EN(G))

S en C (France) société en commandite (limited partnership)

S en NC (France) société en nom collectif (partnership)

Senr Senior

sent. sentence

SEO Senior Executive Officer • Senior Experimental Officer • Society of Education Officers

s.e.o.o. (*or* **s.e. ou o.**) sauf erreur ou omission (French: errors or omissions excepted)

sep. sepal • separable • separate(d) • separation

Sep. (*or* **Sep**) September • Septuagint

SEP (USA) simplified employee pension

SEPM Society of Economic Palaeontologists and Mineralogists

sepn separation

SEPON ('siːpɒn) Stock Exchange Pool Nominees Ltd

sept. septem (Latin: seven) • septembre (French: September) • (*or* **sept^e**) septiembre (Spanish: September)

Sept. (*or* **Sept**) September • Septuagint

seq. sequel • sequence • sequens (Latin: the following (one)) • sequente (Latin: and in what follows) • sequitur (Latin: it follows)

seq. luce *Med.* sequenti luce (Latin: the following day; in prescriptions)

seqq. sequentia (Latin: the following (ones)) • sequentibus (Latin: in the following places)

ser. serial • series • sermon • servant • service

Ser (*or* **ser**) *Biochem.* serine • *Astronomy* Serpens

SER *Cytology* smooth endoplasmic reticulum

SERA Socialist Environment and Resources Association

Serb. Serbia(n)

SERC (sɜːk) Science and Engineering Research Council (formerly SRC)

Serg. (*or* **Sergt**) Sergeant

Serj. (*or* **Serjt**) Serjeant

SERL Services Electronics Research Laboratory

SERLANT ('sɜːlænt) *US Navy* Service Forces, Atlantic

SERPAC ('sɜːpæk) *US Navy* Service Forces, Pacific

SERPS (*or* **Serps**) (sɜːps) State Earnings-Related Pension Scheme

SERT Society of Electronic and Radio Technicians

serv. servant • service

SES (USA) socioeconomic status • Stock Exchange of Singapore

SESCO ('sɛskəʊ) secure submarine communications

SESDAQ ('sɛsdæk) Stock Exchange of Singapore Dealing and Automated Quotation System

SESI Stock Exchange of Singapore Index

SESO Senior Equipment Staff Officer

sess. session

SEST Sweden-ESO Submillimetre Telescope

set. setembro (Portuguese: September) • settlement

SET ('ɛs 'iː 'tiː *or* sɛt) Securities Exchange of Thailand • selective employment tax (1966–73)

SETI ('sɛtɪ) *Astronomy* search for extra-terrestrial intelligence

S-et-L Saône-et-Loire (department of France)

S-et-M Seine-et-Marne (department of France)

S-et-O Seine-et-Oise (department of France)

Sets (sɛts) Stock Exchange electronic trading system

sett. settembre (Italian: September)

sev. sever • several

SEV *Meteorol.* severe (icing, etc.)

sevl several

sew. sewage • sewer • sewerage

SEW safety-equipment worker

sex. sextet • sexual

Sex *Astronomy* Sextans

SEXAFS surface-extended X-ray absorption fine-structure spectroscopy

Sexag. Sexagesima

sext. sextant

sf (*or* **sf.**) *Baseball* sacrifice fly • *Music* sforzando (Italian; strongly accented)

s.f. *Commerce* sans frais (French: no expenses) • (*or* **s-f**) science fiction • *Telecom.* signal frequency • *Finance* sinking fund • sub finem (Latin: towards the end)

Sf *symbol for* Suriname guilder (monetary unit)

SF *British vehicle registration for* Edinburgh • San Francisco • science fiction • senior fellow • *Military* Sherwood Foresters • shipping federation • *Telecom.* signal frequency • *Finance* sinking fund • Sinn Féin • Society of Friends • special facilities • special forces.• *Electrical engineering* standard frequency

SFA Scottish Football Association • Securities and Futures Authority • sulphated fatty alcohol (detergents) • *Slang* sweet Fanny Adams (*or* sweet fuck all; i.e. nothing)

SFAC Statement of Financial Accounting Concepts

SFAS Statement of Financial Accounting Standards

SFB Sender Freies Berlin (German: Broadcasting Station of Free Berlin)

SFBMS Small Farm Business Management Scheme

Sfc (USA) Sergeant first class

SFC specific fuel consumption (of jet engines) • *Meteorol.* surface

SFD *Med.* small for dates (of babies)

SFEU Scottish Further Education Unit

SFF *Computing* small form factor

sfgd safeguard

SFInstE Senior Fellow, Institute of Energy

SFL Scottish Football League • *Aeronautics* sequenced flashing lights

sfm surface feet per minute

SFO Senior Flag Officer • Serious Fraud Office • Superannuation Funds Office

SFOF spaceflight operations facility (in NASA)

S-For (*or* **SFOR**) ('ɛs,fɔ:) (NATO-led) Stabilization Force (in Bosnia)

sfp *Music* sforzato-piano (Italian; strong accent, followed immediately by soft)

SFr Swiss franc

SFR *Finance* sinking fund rate of return

SFT supercritical fluid technology

SFTCD Senior Fellow, Trinity College Dublin

SFU signals flying unit • suitable for upgrade (on airline tickets)

sfz (*or* **sfz.**) *Music* sforzando (*see under* sf)

sg. *Grammar* singular

s.g. specific gravity • steel girder

Sg. Surgeon

SG *British vehicle registration for* Edinburgh • Sa Grâce (French: His (*or* Her) Grace • Sa Grandeur (French: His (*or* Her) Highness • Scots Guards • Seaman Gunner • Secretary General • *Education* (USA) senior grade • *Astronomy* Seyfert galaxy • ship and goods • singular (in transformational grammar) • *Meteorol.* snow grains • Society of Genealogists • Solicitor General • spin-glass (type of crystal) • *postcode for* Stevenage • Surgeon General

SGA *Med.* small for gestational age • (Member of the) Society of Graphic Art

SGB Schweizerischer Gewerkschaftsbund (German: Swiss Federation of Trade Unions)

SGBI Schoolmistresses' and Governesses' Benevolent Institution

SgC Surgeon Captain

SgCr Surgeon Commander

sgd signed

SGD *Freemasonry* Senior Grand Deacon

s.g.d.g. sans garantie du gouvernement (French: without government guarantee; of patents)

Sge *Astronomy* Sagitta

SGF Scottish Grocers' Federation

SGHWR steam-generating heavy-water reactor

sgl. single

S. Glam South Glamorgan

SgLCr Surgeon Lieutenant-Commander

SGM Sea Gallantry Medal

SGML *Computing* standard generalized markup language

SGO Squadron Gunnery Officer

SGOT *Med.* serum glutamic oxaloacetic transaminase

SGP *international vehicle registration for* Singapore

SGPT *Med.* serum glutamic pyruvic transaminase

Sgr *Astronomy* Sagittarius

SgRA Surgeon Rear-Admiral

SGRAM (ˌɛsˈdʒiːˌræm) synchronous graphics random-access memory

Sgt Sergeant

SGT Society of Glass Technology

Sgt Maj. Sergeant Major

SGU Scottish Gliding Union · Scottish Golf Union

SgVA Surgeon Vice-Admiral

SGW *Freemasonry* Senior Grand Warden

sh *Maths.* hyperbolic sine · *Baseball* sacrifice hit

sh. shall · *Stock exchange* share · sheep · *Bookbinding* sheet · shilling · shower

s.h. second-hand · slant (*or* slope) height

Sh. *Military* Shipwright

SH *British vehicle registration for* Edinburgh · *British fishing port registration for* Scarborough · Schleswig-Holstein · school house · *Rugby* scrum-half · *Meteorol.* showers · *Numismatics* small head · southern hemisphere

SHA Scottish Hockey Association · Secondary Heads Association · *Astronomy, Navigation* sidereal hour angle · Special Health Authority · *Maths.* spherical harmonic analysis

SHAC Shelter Housing Aid Centre

SHAEF (ʃeɪf) Supreme Headquarters Allied Expeditionary Forces

Shak. (*or* **Shaks.**) William Shakespeare (1564–1616, English dramatist and poet)

SH&MA Scottish Horse and Motormen's Association

SHAPE (ʃeɪp) Supreme Headquarters Allied Powers Europe (of NATO)

SHC *Physics* specific heat capacity

SHCJ Society of the Holy Child Jesus

shd should

SHE safety, health, and ergonomics

S/HE *Shipping* Sundays and holidays excepted

Shef. (*or* **Sheff.**) Sheffield

SHEFC Scottish Higher Education Funding Council

Shet. (*or* **Shetl.**) Shetland Islands

SHEX *Shipping* Sundays and holidays excepted

Sh.F. shareholders' funds

SHF *Radio, etc.* superhigh frequency

SHHD Scottish Home and Health Department

shipt shipment

SHM *Physics* simple harmonic motion · Society of Housing Managers

SHMIS Society of Headmasters and Headmistresses of Independent Schools

SHMO Senior Hospital Medical Officer

sho *Baseball* shutout

SHO Senior House Officer

shoran (ˈʃɔːræn) short-range navigation

shp shaft horsepower (of an engine)

SHP *Chemical engineering* selective hydrogenation process · *Horticulture* single-flowered hardy perennial (*or* hybrid perpetual) (rose)

shpg shipping

shpt shipment

SHQ station headquarters · supreme headquarters

shr. share(s)

shrap. shrapnel

Shrops Shropshire

SHS Fellow of the Historical Society (Latin *Societatis Historicae Socius*) · Shire Horse Society

sht *Bookbinding* sheet

SHT single-flowered hybrid tea (rose)

shtg. shortage

s.h.v. sub hac voce (*or* hoc verbo) (Latin: under this word)

SHW safety, health, and welfare

s.i. sum insured

Si *Chem., symbol for* silicon

SI *Irish vehicle registration for* Dublin · Sandwich Islands (former name for Hawaii) · seriously ill · Shetland Isles · Smithsonian Institution (Washington, DC) · (USA) Society of Illustrators · South Island (New Zealand) · staff inspector · (Order of the) Star of India · Staten Island (New York) · *Government* statutory instrument · *Chem.* styrene–isoprene (polymer) · Système International (d'Unités; French: International System of Units; in **SI unit(s)**)

SIA Society of Investment Analysts · Spinal Injuries Association

SIAC (USA) Securities Industry Automation Corporation

SIAD Society of Industrial Artists and Designers (now called Chartered Society of Designers, CSD)

SIAM (USA) Society of Industrial and Applied Mathematics

SIAS Statement on Internal Auditing Standards

Sib. Siberia(n)

SIB Securities and Investments Board • *Med.* self-injurious behaviour • Shipbuilding Industry Board • Special Investigation Branch (of the Police)

SIBOR ('si:bɔ:) *Finance* Singapore Inter-Bank Offered Rate

sic. *Pharmacol.* siccus (Latin: dry)

Sic. Sicilian • Sicily

SIC (USA) Scientific Information Center • Standard Industrial Classification

SICAV (France) société d'investissement à capital variable (unit trust)

SICOT Société internationale de chirurgie orthopédique et de traumatologie (French: International Society of Orthopaedic Surgery and Traumatology)

SID Society for International Development • Spiritus in Deo (Latin: his (*or* her) spirit is with God) • (*or* **s.i.d.**) *Radio* sudden ionospheric disturbance

SIDA Swedish International Development Authority

SIDF system independent data format

SIDS (sɪdz) *Med.* sudden infant death syndrome (cot death)

SIE *Physics* secondary ion emission

SIEC Scottish Industrial Estates Corporation

SIESO Society of Industrial and Emergency Service Officers

SIF selective identification feature • *Engineering* stress-intensity factor

SIFS special instructors flying school

sig. signal • signature • signetur (Latin: let it be written *or* labelled) • signification • signifies

Sig. *Med.* signā (Latin: write; in prescriptions, preceding instructions to be written on the label for the patient's use) • Signor (Italian: Mr) • Signore (Italian: Sir)

SIG *Numismatics* signature of engraver present • special-interest group

SIGAC Scottish Industrial Groups Advisory Council

sig. fig. *Maths.* significant figures

sigill. sigillum (Latin: seal)

SIGINT (*or* **Sigint**) ('sɪgɪnt) signals intelligence (gathering network)

SIGMA ('sɪgmə) Science in General Management

Sigmn *Navy* Signalman

sign. signature

sig. n. pro. *Med.* signa nomine proprio (Latin: label with the proper name)

Sig.O. Signal Officer

SIGWEX *Meteorol.* significant weather

SII Société internationale de la lèpre (French: International Leprosy Association)

Sil. Silesia

SIL (sɪl) *Electronics* single in-line (as in **SIL device**)

sim. similar(ly) • simile

SIM *Physics* secondary-ion microscopy • self-inflicted mutilation • Société internationale de musicologie (French: International Musicological Society) • survey information on microfilm

SIMA Scientific Instrument Manufacturers' Association of Great Britain • Steel Industry Management Association • *Psychiatry* system for identifying motivated abilities

SIMC Société internationale pour la musique contemporaine (French: International Society for Contemporary Music)

SIMCA (*or* **Simca**) ('sɪmkə) Société industrielle de mécanique et carrosserie automobiles (French car manufacturers)

SIMD *Computing* single instruction (stream), multiple data (stream)

SIME Security Intelligence Middle East

SIMEX ('saɪmeks) Singapore International Monetary Exchange

SIMG International Society of General Medicine (Latin *Societas Internationalis Medicinae Generalis*)

SIMM *Computing, Electronics* single in-line memory module

SIMPL Scientific, Industrial, and Medical Photographic Laboratories

SIMS *Chem.* secondary-ion mass spectrometry (*or* spectroscopy)

sin (saɪn) *Maths.* sine

sin. sinecure

SIN *Accounting* stores issue note

SinDrs Doctor of Chinese

sing. *Grammar* singular • *Med.* singulorum (Latin: of each; in prescriptions)

Sing. Singapore

sinh (saɪn) *Maths.* hyperbolic sine

Sinh. Sinhalese

SINS (sɪnz) ship's inertial navigation system

SIO Senior Intelligence Officer • *Computing* serial input/output

SIOP ('saɪɒp) (USA) single integrated operations plan (for nuclear war)

SIP *Computing* single in-line package • (USA) supplemental income plan

SIPC (USA) Securities Investor Protection Corporation

SIPO *Computing* serial in, parallel out

SIPRC Society of Independent Public Relations Consultants

SIPRI Stockholm International Peace Research Institute

SIPS Side Impact Protection System

Sir *Computing* serial infrared

SIR *Taxation* small income relief

SIRA Scientific Instrument Research Association

SIRS systemic inflammatory response syndrome

SIRTF Space Infrared Telescope Facility

sis. sister

SIS Secret Intelligence Service (MI6) • (New Zealand) Security Intelligence Service

SISD *Computing* single instruction (stream), single data (stream)

SISO *Electronics* serial in, serial out • *Electronics* single input, single output

SISS submarine integrated sonar system

SISTER ('sɪstə) Special Institutions for Scientific and Technological Education and Research

sit. sitting room • situation

s.i.t. stopping in transit • storing in transit

SIT Society of Industrial Technology • Society of Instrument Technology • *Engineering* spontaneous ignition temperature

SITA Société internationale de télécommunications aéronautiques (French: International Society of Aeronautical Telecommunications) • Students' International Travel Association

SITC Standard International Trade Classification

SITC(R) Standard International Trade Classification (Revised)

SITPRO ('sɪtprəʊ) Simpler Trade Procedures Board (formerly Simplification of International Trade Procedures)

sitt. sitting-room

sit. vac. (pl. **sits vac.**) situation vacant

SIUNA Seafarers International Union of North America

SIV *Microbiol.* simian immunodeficiency virus

SIW self-inflicted wound

s.j. *Law* sub judice

SJ *British vehicle registration for* Glasgow • Society of Jesus • supersonic jet

SJA St John Ambulance (Brigade *or* Association)

SJAA St John Ambulance Association

SJAB St John Ambulance Brigade

SJC Standing Joint Committee • (USA) Supreme Judicial Court

SJD Doctor of Juristic (*or* Juridical) Science (Latin *Scientiae Juridicae Doctor*)

S. J. Res. (USA) Senate joint resolution

sk sack • sick

sk. sketch

SK *British vehicle registration for* Inverness • Saskatchewan • *airline flight code for* SAS-Scandinavian Airline System • *international vehicle registration for* Slovakia • *postcode for* Stockport

SKAMP (skæmp) station keeping and mobile platform (unmanned self-navigating sailing boat)

SKC Scottish Kennel Club • *Meteorol.* sky clear

S. Ken. South Kensington (London)

skid (skɪd) *Med. see* SCID

Skm Stockholm

s.k.p.o. *Knitting* slip one, knit one, pass slipped stitch over

SKr *symbol for* Swedish krona (monetary unit)

Skt (*or* **Skr.**) Sanskrit

SKU *Commerce* stock-keeping unit

sl. sleet • slightly • slip

s.l. *Insurance* salvage loss • secundum legem (*see* sec. leg.) • seditious libel • *Bibliog.* sine loco (Latin: without place (of publication)) • support line

Sl. Slovak(ian)

SL *British vehicle registration for* Dundee • *Insurance* salvage loss • scout leader • sea level • Second Lieutenant • security list • Serjeant-at-Law • *Building trades* short lengths • *postcode for* Slough • Solicitor-at-Law • source language • southern league • south latitude • *Commerce* specification limits • Squadron Leader • supplementary list

SLA Scottish Library Association • special landscape area (in conservation) • (USA) Special Libraries Association • Symbionese Liberation Army

SLAC Stanford Linear Accelerator Center

SLADE (sleɪd) Society of Lithographic Artists, Designers, Engravers, and Process Workers

SLAET Society of Licensed Aircraft Engineers and Technologists

SLAM (*or* **Slam**) (slæm) standoff land-attack missile

s.l.a.n. sine loco, anno, vel nomine (Latin: without place, year, name (of printer))

SLAR (*or* **Slar**) (slɑ:) *Military* side-looking airborne radar

SLAS Society for Latin American Studies

S. Lat. south latitude

Slav. Slavonian • Slavonic (*or* Slavic)

SLBM submarine-launched ballistic missile

SLC Stanford Linear Collider (at SLAC) • Statute Law Committee • Surgeon Lieutenant-Commander

SLCM sea- (*or* ship-, submarine-) launched cruise missile

sld sailed • sealed • sold • solid

SLD self-locking device • (formerly) Social and Liberal Democrats (renamed Liberal Democrats)

S. Ldr. Squadron Leader

SLE *Med.* systemic lupus erythematosus

s.l. et a. sine loco et anno (Latin: without place and year (of publication))

S level *Education* Special (formerly Scholarship) level

s.l.f. *Telecom.* straight-line frequency

SLF Scottish Landowners' Federation

SLFP Sri Lanka Freedom Party

SLIC (USA) (Federal) Savings and Loan Insurance Corporation

SLIM South London Industrial Mission

SLIP *Computing* serial line Internet protocol

Slipar ('slaɪpɑ:) *Military* short light pulse alerting receiver (in an aircraft)

SLLA Scottish Ladies Lacrosse Association

SLLW solid low-level (radioactive) waste

SLM ship-launched missile

SLMA (USA) Student Loan Marketing Association

SLMC Scottish Ladies' Mountaineering Club

s.l.n.d. sine loco nec data (Latin: without indication of date or place (of printing))

SLO Senior Liaison Officer • *international vehicle registration for* Slovenia

SLOC (slɒk) *Computing* source lines of code

SLORC (Burma) State Law and Order Restoration Council

slp slip

s.l.p. sine legitime prole (Latin: without lawful issue)

SLP Scottish Labour Party • (USA) Socialist Labor Party

SLR satellite laser ranging • self-loading rifle • single-lens reflex (camera)

SL Rs *symbol for* Sri Lanka rupee (monetary unit)

SLS sodium lauryl sulphate (detergent) • Stephenson Locomotive Society

SLSC surf life-saving club

SLSI super large-scale integration

SLTA Scottish Licensed Trade Association

Slud (slʌd) *Military slang* salivate, lachrymate, urinate, defecate (effects of chemical weapons)

SLV space (*or* standard) launch vehicle

SLW *Meteorol.* slow • solid low-level (radioactive) waste

sly slowly • southerly

sm. small

s-m (*or* **s/m**) sadomasochism (*or* sadomasochist)

Sm *Chem.*, *symbol for* samarium

SM *British vehicle registration for* Carlisle • *British vehicle registration for* Glasgow • Master of Science (Latin *Scientiae Magister*) • (*or* **S-M**) sadomasochism (*or* sadomasochist) • sales manager • Sa Majesté (French: His (*or* Her) Majesty) • sanctae memoriae (Latin: of holy memory) • Seine Majestät (German: His (*or* Her) Majesty) • senior magistrate • Sergeant Major • (USA) service mark (registered proprietary name) • *Astronautics* service module • shipment memorandum • *British fishing port registration for* Shoreham • *Music* short metre • silver medal(list) • Sisters of Mercy • Society of Miniaturists • Sons of Malta • Staff Major • stage manager • *Astronomy* standard model • (USA) state militia • station master • stipendiary magistrate • strategic missile • Sua Maestà (Italian: His (*or* Her) Majesty) • (officer qualified for) Submarine Duties • Su Magestad (Spanish: His (*or* Her) Majesty) • Surgeon Major • *postcode for* Sutton (Surrey)

SMA *Med.* spinal muscular atrophy • (USA) Surplus Marketing Administration

SMAC Standing Medical Advisory Committee

SMATV satellite (formerly small) master antenna television

SMAW shielded metal-arc welding

SMB Bachelor of Sacred Music • Sa Majesté Britannique (French: Her (*or* His) Britannic Majesty)

SMBA Scottish Marine Biological Association

SMBG *Med.* self-monitoring of blood glucose (for diabetics)

SMC Sa Majesté Catholique (French: His (*or* Her) Catholic Majesty) • Scottish Mountaineering Club • *Astronomy* Small Magellanic Cloud • *Aeronautics* standard mean chord

sm. caps. *Printing* small capitals

SMC(Disp) Dispensing Certificate of the Worshipful Company of Spectacle Makers

SMD Doctor of Sacred Music • *Med.* senile macular degeneration • *Music* short metre double • submarine mine depot • *Electronics* surface mount device

SMDS switched multimegabit data service

SME Sancta Mater Ecclesia (Latin: Holy Mother Church) • seismic-margin earthquake • *international vehicle registration for* Suriname

SMERSH (*or* **Smersh**) Smert Shpionam (Russian: death to spies; section of KGB)

SMEs small and medium enterprises

S.Met.O. Senior Meteorological Officer

SMEX *Astronomy* Small Explorer

SMG Scottish Media Group • submachine gun

SMH *British fishing port registration for* St Margaret's Hope (Orkney)

SMHD Higher Diploma in Ophthalmic Optics of the Worshipful Company of Spectacle Makers

SMHI Swedish Meteorological and Hydrological Institute

SMHO (Malta) Sovereign Military Hospitaller Order

SMI Sa Majesté Impériale (French: His (*or* Her) Imperial Majesty) • Swiss Market Index

SMIA Sheet Metal Industries Association

SMIEEE (USA) Senior Member of the Institute of Electrical and Electronics Engineers

SMIRE (USA) Senior Member of the Institute of Radio Engineers

Smith. Inst. Smithsonian Institution (Washington, DC)

SMJ Sisters of Mary and Joseph

smk. smoke

sml. simulate • simulation • simulator • small

SML Science Museum Library

SMLE short magazine Lee-Enfield (rifle)

SMM Master of Sacred Music • Sancta Mater Maria (Latin: Holy Mother Mary)

SMMB Scottish Milk Marketing Board

SMMT Society of Motor Manufacturers and Traders Ltd

SMO Senior Medical Officer • Sovereign Military Order

SMON (smɒn) *Med.* subacute myelo-opticoneuropathy

smorz. *Music* smorzando (Italian; gradually slower and softer)

s.m.p. sine mascula prole (Latin: without male issue)

SMP Society of Mural Painters • statutory maternity pay • *Computing* symmetric multiprocessing

SMPS Society of Master Printers of Scotland • switched-mode power supply

SMPTE (USA) Society of Motion Picture and Television Engineers

SMR Sa Majesté Royale (French: His (*or* Her) Royal Majesty) • standardized mortality ratio • standard Malaysian rubber • *Med.* standard metabolic rate

SMRE Safety in Mines Research Establishment

SMRTB Ship and Marine Requirements Technology Board

SMS synchronous meteorological satellite

SMSA (USA) Standard Metropolitan Statistical Area

SMSO Senior Maintenance Staff Officer

SMT ship's mean time • *Maths.* Steiner minimum tree • *Computing* surface mount technology

SMTA Scottish Motor Trade Association

SMTF Scottish Milk Trade Federation

SMTO Senior Mechanical Transport Officer

SMTP *Computing* simple mail transfer protocol

SMV *Med.* submentovertical (in radiology)

SMW standard metal window

SMWIA (USA) Sheet Metal Workers' International Association (trade union)

sn. snow

s.n. secundum naturam (Latin: according to nature, naturally) • serial number • series number • service number • sine nomine (Latin: without name) • sub nomine (Latin: under a specified name)

Sn *Chem. symbol for* tin (Latin *stannum*)

SN *British vehicle registration for* Dundee • *British fishing port registration for* North Shields • *airline flight code for* Sabena • (USA) Secretary of the Navy • *international vehicle registration for* Senegal • Sergeant Navigator • shipping note • *Meteorol.* snow • *Astronomy* supernova • *postcode for* Swindon

S/N shipping note • *Electronics* signal-to-noise (in **S/N ratio**) • *Engineering* stress-number (in **S/N curve**)

SNA *Computing* systems network architecture

snafu (snæ'fuː) *Slang* situation normal, all fouled (*or* fucked) up

SNAME (USA) Society of Naval Architects and Marine Engineers

SNAP (snæp) Shelter Neighbourhood Action Project • systems for nuclear auxiliary power

SNB *Stock exchange* sellers no buyers

SNCB Société nationale des chemins de fer belges (Belgian National Railways)

SNCC (snɪk) (USA) Student Nonviolent (later National) Coordinating Committee

SNCF (France) Société nationale des chemins de fer français (state railway authority or system)

Snd Sound (in place names)

SND Sisters of Notre Dame

SNECMA (*or* **Snecma**) ('snɛkmə) (France) Société nationale d'étude et de construction de moteurs d'aviation (state-owned aeroengine company)

SNF solids, nonfat (in nutrition) • spent nuclear fuel • (USA) strategic nuclear forces

SNFA Standing Naval Force, Atlantic

SNFU Scottish National Farmers' Union

Sng. Singapore

SNG substitute (*or* synthetic) natural gas

SNH Scottish National Heritage

SNIF *Finance* short-term note issuance facility

SNIG sustainable noninflationary growth

SNIPEF Scottish and Northern Ireland Plumbing Employers' Federation

SNL Sandia National Laboratories (Albuquerque) • standard nomenclature list

SNLR services no longer required

SNLV strategic nuclear launch vehicle

SNM (USA) Society of Nuclear Medicine • Somali National Movement

SNMP *Computing* simple network management protocol

SNO (formerly) Scottish National Orchestra (renamed Royal Scottish Orchestra, RSO) • Senior Naval Officer • Senior Navigation Officer • Senior Nursing Officer

SNOBOL ('snəʊ,bɒl) *Computing, indicating* a programming language designed for text manipulation

snoRNP *Biochem.* small nucleolar ribonucleoprotein

SNP Scottish National Party

SNPA Scottish Newspaper Proprietors' Association

Snr Senior

SNR *Electronics* signal-to-noise ratio • Society for Nautical Research • *Astronomy* supernova remnant

snRNA *Biochem.* small nuclear RNA

snRNP *Biochem.* small nuclear ribonucleoprotein

s 'n' s *Colloquial* sex and shopping (type of popular fiction)

SNSC Scottish National Ski Council

SNSH *Meteorol.* snow showers

SNTPC Scottish National Town Planning Council

SNTS Society for New Testament Studies

SNU (snjuː) *Astronomy* solar neutrino unit

so. sonata • south(ern)

s.o. seller's option • shipping order • siehe oben (German: see above) • strike out • substance of (specifying weight of paper)

So. south(ern)

SO *British vehicle registration for* Aberdeen • Scientific Officer • Scottish Office • section officer • Senior Officer • shale oil (residue) • Signal Officer • *Irish fishing port registration for* Sligo • *Computing* small outline • *international vehicle registration for* Somalia • sorting office • *postcode for* Southampton • *Meteorol.* southern oscillation • special order • Staff Officer • standing order • Stationery Office •

statistical office • suboffice • Supply Officer • Symphony Orchestra

S/O Section Officer • shipowner

SOA state of the art

SOAD Staff Officer, Air Defence

SOAP (səup) *Med.* subjective, objective, analysis, plan (method of compiling patients' records)

SOAS School of Oriental and African Studies (University of London)

SOB (USA) Senate office building • *Colloquial* silly old bastard (*or* blighter) • (*or* **s.o.b.**) *Colloquial* son of a bitch • state office building

soc. social • socialist • society • sociology

Soc. Socialist • (Italy) società (company *or* partnership) • Society • Socrates (469–399 BC, Greek philosopher)

SOC Scottish Ornithologists' Club • *Colloquial* slightly off colour

SocCE(France) Société des ingénieurs civils de France (Society of Civil Engineers of France)

Soc. Dem. Social Democrat

sociol. sociological • sociologist • sociology

SOCO scene-of-crime officer (in the police)

SOCS Society of County Secretaries

soc. sci. social science (*or* scientist)

socy (*or* **Socy**) society

sod. sodium

SOD superoxide dismutase

SODAC Society of Dyers and Colourists

SODEPAX ('səʊdɪˌpæks) Committee on Society, Development, and Peace

SOE Special Operations Executive (in World War II) • state-owned enterprise

SOED (ital.) Shorter Oxford English Dictionary

SOEID Scottish Office Education and Industry Department

SOES *Finance* (USA) Small Order Execution System (of NASDAQ)

SOF share of freehold (in property advertisements) • *Chem.* soluble organic fraction • *Films* sound on a film

SOFAA Society of Fine Art Auctioneers

sofar ('səʊfɑː) sound fixing and ranging

SOFCS self-organizing flight-control system

SOFFEX ('sɒfɛks) Swiss Options and Financial Futures Exchange

SOFIA Stratospheric Observatory for Infrared Astronomy

S. of S. Secretary of State • *Bible* Song of Songs

SOFS (USA) Small Order Execution System

S. of Sol. *Bible* Song of Solomon

S. of T. Sons of Temperance

S. of TT School of Technical Training

SOGAT ('səʊgæt) (formerly) Society of Graphical and Allied Trades (merged with NGA to form GPMU)

SOH sense of humour (in personal advertisements)

SOHIO (səʊˈhaɪəʊ) Standard Oil of Ohio

SoHo ('səʊˌhəʊ) (USA) South of Houston

SOHO Solar Heliospheric Observatory

Sohyo Nihon Rodo Kumiai So Hygikai (General Council of Japanese Trade Unions)

SO(I) Staff Officer (Intelligence)

SOIC small outline integrated circuit

SO-in-C Signal Officer-in-Chief

sol. solicitor • soluble • solution

s.o.l. *Insurance* shipowner's liability

Sol. Solicitor • *Bible* Song of Solomon

SOL *Slang* (USA) strictly (*or* shit) out of luck

SOLACE (*or* **Solace**) ('sɒlɪs) Society of Local Authority Chief Executives

Sol. Gen. Solicitor General

soln solution

solr (*or* **Solr**) solicitor

solv. solvent

soly solubility

Som. Somerset • (*or* **Som**) Somerville College, Oxford

SOM Society of Occupational Medicine

SOMA ('səʊmə) Society of Mental Awareness

SOME Senior Ordnance Mechanical Engineer

SOMPA ('sɒmpə) (USA) System of Multicultural Pluralistic Assessment (intelligence testing)

Som.Sh. Somali shilling (monetary unit)

SON Spear of the Nation (military arm of African National Congress)

sonar ('səʊnɑː) sound navigation and ranging

SONET ('sɒnɪt) *Computing* synchronous optical network

SO(O) Staff Officer (Operations)

sop. soprano

SOP sleeping-out pass • standard

operating procedure • *Computing* sum of products (in **SOP expression**)

soph. sophomore

Soph. Sophocles (496–406 BC, Greek dramatist and poet)

SOR (*or* **SoR**) sale or return • Society of Radiographers

SORD submerged-object recovery device

SORP *Accounting* Statement of Recommended Practice

s.o.s. *Med.* si opus sit (Latin: if necessary; in prescriptions)

SOS save our souls (also, the clearest letters to transmit and receive in Morse code) • (*or* **SoS**) Secretary of State • senior officers' school • services of supply

SOSc Society of Ordained Scientists

SoSh *symbol for* Somali shilling (monetary unit)

sost. *Music* sostenuto (Italian: sustained)

SOTS Society for Old Testament Study

Sou. (*or* **sou.**) south(ern) • (*or* **Soton**) Southampton

sov. sovereign

s.o.v. shut-off valve

Sov. Soviet

SOV *Linguistics* subject-object-verb (in **SOV language**)

Sov. Un. Soviet Union

sowc senior officers' war course

Soweto (sə'wɛtəʊ) Southwestern Townships (South Africa)

sp. space • special • specie • *Biology* species • specific • specimen • speed • spelling • spirit • sport

s.p. self-propelled • *Music* senza pedale (Italian: without pedal) • sine prole (Latin: without issue) • *Electricity* single phase • starting point • starting price • *Banking* stop payment

Sp. Spain • Spaniard • Spanish

SP *British vehicle registration for* Dundee • *international civil aircraft marking for* Poland • Saint-Père (French: Holy Father) • *postcode for* Salisbury • Sanctissimus Pater (Latin: Most Holy Father) • Self-Propelled (Antitank Regiment) • service pistol • *Military* service police • shore patrol • Sisters of Providence • Socialist Party • *Building trades* soil pipe • spark(ing) plug • Staff Paymaster • standard play (on a VCR) • starting price (odds in a race) • stirrup pump • stop press • stretcher party • submarine patrol • Summus Pontifex (Latin:

Supreme Pontiff; the pope) • supply point • *Finance* supra protest

SpA (Italy) società per azioni (public limited company; plc)

SPA (USA) Society for Personnel Administration

SPAA Scottish Passenger Agents' Association

SPAB Society for the Protection of Ancient Buildings

Sp. Am. Spanish American

Span. Spaniard • Spanish

SPANA Society for the Protection of Animals in North Africa

SPANDAR ('spændɑː) space and range radar (in NASA)

Sp. Ar. Spanish Arabic

SPAR (spɑː) superprecision-approach radar • *indicating* a member of the women's reserve of the US Coast Guard in World War II (from its motto, *Semper paratus* (Latin: always ready))

SPARC (spɑːk) *Computing* scalable processor architecture

SPAS Fellow of the American Philosophical Society (Latin *Societatis Philosophicae Americanae Socius*)

SPATC South Pacific Air Transport Council

SPC Society for the Prevention of Crime • *Commerce* statistical process control • *Telecom.* stored-program control (as in **SPC exchange**)

SPCA (USA) Society for the Prevention of Cruelty to Animals

SPCK Society for Promoting Christian Knowledge

s.p.d. *Finance* subject to permission to deal

SPD Salisbury Plain District • *Astronomy* south polar distance • Sozialdemokratische Partei Deutschlands (Social Democratic Party of Germany)

SPDA single-premium deferred annuity

SPDL *Computing* standard page description language

SPE Society for Pure English • (USA) Society of Petroleum Engineers • *Electronics* solid-phase epitaxy

spec. special(ly) • specific(ally) • specification • specimen • spectrum • speculation

SPEC (Canada) Society for Pollution and Environmental Control • South Pacific Bureau for Economic Cooperation

special. specialized

specif. specifically • specification

SPECmark ('spɛk,mɑːk) Systems Performance Evaluation Cooperative's benchmark

SPECT (*or* **Spect**) (spɛkt) *Med.* single photon emission computed tomography

SPECTRE ('spɛktə) Special Executive for Counter-Intelligence, Revenge, and Extortion (fictional terrorist organization in Ian Fleming's James Bond novels)

Sp.Ed. (USA) Specialist in Education

SPF South Pacific Forum • sun protection factor (of sunscreening preparations)

SPF/DB *Aeronautics* super-plastic forming/diffusion bonding (manufacturing method)

s.p.g. self-propelled gun

SPG Society for the Propagation of the Gospel (*see* USPG) • Special Patrol Group

SPGA Scottish Professional Golfers' Association

SPGB Socialist Party of Great Britain

sp. gr. specific gravity

sp. ht specific heat

SPI *Finance* selected period investment • (USA) Society of the Plastics Industry

SPIRE (spaɪə) *Navigation* spatial inertial reference equipment

spirit. *Music* spiritoso (Italian: in a spirited manner) • spiritualism • spiritualistic

SPIW special-purposes individual weapon

s.p.l. sine prole legitima (Latin: without legitimate issue)

SPLA Sudan People's Liberation Army

s.p.m. sine prole mascula (Latin: without male issue)

SPM scanning proton microscope • *Music* short particular metre

SPMO Senior Principal Medical Officer

SPMU Society of Professional Musicians in Ulster

SPN stop press news

SPNM Society for the Promotion of New Music

SPNR Society for the Promotion of Nature Reserves

SPO Senior Press Officer

SPÖ Sozialistische Partei Österreichs (Austrian Socialist Party)

SPOA Scottish Prison Officers' Association

SPOD Sexual Problems of the Disabled

(department of the Royal Association for Disability and Rehabilitation)

SPOE Society of Post Office Engineers

Spool (spuːl) simultaneous peripheral operation on-line

sport. sporting

SPOT single property ownership trust

spp. *Biology* species (plural)

SPQR Senatus Populusque Romanus (Latin: the Senate and People of Rome) • small profits and quick returns

spr. spring • sprinkle

Spr *Military* Sapper

SPR Society for Psychical Research • strategic petroleum reserve

SPRC Society for the Prevention and Relief of Cancer

SPREd Society of Picture Researchers and Editors

SPRI Scott Polar Research Institute (Cambridge)

SPRINT (sprɪnt) *Military* solid-propellant rocket-intercept missile

sprl (France) société de personnes à responsabilité limitée (private limited company; Ltd)

SPRL Society for the Promotion of Religion and Learning

s.p.s. sine prole supersite (Latin: without surviving issue)

SPS Scottish Painters' Society • *Physics* Super Proton Synchrotron (at CERN) • syndiotactic polystyrene (a plastic)

SPSL Society for the Protection of Science and Learning

SPSO Senior Principal Scientific Officer

SPSP St Peter and St Paul

sp. surf. *Chem.* specific surface

spt seaport • support

sptg sporting

SPTL Society of Public Teachers of Law

SPUC (*or* **Spuc**) (spʌk) Society for the Protection of the Unborn Child

SPURV (spɜːv) self-propelled underwater research vehicle

sq *Military* staff qualified

sq. sequence • sequens (*see under* seq.) • squadron • square

Sq. Squadron • Square (in place names)

SQ sick quarters • *airline flight code for* Singapore Airlines • *Meteorol.* squall • *Building trades* squint quoin • stereophonic-quadraphonic (of audio equipment) • survival quotient

SQA Scottish Qualifications Agency

(formed from a merger of SCOTVEC and SEB) • *Computing* software quality assurance

sq cm square centimetre(s)

sqd squad

sqdn (*or* **Sqdn**) squadron

Sqdn Ldr Squadron Leader

sq ft square feet (*or* foot)

sq in square inch(es)

sq km square kilometre(s)

SQL *Computing* standard query language • *Computing* structured query language

sq m square metre(s)

sq mi square mile(s)

sq mm square millimetre(s)

SQMS Staff Quartermaster Sergeant

sqn (*or* **Sqn**) squadron

Sqn Ldr Squadron Leader

SqnQMS Squadron Quartermaster Sergeant

SqnSM Squadron Sergeant Major

SqO Squadron Officer

sqq. sequentia (*see under* seqq.)

squid (*or* **SQUID**) (skwɪd) *Electronics* superconducting quantum interference device

sq yd square yard

sr *Maths., symbol for* steradian

s.r. self-raising • shipping receipt • short rate

Sr Senhor (Portuguese: Mr, Sir) • Senior (after a name) • Señor (Spanish: Mr, Sir) • Signor (Italian: Mr, Sir) • Sir • Sister (religious) • *Chem., symbol for* strontium

SR *British vehicle registration for* Dundee • *Cytology* sarcoplasmic reticulum • Saudi riyal (Saudi Arabian monetary unit) • Saunders Roe (aircraft) • (USA) Senate resolution • service rifle • *Computing* set-reset • *symbol for* Seychelles rupee (monetary unit) • (formerly, USSR) Socialist Revolutionary (Party) • Society of Radiographers • (USA) Sons of the Revolution • Southern Railway (now Region) • *Physics* special relativity • *Military* Special Reserve • *Taxation* standard rate • (*or* **S–R**) *Psychol.* stimulus–response (as in **SR theory**) • *Chem.* stoichiometric ratio • *British fishing port registration for* Stranraer • *postcode for* Sunderland • Sveriges Radio (Swedish broadcasting corporation) • *airline flight code for* Swissair • *Physics* synchrotron radiation • synthetic rubber

S/R sale or return

Sra Senhora (Portuguese: Mrs) • Señora (Spanish: Mrs)

SRA Squash Rackets Association

SRAM short-range attack missile • *Computing* static random access memory

SR & CC strikes, riot, and civil commotion

SRB solid rocket booster • *Computing* source route bridge • *Microbiol.* sulphate-reducing bacteria

SRBC *Med.* sheep red blood cell(s)

SRBM short-range ballistic missile

SRBP synthetic resin-bonded paper

SRC sample return container • Science Research Council (*see* SERC) • (Spain) sociedad regular colectiva (partnership) • solvent-refined coal • *Chem.* standard reference compound • *Engineering* steam Rankine cycle • Students' Representative Council • Swiss Red Cross

SRCh State Registered Chiropodist

SRCN (formerly) State Registered Children's Nurse (*see* RSCN)

SRD service rum diluted

SRDE Signals Research and Development Establishment

SRE Sancta Romana Ecclesia (Latin: Holy Roman Church)

S. Rept (USA) Senate report

S. Res. (USA) Senate resolution

SRG standard reformed gas • Strategic Research Group (marketing research company)

SRHE Society for Research into Higher Education

SRI Sacrum Romanum Imperium (Latin: Holy Roman Empire)

SRIS Science Reference Information Service

Srl. (Italy) società a responsabilità limitata (private limited company; Ltd)

SRls *symbol for* Saudi riyal (monetary unit of Saudi Arabia)

SRM short-range missile • speed of relative movement

SRN (formerly) State Registered Nurse (*see* RGN) • *Accounting* stores return note

sRNA *Biochem.* soluble RNA

SRNA Shipbuilders and Repairers National Association

SRO *Finance* self-regulatory organization • *Microbiol.* sex-ratio organism • (USA) single-room occupancy (as in **SRO hotel**) • standing room only • Statutory

Rules and Orders • Supplementary Reserve of Officers

SRP State Registered Physiotherapist • suggested retail price • supply refuelling point

SRS Fellow of the Royal Society (Latin *Societatis Regiae Sodalis*)

SRSA Scientific Research Society of America • *Immunol.* slow-reacting substance A (of anaphylaxis)

SRSS *Maths.* square-root sum of squares

Srta Senhorita (Portuguese: Miss) • Señorita (Spanish: Miss)

SRU Scottish Rugby Union

SRY Sherwood Rangers Yeomanry

ss. sections • *Med.* semis (Latin: half; in prescriptions) • *Baseball* shortstop • subsection

s.s. screw steamer • sensu stricto (Latin: in the strict sense) • *Music* senza sordini (Italian: without mutes) • steamship • supra scriptum (Latin: written above)

SS *British vehicle registration for* Aberdeen • Sacra Scriptura (Latin: Holy Scripture) • Saints • Santa Sede (Italian: Holy See) • Sa Sainteté (French: His Holiness) • Schutzstaffel (German: protection squad; Nazi paramilitary organization) • secondary school • Secretary of State • secret service • security service • short sleeves • Sidney Sussex College, Cambridge • *Computing* single-sided • social security • *Chem.* sodium sulphate • *postcode for* Southend-on-Sea • Staff Surgeon • stainless steel • standard size • steamship • *British fishing port registration for* St Ives • (Italy) Strada Statale (National Highway) • Straits Settlements • *Chem.* styrylstilbene • Sunday school • surface to surface (missile)

SS. Saints • sanctissimus (Latin: most holy)

S/S same size (of illustrations) • *Printing* silk screen • steamship

SSA Scottish Schoolmasters' Association • *Computing* serial storage architecture • (USA) Social Security Administration • Society of Scottish Artists • standard spending assessment (in local government)

SSAC Scottish Sub-Aqua Club • Social Security Advisory Committee

SSADM (*or* **SSadm**) *Computing* structured systems analysis and design method

SSAE stamped self-addressed envelope

SSAFA (*or* **SS&AFA**) Soldiers', Sailors', and Airmen's Families Association

SSAP Statement of Standard Accounting Practice

SSB Bachelor of Sacred Scripture (Latin *Sacrae Scripturae Baccalaureus*) • *Telecom.* single-sideband (transmission) • *Genetics* single-strand binding (as in **SSB protein**) • (USA) Social Security Board • *Physics* spontaneous symmetry breaking

SSBN *US Navy* strategic submarine, ballistic nuclear

SSC Scottish Ski Club • Scottish Sports Council • Sculptors' Society of Canada • (India) Secondary School Certificate • Short Service Commission • (USA) small-saver certificate • Society of the Holy Cross (Latin *Societas Sanctae Crucis*) • (Scotland) Solicitor before the Supreme Court • Species Survival Commission • *Physics* Superconducting Super Collider

SScD Doctor of Social Science

SSD Doctor of Sacred Scripture (Latin *Sacrae Scripturae Doctor*) • Social Services Department

SS.D Sanctissimus Dominus (Latin: Most Holy Lord; the pope)

ssDNA *Biochem.* single-stranded DNA

SSE Society of St Edmund • south-southeast

SSEB (formerly) South of Scotland Electricity Board

SSEC Secondary School Examinations Council

S-SEED *Optics* symmetric self-electro-optic-effect device

SSEES School of Slavonic and East European Studies (University of London)

SSF *Military* single-seater fighter (aircraft) • Society of St Francis

SSFA Scottish Schools' Football Association • Scottish Steel Founders' Association

S/Sgt (*or* **S.Sgt**) Staff Sergeant

SSHA Scottish Special Housing Association

SSI Scottish Symphony Orchestra • site of scientific interest • *Electronics* small-scale integration • Social Services Inspectorate • Society of Scribes and Illuminators • (USA) supplemental security income

SSJE Society of St John the Evangelist

SSL Licentiate in Sacred Scripture (Latin *Sacrae Scripturae Licentiatus*)

SSM *Telecom.* single-sideband modulation · Society of the Sacred Mission · *Commerce* soft systems methodology · Staff Sergeant Major · *Military* surface-to-surface missile

SSMA Stainless Steel Manufacturers' Association

SSN severely subnormal · (USA) Social Security number · Standard Serial Number

SSO Senior Scientific Officer · Senior Supply Officer · Staff Signal Officer · Station Staff Officer

ssp. *Biology* subspecies

SSP statutory sick pay

SSPCA Scottish Society for the Prevention of Cruelty to Animals

SSPE *Med.* subacute sclerosing panencephalitis

sspp. *Biology* subspecies (plural)

SS.PP. Sancti Patres (Latin: Holy Fathers)

SSQ *Military* station sick quarters

SSR secondary surveillance radar · Soviet Socialist Republic

SSRA Scottish Squash Rackets Association

SSRC Social Science Research Council (*see under* ESRC)

SSRI *Med.* selective serotonin reuptake inhibitor · (USA) Social Science Research Institute

SSS Secretary of State for Scotland · (USA) Selective Service System (for mobilizing military forces) · *Med.* sick sinus syndrome · single-screw ship · *British fishing port registration for* South Shields · *Golf* standard scratch score

SSSI site of special scientific interest

SSSR Soyuz Sovietskikh Sotsialisticheskikh Respublik (Russian: Union of Soviet Socialist Republics)

SSStJ Serving Sister, Order of St John of Jerusalem

SST *Telecom.* single-sideband transmission · Society of Surveying Technicians · supersonic transport

SSTA Scottish Secondary Teachers' Association

SSU Sunday School Union

SSV *Med.* simian sarcoma virus

SSW Secretary of State for War · southwest · special security wing (of a prison)

SSWA Scottish Society of Women Artists

st. stanza · state · statement · statute · stem · *Printing* stet · *Knitting* stitch · stone (weight) · strait · street · *Prosody* strophe · *Cricket* stumped by

s.t. select time · short ton · static thrust · steam trawler

St Saint · (ital.) *Physics, symbol for* Stanton number · *Physics, symbol for* stokes · (ital.) *Meteorol., symbol for* stratus

St. Statute · Strait · Street

ST *British vehicle registration for* Inverness · septic tank · shipping ticket · (Hubble) Space Telescope · speech therapist · spring tide · Standard Time · *British fishing port registration for* Stockton · *postcode for* Stoke-on-Trent · *Meteorol.* stratus · *international civil aircraft marking for* Sudan · Summer Time · (ital.) Sunday Times · *Building trades* surface trench · surtax

sta. (*or* **Sta.**) station · stationary

Sta Santa (Italian, Spanish, Portuguese: Saint (female))

STA Sail Training Association · (USA) Science and Technology Agency · Scottish Typographical Association · Society of Typographic Arts · Swimming Teachers' Association

stab. stabilization · stabilizer · stable

stacc. *Music* staccato

Staffs Staffordshire

STAGS (stægz) Sterling Transferable Accruing Government Securities

STANAG ('stænæg) Standard NATO Agreement

St And. St Andrews (Scotland)

stand. standard

STARCAT ('staː,kæt) Space Telescope Archive and Catalogue

START (staːt) Strategic Arms Reduction Talks

stat. statics · *Med.* statim (Latin: immediately; in prescriptions) · stationary · statistical · statistics · statuary · statue · statute

STATE (steɪt) *Military* simplified tactical approach and terminal equipment

Stat. Hall Stationers' Hall (London)

STAUK Seed Trade Association of the United Kingdom

STB Bachelor of Sacred Theology (Latin

Sacrae Theologiae Baccalaureus) • Scottish Tourist Board

stbd starboard

stbt steamboat

STC Samuel Taylor Coleridge (1772– 1834, British poet) • Senior Training Corps • Short-Title Catalogue • Standard Telephones and Cables Ltd • (India) State Trading Corporation • subject to contract (in property advertisements)

std standard • started

Std. Stunde (German: hour)

STD Doctor of Sacred Theology (Latin *Sacrae Theologiae Doctor*) • salinity- temperature-depth (sensor system) • *Med.* sexually transmitted disease • Society of Typographic Designers • *Computing* state transition diagram • (New Zealand) subscriber toll dialling • subscriber trunk dialling

Ste Sainte (French: Saint (female))

Sté (*or* **S^té**) société (French: company; Co.)

STE Society of Telecom Executives (trade union)

STEL short-term exposure level *or* limit (of radiation)

STEM (stɛm) scanning transmission electron microscope (*or* microscopy)

sten. stenographer • stenography

Sten Shepherd and Turpin (inventors), Enfield (in **Sten gun**)

steno. (*or* **stenog.**) stenographer • stenographic • stenography

STEP (stɛp) Solar/Terrestrial Energy Programme • Special Temporary Employment Programme

ster. (*or* **stereo.**) stereophonic • (*or* **stereo.**) stereotype • sterling

St. Ex. Stock Exchange

STF *Meteorol.* stratiform

stg sterling

StGB Strafgesetzbuch (German: Penal Code)

stge storage

STGWU Scottish Transport and General Workers' Union

Sth South

STh Scholar in Theology

STH *Biochem.* somatotrophic hormone

sthn southern

STI Straits Times Index (of the Singapore Stock Exchange)

STIM scanning transmission ion microscope

STINGS (stɪŋz) *Aeronautics* stellar inertial guidance system

stip. stipend • (*or* **Stip.**) stipendiary • stipulation

s.t.i.r. surplus to immediate requirements

Stir. Stirling

stk stock

STL Licentiate in Sacred Theology (Latin *Sacrae Theologiae Licentiatus*) • Reader (*or* Professor) of Sacred Theology (Latin *Sacrae Theologiae Lector*) • Standard Tele- communications Laboratories • *Telecom.* studio-to-transmitter link

stlg sterling

STLO Scientific Technical Liaison Office(r)

STLV *Med.* simian T-lymphotropic virus

STM Master of Sacred Theology (Latin *Sacrae Theologiae Magister*) • scanning tunnelling microscope (*or* microscopy) • *Med., Psychiatry* short-term memory • *Computing* synchronous transfer module

STMS *Finance* short-term monetary support (within the EMS)

stmt statement

stn stain • (*or* **Stn**) station

STNR *Meteorol.* stationary

St° Santo (Portuguese: Saint)

STO Sea Transport Officer • Senior Tech- nical Officer • standing order

STOL (stɒl) *Aeronautics* short takeoff and landing

STOLVCD *Aeronautics* short takeoff and landing, vertical climb and descent

S'ton Southampton

S to S ship to shore • station to station

stp (*or* **s.t.p.**) *Physics* standard tempera- ture and pressure

STP Professor of Sacred Theology (Latin *Sacrae Theologiae Professor*) • *Trademark* scientifically treated petroleum (an oil substitute; refers colloquially to a hallu- cinogenic drug) • sewage treatment plant • *Computing* shielded twisted pair • *Physics* standard temperature and pressure

STPMA Scottish Theatrical Proprietors' and Managers' Association

str seater • steamer

str. straight • (*or* **Str.**) strait • (*or* **Str.**) street • strength • *Journalism* stringer • *Music* strings (*or* stringed) • *Rowing* stroke (oar) • strong • structural • structure

s.t.r. surplus to requirements

STRAC (stræk) (USA) strategic air command • strategic army corps

STRAD (stræd) signal transmitting, receiving, and distribution

stratig. stratigraphy

strd stranded

STRICOM ('straɪ,kɒm) US Strike Command

string. *Music* stringendo (Italian; intensifying)

STRIVE (straɪv) Society for the Preservation of Rural Industries and Village Enterprises

STROBE (strəʊb) satellite tracking of balloons and emergencies

Sts Saints

STS Scottish Text Society • *Astronautics* space transportation system

STSO Senior Technical Staff Officer

st. st. *Knitting* stocking stitch

STTA Scottish Table Tennis Association

STTL Schottky transistor-transistor logic

STUC Scottish Trades Union Congress

stud. student

Stuka ('stuːkə) Sturzkampfflugzeug (German: dive bomber)

STV Scottish Television • single transferable vote • standard test vehicle • (USA) subscription television

stvdr. stevedore

stwy stairway

s.u. set up • siehe unten (German: see below)

Su. Sudan(ese) • Sunday

SU *airline flight code for* Aeroflot-Russian International • *international civil aircraft marking for* Egypt • *British vehicle registration for* Glasgow • Scripture Union • *British fishing port registration for* Southampton • *Physics, Maths.* special unitary (group; as in **SU₃**) • *Physics* strontium unit (of radioactive strontium)

SUA (USA) State Universities Association

sub. subaltern • subeditor • *Music* subito (Italian: immediately, suddenly) • subject • subjunctive • submarine • subordinated • subscription • subsidiary • subsidy • subsistence • *Grammar* substantive • substitute • suburb(an) • subvention • subway

SUB (USA) supplemental unemployment benefits

subd. subdivision

sub-ed. subeditor

subj. subject • subjective(ly) • subjunctive

SUBLANT ('sʌblænt) US Navy Submarine Forces, Atlantic

Sub-Lt (*or* **Sub-Lieut.**) Sub-Lieutenant

subord. cl. subordinate clause

SUBPAC ('sʌb,pæk) US Navy Submarine Forces, Pacific

SUBROC ('sʌb,rɒk) submarine rocket

subs. subsidiary • subsistence

subsc. subscription

subsec. subsection

subseq. (*or* **subsq.**) subsequent(ly)

subsp. *Biology* (pl. **subspp.**) subspecies

subst. substantive(ly) • substitute

substand. substandard

suc. (*or* **succ.**) succeed • (*or* **succ.**) success • (*or* **succ.**) successor • suction

SUD *international vehicle registration for* Sudan

suff. sufficient • (*or* **suf.**) suffix

Suff. Suffolk • (*or* **Suffr.**) Suffragan

sug. suggest(ion)

SUIT Scottish and Universal Investment Trust

suiv. suivant (French: following)

Sult. Sultan

sum. *Med.* sumat *or* sumendum (Latin: let him (*or* her) take *or* let it be taken; in prescriptions) • summary • summer

SUM surface-to-underwater missile

sums. summons

Sun. (*or* **Sund.**) Sunday

SUNS (sʌnz) sonic underwater navigation system

SUNY State University of New York

sup *Maths.* supremum

sup. superficial • superfine • superior • superlative • *Grammar* supine (noun) • supplement(ary) • supply • supra (Latin: above) • supreme

sup. ben. supplementary benefit

Sup. Ct Superior Court • Supreme Court

Supdt Superintendent

super. superficial • superfine • superior • supernumerary

superhet ('suːpə,hɛt) *Telecom.* supersonic heterodyne

superl. superlative

SUPLO Scottish Union of Power-loom Overlookers

supp. (*or* **suppl.**) supplement(ary)

Supp. Res. Supplementary Reserve (of officers)

supr. superior • (*or* **Supr.**) supreme

Supt Superintendent

supvr (*or* **supr**) supervisor

sur. surface • surplus

Sur. Surrey

SUR *Commerce* set-up reduction

surg. surgeon • surgery • surgical

Surg. Cdr (*or* **Surg. Comdr**) Surgeon Commander

Surg. Gen. Surgeon General

Surg. Lt-Cdr Surgeon Lieutenant-Commander

Surg. Maj. Surgeon Major

surr. surrender • (*or* **surro.**) surrogate

Surr. Surrey

surv. survey • (*or* **survey.**) surveying • surveyor • survive • surviving

SURV standard underwater research vessel

Surv. Gen. Surveyor General

Sus. Sussex • Susanna (in the Apocrypha)

SUS Scottish Union of Students

SUSI Sydney University Stellar Interferometer

susp. suspend • suspension

Suss. Sussex

Sustrans ('sʌs,træns) Sustainable Transport

SUSY *Physics* supersymmetry

SUT Society for Underwater Technology

Suth. Sutherland (former Scottish county)

SUV (USA) sport-utility vehicle

sv *Baseball* save

sv. *Microbiol.* serovar

s.v. sailing vessel • side valve • sub verbo (*or* voce) (Latin: under the word *or* heading) • *Insurance* surrender value

Sv *Physics, symbol for* sievert

SV safety valve • Sancta Virgo (Latin: Holy Virgin) • Sanctitas Vestra (Latin: Your Holiness) • *airline flight code for* Saudi Arabian Airlines • *Microbiol.* simian virus (as in **SV40**) • stroke volume (of an engine)

svc. (*or* **svce**) service

SVC *Med.* superior vena cava • *Computing* supervisor call

SVD swine vesicular disease

SVGA super video graphics array

svgs savings

S-VHS super-VHS

SVO Scottish Variety Orchestra • *Linguistics* subject-verb-object (in **SVO language**) • Superintending Veterinary Officer

s.v.p. s'il vous plaît (French: if you please)

SVP saturated vapour pressure

SVQ Scottish Vocational Qualification

s.v.r. *Med.* spiritus vini rectificatus (Latin: rectified spirit of wine; in prescriptions)

SVS still-camera video system

SVTP sound velocity, temperature, pressure

s.vv. sub verbis (Latin: under the words or headings)

svy survey

sw. switch

s.w. salt water • sea water

s/w sea water • seaworthy

Sw. Sweden • Swedish • Swiss

SW *airline flight code for* Air Namibia • *British vehicle registration for* Carlisle • *British vehicle registration for* Glasgow • senior warden • shipper's weight • shock wave • *Radio* short wave • small women (clothing size) • South Wales • south-west(ern) • *postcode for* southwest London • standard weight

S/W *Computing* software

Swab. Swabia(n)

SWACS space warning and control system

SWA(L)K (swɔːlk *or* swæk) sealed with a (loving) kiss (on envelopes)

SWANU ('swɑːuː) South West Africa National Union

SWANUF ('swɑːnuf) South West Africa National United Front

SWAPO (*or* **Swapo**) ('swɑːpəu) South-West Africa People's Organization

S/WARE *Computing* software

SWAS Submillimetre Wave Astronomy Satellite

SWAT (swɒt) (USA) Special Weapons and Tactics (police unit)

SWB short wheel base • *Military* South Wales Borderers

swbd switchboard

SWCI *Computing* software configuration item

swd *Bookbinding* sewed

SWE (USA) Society of Women Engineers

SWEB South Wales Electricity Board • Southwest Electricity Board

Swed. Sweden • Swedish

SWET Society of West End Theatre

SwF *symbol for* Swiss franc (monetary unit)

SWG standard wire gauge

SWH solar water heating

SWIE South Wales Institute of Engineers

SWIFT (swɪft) Society for Worldwide Interbank Financial Transmission

Swing (swɪŋ) *Finance* Sterling Warrant into Gilt-edged Stock

Switz. (*or* **Swit.**) Switzerland

SWL safe working load

SWLA (*or* **SWIA**) Society of Wildlife Artists

SWMF South Wales Miners' Federation

SWO Station Warrant Officer

SWOA Scottish Woodland Owners' Association

SWOPS single-well oil-production system

SWOT (swɒt) *Marketing* strengths, weaknesses, opportunities, and threats (of a new product)

SWP safe working pressure • Socialist Workers' Party

SWPA South-West Pacific Area

SWR *Telecom.* standing-wave ratio

SWRB Sadler's Wells Royal Ballet (now the Royal Ballet)

SWS static water supply

SWSWU Sheffield Wool Shear Workers' Union

Swtz. Switzerland

SWWJ Society of Women Writers and Journalists

Sx Sussex

SX *British vehicle registration for* Edinburgh • *international civil aircraft marking for* Greece • soft X-rays • *Shipping* Sundays excepted

SXES soft X-ray electron spectroscopy

SXR soft X-rays (*or* radiation)

SXT sextant

Sy. Seychelles • supply • Surrey • Syria

SY *international vehicle registration for* Seychelles • *postcode for* Shrewsbury • steam yacht • *British fishing port registration for* Stornoway

SYB (ital.) The Statesman's Year-Book

Syd. Sydney

S. Yd Scotland Yard

SYHA Scottish Youth Hostels Association

syl. (*or* **syll.**) syllable • syllabus

Sylk (sɪlk) *Computing* symbolic link

sym. symbol • symbolic • *Chem.* symmetrical • symmetry • symphonic • symphony • symptom

symp. symposium

syn. synchronize • synonym(ous) • synonymy • synthetic

sync *Image technol.* synchronization (*or* synchronize)

Syncom ('sɪnkɒm) (USA) synchronous communications satellite

synd. syndicate • syndicated

synon. synonymous

synop. synopsis

synth. *Music* synthesizer • synthetic

Sy. PO Supply Petty Officer

syr. *Pharmacol.* syrup

Syr. Syria(n) • Syriac

SYR *international vehicle registration for* Syria

syst. system • systematic

sz. size

SZ *British vehicle registration for* Down

T

t (ital.) *symbol for* Celsius temperature • *Astronomy, symbol for* hour angle • (ital.) *Statistics, symbol for* Student's t distribution • *Music* te (in tonic sol-fa) • (ital.) *Chem.* tertiary (isomer; as in **t-butane**) • *symbol for* tonne(s) • *Physics* top (a quark flavour) • (ital.) *Chem., symbol for* transport number

t. table • tabulated • *Sport* tackle • taken (from) • *Commerce* tare • teaspoon(ful) • teeth • *Music* tempo • tempore (Latin: in the time of) • tenor • *Grammar* tense • terminal • territorial • territory • thunder • time • tome (French: volume) • tonneau (French: ton) • ton(s) *or* tonne(s) • town • township • transit • *Grammar* transitive • troy • tun • turn

T (ital.) *Physics, symbol for* kinetic energy • (ital.) *Physics, symbol for* period • *Biochem., symbol for* ribosylthymine •

Chem., *symbol for* tautomeric effect • *Meteorol.* temperature • tera- (prefix indicating 10^{12} as in **TJ**, terajoule) • *Physics*, *symbol for* tesla(s) • *international vehicle registration for* Thailand • (ital.) *symbol for* thermodynamic temperature • *Biochem.*, *symbol for* threonine • *Biochem.*, *symbol for* thymine • *Immunol.* thymus (in **T cell** *or* **lymphocyte**) • *Physics* time reversal (as in **T invariance**) • (bold ital.) *Physics, Engineering, symbol for* torque • *Photog.* total light transmission (in **T-number**) • trainer (aircraft; as in **T-37**) • *Irish fishing port registration for* Tralee • *Chem., symbol for* tritium • *Shipping* tropical loading (on load line) • *Logic, Computing, symbol for* true • (ital.) *Biochem., symbol for* twisting number

T. (*or* **T**) tablespoon(ful) • *Music* tace (Italian: be silent) • tanker • target • tea (rose) • teacher • telegraph(ic) • telephone • *Music* tempo • temporary • *Music* tenor • Territorial • Territory • Testament • thermometer • *Advertising* third of a page (type area) • Thursday • time • torpedo • transaction(s) • translation • transport(ation) • *Obstetrics* transverse (presentation) • Treasury • Trinity • Tuesday • Turkish • *Knitting* twist

T3 *airline flight code for* Tri Star Airlines

T₃ *Biochem.* triiodothyronine

T4 *airline flight code for* Transeast Airlines

T₄ *Biochem.* thyroxine (tetraiodothyronine)

2,4,5-T 2,4,5-trichlorophenoxyacetic acid (herbicide)

5T *international civil aircraft marking for* Mauritania

6T *airline flight code for* Air Mandalay

7T *international civil aircraft marking for* Algeria • *airline flight code for* Trans Cote

8T *airline flight code for* Travelair

9T *airline flight code for* Athabaska Airways

ta. tableau • tablet

t.a. target area • time and attendance • true altitude

Ta *Chem., symbol for* tantalum

TA *British vehicle registration for* Exeter • *Taxation* table of allowances • *postcode for* Taunton • (USA) teaching assistant • telegraphic address • temporary admission • *Computing* terminal adapter • Territorial Army • *Physics* thermal analysis • tithe annuity • training adviser • *Psychol.* transactional analysis • (USA) transit authority • travelling allowance

T/A temporary assistant

TAA Territorial Army Association • *Chem.* tertiary amyl alcohol • test of academic aptitude • Trans-Australia Airlines

TA&VRA Territorial Auxiliary and Volunteer Reserve Association

tab. table (list *or* chart) • tablet • tabulation • tabulator

TAB tabulator (on a typewriter) • Technical Assistance Board (of the UN) • (Australia, New Zealand) Totalizator Administration (*or* Agency) Board • *Med.* typhoid, paratyphoid A, paratyphoid B (vaccine)

TABA Timber Agents' and Brokers' Association of the United Kingdom

Tac. Publius Cornelius Tacitus (*c.* AD 55– *c.* 120, Roman historian)

TAC *US Air Force* Tactical Air Command • Technical Assistance Committee (of the UN) • Television Advisory Committee • The Athletics Congress • Tobacco Advisory Committee • Trades Advisory Council

TACAN ('tækæn) tactical air navigation

TACL Training for Action-Centred Leadership

TACMAR ('tækmaː) tactical multifunction array radar

TACS tactical air-control system

TACV tracked air-cushion vehicle

Tads (tædz) *Military* target acquisition and designation sight

TAF Tactical Air Force

TAFE ('tæfɪ) technical and further education

tafu ('tæfuː) *Slang* things are fouled (*or* fucked) up

tafubar ('tæfuːˌbaː) *Slang* things are fouled (*or* fucked) up beyond all recognition

Tag. Tagalog (language)

TAG Taxon Advisory Group • (USA) The Adjutant-General

T/Agt (*or* **T Agt**) (USA) transfer agent

Tai. Taiwan

TAI temps atomique international (French: International Atomic Time; IAT)

tal. talis (Latin: such)

Tal. *Judaism* Talmud

TAL *Insurance* traffic and accident loss

TALISMAN (*or* **Talisman**) ('tælɪzmən) *Stock exchange* Transfer Accounting Lodgement for Investors and Stock

Management (now replaced by CREST system)

tal. qual. talis qualis (Latin: average quality)

Tam. Tamil (language)

TAM tactical air missile • (tæm) Television Audience Measurement (as in **TAM rating**)

Tamba ('tæmbə) Twins and Multiple Births Association

tan (tæn) *Maths.* tangent

T & A (*or* **T and A**) *Slang* (USA) tits and ass • *Med.* (USA) tonsils and adenoids (*or* tonsillectomy and adenoidectomy)

T&AFA Territorial and Auxiliary Forces Association

T & AVR Territorial and Army Volunteer Reserve

t. & b. top and bottom

T & E test and evaluation • *Colloquial* tired and emotional (i.e. drunk) • travel and entertainment • trial and error

t. & g. *Carpentry* tongued and grooved

T & G Transport and General Workers' Union

t. & o. *Bookmaking* taken and offered

t. & p. *Insurance* theft and pilferage

t. & s. toilet and shower

T & S transport and supply

T&SG Television and Screen Writers' Guild

T & T Trinidad and Tobago

Tang. Tangier

tanh *Maths.* hyperbolic tangent

TANS terminal-area navigation system • Territorial Army Nursing Service (now merged with QARANC)

TANU ('tænuː) Tanganyika African National Union

TAO Technical Assistance Operations (of the UN)

TAOC (USA) Tactical Air Operations Center

TAP (USA) Technical Assistance Program • Transportes Aéreos Portugueses (Portuguese Airlines)

Tapi *Trademark* telephony application programming interface

Tapline ('tæp,laɪn) (Saudi Arabia) Trans-Arabian Pipeline Company

TAPPI (USA) Technical Association of the Pulp and Paper Industry

TAPS Trans-Alaska Pipeline System

tar. tariff • tarpaulin

TAR terrain-avoidance radar • Territorial

Army Regulations • *Accounting* throughput accounting ratio • thrust-augmented rocket

TARA Technical Assistant, Royal Artillery • Territorial Army Rifle Association

TARAN test and replace as necessary

tarfu ('taːfuː) *Slang* things are really fouled (*or* fucked) up

TARO Territorial Army Reserve of Officers

TARS Technical Assistance Recruitment Service (of the UN)

Tas. Tasmania(n)

TAS torpedo antisubmarine (course) • *Aeronautics* true air speed

TASI *Telecom.* time-assignment speech interpolation

Tasm. Tasmania(n)

TASMO tactical air support of maritime operations

TASR terminal area surveillance radar

Tass (tæs) Telegrafnoye Agentstvo Sovetskovo Soyuza (Russian news agency)

TASS Transport Aircraft Servicing Specialist (in the RAF)

TAT *Psychol.* thematic apperception test • *Astronautics* thrust-augmented Thor • *Med.* tired all the time • transatlantic telephone cable

TATSA transportation aircraft test and support activity

Tau *Astronomy* Taurus

TAUN Technical Assistance of the United Nations

TAURUS ('tɔːrəs) (formerly) *Stock exchange* Transfer and Automated Registration of Uncertified Stock

taut. tautology

t.-à-v. tout-à-vous (French: yours ever; in correspondence)

tav. tavern

TAVR Territorial and Army Volunteer Reserve (1967–79)

TAVRA Territorial Auxiliary and Volunteer Reserve Association

t.a.w. twice a week

tax. (*or* **taxn**) taxation

tb *Baseball* total bases

t.b. temporary buoy • *Book-keeping* trial balance • true bearing • tubercle bacillus • tuberculosis

Tb *symbol for* terabyte(s) • *Chem., symbol for* terbium

TB *British vehicle registration for* Liverpool • *symbol for* terabyte(s) • torpedo boat •

torpedo bomber • training battalion • training board • Treasury bill • *Bookkeeping* trial balance • tuberculosis

t.b.a. *Commerce* to be advised • to be agreed • to be announced

TBA tyres, batteries, and accessories

t.b.&s. top, bottom, and sides

TBC *Image technol.* time-based corrector

TBCEP *Chem.* tri-beta-chloroethyl phosphate (flame retardant)

t.b.c.f. to be called for

t.b.d. to be determined

TBD torpedo-boat destroyer

TBF Teachers Benevolent Fund

TBG *Immunol.* thyroxine-binding globulin

TBI *Engineering* throttle-body injection • *Med.* total body irradiation

T-bill Treasury bill

t.b.l. *Knitting* through back of loop

TBL *Commerce* through bill of lading

TBM *Military* tactical ballistic missile • *Surveying* temporary benchmark • *Computing* terabit memory • tunnel-boring machine

TBO *Aeronautics* time between overhauls • *Theatre* total blackout

t-BOC *Chem.* tertiary butoxycarbonyl

T-bond Treasury bond

TBP *Chem.* tertiary butyl peroxide • *Chem.* tributyl phosphate

tbs. (*or* **tbsp.**) tablespoon(ful)

TBS talk between ships (radio apparatus) • *Computing* tape backup system • tight building syndrome • training battle simulation

TBSV *Microbiol.* tomato bushy stunt virus

TBT tributyl tin (used in marine paints)

tc. *Music* tierce (organ stop)

t.c. temperature control • till cancelled • time check • true course

Tc *Chem., symbol for* technetium

TC *airline flight code for* Air Tanzania • *British vehicle registration for* Bristol • *Military* Tank Corps • (USA) Tariff Commission • *Law* Tax Cases • technical college • temporary clerk • Temporary Constable • tennis club • touring club • town clerk • town council(lor) • training centre • training college • training corps • *Military* Transport Command • traveller's cheque • *Music* tre corde (Italian: three strings; i.e. release soft pedal) • Trinity College • (Trinidad and Tobago) (Order of the) Trinity Cross • *Meteorol.* tropical cyclone • Trusteeship Council (of the

UN) • *Chem.* tungsten carbide • *international civil aircraft marking for* Turkey • twin carburettors (on motor vehicles)

TCA *Biochem.* tricarboxylic acid (in **TCA cycle**) • trichloroacetic acid (herbicide) • tricyclic antidepressant (drug)

TCB *Slang* (USA) take care of business • *Chem.* tetrachlorobiphenyl • Thames Conservancy Board

TCBM transcontinental ballistic missile

TCC *Telecom.* time compression coding • Transport and Communications Commission (of the UN) • Trinity College, Cambridge • (USA) Troop Carrier Command

TCCB Test and County Cricket Board

TCD Trinity College, Dublin

TCDD tetrachlorodibenzodioxin (dioxin; environmental pollutant)

TCE *Chem.* trichloroeth(yl)ene (solvent)

tcf trillion cubic feet (measure of natural gas)

TCF Temporary Chaplain to the Forces • time-correction factor • Touring Club de France

TCFB Transcontinental Freight Bureau

TCGF *Immunol.* T-cell growth factor

TCH *international vehicle registration for* Chad

tchg teaching

tchr teacher

TCI Touring Club Italiano (Italian Touring Club)

TCL *Chem.* trichloroeth(yl)ene (solvent) • Trinity College (of Music) London

TCM Trinity College of Music (London)

TCMA Telephone Cable Makers' Association

TCNE *Chem.* tetracyanoeth(yl)ene

TCNQ *Chem.* tetracyanoquinodimethane

TCO test control office • (Sweden) Tjänstemännens Centralorganisation (Central Organization of Salaried Employees; trade union) • Trinity College, Oxford

TCP *Computing* transmission control protocol • *Trademark* trichlorophenylmethyliodisalicyl (an antiseptic) • *Chem.* tricresyl phosphate

TCPA Town and Country Planning Association

TCP/IP transmission control protocol/Internet protocol

TCR *Immunol.* T-cell receptor • *Physics* temperature coefficient of resistance

TCRE *Med.* transcervical resection of the endometrium

TCS target cost system • (USA) traffic control station

tctl tactical

TCU *Meteorol.* towering cumulus • (USA) Transportation, Communications, International Union

td. touchdown

t.d. technical data • *Med.* ter in die (Latin: three times a day; in prescriptions) • test data • time delay • tractor-drawn

TD *postcode for* Galashiels • *British vehicle registration for* Manchester • *Military* Tactical Division • tank destroyer • *Med.* tardive dyskinesia • Teaching Diploma • (Ireland) Teachta Dála (Gaelic: Member of the Dáil) • technical development • technical drawing • Territorial (Efficiency) Decoration (in the Territorial Army) • Tilbury Docks • torpedo depot • *Football* (USA, Canada) touchdown(s) • (USA) traffic director • (USA) Treasury Department • trust deed • Tunisian dinar (monetary unit)

TDA tax-deferred annuity • 2,4-toluene diamine (possible carcinogen released by breast implants)

TDAL *Chem.* tetradecenal

TDB temps dynamique barycentrique (French: barycentric dynamical time) • total disability benefit

TDC Temporary Detective Constable • through-deck cruiser • (*or* **t.d.c.**) *Engineering* top dead centre

TDD telecommunications device for the deaf • Tubercular Diseases Diploma

TDDA *Chem.* tetradecadien-1-yl acetate

TDDL *Telecom.* time-division data link

TDG twist drill gauge

TDH tall, dark, and handsome (in personal advertisements)

TDI *Chem.* toluene-2,4-diisocyanate

TDL tunable diode laser

TDM telemetric data monitor • *Telecom.* time-division multiplexing

TDMA *Telecom.* time-division multiple access

TDN total digestible nutrients

T-DNA *Genetics* transferred DNA

TDO *Meteorol.* tornado

TDP technical development plan

t.d.r. tous droits réservés (French: all rights reserved)

TDR *Computing* time domain reflectometer • *Finance* Treasury deposit receipt

TDRSS (*or* **TDRS**) tracking and data-relay satellite system

t.d.s. *Med.* ter die sumendum (Latin: to be taken three times a day; in prescriptions)

TDS *Computing* tabular data stream • *Physics* thermal desorption spectroscopy (*or* spectrum) • *Nuclear physics* thermal diffuse scattering

TDT *Astronomy* terrestrial dynamical time

t.e. thermal efficiency • tinted edge (of paper) • trailing edge • turbine engine

t/e twin-engined

Te *Chem., symbol for* tellurium

TE *airline flight code for* Lithuanian Airlines • *British vehicle registration for* Manchester • telecommunications engineering • *Biochem.* trace element • trade expenses • *Telecom.* transverse electric (as in **TE wave**)

TEAC Technical Educational Advisory Council

TEC *Physics* thermal expansion coefficient • *Physics* thermionic energy conversion • (tɛk) Training and Enterprise Council

tech. (*or* **techn.**) technical(ly) • technician • technique • technology

Tech(CEI) Technician (Council of Engineering Institutions)

technol. technological • technology

TED *Chem.* trieth(yl)enediamine

TEE Telecommunications Engineering Establishment • Torpedo Experimental Establishment • Trans-Europe Express (train)

TEF toxicity equivalence factor

TEFL ('tɛfᵊl) teaching (of) English as a foreign language

t.e.g. *Bookbinding* top edges gilt

Teh. Tehran

tel. telegram • telegraph(ic) • telephone

Tel *Astronomy* Telescopium

TEL tetraethyl lead (petrol additive) • *Astronautics* transporter-erector-launcher

telecom. telecommunication(s)

teleg. telegram • telegraph(ic) • telegraphy

teleph. telephone • telephony

telex ('tɛlɛks) teleprinter exchange

TELNET ('tɛlnɛt) *Computing* teletype network

tel. no. telephone number

TEM Territorial Efficiency Medal •

transmission electron microscope (*or* microscopy) · *Telecom.* transverse electromagnetic

TEMA Telecommunications Engineering and Manufacturing Association

temp. temperance · temperate · temperature · *Music* tempo · temporal · temporary · tempore (Latin: in the time of)

Templar ('templə) tactical expert mission-planner (military computer)

temp. prim. *Music* tempo primo (Italian; at the original pace)

ten. tenant · tenement · tenor · *Music* tenuto (Italian: held, sustained)

tency tenancy

Tenn. Tennessee

TENS (tɛnz) *Med.* transcutaneous electrical nerve stimulation

TeolD Doctor of Theology

TEPP tetraethyl pyrophosphate (pesticide)

ter. (*or* **Ter.**) terrace · territorial · territory

terat. teratology

TERCOM ('tɜːkɒm) *Aeronautics* terrain contour matching (*or* mapping)

term. terminal · terminate · termination · terminology

terr. (*or* **Terr.**) terrace · territorial · territory

tert. tertiary

TES thermal-energy storage · (ital.) Times Educational Supplement

TESL ('tɛsᵊl) teaching (of) English as a second language

TESOL ('tiːsɒl) teaching (*or* teacher) of English to speakers of other languages

Tessa ('tɛsə) Tax-Exempt Special Savings Account

test. testament · testator (*or* testatrix) · testimonial · testimony

TET Teacher of Electrotherapy

T.-et-G. Tarn-et-Garonne (department of France)

TETOC technical education and training for overseas countries

tet. tox. tetanus toxin

TEU *Shipping* twenty-foot equivalent unit

Teut. Teuton(ic)

TeV *Physics, symbol for* teraelectronvolt

TEWT (*or* **Tewt**) (tjuːt) *Military* tactical exercise without troops

Tex. Texan · Texas

text. textile

text. rec. textus receptus (Latin: the received text)

t.f. tabulating form · tax-free · travaux forcés (French: hard labour)

TF *British vehicle registration for* Reading · *postcode for* Telford · Territorial Force · *Electronics* thin film · *Biochem.* transcription factor · *Shipping* tropical freshwater (on load lines) · *international civil aircraft marking for* Iceland

TFA Tenant Farmers' Association · *Biochem.* total fatty acids · *Chem.* trifluoroacetic acid

TFAA *Chem.* trifluoroacetic anhydride

TFAP Tropical Forestry Action Plan

tfc traffic

TFD *Electronics* thin-film detector

TFECG Training and Further Education Consultative Group

TFEL *Computing* thin-film electroluminescent display

TFMS *Physical chem.* time-of-flight mass spectrometer

tfr transfer

TFR Territorial Force Reserve · total fertility rate

TFSC (*or* **TFSK**) Turkish Federated State of Cyprus (Turkish *Kibris*)

TFT *Electronics* thin-film transistor

TFT LCD *Electronics* thin-film transistor liquid-crystal display

TFTR *Nuclear engineering* Tokamak Fusion Test Reactor (Princetown, USA)

TFU telecommunications flying unit

TFW *Military* tactical fighter wing

TFX tactical fighter experimental (aircraft)

tg *Maths.* tangent

t.g. tail gear · *Biology* type genus

TG *British vehicle registration for* Cardiff · *international civil aircraft marking for* Guatemala · Tate Gallery (London) · temporary gentleman · *airline flight code for* Thai Airways International · *Colloquial* thank God · (USA) Theater Guild · *international vehicle registration for* Togo · training group · transformational-generative (grammar) · transformational grammar · Translators' Guild (London)

TGA *Chem.* thermal gravimetric analysis

TGAT ('tiːgæt) *Education* Task Group on Assessment and Testing

T-gate *Computing* ternary selector gate

t.g.b. *Carpentry* tongued, grooved, and beaded

TGB thyroglobulin · (France) Très Grande Bibliothèque (*or* Bibliothèque de France; proposed French national library)

TGEW Timber Growers England and Wales Ltd

TGF *Med.* transforming growth factor

TGI *Marketing* Target Group Index

TGIF *Colloquial* thank God it's Friday

TGM torpedo gunner's mate

TGMV *Microbiol.* tomato golden mosaic virus

T-group training group

tgt target

TGT *Physics* temperature gradient technique • *Aeronautics* turbine gas temperature

TGV (France) train à grande vitesse (high-speed passenger train)

TGWU Transport and General Workers' Union

th *Maths.* hyperbolic tangent

th. thermal

Th *Chem., symbol for* thorium

Th. Theatre • Thursday

TH *British vehicle registration for* Swansea • Technische Hochschule (German: technical university *or* college) • *British fishing port registration for* Teignmouth • Territory of Hawaii • *Numismatics* toothed border • town hall (on maps) • Toynbee Hall • Transport House • Trinity House

THA *Biochem.* tetrahydroaminoacridine

Thai. Thailand

thanat. thanatology

ThB Bachelor of Theology (Latin *Theologicae Baccalaureus*)

THC tetrahydrocannabinol (cannabis component) • (New Zealand) Tourist Hotel Corporation

ThD Doctor of Theology (Latin *Theologicae Doctor*)

THD total harmonic distortion (in sound recording)

THE Technical Help to Exporters (division of the British Standards Institute)

theat. theatre • theatrical

THELEP Therapy of Leprosy

Theoc. Theocritus (*c.* 310–250 BC, Greek poet)

theol. theologian • theological • theology

Theoph. Theophrastus (*c.* 370–*c.* 286 BC, Greek philosopher)

theor. theorem • (*or* **theoret.**) theoretical • theory

theos. theosophical • theosophist • theosophy

therap. (*or* **therapeut.**) therapeutic(s)

therm. thermometer • thermometry

thermochem. thermochemistry

thermodyn. thermodynamics

thermom. thermometer • thermometry

thes. thesis

THES (ital.) Times Higher Education Supplement

thesp. thespian

Thess. *Bible* Thessalonians • Thessaly

THF *Chem.* tetrahydrofuran • Trusthouse Forte plc

THI temperature–humidity index

thk thick

ThL Theological Licentiate

ThM Master of Theology (Latin *Theologiae Magister*)

THN *Chem.* tetrahydronaphthalene

thor. thorax

thoro. thoroughfare

Thos. Thomas

thou. thousand

thp thrust horsepower

thr. their • through • thrust

Thr (*or* **thr**) *Biochem.* threonine

THR *Med.* total hip replacement

3i Investors in Industry

3M Minnesota Mining and Manufacturing Company

three Rs (*or* **3 Rs**) reading, (w)riting, and (a)rithmetic

throt. throttle

ThSchol Scholar in Theology

Thuc. Thucydides (*c.* 460–*c.* 400 BC, Greek historian)

Thurs. (*or* **Thur.**) Thursday

THWM Trinity (House) high-water mark

THz *Physics, symbol for* terahertz

Ti *Chem., symbol for* titanium • *Genetics* tumour-inducing (in **Ti plasmid**)

Ti. Tiberius (42 BC–AD 37, Roman emperor) • Tibet

TI *international civil aircraft marking for* Costa Rica • *Irish vehicle registration for* Limerick • technical inspection • technical institute • temperature indication (*or* indicator) • (USA) Texas Instruments • *Med.* thermal imaging • *Chem.* toluene-insoluble

T/I target identification • target indicator

TIA Tax Institute of America • *Med.* transient ischaemic attack

Tib. Tibet(an)

TIBC *Med.* total iron-binding capacity

TIBOR ('tiːbɔː) *Finance* Tokyo InterBank Offered Rate

TIC (*or* **tic**) *Law* taken into consideration (of an offence) • total inorganic carbon (in chemical analysis) • tourist information centre

t.i.d. *Med.* ter in die (Latin: three times a day; in prescriptions)

TIE theatre in education

tier. *Music* tierce (organ stop)

TIF *Computing* tagged image format • telephone interference (*or* influence) factor • *Biochem.* transcription initiation factor • Transports Internationaux par Chemin de Fer (French: International Rail Transport)

TIFF (tɪf) *Computing* tagged image file format

TIG tungsten inert gas (welding)

TIGR ('taɪgə) Treasury Investment Growth Receipts (type of bond)

TIH Their Imperial Highnesses

TIL *Med.* tumour-infiltrating lymphocyte

TILS Technical Information and Library Service

t.i.m. time is money

Tim. *Bible* Timothy

TIM transient intermodulation distortion (in sound recording)

timp. *Music* timpani

TIMP *Med.* tissue inhibitor of metalloproteinase

TIMS The Institute of Management Sciences • *Physical chem.* thermal ionization mass spectrometry

TIN (USA) taxpayer identification number

TINA ('tiːnə) *Politics, colloquial* there is no alternative (usually referring to Margaret Thatcher)

tinct. tincture

TIO Technical Information Officer

Tip. Tipperary

TIP temperature-independent paramagnetism • *Computing* terminal interface processor

TIPSS *Med.* transcutaneous intrahepatic porto-systemic shunt

TIR *Optics* total internal reflection • Transport International Routier (French: International Road Transport; on continental lorries)

TIRC Tobacco Industry Research Committee

TIROS ('taɪrəs) television and infrared observation satellite

tis. tissue

TIS technical information service

tit. title • titular

Tit. *Bible* Titus

TJ *international civil aircraft marking for* Cameroon • *British vehicle registration for* Liverpool • *international vehicle registration for* Tajikistan • (*or* **t.j.**) talk jockey • *Physics, symbol for* terajoule • *Athletics* triple jump

tk tank • truck

Tk *symbol for* taka (monetary unit of Bangladesh)

TK *British vehicle registration for* Exeter • *airline flight code for* Turkish Airlines

TKO (*or* **t.k.o.**) *Boxing* technical knockout

tkr tanker

tks thanks

tkt ticket

tl *Meteorol., symbol for* thunderstorm

t.l. test link • *Engineering* thrust link • time length • total load • *Insurance* total loss • trade list

Tl *Chem., symbol for* thallium

TL *international civil aircraft marking for* Central African Republic • *British vehicle registration for* Lincoln • target language • *Physics* thermoluminescence • thermoluminescent (as in **TL-dating**) • Torpedo Lieutenant • *Insurance* total loss • *Colloquial* (USA) trade-last • transmission line • Turkish lira (monetary unit)

T/L *Banking* time loan • *Insurance* total loss

TLA three-letter abbreviation (*or* acronym)

t.l.b. temporary lighted buoy

TLB *Computing* translation look-aside buffer

TLC tender loving care • *Chem.* thin-layer chromatography • *Med.* total lung capacity • (Australia) Trades and Labour Council

tld tooled

TLD thermoluminescent dosimeter

TLF transferable loan facility

TLG Theatrical Ladies' Guild

t.l.o. *Insurance* total loss only

TLO Technical Liaison Officer

TLP *Astronomy* transient lunar phenomenon

tlr tailor • trailer

TLR Times Law Reports • *Photog.* twin-lens reflex

TLS (ital.) Times Literary Supplement • typed letter, signed

tltr translator

TLU *Computing* table look-up

TLV *Electronics* threshold-limit value

TLWM Trinity (House) low-water mark

t.m. temperature meter • true mean

Tm *Chem., symbol for* thulium

TM *British vehicle registration for* Luton • tactical missile • technical manual • technical memorandum • test manual • Their Majesties • *Telecom.* tone modulation • trademark • training manual • transcendental meditation • *Telecom.* transverse magnetic (as in **TM wave**) • trench mortar • tropical medicine • *Computing* Turing machine • *international vehicle registration for* Turkmenistan

TMA Theatrical Management Association • Trans-Mediterranean Airways (Lebanese national airline) • *Chem.* trimellitic acid

TMB *Chem.* tetramethylbenzidine • travelling medical board

tmbr timber

TMC Tanglewood Music Center (USA) • The Movie Channel

TMCS *Chem.* trimethylchlorosilane

TMD *Military* theatre missile defence

tme time

TMEDA *Chem.* tetramethylethylenediamine

TMI (USA) Three Mile Island

TMJ *Med.* temporomandibular joint (as in **TMJ syndrome**)

tmkpr timekeeper

TML *Chem.* tetramethyl lead • *Shipping* three-mile limit

TMMG Teacher of Massage and Medical Gymnastics

TMO telegraph(ic) money order

TMP thermomechanical pump

tmpry temporary

tmr timer

TMS *Chem.* tetramethylsilane

TMSDEA *Chem.* trimethylsilyldiethylamine

TMT turbine motor train

TMV *Microbiol.* tobacco mosaic virus • true mean value

tn ton (*or* tonne) • town • train • transportation

t.n. technical note • telephone number

Tn *Genetics* transposon (followed by a number or letter)

TN *British vehicle registration for* Newcastle upon Tyne • *international civil aircraft marking for* Republic of the Congo • *US postcode for* Tennessee • *postcode for* Tonbridge • *British fishing port registration for* Troon • true north • *international vehicle registration for* Tunisia

TNBT *Chem.* titanium(IV) butoxide

TNC Theatres National Committee • total numerical control • transnational corporation

TNF *Military* theatre nuclear forces • *Trademark* The North Face (outdoor equipment and clothing) • *Med.* tumour necrosis factor

tng training • turning

TNG (USA) The Newspaper Guild

TNIP (South Africa) Transkei National Independence Party

TNM tactical nuclear missile • *Med.* tumour (size), (lymph) node (involvement), metastasis (in **TNM classification**)

T-note *Finance* Treasury note

TNP (France) Théâtre National Populaire (national theatre)

TNPG The Nuclear Power Group

tnpk. turnpike

TNT 2,4,6-trinitrotoluene (explosive)

TNW *Military* theatre nuclear weapon

TNX *Chem.* trinitroxylene

t.o. takeoff • turnover

To. Togo

TO *British vehicle registration for* Nottingham • *Management* table of organization • Technical Officer • telegraphic order • telegraph office • telephone office • telephone order • Torpedo Officer • trained operator • Transport Officer • *British fishing port registration for* Truro • turn over (page)

T/O turnover

tob. tobacco(nist)

Tob. *Old Testament* Tobit

ToB *Cycling* Tour of Britain

TOB temporary office building

TobRV *Microbiol.* tobacco ringspot virus

TOC total organic carbon (in chemical analysis)

Toc H Talbot House (obsolete telegraphic code for its initials; original headquarters of the movement)

TOD time of delivery

t.o.e. ton oil equivalent

TOE *Physics* theory of everything

TOEFL test(ing) of English as a foreign language

TOET test of elementary training

TOF *Physics* time of flight

TOFC trailer on flat car (type of freight container)

TOFMS *Chem.* time-of-flight mass spectroscopy

TOFPET ('tɒf,pet) *Med.* time-of-flight positron-emission tomography

tog. (*or* **togr**) together

Tok. Tokyo

TOL Tower of London

tom. (*or* **tomat.**) tomato • tomus (Latin: volume)

TOM territoire d'outre mer (French: overseas territory) • total organic matter (in chemical analysis)

TOMCAT ('tɒm,kæt) *Military* theatre of operations missile continuous-wave antitank weapon

TOMS Total Ozone Mapping Spectrometer

TON total organic nitrogen (in chemical analysis)

tonn. tonnage

TOO (*or* **t.o.o.**) time of origin • *Commerce* to order only

TOP (tɒp) *Computing* technical office protocol • temporarily out of print

TOPIC ('tɒpɪk) *Stock exchange* Teletext Output Price Information Computer

topog. topographer • topographical • topography

topol. togological • topology

TOPS (tɒps) Training Opportunities Scheme

t.o.r. time of receipt (*or* reception)

Tor. Toronto

TOR Tertiary Order Regular of St Francis

torn. tornado

torp. torpedo

t.o.s. temporarily out of stock • terms of service

TOSD Tertiary Order of St Dominic

TOSF Tertiary Order of St Francis

Toshiba (tɒ'ʃiːbə) Tokyo Shibaura Denki KK (Japanese corporation)

tot. total

t.o.t. time on (*or* over) target

TOTC time-on-target computation (military computer)

TOTP Top of the Pops (television programme)

Tou. Toulon (France)

tour. tourism • tourist

tourn. tournament

TOW (təʊ) tube-launched optically tracked wire-guided (antitank missile) • tug-of-war

toxicol. toxicological • toxicologist • (*or* **tox.**) toxicology

tp township • troop

t.p. target practice • teaching practice • title page • to pay • *Commerce* tout payé (French: all expenses paid)

TP *British vehicle registration for* Portsmouth • *airline flight code for* TAP Air Portugal • taxpayer • technical paper (*or* publication) • teleprinter • *Computing* teleprocessing • *Music* tempo primo (Italian; at the original tempo) • tempore Paschale (Latin: at Easter) • test panel • *Insurance* third party • town planner (*or* planning) • *Computing* transaction processing • Transvaal Province • treaty port • *Surveying* trigonometric point • true position • *Surveying* turning point

TPA *Chem.* tetradecanoylphorbol-13-acetate • (*or* **tPA**) *Med.* tissue plasminogen activator

TPC (USA) The Peace Corps • (Australia) Trade Practices Commission • Transaction Processing Council

tpd (*or* **TPD**) tons per day

tph (*or* **TPH**) tons per hour

tpi *Engineering* teeth per inch • *Computing* tracks per inch • *Engineering* turns per inch

TPI tax and price index • *Computing* terminal phase initiation • *Engineering* threads per inch • *Shipping* tons per inch (immersion) • (Australia) totally and permanently incapacitated • *Computing* tracks per inch • transpolyisoprene (synthetic gutta percha) • Tropical Products Institute

tpk. turnpike

TPLF Tigrean People's Liberation Front

TPM *Computing, etc.* third-party maintenance • (*or* **tpm**) tons per minute • *Commerce* total productive management • *Chem.* triphenylmethane

TPN *Med.* total parenteral nutrition • *Biochem.* triphosphopyridine nucleotide (now called NADP)

TPO travelling post office • Tree Preservation Order

TPP *Chem.* thiamine pyrophosphate

Tpr *Military* Trooper

TPR *Med.* temperature, pulse, respiration

TPS toughened polystyrene

tpt transport • *Music* trumpet

tptr *Music* trumpeter

TQ (*or* **t.q.**) *Banking* tel quel (exchange rate) • *Computing* text quality • *postcode for* Torquay • total quality (quality control) • *airline flight code for* Transwede Airways

TQM total quality management

tr. *Med.* tinctura (Latin: tincture) • trace • track • tragedy • train • transaction • transfer • transitive • translate(d) • translation • translator • transport(ation) • *Printing* transpose • transposition • treasurer • *Music* treble • *Music* trill • troop • truck • *Music* trumpet(er) • trust • trustee

Tr *Chem., former symbol for* terbium • *Geology* Triassic • *Chem., symbol for* trityl (triphenylmethyl) group (in formulae)

TR *international civil aircraft marking for* Gabon • *British vehicle registration for* Portsmouth • target rifle • tariff reform • Telephone Rentals plc • tempore regis *or* reginae (Latin: in the time of the king *or* queen) • Territorial Reserve • test run • Theodore Roosevelt (1858–1919, US president 1901–09) • tons registered • tracking radar • *airline flight code for* Transbrasil • *Telecom.* transmit–receive (as in **TR switch**) • *postcode for* Truro • trust receipt • *international vehicle registration for* Turkey

T/R transmitter-receiver

TrA *Astronomy* Triangulum Australe

TRA (USA) Thoroughbred Racing Association

trac. tracer • tractor

TRACALS ('trækælz) *Aeronautics* traffic control and landing system

TRACE (treɪs) task reporting and current evaluation • *Aeronautics* test equipment for rapid automatic checkout evaluation

trad. tradition(al) • traduttore (Italian: translator) • traduzione (Italian: translation)

TRADA Timber Research and Development Association

trag. tragedy • tragic

TRAM *Med.* transverse rectus abdominus myocutaneus (breast reconstruction)

TRAMPS (træmps) temperature regulator and missile power supply

trans. transaction • transfer(red) • transit • transitive • transitory • translate(d) •

translation • translator • transparent • transport(ation) • *Printing* transpose • transposition • transverse

Trans. Transvaal

transcr. transcribed (by *or* for) • transcription

transf. transferred

transl. translate(d) • translation • translator

translit. transliterate • transliteration

transp. transport(ation)

trany transparency

trav. traveller • travels

trbn. *Music* trombone

TRC Thames Rowing Club • Tobacco Research Council

tr. co. trust company

tr. coll. training college

Trd Trinidad

TRDA Timber Research and Development Association

treas. treasurer • (*or* **Treas.**) treasury

tree. trustee

trem (trem) transport emergency (in **trem card**; carried by chemical tank vehicles)

trem. *Music* tremolando (Italian: trembling) • *Music* tremulant (device in an organ)

tren *Chem.* triaminotriethylamine (used in formulae)

trf. tariff • transfer

TRF *Biochem.* thyrotrophin-releasing factor • tuned radio frequency

trg training

trg. (*or* **trge**) *Music* triangle

TRG Tory Reform Group

TRH Their Royal Highnesses • *Biochem.* thyrotrophin-releasing hormone

Tri *Astronomy* Triangulum

TRI Television Reporters International • (USA) Textile Research Institute

trib. tribal • tributary

TRIC Television and Radio Industries Club • *Med.* trachoma inclusion conjunctivitis

trid. *Med.* triduum (Latin: three days; in prescriptions)

trien *Chem.* trimethylenetetramine (used in formulae)

trig. triangulation (station *or* point) • trigger • trigonometric(al) • trigonometry

TRIGA ('trɪɡə) *Engineering* training, research, and isotope-production reactors – General Atomic (in **TRIGA reactor**)

trike (traɪk) *Chem.* trichloroeth(yl)ene (solvent)

trim. trimester

Trin. Trinidad • Trinity • Trinity College (Oxford) • Trinity Hall (Cambridge)

Trip. Tripos

tripl. triplicate

triple A (*or* **AAA**) anti-aircraft artillery

trit. triturate

TRITC *Immunol.* tetramethylrhodamine isothiocyanate (fluorescent dye)

TRJ turboramjet (engine)

TRLFSW tactical range landing-force support weapon

trlr trawler

TRM trademark

trml terminal

t.r.n. technical research note

tRNA *Biochem.* transfer RNA

TRNC Turkish Republic of Northern Cyprus

trng training

TRO *Law* temporary restraining order

trom. (*or* **tromb.**) trombone

trombst trombonist

Tron (trɒn) *Computing* the real-time operating system nucleus

trop. tropic(al)

Trop. Can. Tropic of Cancer

Trop. Cap. Tropic of Capricorn

trop. med. tropical medicine

Trp (*or* **trp**) *Military* troop • *Biochem.* tryptophan

TRPGDA *Chem.* tripropylene glycol diacrylate

TRRL Transport and Road Research Laboratory

trs. transfer • *Printing* transpose • trustees

TRS *Computing* term rewriting system • Torry Research Station (Scotland)

TRSB time reference scanning beam

trsd transferred • transposed

trsp. transport

TRSR taxi and runway surveillance radar

TRSSGM tactical range surface-to-surface guided missile

TRSSM tactical range surface-to-surface missile

trt turret

Truron. Truronensis (Latin: (Bishop) of Truro)

try. truly

t.s. temperature switch • tensile strength • test summary • till sale • turbine ship • twin screw • type specification

Ts *Chem., symbol for* tosyl group (in formulae)

TS *postcode for* Cleveland • *British vehicle registration for* Dundee • *Music* tasto solo (Italian: one key alone; in figured-bass playing) • Television Society • Theosophical Society (in England) • tool steel • *Slang* tough shit • *Med.* Tourette's syndrome • training ship • *Chem.* transition state • Treasury Solicitor • *Med.* tuberous sclerosis • tub-sized (paper) • *international civil aircraft marking for* Tunisia • typescript

T/S trans-shipment

TSA The Securities Association Ltd • *Statistics* time-series analysis • total surface area • Training Services Agency (now called Training Agency)

TSAPI telephony services application programming interface

TSB Trustee Savings Bank

tsc *indicating* passed a Territorial Army course in staff duties

TSD Tertiary of St Dominic

TSDS two-speed destroyer sweeper

TSE Tokyo Stock Exchange • Toronto Stock Exchange • *Med.* transmissible spongiform encephalopathy

TSF two-seater fighter

tsfr transfer

T.Sgt Technical Sergeant

TSh *symbol for* Tanzanian shilling (monetary unit)

TSH Their Serene Highnesses • *Biochem.* thyroid-stimulating hormone

TSH-RH *Biochem.* thyroid-stimulating-hormone-releasing hormone

tsi tons per square inch

TSO Trading Standards Officer

tsp. teaspoon(ful)

TSP *Computing* travelling salesman problem

TSR tactical strike reconnaissance • *Computing* terminate but stay resident (program) • torpedo-spotter reconnaissance • Trans-Siberian Railway

TSRB Top Salaries Review Body

TSS time-sharing system • *Med.* toxic shock syndrome • turbine steamship • twin-screw steamer (*or* steamship, ship) • typescripts

TSSA Transport Salaried Staffs' Association

tstr tester

t.s.u. this side up

TSV *Microbiol.* tobacco streak virus

t.s.v.p. tournez s'il vous plaît (French: please turn over; PTO)

TSW Television South West

TT *international civil aircraft marking for* Chad • *British vehicle registration for* Exeter • *British fishing port registration for* Tarbert • technical training • teetotal(ler) • *Banking* telegraphic transfer • *Geology* Tertiary • tetanus toxoid • *Cycling* time trial • torpedo tube • *Motorcycling* Tourist Trophy (races) • *international vehicle registration for* Trinidad and Tobago • Trust Territories (abbrev. *or* postcode) • tuberculin-tested (as in **TT milk**)

TTA Travel Trade Association

TTAPS Turco, Toon, Ackerman, Pollack, and Sagan (in **TTAPS study** on effects of nuclear winter; from authors' names)

TTB tetragonal tungsten bronze

TTBT Threshold Test Ban Treaty

TTC teachers' training course • technical training centre • *Military* Technical Training Command • *Med.* transcutaneous transhepatic cholangiography

TTF Timber Trade Federation

TTFN *Colloquial* ta-ta for now

TTL *Photog.* through the lens • (*or* **t.t.l.**) to take leave • *Electronics* transistor-transistor logic • (Zimbabwe) tribal trust land

TTM *Commerce* time to market

Tto Toronto

TTS teletypesetter (*or* teletypesetting) • *Acoustics* temporary threshold shift • *Computing* text-to-speech

TTT *Cycling* team time trial • *Banking* telegraphic transfer • Tyne Tees Television Limited

TTY (USA) teletypewriter

Tu. Tudor • Tuesday

TU *British vehicle registration for* Chester • *international civil aircraft marking for* Côte d'Ivoire (Ivory Coast) • thermal unit • toxic unit • trade union • *Telecom.* traffic unit • training unit • *Acoustics, Telecom.* transmission unit • *airline flight code for* Tunis Air • Tupolev (aircraft; as in **TU-104**)

TUAC Trade Union Advisory Committee

tub. *Cycling* tubular (tyre)

tuberc. tuberculosis

Tuc *Astronomy* Tucana

TUC Trades Union Congress

TUCC Transport Users' Consultative Committee (*or* Council)

TUCGC Trades Union Congress General Council

Tues. Tuesday

TUG Telephone Users' Group

TULRA Trade Union and Labour Relations Act (1974)

TUPE Transfer of Undertakings (Protection of Employment) Regulations

TUR *Med.* transurethral resection

turb. turbine

TURB *Meteorol.* turbulence

Turk. Turkey • Turkish

t.v. terminal velocity • test vehicle

TV *British vehicle registration for* Nottingham • television • *Finance* terminal value • (USA) *Colloquial* transvestite

TVA taxe à (*or* sur) la valeur ajoutée (French: value-added tax; VAT) • (USA) Tennessee Valley Authority

TVEI Technical and Vocational Educational Initiative

tvl travel

Tvl Transvaal

TVO tractor vaporizing oil

TVP textured vegetable protein (a meat substitute)

TVR television rating • *Physics* temperature variation of resistance

TVRO television receive only (type of antenna)

t.w. tail wind

TW *British vehicle registration for* Chelmsford • *airline flight code for* Trans World Airlines • *Telecom.* travelling wave (in **TW antenna**) • *postcode for* Twickenham

T-W (USA) three-wheeler (motorcycle)

TWA Thames Water Authority • time-weighted average • Trans-World Airlines

TWh (*or* **TW h**) *Electricity, symbol for* terawatt hour(s)

TWIMC to whom it may concern

TWN teleprinter weather network

twocing ('twʊkɪŋ) *Colloquial* taking without owner's consent (car theft)

TWT transonic wind tunnel • *Electronics* travelling-wave tube

TWU Transport Workers' Union of America

TWX (USA) teletypewriter exchange service

twy twenty

tx tax(ation)

Tx *Broadcasting* transmission

TX *British vehicle registration for* Cardiff • *US postcode for* Texas

Ty Territory • truly (in correspondence)

TY *international civil aircraft marking for* Benin • *British vehicle registration for* Newcastle upon Tyne

TYC *Horse racing* two-year-old course

TYMV *Microbiol.* turnip yellow mosaic virus

t.y.o. *Horse racing* two-year-old

typ. typical • typing • typist • typographer • typographic(al) • typography

typh. typhoon

typo. (*or* **typog.**) typographer • typographic(al) • typography

typw. typewriter • typewriting • typewritten

Tyr (*or* **tyr**) *Biochem.* tyrosine

Tyr. Tyrone

Tyrol. Tyrolean • Tyrolese

TZ *British vehicle registration for* Belfast • *international civil aircraft marking for* Mali

U

u (*ital.*) *Electricity, symbol for* instantaneous potential difference • (*ital.*) *Optics, symbol for* object distance • (*ital.*) *Thermodynamics* specific internal energy • *Meteorol., symbol for* ugly threatening sky • *Physics* ungerade (German: odd; in spectroscopy) • *Chem., symbol for* unified atomic mass unit • *Physics* up (a quark flavour) • (*ital.*) *Physics, symbol for* a velocity component or speed

u. uncle • und (German: and) • unit • unsatisfactory • unter (German: under *or* among) • upper • utility

U (*ital.*) *Thermodynamics, symbol for* internal energy • (*ital.*) *Electricity, symbol for* potential difference • *symbol for* rate of heat loss (measured in British thermal units; in **U value**, a measure of insulating power) • *Films* universal (certification) • *Colloquial* upper class (of characteristics, language habits, etc.; also in **non-U**) • *Biochem., symbol for* uracil • *Chem., symbol for* uranium • *Biochem., symbol for* uridine • you (from phonetic spelling, as in **IOU, while U wait**)

U. (*or* **U**) Union • Unionist • unit • United • University • unsatisfactory • upper

2U *airline flight code for* Western Pacific Air Services

3U *airline flight code for* Sichuan Airlines

5U *international civil aircraft marking for* Niger

6U *airline flight code for* Air Ukraine

9U *airline flight code for* Air Moldova • *international civil aircraft marking for* Burundi

u.a. under age • unter anderem (German: among other things) • usque ad (Latin: as far as)

U/a *Insurance* underwriting account

UA *British vehicle registration for* Leeds • *international vehicle registration for* Ukraine • Ulster Association • *airline flight code for* United Airlines • United Artists Corporation • University of Alabama

UAB Unemployment Assistance Board • Universities Appointments Board • University of Alabama in Birmingham

UABS Union of American Biological Societies

UAC Ulster Automobile Club

UADW Universal Alliance of Diamond Workers

UAE United Arab Emirates

UAI Union des associations internationales (French: Union of International Associations)

UAL United Airlines

UAM underwater-to-air missile

u. & l.c. *Printing* upper and lower case

u. & o. use and occupancy

UAOD United Ancient Order of Druids

UAOS Ulster Agricultural Organization Society

UAP United Australia Party

UAPT Union africaine des postes et

télécommunications (French: African Postal and Telecommunications Union)

UAR United Arab Republic (1958–71)

UARS upper-atmosphere research satellite

UART ('juː,ɑːt) *Electronics, Computing* universal asynchronous receiver/transmitter

u.a.s. upper airspace

UAS University Air Squadron

UAU Universities Athletic Union

UAV unmanned air vehicle

UAW (USA) United Automobile Workers (United International Union of Automobile, Aerospace, and Agricultural Implement Workers of America; trade union)

u.A.w.g. um Antwort wird gebeten (German: an answer is requested)

UB *British vehicle registration for* Leeds • *airline flight code for* Myanmar Airways International • *postcode for* Southall • United Brethren

UB40 *indicating* the index card used for unemployment benefit

UBA *Med.* ultrasonic bone analysis

UBC University of British Columbia (Vancouver)

UBF Union of British Fascists

UBI Understanding British Industry (organization)

U-boat *indicating* a German submarine during the World Wars (German *Unterseeboot*)

UBR *Taxation* Uniform Business Rate • University Boat Race

UBS United Bible Societies

UBV *Astronomy* ultraviolet, blue, visual (green-yellow)

u.c. *Music* una chorda (*see under* UC) • *Printing* upper case

u/c *Commerce* undercharge

UC *British vehicle registration for* London (central) • *Music* una corda (Italian: on one string; i.e. use soft pedal) • under construction • undercover • University College • *Civil engineering* upcast shaft • *Theatre* up centre (of stage) • Upper Canada • urban council • urbe condita (Latin: the city being built) • *Obstetrics* uterine contraction

U$_c$ *Films* universal, particularly suitable for children (certification)

UCA United Chemists' Association

UCAE Universities' Council for Adult Education

UCAS ('juːkæs) Universities and Colleges Admissions Service (formed from a merger of UCCA and PCAS)

UCATT (*or* **Ucatt**) ('juːkət) Union of Construction, Allied Trades, and Technicians

u.c.b. unless caused by

UCBSA United Cricket Board of South Africa

UCC Union Carbide Corporation • Universal Copyright Convention • University Computing Company

UCCA (*or* **Ucca**) ('ʌkə) (formerly) Universities Central Council on Admissions (merged with PCAS to form UCAS)

UCCD United Christian Council for Democracy

UCD University College, Dublin • *Oceanog.* upper critical depth

UCET Universities Council for Education of Teachers

UCG underground coal gasification

UCH University College Hospital (London)

UCI Union cycliste internationale (French: International Cyclists' Union) • United Cinemas International • University of California, Irvine

UCITS *Finance* Undertakings for Collective Investment in Transferable Securities

UCJG Alliance universelle des unions chrétiennes de jeunes gens (French: World Alliance of Young Men's Christian Associations)

u.c.l. *Engineering* upper cylinder lubricant

UCL University College London • upper control limit

UCLA University of California at Los Angeles

UCM University Christian Movement

UCMJ (USA) uniform code of military justice

UCMSM University College and Middlesex School of Medicine

UCNS Universities' Council for Nonacademic Staff

UCNW University College of North Wales

UCR *Physiol., Psychol.* unconditioned reflex (*or* response) • *Computing* under-colour removal • (USA) Uniform Crime Report

UCS *Physiol., Psychol.* unconditioned stimulus • Union of Concerned Scientists • *Computing* universal multiple-octet coded character set • University College School (London) • Upper Clyde Shipbuilders

UCSB University of California, Santa Barbara

UCSD University of California, San Diego

UCSW University College of South Wales

UCTA United Commercial Travellers' Association (now part of the MSF union)

UCV (USA) United Confederate Veterans

UCW Union of Communication Workers • University College of Wales

UCWRE Underwater Countermeasures and Weapons Research Establishment

u.d. unfair dismissal • *Med.* ut dictum (Latin: as directed; in prescriptions)

UD *British vehicle registration for* Oxford • United Dairies

U/D (USA) under deed

UDA Ulster Defence Association

UDAG ('juːdæg) (USA) Urban Development Action Grant

u.d.c. upper dead centre

UDC (USA) United Daughters of the Confederacy • universal decimal classification • Urban Development Corporation • Urban District Council

UDCA Urban District Councils' Association

UDE Underwater Development Establishment

UDEAC Union douanière et économique de l'Afrique centrale (French: Central African Customs and Economic Union)

UDEAO Union douanière des états d'Afrique d'Ouest (French: Customs Union of West African States)

UDF Ulster Defence Force • (South Africa) Union Defence Force • (Bulgaria) Union of Democratic Forces • (France) Union pour la démocratie française (French Democratic Union; political party) • (South Africa) United Democratic Front

u. dgl. und dergleichen (German: and the like)

UDI unilateral declaration of independence

UDM Union of Democratic Mineworkers

UDMH *Chem.* unsymmetrical dimethyl hydrazine

UDN ulcerative dermal necrosis (fish disease)

UDP United Democratic Party • *Biochem.* uridine 5'-diphosphate

UDR Ulster Defence Regiment • (France) Union des démocrates pour la république (French: Union of Democrats for the Republic; former name of the RPR)

UDSR (France) Union démocratique et socialiste de la résistance (Democratic and Socialist Union of the Resistance; 1946–58)

UE *British vehicle registration for* Dudley • (New Zealand) university entrance (examination)

UEA Universal Esperanto Association • University of East Anglia

UED University Education Diploma

u.e.f. universal extra fine (screw)

UEF Union européenne des fédéralistes (French: European Union of Federalists) • Union européenne féminine (French: European Union of Women)

UEFA (juːˈɜːfə) Union of European Football Associations

UEI Union of Educational Institutions

UEIC United East India Company

UEL United Empire Loyalists

UEMOA Union économique et monetaire ouest-africaine (French: West African Economic and Monetary Union)

UEO Union de l'Europe occidentale (French: Western European Union; WEU) • Unit Education(al) Officer

UEP Union européenne de paiements (French: European Payments Union) • *Genetics* unit evolutionary period

UEPS Union européenne de la presse sportive (French: European Sports Press Union)

UER Union européenne de radiodiffusion (French: European Broadcasting Union; EBU) • university entrance requirements

UETA (USA) universal engineer tractor, armoured

UETRT (USA) universal engineer tractor, rubber-tyred

u/f unfurnished (in property advertisements)

UF *British vehicle registration for* Brighton • United Free (Church, of Scotland) • (India) United Front • urea–formaldehyde (as in **UF resin**) • utilization factor (of an electric lamp)

UFA (formerly, in Germany) Universum Film-Aktiengesellschaft (German: Universal Film Company) • *Biochem.* unsaturated fatty acid

UFAW Universities Federation for Animal Welfare

UFC United Free Church (of Scotland) • University Funding Council

UFCW (USA) United Food and

Commercial Workers International Union (trade union)

UFF Ulster Freedom Fighters

UFFI urea–formaldehyde foam insulation

UFO ('ju: 'ɛf 'əʊ *or* 'ju:fəʊ) unidentified flying object

UFT *Physics* unified field theory · (USA) United Federation of Teachers

UFTAA Universal Federation of Travel Agents' Association

UFTU (Romania) Union of Free Trades Union

UFU Ulster Farmers' Union

UFW United Farm Workers of America (trade union)

u/g underground

Ug. (*or* **Ugan.**) Uganda(n)

UG *British vehicle registration for* Leeds

UGC underground gasification of coal · University Grants Committee (now Funding Council, UFC)

UGLE *Freemasonry* United Grand Lodge of England

UGT (Spain) Unión General de Trabajadores (General Union of Workers)

UGWA United Garment Workers of America (trade union)

UH *British vehicle registration for* Cardiff

UHA (South Africa) Union House of Assembly

UHB *Colloquial* urban haute bourgeoisie

UHCC Upper House of the Convocation of Canterbury

UHCY Upper House of the Convocation of York

UHF ultrahigh frequency

UHT ultra-heat-treated (as in **UHT milk**) · ultrahigh temperature

UHV ultrahigh vacuum

u.i. ut infra (Latin: as below)

u/i under instruction

UI *British vehicle registration for* Londonderry · (USA) unemployment insurance · *Computing* user interface

UIA Union of International Associations

UIAA Union internationale des associations d'alpinisme (French: International Union of Alpine Associations)

UIC Union internationale des chemins de fer (French: International Union of Railways)

UICC Union internationale contre le cancer (French: International Union against Cancer)

UICN Union internationale pour la conservation de la nature et de ses resources (French: International Union for Conservation of Nature and Natural Resources; IUCNN)

UICPA Union internationale de chimie pure et appliquée (French: International Union of Pure and Applied Chemistry; IUPAC)

UIE Union internationale des étudiants (French: International Union of Students)

UIEO Union of International Engineering Organizations

UIHPS Union internationale d'histoire et de philosophie des sciences (French: International Union of the History and Philosophy of Science)

UIJS Union internationale de la jeunesse socialiste (French: International Union of Socialist Youth)

UIL Unione Italiana del Lavoro (Italian Federation of Trade Unions) · United Irish League

UIMS *Computing* user-interface management system

UIP Union internationale de patinage (French: International Skating Union) · Union internationale de physique pure et appliquée (French: International Union of Pure and Applied Physics; IUPAP) · Union interparlementaire (French: Inter-Parliamentary Union)

UIPC Union internationale de la presse catholique (French: International Catholic Press Union)

UIPM Union internationale de pentathlon moderne (French: International Modern Pentathlon Union)

UISB Union internationale des sciences biologiques (French: International Union of Biological Sciences)

UISM Union internationale des syndicats des mineurs (French: Miners' Trade Unions International)

UISPP Union internationale des sciences préhistoriques et prohistoriques (French: International Union of Prehistoric and Prohistoric Sciences)

UIT Union internationale des télécommunications (French: International Telecommunications Union; ITU) · unit investment trust

UITF *Accounting* Urgent Issues Task Force

UITP Union internationale des transports publics (French: International Union of Public Transport)

UIU Upholsterers' International Union of North America

UJ *British vehicle registration for* Shrewsbury • Union Jack • *Engineering* universal joint

UJC Union Jack Club (London)

UJD Utriusque Juris Doctor (Latin: Doctor of Civil and Canon Law)

UK *airline flight code for* Air UK • *British vehicle registration for* Birmingham • United Kingdom • *international civil aircraft marking for* Uzbekistan

UK(A) *Athletics* United Kingdom Allcomers

UKA Ulster King of Arms • United Kingdom Alliance

UKAC United Kingdom Automation Council

UKADGE United Kingdom Air Defence Ground Environment

UKAEA United Kingdom Atomic Energy Authority

UKAPE United Kingdom Association of Professional Engineers

UKBG United Kingdom Bartenders' Guild

UKCC United Kingdom Central Council for Nursing, Midwifery, and Health Visiting

UKCIS United Kingdom Chemical Information Service

UKCOSA (juːˈkəʊzə) United Kingdom Council for Overseas Students' Affairs

UKCSBS United Kingdom Civil Service Benefit Society

UKCTA United Kingdom Commercial Travellers' Association

UKDA United Kingdom Dairy Association

UKFBPW United Kingdom Federation of Business and Professional Women

UKgal *symbol for* UK gallon

UKIAS (*or* **Ukias**) (juːˈkaɪəs) United Kingdom Immigrants' Advisory Service

UKIRT UK Infrared Telescope (Mauna Kea, Hawaii)

UKISC United Kingdom Industrial Space Committee

UKLF United Kingdom Land Forces

UKMF(L) United Kingdom Military Forces (Land)

UKMIS United Kingdom Mission

UK(N) *Athletics* United Kingdom National

UKOOA United Kingdom Offshore Operators Association

UKOP United Kingdom Oil Pipelines

UKPA United Kingdom Pilots' Association

UKPIA United Kingdom Petroleum Industry Association Ltd

Ukr. Ukraine • Ukrainian

UKSATA United Kingdom South Africa Trade Association

UKSLS United Kingdom Services Liaison Staff

UKSMA United Kingdom Sugar Merchant Association Ltd

UKST UK Schmidt Telescope (Siding Spring, Australia)

UKW Ultrakurzwelle (German: ultrashort wave, i.e. VHF)

u.l. upper limit

UL *airline flight code for* Air Lanka • *British vehicle registration for* London (central) • *British fishing port registration for* Ullapool • (USA) Underwriters' Laboratories (on labels for electrical appliances) • university library • *Biochem.* unsaturated lipid • *Theatre* up left (of stage) • *Med.* upper limb

ULA *Computing* uncommitted logic array

ULC *Theatre* up left centre (of stage)

ULCC ultralarge crude carrier (oil tanker) • University of London Computer Centre

ULCI Union of Lancashire and Cheshire Institutes

ULF *Radio, etc.* ultralow frequency • upper limiting frequency

ULICS University of London Institute of Computer Science

ULM ultrasonic light modulator • universal logic module

ULMS underwater long-range missile system

u.l.s. *Finance* unsecured loan stock

ULS unsecured loan stock

ULSEB University of London School Examinations Board

ult. ultimate(ly) • (*or* **ulto.**) ultimo (Latin: in (*or* of) the last (month); in correspondence)

ULT United Lodge of Theosophists

ult. praes. *Med.* ultimum praescriptum (Latin: last prescribed)

ULV ultralow volume (as in **ULV sprayer**)

um. unmarried

u/m undermentioned

UM *airline flight code for* Air Zimbabwe • *British vehicle registration for* Leeds •

symbol for Mauritanian ouguiya (monetary unit) • University of Minnesota

UMa *Astronomy* Ursa Major

UMA Union du Maghreb Arabe (French: Arab Maghreb Union; customs union)

UMB Union mondiale de billard (French: World Billiards Union) • *Computing* upper memory block

umbl. umbilical

UMC University of Missouri, Columbia

UMCP University of Maryland, College Park

UMDS United Medical and Dental Schools

UMEJ Union mondiale des étudiants juifs (French: World Union of Jewish Students)

UMF Umbrella Makers' Federation

UMFC United Methodist Free Churches

UMi *Astronomy* Ursa Minor

UMIST (*or* **Umist**) ('juː‚mɪst) University of Manchester Institute of Science and Technology

UMNO United Malays (later Malaysia) National Organization

ump. umpire

UMP *Biochem.* uridine monophosphate

UMT universal military training

UMTS universal military training service (*or* system)

UMW (*or* **UMWA**) United Mine Workers (of America; trade union)

un. unified • union • united • unsatisfactory

UN *British vehicle registration for* Exeter • *international civil aircraft marking for* Kazakhstan • *airline flight code for* Transaero Airlines • United Nations

UNA United Nations Association

UNAA United Nations Association of Australia

unab. unabridged

unacc. (*or* **unaccomp.**) unaccompanied

UNACC United Nations Administrative Committee and Coordination

UNACOM universal army communication system

unan. unanimous

UNARCO United Nations Narcotics Commission

unasgd unassigned

unatt. unattached

unattrib. unattributed

UNAUS United Nations Association of the United States

unauthd unauthorized

unb. (*or* **unbd**) unbound

UNB universal navigation beacon • University of New Brunswick

UNBRO (*or* **Unbro**) ('ʌnbrəʊ) United Nations Border Relief Operation

unc. uncertain • uncle

UNC *Numismatics* uncirculated • Union Nationale Camerounaise (French: Cameroon National Union *or* Cameroon People's Democratic Movement) • United Nations Command • University of North Carolina

UNCAST United Nations Conference on the Applications of Science and Technology

UNCC United Nations Cartographic Commission

UNCCP United Nations Conciliation Commission for Palestine

UNCDF United Nations Capital Development Fund

UNCED United Nations Conference on Environment and Development

UNCID Uniform Rules of Conduct for Interchange of Trade Data by Teletransmission

UNCIO United Nations Conference on International Organization

UNCITRAL United Nations Commission on International Trade Law

unclas. (*or* **unclass.**) unclassified

UNCLE ('ʌŋkᵊl) United Network Command for Law Enforcement (fictional organization in TV series 'The Man from UNCLE')

UNCLOS United Nations Conference on the Law of the Sea

UNCOK United Nations Commission on Korea

uncond. unconditional

uncor. uncorrected

UNCSTD United Nations Conference on Science and Technology for Development

UNCTAD (*or* **Unctad**) ('ʌŋk‚tæd) United Nations Conference on Trade and Development

UNCURK United Nations Commission for Unification and Rehabilitation of Korea

UND University of North Dakota

UNDP United Nations Development Programme

UNDRO (*or* **Undro**) ('ʌn,drəʊ) United Nations Disaster Relief Organization

undsgd undersigned

undtkr undertaker

UNE underground nuclear explosion

UNEC United Nations Education Conference

UNECA United Nations Economic Commission for Asia

UNEDA United Nations Economic Development Administration

UNEF ('juː,nɛf) United Nations Emergency Force

UNEP ('juː,nɛp) United Nations Environment Programme

UNESCO (*or* **Unesco**) (juːˈnɛskəʊ) United Nations Educational, Scientific, and Cultural Organization

UNETAS United Nations Emergency Technical Aid Service

unexpl. unexplained • unexploded • unexplored

UNFAO United Nations Food and Agriculture Organization

UNFB United Nations Film Board

UNFC United Nations Food Conference

UNFICYP United Nations (Peace-Keeping) Force in Cyprus

UNFPA United Nations Fund for Population Activities

ung. *Med.* unguentum (Latin: ointment)

UNGA United Nations General Assembly

Unh *Chem., symbol for* unnilhexium (element 106)

UNH *Chem.* uranyl nitrate hexahydrate

UNHCR United Nations High Commissioner for Refugees

UNHQ United Nations Headquarters

Uni. University College, Oxford

UNI Ente Nazionale Italiano di Unificazione (Italian Standards Association)

UNIA (USA) Universal Negro Improvement Association

UNIC United National Information Centre

UNICA Union internationale du cinéma d'amateurs (French: International Union of Amateur Cinema)

UNICE Union des industries de la communauté européenne (French: Union of Industries of the European Community)

UNICEF (*or* **Unicef**) ('juː,nɪ,sɛf) United Nations Children's Fund (formerly United Nations International Children's Emergency Fund)

UNICOM (*or* **Unicom**) ('juː,nɪ,kɒm) universal integrated communication system

Unics ('juː,nɪks) *Trademark, indicating* a type of computer operating system

UNIDO (*or* **Unido**) (juːˈniːdəʊ) United Nations Industrial Development Organization

UNIDROIT Institut international pour l'unification du droit privé (French: International Institute for the Unification of Private Law)

unif. uniform

UNIFIL (*or* **Unifil**) ('juː,nɪ,fɪl) United Nations Interim Force in Lebanon

UNIMA Union internationale des marionnettes (French: International Union of Puppeteers)

UNIO United Nations Information Organization

UNIP (*or* **Unip**) ('juː,nɪp) (Zambia) United National Independence Party

UNIPEDE Union internationale des producteurs et distributeurs d'énergie électrique (French: International Union of Producers and Distributors of Electrical Energy)

unis. *Music* unison

UNIS United Nations International School

UNISCAT (*or* **Uniscat**) ('juː,nɪ,skæt) United Nations Expert Committee on the Application of Science and Technology

UNISIST (*or* **Unisist**) ('juː,nɪ,sɪst) Universal System for Information in Science and Technology

Unit. Unitarian(ism)

UNITA (*or* **Unita**) (juːˈniːtə) União Nacional para a Independência Total de Angola (Portuguese: National Union for the Total Independence of Angola)

UNITAR United Nations Institute for Training and Research

univ. universal(ly) • university

Univ. Universalist • University • University College, Oxford

UNIVAC ('juː,nɪ,væk) universal automatic computer

UNIX ('juː,nɪks) *Trademark, indicating* a type of computer operating system

UNJSPB United Nations Joint Staff Pension Board

unkn. unknown

UNKRA (*or* **Unkra**) United Nations Korean Reconstruction Agency

UNLC United Nations Liaison Committee

unm. unmarried

UNM University of New Mexico

UNMC United Nations Mediterranean Commission

UNO United Nations Organization · (Nicaragua) *indicating* National Opposition Union (Spanish; political party)

unop. unopposed

unp. unpaged · unpaid

Unp *Chem., symbol for* unnilpentium (element 105)

UNP (Sri Lanka) United National Party

UNPA United Nations Postal Administration

UNPC United Nations Palestine Commission

UNPCC United Nations Conciliation Commission for Palestine

unpd unpaid

UNPROFOR ('ʌnprʊə,fɔ:) United Nations Protection Force in Yugoslavia

unpub. (*or* **unpubd**) unpublished

Unq *Chem., symbol for* unnilquadium (element 104)

UNREF (*or* **Unref**) ('ʌnrɛf) United Nations Refugee Emergency Fund

UNRISD United Nations Research Institute for Social Development

UNRPR United Nations Relief for Palestine Refugees

UNRRA (*or* **Unrra**) ('ʌnrə) United Nations Relief and Rehabilitation Administration

UNRWA (*or* **Unrwa**) ('ʌnrə) United Nations Relief and Works Agency

unsat. unsatisfactory · unsaturated

UNSC United Nations Security Council · United Nations Social Commission

UNSCC United Nations Standards Coordinating Committee

UNSCCUR United Nations Scientific Conference on the Conservation and Utilization of Resources

UNSCOB United Nations Special Committee on the Balkans

UNSCOP United Nations Special Committee on Palestine

UNSF United Nations Special Fund for Economic Development

UNSG United Nations Secretary-General

UNSR United Nations Space Registry

UNSW University of New South Wales

UNTAA United Nations Technical Assistance Administration

UNTAB United Nations Technical Assistance Board

UNTAC (*or* **Untac**) ('ʌntæk) United Nations Transitional Authority for Cambodia

UNTAG United Nations Transition Assistance Group (in Namibia)

UNTAM United Nations Technical Assistance Mission

UNTC United Nations Trusteeship Council

UNTT United Nations Trust Territory

UNWCC United Nations War Crimes Commission

UO *British vehicle registration for* Exeter

u.o.c. ultimate operating capability

UOD ultimate oxygen demand (in water conservation)

U of A University of Alaska

U of NC University of North Carolina

U of S University of Saskatchewan

U of T University of Toronto

up. (*or* **u.p.**) underproof (of alcohol) · upper

UP *airline flight code for* Bahamasair · *British vehicle registration for* Newcastle upon Tyne · Ulster Parliament · Union Pacific · United Party (esp. in South Africa) · United Presbyterian · United Press (news agency) · University of Paris · University of Pennsylvania · University of Pittsburgh · University Press · unsaturated polyester (polymer) · (USA) Upper Peninsula · Uttar Pradesh (formerly United Provinces)

UPA Union postale arabe (French: Arab Postal Union) · United Productions of America (cartoon film producer)

UPC Uganda People's Congress · Union des Populations Camerounaises (French: Union of the Populations of Cameroon) · United Presbyterian Church · Universal Postal Convention · universal product code (bar code)

upd unpaid

UPD united port district · Urban Planning Directorate

UPF untreated polyurethane foam

UPGC University and Polytechnic Grants Committee

UPGWA United Plant Guard Workers of America (trade union)

uphd uphold

uphol. upholsterer · upholstery

UPI United Press International

UPIU (USA) United Paperworkers International Union

UPIGO Union professionnelle internationale des gynécologues et obstétriciens (French: International Union of Professional Gynaecologists and Obstetricians)

UPNI Unionist Party of Northern Ireland

UPOA Ulster Public Officers' Association

UPOW Union of Post Office Workers (now merged with UCW)

UPP (Poland) United Peasant Party

UPPP *Med.* uvulopalatopharyngoplasty

UPR *Insurance* unearned premiums reserve • Union Pacific Railroad

UPS *Chem.* ultraviolet photoelectron spectroscopy • uninterruptible power supply (to computers, etc.) • *Trademark* (USA) United Parcel Service • United Publishers' Services

UPU Universal Postal Union (UN agency; formerly General Postal Union, GPU)

UPUP Ulster Popular Unionist Party

uPVC unplasticized polyvinyl chloride

UPW Union of Post Office Workers (now merged with UCW)

UPWA United Packinghouse Workers of America (former trade union)

ur. *Med.* urine

Ur. Urdu • Uruguay(an)

UR *British vehicle registration for* Luton • *international civil aircraft marking for* Ukraine • *Physiol., Psychol.* unconditioned reflex (*or* response) • uniform regulations • *Theatre* up right (of stage)

URA (USA) Urban Renewal Administration

Uran. *Astronomy* Uranus

urb. urban

URBM ultimate-range ballistic missile

URC United Reformed Church • *Theatre* up right centre (of stage)

Urd. Urdu

URF Union des services routiers des chemins de fer européens (French: Union of European Railways Road Services)

urg. urgent

URI *Med.* upper respiratory infection

URL *Computing* Uniform Resource Locator • Unilever Research Laboratory

UR-NAP *Med.* urea-resistant neutrophil alkaline phosphatase (enzyme)

urol. urology

URSI Union radio scientifique internationale (French: International Scientific Radio Union)

URTI *Med.* upper respiratory tract infection

URTU United Road Transport Union

Uru. Uruguay(an)

Urupabol Uruguay, Paraguay, and Bolivia (commission for cooperation, trade, and integration)

URW United Rubber, Cork, Linoleum, and Plastic Workers of America (trade union)

u.s. ubi supra (Latin: where (mentioned *or* cited) above) • ut supra (Latin: as above)

u/s unserviceable • useless

US *British vehicle registration for* Glasgow • (Italy) ufficio stampa (press office) • *Med.* ultrasound (*or* ultrasonic) scanning • (USA) Uncle Sam • *Physiol., Psychol.* unconditioned stimulus • (USA) United Service • United States • United States highway (as in **US66**) • *airline flight code for* USAir

U/S unserviceable • useless

USA United States Army • United States of America (abbrev. *or* IVR) • United Synagogue of America

USA/ABF United States of America Amateur Boxing Federation

USAAC United States Army Air Corps (forerunner of USAF)

USAAF United States Army Air Force (forerunner of USAF)

USAC United States Air Corps • United States Auto Club

USAEC United States Atomic Energy Commission

USAF United States Air Force

USAFA United States Air Force Academy

USAFC United States Army Forces Command

USAFE United States Air Forces in Europe

USAFI United States Armed Forces Institute

USAFR United States Air Force Reserve

USAID United States Agency for International Development

USAMC United States Army Materiel Command

USAMedS United States Army Medical Service

USAR United States Army Reserve

USAREUR United States Army, Europe

USASA United States Army Security Agency

USAT United States Army Transport

USATDC United States Army Training and Doctrine Command

USB *Computing* universal serial bus

USBC United States Bureau of the Census

USBM United States Bureau of Mines

USC Ulster Special Constabulary · United Services Club · United Somali Congress · United States Code · United States Congress · United States of Colombia · University of South Carolina · University of Southern California

USCA United States Code Annotated

USCC United States Circuit Court

USCCA United States Circuit Court of Appeals

USCG United States Coast Guard

USCGA United States Coast Guard Academy

USCGR United States Coast Guard Reserve

USCGS (*or* **USC&GS**) United States Coast and Geodetic Survey

USCL United Society for Christian Literature

USCRC United States Citizens Radio Council

USCSC United States Civil Service Commission

USC Supp. United States Code Supplement

USDA United States Department of Agriculture

USDAW (*or* **Usdaw**) ('ʌz,dɔ:) Union of Shop, Distributive, and Allied Workers

USDOE United States Department of Energy

USEA United States Energy Association

U/sec. Undersecretary

USECC United States Employees' Compensation Commission

USES United States Employment Service

usf. und so fort (German: and so on; etc.)

USF United States Forces

USFL United States Football League

USG United States Government · *Railways* United States Standard Gauge

USGA United States Golf Association

USgal *symbol for* US gallon

USGPO United States Government Printing Office

USGS United States Geological Survey

USh *symbol for* Uganda shilling (monetary unit)

USHA United States Housing Authority

USI United Schools International · United

Service Institution · United States Industries

USIA United States Information Agency

USIS United States Information Service

USITC United States International Trade Commission

USL United States Legation

USLTA United States Lawn Tennis Association

USM ultrasonic machining · underwater-to-surface missile · United States Mail · United States Marines · United States Mint · *Stock exchange* unlisted securities market

USMA United States Military Academy

USMC United States Marine Corps · United States Maritime Commission

USMH United States Marine Hospital

USMS United States Maritime Service

USN United States Navy

USNA United States National Army · United States Naval Academy

USNC United States National Committee

USNG United States National Guard

USNI United States Naval Institute

USNO United States Naval Observatory

USNR United States Naval Reserve

USNRC United States Nuclear Regulatory Commission

USNS United States Navy Ship

USO (USA) United Service Organization

US of A *Colloquial* United States of America

USP *Papermaking* unbleached sulphite pulp · *Advertising* unique selling proposition (of a product) · (*or* **USPat.**) United States Patent · United States Pharmacopeia

USPC Ulster Society for the Preservation of the Countryside

USPCA Ulster Society for the Prevention of Cruelty to Animals

USPG United Society for the Propagation of the Gospel (formerly SPG)

USPHS United States Public Health Service

USPO United States Post Office

USPS United States Postal Service

USR United States Reserves · Universities' Statistical Record

USRC United States Reserve Corps

USS Undersecretary of State · Union Syndicale Suisse (French: Swiss Federation of Trade Unions) · United States Senate · United States Service · United

States Ship • United States Steamer (*or* Steamship*) • Universities Superannuation Scheme

USSB United States Shipping Board

USSC (*or* **USSCt**) United States Supreme Court

USSR Union of Soviet Socialist Republics

USSS United States Steamship

UST undersea technology

USTA United States Tennis Association • United States Trademark Association

USTC United States Tariff Commission

USTS United States Travel Service

usu. usual(ly)

USV United States Volunteers

USVA United States Volleyball Association

USVB United States Veterans' Bureau

USVI United States Virgin Islands

usw. und so weiter (German: and so forth; etc.)

USW *Radio, etc.* ultrashort wave • underwater sea warfare

USWA United Steelworkers of America (trade union)

USWB United States Weather Bureau

USWI United States West Indies • urban solid-waste incinerator

ut. utility

u.t. universal trainer • user test

Ut. Utah

UT *British vehicle registration for* Leicester • ultrasonic testing • ultrasonic transducer • (India) Union Territory • unit trust • *Astronomy* universal time • University of Texas at Austin • *Med.* urinary tract • *airline flight code for* UTA French Airlines • *US postcode for* Utah

U/T under trust

UTA Ulster Transport Authority • Union de Transports Aériens (French airline) • Unit Trust Association

UTC Coordinated Universal Time (French *universel temps coordonné*) • University Training Corps

Utd United

UTDA Ulster Tourist Development Association

ut dict. *Med.* ut dictum (Latin: as directed; in prescriptions)

utend. *Med.* utendus (Latin: to be used)

U3A University of the Third Age

UTI *Med.* urinary-tract infection

ut inf. ut infra (Latin: as below)

UTK University of Tennessee, Knoxville

UTP *Computing* unshielded twisted pair • *Biochem.* uridine 5′-triphosphate

UTS ultimate tensile strength

ut sup. ut supra (Latin: as above)

UTU Ulster Teachers' Union • (USA) United Transportation Union

UTW Union of Textile Workers

UTWA United Textile Workers of America (trade union)

u.U. unter Umständen (German: circumstances permitting)

UU *airline flight code for* Air Austral • *British vehicle registration for* London (central) • Ulster Unionist

UUA (USA) Unitarian Universalist Association

UUCP *Computing* UNIX to UNIX copy

UUM underwater-to-underwater missile

UUUC United Ulster Unionist Coalition (*or* Council)

UUUP United Ulster Unionist Party

u.u.V. unter üblichem Vorbehalt (German: errors and omissions excepted)

UV *British vehicle registration for* London (central) • ultraviolet

UV-A (*or* **UVA**) *indicating* ultraviolet radiation of wavelength 320–400 nm

UVAS ultraviolet astronomical satellite

uvaser (juːˈveɪzə) ultraviolet amplification by stimulated emission of radiation

UV-B (*or* **UVB**) *indicating* ultraviolet radiation of wavelength 290–320 nm

UVF Ulster Volunteer Force

UVL ultraviolet light

UW *British vehicle registration for* London (central) • underwater • University of Washington • unladen weight

U/W *Law* under will • (*or* **U/w**) underwriter

UWA University of Western Australia

UWC Ulster Workers' Council

UWCE Underwater Weapons and Countermeasures Establishment

UWIST (*or* **Uwist**) (ˈjuːˌwɪst) University of Wales Institute of Science and Technology

UWT Union of Women Teachers (now merged with NAS/UWT)

UWUA Utility Workers Union of America

UWUSA United Workers' Union of South Africa

ux. uxor (Latin: wife)

UX *airline flight code for* Air Europa • *British vehicle registration for* Shrewsbury

UXB unexploded bomb

UY *airline flight code for* Cameroon

Airlines • Universal Youth • *British vehicle registration for* Worcester

Uz. Uzbek(istan)

UZ *British vehicle registration for* Belfast

V

v (ital.) *Optics, symbol for* image distance • (ital.) *Physics, symbol for* instantaneous potential difference • *Physics, symbol for* instantaneous voltage • (ital.) *Chem., symbol for* specific volume • *Spectroscopy, symbol for* variable absorption • (bold ital.) *Physics, symbol for* velocity • (ital.) *Physics, symbol for* a velocity component *or* speed • (ital.) *Chem., symbol for* vibrational quantum number • *Meteorol., symbol for* (abnormally good) visibility

v. vacuum • vagrant • vale • valley • valve • vein • vel (Latin: or) • ventilator • ventral • verb(al) • verse • version • *Printing* verso • versus • vertical • very • via • vicarage • vice (Latin: in place of) • vide (Latin: see) • village • violin • virus • visibility • *Med.* vision • *Grammar* vocative • *Music* voice • volcano • volume • von (German: of; in names) • votre (French: your) • vowel

V (ital.) *Electricity, symbol for* electric potential • *Roman numeral for five* • *(ital.) Physics, symbol for* luminous efficiency • (ital.) *Electricity, symbol for* potential difference • (ital.) *Physics, symbol for* potential energy • *Biochem. symbol for* valine • *Chem., symbol for* vanadium • *Immunol.* variable region (of an immunoglobulin chain) • *international vehicle registration for* Vatican City • verb (in transformational grammar) • Vergeltungswaffe (German: reprisal weapon, as in **V-1** and **V-2**, World War II missiles) • victory (as in **V-Day**, **V-sign**) • *symbol for* volt • (ital.) *symbol for* volume (capacity) • *indicating* types of aircraft (*see* V bomber)

V. (*or* **V**) Venerable • version • Very (in titles) • vespers • Via (Italian: Street) • Vicar • Vice (in titles) • Viscount (*or* Viscountess) • *Music* volti (Italian: turn over) • voltmeter • Volunteer(s)

V1 (*or* **V-1**) Vergeltungswaffe 1 (*see under* V)

V¹ *Music* violino primo (Italian: first violin)

V2 (*or* **V-2**) Vergeltungswaffe 2 (*see under* V)

V² *Music* violino secondo (Italian: second violin)

V3 *international civil aircraft marking for* Belize

V4 *airline flight code for* Venus Airlines

V5 *airline flight code for* Vnukovo Airlines

V8 *international civil aircraft marking for* Brunei

V9 *airline flight code for* Bashkir Airlines

4V *airline flight code for* Voyageur Airways

5V *international civil aircraft marking for* Togo

6V *airline flight code for* Air Vegas • *international civil aircraft marking for* Senegal

9V *international civil aircraft marking for* Singapore

va *Music* viola

v.a. value analysis • verb active • verbal adjective

Va. Virginia (USA)

VA *British vehicle registration for* Peterborough • value-added • *Commerce* value analysis • *Med.* ventricular arrhythmia • (USA) Veterans' Administration • *airline flight code for* VIASA • Vicar Apostolic • Vice-Admiral • (Order of) Victoria and Albert • *US postcode for* Virginia • *Med.* visual acuity • *Radio* Voice of America • Volunteer Artillery • Volunteers of America • Vostra Altezza (Italian: Your Highness) • Votre Altesse (French: Your Highness) • Vuestra Alteza (Spanish: Your Highness)

V/A voucher attached

v.a. & i. verb active and intransitive

VAB *Astronautics* vehicle assembly building (of NASA)

VABF Variety Artists' Benevolent Fund

VABM *Surveying* vertical angle benchmark

vac. vacancy • vacant • vacation • vacuum

vacc. vaccination • vaccine

vac. dist. vacuum distilled (*or* distillation)

vac. pmp vacuum pump

VAD (member of the) Voluntary Aid Detachment

VADAS *Med.* voice-activated domestic appliance system

V-Adm Vice-Admiral

VAF Variety Artists' Federation

VAFC *Computing* VESA advanced feature connector

vag. vagabond • vagina • vagrancy • vagrant

VAH (USA) Veterans' Administration Hospital

val. valley • valuation • value(d)

Val (*or* **val**) *Biochem.* valine

valid. validate • validation

valn valuation

Valpo Valparaiso

VAM *Botany* vesicular-arbuscular mycorrhiza

vamp. vampire

van (væn) *Tennis* advantage

van. vanguard • vanilla

VAN (væn) *Computing* value-added network

Vanc. Vancouver

V&A Victoria and Albert Museum

V & M Virgin and Martyr

V & V *Computing* verification and validation

VAPI *Aeronautics* visual approach path indicator

vapor. vaporization

vap. prf vapour proof

var. variable • variant • variation • *Botany* variety • variometer • various

VAR *Computing* value-added reseller • visual aural range • volunteer air reserve • Votre Altesse Royale (French: Your Royal Highness)

varactor ('vɛə,ræktə) *Electronics* variable reactor

Varig (*or* **VARIG**) ('værɪg) (Empresa de) Viaçäo Aérea Rio Grandense (Brazilian airline)

varistor (və'rɪstə) *Electronics* variable resistor

var. lect. varia lectio (Latin: a variant reading)

varn. varnish

vas. vasectomy

VAS *Med.* ventricular-atrial shunt

vasc. vascular

Vascar (*or* **VASCAR**) ('væskɑː) visual average speed computer and recorder

VASI *Aeronautics* visual approach slope indicator

VASP Viaçäo Aérea Säo Paulo (Brazilian airline)

Vat. Vatican

VAT ('viː 'eɪ 'tiː *or* væt) value-added tax

VATE versatile automatic test equipment

Vat. Lib. Vatican Library

vaud. vaudeville

v. aux. auxiliary verb

VAV variable air volume (in **VAV** (air-conditioning) **system**)

VAWT vertical-axis wind turbine

VAX *Computing, trademark* virtual address extension (range of computers manufactured by DEC)

VAX/VMS *Trademark, indicating* the standard operating system for VAX processors (*see under* VAX; VMS)

vb (*or* **vb.**) verb(al)

v.b. vehicle borne • vertical bomb

VB *airline flight code for* Birmingham European Airways • *British vehicle registration for* Maidstone • *Chem.* valence bond • verbal constituent (in transformational grammar) • volunteer battalion

VBI *Med.* vertebrobasilar insufficiency

vbl verbal

V bomber *indicating* various types of aircraft (named after the types Victor, Vulcan, and Valiant)

VBR *Computing* variable bit rate

vc. *Music* cello (from *violoncello*)

v.c. valuation clause • vehicular communication • visual communication

VC *British vehicle registration for* Coventry • *Engineering* vapour compression • Vatican City • *Cycling* velo club • *Finance* venture capital • (USA) Veterinary Corps • Vice-Chairman • Vice-Chamberlain • Vice-Chancellor • Vice-Consul • Vickers Commercial (aircraft, as in **VC10**) • Victoria Cross • Viet Cong • *Chem.* vinyl chloride • *Med., Physiol.* vital capacity • *indicating* acuity of colour vision

VCA *Chem.* vinyl carbonate • *Microbiol., Med.* viral capsid antigen • (USA) Volunteer Civic Association

VCAS Vice-Chief of the Air Staff

VCC Veteran Car Club of Great Britain • Vice-Chancellors' Committee

VCCS *Electronics* voltage-controlled current source

VCDS Vice-Chief of the Defence Staff

Vce Venice

VCE variable-cycle engine

VCG vertical centre of gravity • Vice-Consul-General

VCGS Vice-Chief of the General Staff

VCH Victoria County History (reference book) • *Chem.* vinyl cyclohexene

v.Chr. vor Christus (German: before Christ; BC)

VCI volatile corrosion inhibitor

vcl. *Music* cello (from *violoncello*) • vehicle

VCM vinyl chloride monomer (a plastic)

VCNS Vice-Chief of the Naval Staff

vcnty vicinity

VCO (India) Viceroy's Commissioned Officer • *Electronics* voltage-controlled oscillator • voluntary county organizer (in the Women's Institute)

VCPI *Computing* virtual control program interface

Vcr Vancouver

VCR video-cassette recorder • visual control room (at an airfield)

VCT venture capital trust

vcs voices

V. Cz. Vera Cruz (Mexico)

vd void

v.d. vapour density • various dates

VD *Med.* vascular dementia • venereal disease • Victorian Decoration • Volunteer Decoration (formerly awarded in the Territorial Army or the Royal Naval Volunteer Reserve)

V-Day Victory Day

Vdc volts direct current

VDC Volunteer Defence Corps

v. def. verb defective

v. dep. verb deponent

VDF *Computing* voice data fax

VDH valvular disease of the heart

VDI *Computing* virtual device interface

VDJ video disc jockey

VDM Verbi Dei Minister (Latin: Minister of the Word of God) • *Computing* Vienna Development Method (notation)

VDQS vin délimité de qualité supérieure (French; superior-quality wine; wine classification)

VDR variable-diameter rotor • video-disc recording

VDRL venereal disease research laboratory (in **VDRL test**, for syphilis)

VDS variable-depth sonar

VDT *Computing* visual display terminal

VDU *Computing* visual display unit

VDW *Chem.* van der Waals (as in **VDW force**)

ve. (*or* v^e) veuve (French: widow)

VE *airline flight code for* Avensa • *British vehicle registration for* Peterborough • *Chem.* valence electron • Victory in Europe (in **VE Day**, 8 May 1945) • vocational education • Vostra Eccellenza (Italian: Your Excellency) • Votre Éminence (French: Your Eminence) • Vuestra Excelencia (Spanish: Your Excellency)

VEB (formerly, in East Germany) Volkseigener Betrieb (German: People's Concern; state-owned company)

vec. vector

ved. vedova (Italian: widow)

veg. vegetable • vegetarian • vegetation

veh. vehicle • vehicular

vel. vellum • velocity • velvet

Vel *Astronomy* Vela

ven. vendredi (French: Friday) • veneer • venerdì (Italian: Friday) • venereal • venery • venison • venom(ous) • ventral • ventricle

Ven. Venerable • (*or* **Venet.**) Venetian • (*or* **Venez.**) Venezuela(n) • Venice • *Astronomy* Venus

vent. ventilate • ventilation • ventriloquist

ver. verification • verify • vermilion • verse • version

Ver. Verein (German: association *or* company)

VER *Meteorol.* vertical

VERA ('vɪərə) versatile reactor assembly • vision electronic recording apparatus

verb. verbessert (German: improved *or* revised)

verb. et lit. *Law* verbatim et literatim (Latin: word for word and letter for letter)

verb. sap. (*or* **sat.**) verbum sapienti satis (Latin: a word is enough to the wise)

verdt verdict

Verf. Verfasser (German: author)

Verl. Verlag (German: publisher)

verm. vermiculite • vermilion

vern. vernacular

vers *Maths.* versed sine (*or* versine)

vers. version

vert. vertebra(l) • vertical • vertigo

Very Rev. (*or* **Very Revd**) Very Reverend

ves. (*or* **vesp.**) *Med.* vespere (Latin: in the evening; in prescriptions) • vessel • vestry

VESA Video Electronics Standards Association

VESPER ('vespə) Voluntary Enterprises and Services and Part-time Employment for the Retired

vet. veteran • (*or* **veter.**) veterinarian • (*or* **veter.**) veterinary

Vet. Admin. (USA) Veterans' Administration

VetMB Bachelor of Veterinary Medicine

vet. sci. veterinary science

vet. surg. veterinary surgeon

v.f. very fair • very fine

VF *British vehicle registration for* Norwich • *airline flight code for* Tropical Airlines • *Med.* ventricular fibrillation • *RC Church* Vicar Forane • video frequency • *Telecom.* voice frequency

VFA (Australia) Victorian Football Association • *Chem.* volatile fatty acid

Vfat *Computing* virtual file allocation table

VFD *Computing* vacuum fluorescent display • verified free distribution • (USA) volunteer fire department

VFL (Australia) Victorian Football League

VFM *Accounting* value for money (audit)

VFO *Electronics* variable-frequency oscillator

VFOAR *US Air Force* Vandenberg Field Office of Aerospace Research

VFR *Aeronautics* visual flight rules

VFT (Australia) very fast train

VFU *Computing* vertical format unit

VFW (USA) Veterans of Foreign Wars

vfy verify

v.g. verbigracia (Spanish: for example) • verbi gratia (Latin: for example) • very good

Vg. *Ecclesiast.* Virgin

VG *British vehicle registration for* Norwich • vaisseau de guerre (French: warship) • very good • Vicar General • *Freemasonry* Vice Grand • Votre Grâce (French: Your Grace) • Votre Grandeur (French: Your Highness)

VGA *Computing* video graphics array (*or* adapter)

v.g.c. very good condition

vgl. vergleiche (German: compare)

VGPI *Aeronautics* visual glide path indicator (in World War II)

VGSOH very good sense of humour (in personal advertisements)

v.h. vertical height • very high

VH *airline flight code for* Air Burkina • *international civil aircraft marking for* Australia • *British vehicle registration for* Huddersfield • Votre Hautesse (French: Your Highness)

v.h.b. very heavy bombardment

VHC (*or* **v.h.c.**) very highly commended • *Chem.* volatile halocarbon compound

VHD video high density (system)

VHDF *Computing* very high-density floppy

VHE very high energy

VHF very high fidelity (in sound recording) • *Radio, etc.* very high frequency

VHLW vitrified high-level (radioactive) waste

VHN *Engineering* Vickers hardness number

VHO very high output

VHP (India) Vishwa Hindu Parishad (militant Hindu group)

VHS *Trademark* Video Home System

VHT very high temperature

VHV very high vacuum

v.i. verb intransitive • vide infra (Latin: see below)

VI Vancouver Island • *Cartography* vertical interval • Virgin Islands • viscosity index • volume indicator

VIA *Computing* versatile interface adapter • Visually Impaired Association

viad. viaduct

VIASA Venezolana Internacional de Aviácion, SA (Venezuelan International Airways)

vib. *Music* vibraphone • vibrate • vibration

VIB vertical integration building

vic. vicar • vicarage • vicinity • victory

Vic. Victoria (Australia)

VIC Victoria Institute of Colleges

Vic. Ap. (*or* **Vic. Apos.**) Vicar Apostolic

Vice-Adm. Vice-Admiral

Vic. Gen. Vicar General

vid. vide (Latin: see) • *Law* vidua (Latin: widow)

VID *Computing* virtual image display

Vien. Vienna

VIF variable import fee

vig. (*or* **vign.**) vignette

vil. village

v. imp. verb impersonal

v. imper. verb imperative

vin. vinegar

VIN (USA) vehicle identification number

vind. vindicate • vindication

vini. viniculture

VIO Veterinary Investigation Officer

VIP *Med.* vasoactive intestinal peptide • very important person

vir. viridis (Latin: green)

Vir *Astronomy* Virgo

Vir. (*or* **Virg.**) Virgil (70–19 BC, Roman poet)

VIR Victoria Imperatrix Regina (Latin: Victoria, Empress and Queen)

Virg. Virginia (USA)

v. irr. verb irregular

vis. viscosity • visibility • visible • visual

Vis. Viscount (*or* Viscountess)

VIS Veterinary Investigation Service • *Meteorol.* visibility

visc. viscosity

Visc. (*or* **Visct.**) Viscount (*or* Viscountess)

VISS *Computing* VHS index search system

VISTA ('vɪstə) Volunteers in Service to America

vit. vitreous

VITA Volunteers for International Technical Assistance

viti. viticulture

vitr. vitreum (Latin: glass)

vit. stat. vital statistics

viv. *Music* vivace (Italian: lively, animated)

vivi. vivisection

vix. vixit (Latin: he *or* she lived)

viz videlicet (Latin: namely; *z* is medieval Latin symbol of contraction)

v.J. vorigen Jahres (German: of last year)

VJ *British vehicle registration for* Gloucester • *airline flight code for* Royal Air Cambodge • (Australia) Vaucluse Junior (yacht) • Victory over Japan (in **VJ Day**, 15 Aug. or 2 Sept. 1945) • video jockey

v.k. vertical keel

VK *British vehicle registration for* Newcastle upon Tyne

vl. violin

v.l. varia lectio (Latin: a variant reading)

VL *British vehicle registration for* Lincoln • Vulgar Latin

vla *Music* viola

VLA *Astronomy* Very Large Array (system of radio telescopes, New Mexico)

Vlad. Vladivostok

VLB vertical-lift bridge • *Computing* VESA local bus

VLBA *Astronomy* very long baseline array

VLBC very large bulk carrier (ship)

VLBI *Astronomy* very long-baseline interferometry

VL-Bus *Computing* VESA's local bus

VLBW *Med.* very low birth weight

VLCC very large crude carrier (oil tanker)

VLCFA *Biochem.* very long-chain fatty acid

VLDL *Biochem.* very low-density lipoprotein

vle *Music* violone (double-bass viol)

VLE *Chem.* vapour–liquid equilibrium

VLF *Radio, etc.* very low frequency

VLFMF very low-frequency magnetic field

VLIW *Computing* very long instruction word

VLLW very low-level (radioactive) waste

vln violin

VLR very long range (aircraft) • Victoria Law Reports

VLS *Physics* vapour–liquid–solid (as in **VLS system**)

VLSI *Electronics* very large-scale integration

VLT very large telescope

vltg. voltage

vlv. valve • valvular

v.M. vorigen Monats (German: last month)

VM *British vehicle registration for* Manchester • *airline flight code for* Regional Airlines • *Physics* velocity modulation • Victory Medal • Viet Minh • Virgin Mary • *Computing* virtual machine (*or* memory) • *Chem.* volatile matter • Votre Majesté (French: Your Majesty)

V-Mail *Military* (USA) victory mail

VMC *Computing* VESA medium channel • *Aeronautics* visual meteorological conditions

VMCCA Veteran Motor Car Club of America

VM/CMS *Computing, trademark* virtual machine, conversational monitor system

VMD Doctor of Veterinary Medicine (Latin *Veterinariae Medicinae Doctor*)

VME *Computing, trademark* Versa Module Eurocard (in **VME bus**)

VMH Victoria Medal of Honour (awarded by the Royal Horticultural Society)

VMO *Astronomy* very massive object

VMS vertical marketing system • *Computing* virtual machine system (*see also* VAX/VMS) • Voluntary Medical Services

v.m.t. very many thanks

vn violin

v.n. verb neuter

VN *British vehicle registration for* Middlesbrough • Vietnam (abbrev. *or* IVR) • *airline flight code for* Vietnam Airlines • Vietnamese

Vna Vienna

VNM (Canada) Victoria National Museum

VNTR *Genetics* variable number tandem repeat

vo. verso

VO *British vehicle registration for* Nottingham • *airline flight code for* Tyrolean Airways • valuation officer • verbal order • very old (of brandy, whisky, etc.) • Veterinary Officer • (Royal) Victorian Order • voice-over (commentary)

VOA *Radio* Voice of America • Volunteers of America

voc. vocal(ist) • vocation • *Grammar* vocative

VOC Vehicle Observer Corps • volatile organic compound (*or* chemical)

vocab. vocabulary

VOCAL ('vəuk³l) Voluntary Organizations Communication and Language

vocat. *Grammar* vocative

voc-ed (ˌvɒk'ɛd) vocational education

voctl vocational

VOD velocity of detonation

VODAT voice-operated device for automatic transmission

vol. volatile • volcanic • volcano • (*or* **Vol.**) volume • voluntary • (*or* **Vol.**) volunteer

Vol *Astronomy* Volans

volc. volcanic • volcano

vols (*or* **Vols**) volumes

volum. *Chem.* volumetric

voly voluntary

VONA vehicle of the new age (rapid transport shuttle)

VOP very oldest procurable (of brandy, port, etc.)

VOR very-high-frequency omnirange *or* omnidirectional radio range (navigation aid)

vorm. vormals (German: formerly) • vormittags (German: in the morning; a.m.)

Vors. Vorsitzender (German: chairman)

vou. voucher

vox pop (vɒks pɒp) vox populi (Latin: voice of the people)

v.p. *Photog.* vanishing point • vapour pressure • variable pitch • verb passive

VP *British vehicle registration for* Birmingham • *international civil aircraft marking for* United Kingdom Colonies and Protectorates • *Chem., Physics* vapour phase • *Building trades* vent pipe • verb phrase (in transformational grammar) • Vice-President • Vice-Principal • victory points • vita patris (Latin: during the life of his (*or* her) father)

VPC *Chem.* vapour-phase chromatography • vente par correspondence (French: mail order)

v.p.d. vehicles per day

VPE *Electronics* vapour-phase epitaxy

v.ph. vertical photography

v.p.h. vehicles per hour

VPI *Chem.* vapour-phase inhibitor

VPL visible panty line

v.p.m. vehicles per mile

Vpo Valparaiso

VPO Vienna Philharmonic Orchestra

VPop *Computing* virtual point of presence

VPP (India) value payable post • *Vet. science* virus pneumonia of pigs • Volunteer Political Party

V.Pres. Vice-President

VPRGS Vice-President of the Royal Geographical Society

VPRP Vice-President of the Royal Society of Portrait Painters

VPRS Vice-President of the Royal Society

v.p.s. vibrations per second

VPZS Vice-President of the Zoological Society

VQ *international civil aircraft marking for* United Kingdom Colonies and Protectorates

VQMG Vice-Quartermaster-General

v.r. variant reading • vedi retro (Italian: please turn over; PTO) • verb reflexive

VR *British vehicle registration for* Manchester • *airline flight code for* TACV-Cabo Verde Airlines • *international civil aircraft marking for* United Kingdom Colonies and Protectorates • variant reading • *Physics* velocity ratio • *Med.* ventricular reflux • Vicar Rural • Victoria Regina (Latin: Queen Victoria) • *Computing* virtual reality • voltage regulator • Volunteer Reserve • vulcanized rubber

VRA (USA) Vocational Rehabilitation Administration

VRAM *Computing* video random access memory

VRB *Meteorol.* variable

VRC Vehicle Research Corporation • Volunteer Rifle Corps

VRD Royal Naval Volunteer Reserve Officers' Decoration

v. refl. verb reflexive

V. Rev. Very Reverend

VRG *Biochem.* vaccinia rabies glycoprotein

VRI Victoria Regina et Imperatrix (Latin: Victoria, Queen and Empress) • *Aeronautics* visual rule instrument (landing)

VRM (USA) variable rate mortgage

VRN *Finance* variable-rate note

VRO vehicle registration office

Vry Viceroy

vs. versus

v.s. variable speed • vide supra (Latin: see above) • *Music* volti subito (Italian: turn over quickly)

VS *British vehicle registration for* Luton • *Freemasonry* Venerable Sage • Veterinary Surgeon • vieux style (French: Old Style; method of reckoning dates) • *airline flight code for* Virgin Atlantic Airways • *Chem.* volatile solid • *Music* volti subito (Italian: turn over quickly) • *Chem.* volumetric solution • Vostra Santità (Italian: Your Holiness) • Votre Sainteté (French: Your Holiness)

VSAM *Computing* virtual storage access method

VSB *Telecom.* vestigial sideband

vsby visibility

VSC Volunteer Staff Corps

VSCC Vintage Sports Car Club

VSD vendor's shipping document • *Med.* ventricular septal defect

VSG *Immunol.* variable (cell-)surface glycoprotein

VSI *Navigation* vertical speed indicator

V-sign victory sign

VSL venture scout leader

VSM *Telecom.* vestigial sideband modulation

VSO *Linguistics* verb-subject-object (as in **VSO language**) • *Computing* very small outline • very superior old (of brandy, port, etc.) • Vienna State Opera • Voluntary Service Overseas

VSOP very special (*or* superior) old pale (of brandy, port, etc.)

VSP *Meteorol.* vertical speed

VSR very short range • very special reserve (of wine)

vst violinist

V/STOL (*or* **V-STOL**, **VSTOL**) ('vi:ˌstɒl) vertical and short takeoff and landing (aircraft)

VSV *Med.* vesicular stomatitis virus

VSW vitrified stoneware

VSWR *Telecom.* voltage standing-wave ratio

v.t. vacuum technology • variable transmission • verb transitive

Vt Vermont

VT *airline flight code for* Air Tahiti • *international civil aircraft marking for* India • *British vehicle registration for* Stoke-on-Trent • variable time (as in **VT fuse**) • *symbol for* vatu (monetary unit of Vanuatu) • *Med.* ventricular tachycardia • *US postcode for* Vermont

VTC Volunteer Training Corps • voting trust certificate

Vte Vicomte (French: Viscount)

VTE *Psychol.* vicarious trial and error

Vtesse Vicomtesse (French: Viscountess)

VTFL *Computing* variable-to-fixed-length (in **VTFL code**)

vtg voting

VTL *Computing* variable threshold logic

VTMoV *Microbiol.* velvet tobacco mottle virus

VTO vertical takeoff (aircraft)

VTOHL vertical takeoff, horizontal landing (aircraft)

VTOL ('vi:ˌtɒl) vertical takeoff and landing (aircraft)

VTOVL vertical takeoff, vertical landing (aircraft)

VTR videotape recorder

VTVL *Computing* variable-to-variable length (in **VTVL code**)

VU *airline flight code for* Air Ivoire • *British vehicle registration for* Manchester • *Acoustics* volume unit (as in **VU meter**)

Vul *Astronomy* Vulpecula

Vul. (*or* **Vulg.**) Vulgate

vulg. vulgar(ly)

vv. verbs • verses • *Music* (first and second) violins • *Music* voices • volumes

v.v. vice versa • viva voce

v/v *Chem.* volume in volume

VV *British vehicle registration for* Northampton

VVD (Netherlands) Volkspartij voor Vrijheid en Democratie (People's Party for Freedom and Democracy)

vve veuve (French: widow)

vv.ll. variae lectiones (Latin: variant readings)

VV.MM. Vos Majestés (French: Your Majesties)

VVO very very old (of brandy, port, etc.)

VW British vehicle registration for Chelmsford • Computing van Wijngaarden (in

VW-grammar) • Very Worshipful • Volkswagen

VWH Vale of the White Horse

vx vertex

VX British vehicle registration for Chelmsford

vy very

v.y. Bibliog. various years

VY British vehicle registration for Leeds

VZ British vehicle registration for Tyrone

VZIG Immunol. varicella zoster hyper-immune globulin

W

w Meteorol., symbol for dew • (ital.) Chem., symbol for mass fraction • (ital.) Physics, symbol for a velocity component • Spectroscopy, symbol for weak absorption

w. war • warm • water • weather • week • weight • wet • white • Cricket wicket • Cricket wide • width • wife • win • wind • wire • with • won • word • work

W (ital.) Engineering, symbol for load • Biochem., symbol for tryptophan • Chem., symbol for tungsten (formerly wolfram) • Colloquial water closet; WC • Irish fishing port registration for Waterford • symbol for watt • (ital.) Physics, symbol for weight • Genetics Weigle (in **W reactivation**) • west(ern) • postcode for west London • Shipping winter loading (on load line) • women's (clothing size) • symbol for won (South Korean monetary unit) • (ital.) Physics, symbol for work • (ital.) Biochem., symbol for writhing number • Genetics, indicating a sex chromosome in birds and some insects • Physics, indicating a type of fundamental particle

W. (or **W**) Wales • Warden • Wednesday • Welsh • Wesleyan • west(ern) • white • wide • widow(er) • widowed

W2 airline flight code for World Airlines

W3 airline flight code for Swiftair

W7 airline flight code for Western Pacific Airlines

2W airline flight code for Wairarapa Airlines

4W international civil aircraft marking for Yemen

5W airline flight code for Interline Aviation • international civil aircraft marking for Samoa

6W international civil aircraft marking for Senegal • airline flight code for Wilderness Airline

7W airline flight code for Air Sask Aviation

9W airline flight code for Jet Airways (India)

w.a. with answers

WA airline flight code for Newair • British vehicle registration for Sheffield • postcode for Warrington • US postcode for Washington (state) • (USA) Welfare Administraion • West Africa(n) • Western Australia • Westminster Abbey • Banking (USA) withholding agent • British fishing port registration for Whitehaven • Netball wing attack • Insurance with average • Woodworkers of America (trade union)

WAA Women's Auxiliary Association

WAAA Women's Amateur Athletic Association

WAAAF (wæf) Women's Auxiliary Australian Air Force

WAAC (wæk) Women's Army Auxiliary Corps

WAAE World Association for Adult Education

WAAF (wæf) Women's Auxiliary Air Force • Women's Auxiliary Australian Air Force

WAAS Women's Auxiliary Army Service ·
World Academy of Art and Science

WAC (wæk) (USA) Women's Army Corps
(in World War II) · World Aeronautical
Chart

WACC *Accounting* weighted average cost of
capital · World Association for Christian
Communications

WACCC Worldwide Air Cargo Commodity Classification

WACSM (USA) Women's Army Corps
Service Medal

Wad (*or* **Wadh**) Wadham College, Oxford

WADEX (*or* **Wadex**) ('wɒdɛks) word and
author index

WADF (USA) Western Air Defense

WADS wide-area data service

w.a.e. when actually employed

WAF West African Forces · (*or* **w.a.f.**)
with all faults · (USA) Women in the Air
Force

WAFC West African Fisheries Commission (of the FAO)

W. Afr. West Africa(n)

WAFS (USA) Women's Auxiliary Ferrying
Squadron

WAG *international vehicle registration for*
(West Africa) Gambia · Writers' Action
Group

WAGBI Wildfowl Association of Great
Britain and Ireland

WAGGGS World Association of Girl
Guides and Girl Scouts

WAHUHA Waugh, Huber, and Haberlen
(as in **WAHUHA sequence**)

WAIF World Adoption International Fund

WAIS *Psychol.* Wechsler Adult Intelligence
Scale · *Computing* wide-area information
service

WAIS-R *Psychol.* Wechsler Adult Intelligence Scale – Revised

Wal. Walloon

WAL *international vehicle registration for*
(West Africa) Sierra Leone

WAM work analysis and measurement ·
(USA) wrap around mortgage

WAN (wæn) *Computing* wide-area network

Wand. *Football* Wanderers

w & i weighing and inspection

W & L (USA) Washington and Lee University

W & M William and Mary (English joint
sovereigns 1689–1702 and 1689–94,
respectively)

W & S whisky and soda

W & T *Taxation* wear and tear

WANS Women's Australian National
Service

WAOS Welsh Agricultural Organization
Society

WAP work assignment plan (*or* procedure)

WAPC Women's Auxiliary Police Corps

WAPOR World Association for Public
Opinion Research

war. warrant

War. Warsaw · Warwickshire

WAR West Africa Regiment

WARC (USA) Western Air Rescue
Center · World Administrative Radio
Conference · World Alliance of
Reformed Churches

WARI (Australia) Waite Agricultural
Research Institute

Warks Warwickshire

warn. warning

warr. (*or* **warrty**) warranty

WASA Welsh Amateur Swimming Association

Wash. Washington

WASP (wɒsp) (*or* **Wasp**) (USA) white
Anglo-Saxon Protestant · (USA) Women
Airforce Service Pilots

WAST Western Australia Standard Time

Wat. Waterford

WAT *Aeronautics* weight, altitude, temperature (as in **WAT curves**) · *Psychol.*
word association test

WATA World Association of Travel
Agencies

WATFOR ('wɒtfɔː) *Computing* University
of Waterloo Fortran (Fortran compiler)

WATS (USA) Wide Area Telephone
Service

W. Aus. (*or* **W. Aust.**) Western Australia

WAV *Computing* waveform

WAVES (*or* **Waves**) (weɪvz) *US Navy*
Women Accepted for Volunteer Emergency Service

WAWF World Association of World Federalists

WAY World Assembly of Youth

WAYC Welsh Association of Youth Clubs

w.b. wage board · waste ballast · water
ballast · waybill · westbound · wheel
base · *Knitting* wool back

Wb *Magnetism, symbol for* weber

WB *British vehicle registration for* Sheffield ·
(*or* **W/B**) warehouse book · Warner
Brothers (Pictures, Incorporated) · water

board • (*or* **W/B**) *Telecom.* waveband • (*or* **W/B**) *Commerce* waybill • weather bureau • *Psychol.* Wechsler–Bellevue (Intelligence Scale) • *Insurance* weekly benefits • World Bank

WBA West Bromwich Albion (Football Club) • World Boxing Association

WBAFC (USA) Weather Bureau Area Forecast Center

WBAN (USA) Weather Bureau, Air Force and Navy

WBC *Med.* white blood cell • *Med.* white blood (cell) count • World Boxing Council

WBF World Bridge Federation

WBGT *Meteorol.* wet-bulb globe temperature (*or* thermometer)

w.b.i. will be issued

WBI *Med.* whole-body irradiation

WbN west by north

w.b.s. *Engineering* walking-beam suspension • *Insurance* without benefit of salvage

WbS west by south

WBS *Med.* whole-body scan • *Physics* wide-band spectrometer (*or* spectrum)

WBT *Meteorol.* wet-bulb temperature

w.c. watch committee • water closet • water cock • without charge

WC *British vehicle registration for* Chelmsford • *airline flight code for* Islena Airlines • war cabinet • war council • water closet • Wesleyan chapel • *postcode for* west central London • *Military* Western Command • Whitley Council • working capital • *Computing* world coordinates

W/C Wing Commander

WCA Wholesale Confectioners' Alliance • Wildlife and Countryside Act • Women's Christian Association

WCAT Welsh College of Advanced Technology

WCC War Crimes Commission • World Council of Churches

W/Cdr (*or* **W. Cdr**) Wing Commander

WCEU World Christian Endeavour Union

WCF World Congress of Faiths

WCL World Confederation of Labour

WCP World Climate Programme • World Council of Peace

WCRA (USA) Weather Control Research Association • Women's Cycle Racing Association

WCT World Championship Tennis

WCTU (USA, Canada) Women's Christian Temperance Union

WCWB World Council for the Welfare of the Blind

wd ward • warranted • weed • wood • word • would • wound

w/d warranted • well developed

WD *airline flight code for* Halisa Air • *international vehicle registration for* (Windward Islands) Dominica • *British vehicle registration for* Dudley • War Department • *postcode for* Watford • *Irish fishing port registration for* Wexford • *Astronomy* white dwarf • *Netball* wing defence • Works Department

4WD (USA) four-wheel drive

W/D *Banking* withdrawal

WDA Welsh Development Agency • *Taxation* writing-down allowance

WDC War Damage Commission • (USA) War Damage Corporation • Woman Detective Constable • Women's Diocesan Association • World Data Centre

w.d.f. wood door and frame

wdg winding

WDM *Telecom.* wavelength division multiplex

Wdr *Navy* Wardmaster

WDS Woman Detective Sergeant

wd sc. wood screw

WDSPR *Meteorol.* widespread

wdth width

WDV *Taxation* written-down value

w/e weekend • week ending

WE *British vehicle registration for* Sheffield

wea. weapon • weather

WEA Royal West of England Academy • Workers' Educational Association

WE&FA Welsh Engineers' and Founders' Association

WEARCON ('weə,kɒn) weather observation and forecasting control system

WEC wave (*or* wind) energy converter • World Energy Conference

WECOM ('wekɒm) (USA) Weapons Command

Wed. Wednesday

WEDA Wholesale Engineering Distributors' Association

Weds. Wednesday

w.e.f. with effect from

WEFC West European Fisheries Conference

WEFT (weft) *Aeronautics* wings, engine, fuselage, tail

Wel. Welsh

weld. welding

Well. Wellington (New Zealand)

WEN (*or* **Wen**) (wen) Women's Environmental Network

Wes. Wesleyan

WES Women's Engineering Society • World Economic Survey

WES/PNEU Worldwide Education Service of Parents' National Educational Union

west. western

Westm. Westmeath • Westminster • Westmorland (former English county)

WET West(ern) European Time

WETUC Workers' Educational Trade Union Committee

WEU Western European Union (for defence policy)

Wex. (*or* **Wexf.**) Wexford

wf (*or* **w.f.**) *Printing* wrong fount

WF *British vehicle registration for* Sheffield • *postcode for* Wakefield • *Physics* wave function • (USA) Wells Fargo and Company • (USA) white female • *Sport* wing forward

WFA White Fish Authority • Women's Football Association • World Friendship Association

w factor *Psychol.* will factor

WFB World Fellowship of Buddhists

WFC World Food Council

w.fd *Knitting* wool forward

WFD World Federation of the Deaf

WFDY World Federation of Democratic Youth

WFEO World Federation of Engineering Organizations

wff *Logic* well-formed formula

WFF World Friendship Federation

WFGA Women's Farm and Garden Association

Wfl Worshipful

WFL Women's Freedom League

WFMH World Federation for Mental Health

WFMW World Federation of Methodist Women

WFN World Federation of Neurology

WFP World Food Programme (of the FAO)

WFPA World Federation for the Protection of Animals

WFSW World Federation of Scientific Workers

WFTU World Federation of Trade Unions

WFUNA World Federation of United Nations Associations

w.fwd *Knitting* wool forward

wg weighing • wing

WG *international vehicle registration for* (Windward Islands) Grenada • *British vehicle registration for* Sheffield • (*or* **w.g.**) water gauge • (*or* **w.g.**) weight guaranteed • Welsh Guards • West German(y) • West Germanic (language group) • *Colloquial* W(illiam) G(ilbert) Grace (1848–1915, English cricketer) • (*or* **w.g.**) wire gauge • Working Group

WGA Writers' Guild of America

WGC Welwyn Garden City • World Gold Council • *Freemasonry* Worthy Grand Chaplain

Wg/Cdr (*or* **Wg Comdr**) Wing Commander

W. Ger. West German(y) • West Germanic (language group)

WGG *Freemasonry* Worthy Grand Guardian (*or* Guide)

WGGB Writers' Guild of Great Britain

WGI world geophysical interval

w. gl. wired glass

W. Glam. West Glamorgan

WGM *Freemasonry* Worthy Grand Master

WGmc West Germanic (language group)

WGPMS warehouse gross performance measurement system

WGS *Chem.* water-gas shift (reaction) • *Freemasonry* Worthy Grand Sentinel

WGU Welsh Golfing Union

wh. wharf • which • whispered • white

W h *Electricity, symbol for* watt hour(s)

WH *airline flight code for* China Northwest Airlines • *British vehicle registration for* Manchester • water heater • *British fishing port registration for* Weymouth • *Cycling* wheelers • (USA) White House • *Sport* wing half • (*or* **w/h**) *Banking* withholding

WHA World Health Assembly (of WHO) • World Hockey Association

W'hampton Wolverhampton

w.h.b. wash-hand basin

whd warhead

whf. wharf

WhF Whitworth Fellow

whfg. wharfage

whfr wharfinger

Whi. Whitehall (London)

whis. whistle

whmstr weighmaster

WHO (USA) White House Office • World Health Organization

WHOA! (wəu) (USA) Wild Horse Organized Assistance

WHOI (USA) Woods Hole Oceanographic Institution

whr whether

WHRA World Health Research Centre

whs. (or **whse**) warehouse

WhSch Whitworth Scholar

whsle wholesale

whsng warehousing

whs. rec. warehouse receipt

whs. stk warehouse stock

WHT William Herschel Telescope (La Palma)

WHTSO Welsh Health Technical Services Organization

w.i. *Finance* when issued

WI *Irish vehicle registration for* Waterford • West Indian • West Indies • Windward Islands • *British fishing port registration for* Wisbech • *US postcode for* Wisconsin • Women's Institute • wrought iron

WIA wounded in action

WIBC (USA) Women's International Bowling Congress

WICA Warsaw International Consumer Association

Wich. *Horticulture* Wichuraiana (rose)

Wick. Wicklow

wid. widow(er)

WID West India Docks (London)

WIDF Women's International Democratic Federation

WIF West Indies Federation

Wig. Wigtown (Scotland) • Wigtownshire (former Scottish county)

Wigorn. Wigorniensis (Latin: (Bishop) of Worcester)

wilco ('wɪl,kəu) will comply (in radio communications, etc.)

WILPF Women's International League for Peace and Freedom

Wilts Wiltshire

WIMP (wɪmp) *Physics* weakly interacting massive particle • (or **wimp**) *Computing* windows, icons, menus, and pointers

WIN (wɪn) (USA) Work Incentive

Winch. Winchester

W. Ind. West Indian • West Indies

Wind. I. Windward Islands

Wing Cdr Wing Commander

Wings (wɪŋz) *Finance* warrants in negotiable government securities

Winn. Winnipeg

wint. winter

Winton. Wintoniensis (Latin: (Bishop) of Winchester)

WIP waste incineration plant • work in progress

WIPO (or **Wipo**) ('waɪpəu) World Intellectual Property Organization

WIRA (or **Wira**) Wool Industries Research Association

WIRDS weather information reporting and display system

Wis. Wisconsin

WISC *Psychol.* Wechsler Intelligence Scale for Children

WISC-R *Psychol.* Wechsler Intelligence Scale for Children – Revised

Wisd. *Old Testament* Wisdom of Solomon

WISP (wɪsp) wide-range imaging spectrometer

wit. witness

WITA Women's International Tennis Association

withdrl withdrawal

witht without

Wits. Witwatersrand

WIZO (or **Wizo**) ('waɪtzəu) Women's International Zionist Organization

WJ *airline flight code for* Labrador Airways • *British vehicle registration for* Sheffield

WJC World Jewish Congress

WJEC Welsh Joint Education Committee

wk weak • week • work • wreck

w.k. warehouse keeper • well-known

Wk Walk (in place names)

WK *British vehicle registration for* Coventry • *British fishing port registration for* Wick

WKB *Physics* Wentzel–Kramers–Brillouin (as in **WKB solution** of the Schrödinger equation)

wkds weekdays

wkg working

wkly weekly

WKN *Meteorol.* weaken(ing)

wkr worker • wrecker

wks weeks • works

wks. workshop

wkt *Cricket* wicket

wl wool

WL *British vehicle registration for* Oxford • *international vehicle registration for* (Windward Islands) St Lucia • wagon-lit

(French: sleeping car) • waiting list • (*or*
w.l.) water line • (*or* **W/L**) wavelength •
West Lothian • Women's Liberation

WLA Women's Land Army

WLAN *Computing* wireless local area
network

WLB (USA) War Labor Board

wld would

Wld Ch. World Championship

wldr welder

WLF Women's Liberal Federation

wl fwd *Knitting* wool forward

WLGS Women's Local Government
Society

WLHB Women's League of Health and
Beauty

WLI workload index

WLM Women's Liberation Movement

Wln Wellington (New Zealand)

WLN Wiswesser line notation

W. long. west longitude

W. Loth. West Lothian

WLPSA Wild Life Preservation Society of
Australia

WLR Weekly Law Reports

WLRI World Life Research Institute

WLS *Statistics* weighted least squares

WLTBU Watermen, Lightermen,
Tugmen, and Bargemen's Union

WLTM would like to meet (in personal
advertisements)

WLU World Liberal Union

WLUS World Land Use Survey

wly westerly

wlz waltz

Wm. William

WM *British vehicle registration for* Liverpool • war memorial • wattmeter • (USA)
white male • *airline flight code for*
Windward Island Airways • wire mesh •
Freemasonry Worshipful Master

W/M (*or* **w/m**) *Shipping* weight or measurement

WMA Working Mothers' Association •
World Medical Association

WMAA Whitney Museum of American Art

WMC (USA) War Manpower Commission
(in World War II) • Ways and Means
Committee • working men's club •
Working Men's College • World
Meteorological Centre • World
Methodist Council

WMCIU Working Men's Club and Institute Union Ltd

WMF *Trademark* Windows metafile format

wmk watermark

WMM World Movement of Mothers

WMO World Meteorological Organization

WMP with much pleasure

WMS Wesleyan Missionary Society •
World Magnetic Survey

WMTC Women's Mechanized Transport
Corps

WN *airline flight code for* Southwest
Airlines • *British vehicle registration for*
Swansea • *postcode for* Wigan • *British
fishing port registration for* Wigtown

WNA *Shipping* winter North Atlantic
loading (on load line)

w.n.d.p. with no down payment

WNE Welsh National Eisteddfod

WNL within normal limits

WNLF Women's National Liberal Federation

WNO Welsh National Opera

WNP Welsh Nationalist Party

WNW west-northwest

w.o. walkover • wie oben (German: as
mentioned above) • *Commerce* written
order

w/o without • *Accounting* written off

WO *British vehicle registration for* Cardiff •
walkover • War Office • Warrant Officer •
welfare officer • wireless operator • *British
fishing port registration for* Workington •
airline flight code for World Airways •
written order

w.o.a. without answers

WOA Wharf Owners' Association

WOAR Women Organized Against Rape

w.o.b. washed overboard

w.o.c. without compensation

WOCA ('wəʊkə) world outside centrally
planned economic areas

WOC(S) *Building trades* waiting on cement
(to set)

w.o.e. without equipment

WOF (New Zealand) Warrant of Fitness
(for vehicles)

W/offr welfare officer

w.o.g. water, oil, or gas • with other goods

w.o.l. wharf-owner's liability

Wolfs. Wolfson College, Oxford

WOMAN ('wʊmən) World Organization
for Mothers of All Nations

w.o.n. *Knitting* wool on needle

WOO *Navy* Warrant Ordnance Officer •
World Oceanographic Organization

w.o.p. with other property • without personnel

w.o.p.e. without personnel or equipment

Wor. Worshipful

WOR without our responsibility

Worc. Worcester College, Oxford

WORC (USA) Washington Operations Research Council

Worcs Worcestershire

WORM (*or* **worm**) (wɜːm) *Computing* write once, read many (times)

WOSAC ('wəʊsæk) worldwide synchronization of atomic clocks

WOSB ('wɒzbɪ) War Office Selection Board

WOW waiting on weather • Women Against the Ordination of Women

wp. *Baseball* wild pitch(es)

w.p. waste paper • waste pipe • weather permitting • will proceed • *Law* without prejudice • word processing (*or* processor)

Wp. Worship • Worshipful

WP *airline flight code for* Aloha Islandair • weather permitting • Western Province (South Africa) • (USA) West Point (military academy) • *Government* White Paper • (USA) wire payment • *Law* without prejudice • *British vehicle registration for* Worcester • word processing (*or* processor) • working paper • working party • working pressure • *Freemasonry* Worthy Patriarch • *Freemasonry* Worthy President

WPA Water Polo Association • Western Provident Association • *Insurance* with particular average • (USA) Work Projects (originally Progress) Administration (1935–43) • World Parliament Association • World Pool-Billiard Association • World Presbyterian Alliance

WPB (USA) War Production Board (in World War II) • (*or* **w.p.b.**) wastepaper basket

WPBL (USA) Women's Professional Basketball League

WPBSA World Professional Billiards and Snooker Association

WPC War Pension(s) Committee • Woman Police Constable • wood–plastic (*or* –polymer) composite • World Petroleum Congress

WPCA (USA) Water Pollution Control Administration

WPCF (USA) Water Pollution Control Federation

WPE white porcelain enamel

WPESS within pulse electronic sector scanning (in **WPESS sonar**)

WPFC West Pacific Fisheries Commission

Wpfl Worshipful

wpg waterproofing

WPG *Computing, trademark* WordPerfect graphic

WPGA Women's Professional Golfers' Association

WPHC Western Pacific High Commission

WPI wholesale price index • World Press Institute

WPL warning-point level

w.p.m. words per minute

WPMSF World Professional Marathon Swimming Federation

wpn weapon

WPPSI *Psychol.* Wechsler Preschool and Primary Scale of Intelligence

WPRL Water Pollution Research Laboratory

w.p.s. with prior service

WPT Women's Playhouse Trust

w.r. warehouse receipt • *Insurance* war risk • water repellent

WR *British vehicle registration for* Leeds • *airline flight code for* Royal Tongan Airlines • ward room • warehouse receipt • war reserve • *Med.* Wassermann reaction • *Railways* Western Region • (formerly) West Riding (Yorkshire) • Willelmus Rex (Latin: King William) • (*or* **W–R**) *Astronomy* Wolf–Rayet (in **WR stars**) • *postcode for* Worcester

WRA (USA) War Relocation Authority • Water Research Association • *Horticulture* Wisley Rose Award

WRAAC (ræk) Women's Royal Australian Army Corps

WRAAF (ræf) Women's Royal Australian Air Force

WRAC (ræk) Women's Royal Army Corps

WRAF (ræf) Women's Royal Air Force

WRAM *Computing* Windows random-access memory

WRANS (rænz) Women's Royal Australian Naval Service

WRC Water Research Centre • (USA) Welding Research Council

WRE (Australia) Weapons Research Establishment

w. ref. with reference (to)

WRI war risks insurance • Women's Rural Institute

w.r.n. *Knitting* wool round needle

WRNG *Meteorol.* warning

WRNR Women's Royal Naval Reserve

WRNS Women's Royal Naval Service

wrnt warrant

w.r.o. *Insurance* war risks only

WRO Weed Research Organization (of the AFRC)

WRP Worker's Revolutionary Party

WRRA *Cycling* Women's Road Records Association

wrt wrought (of iron)

w.r.t. with respect to

WRU Welsh Rugby Union • Wesleyan Reform Union

WRVS Women's Royal Voluntary Service (formerly WVS)

WS *British vehicle registration for* Bristol • *postcode for* Walsall • water-soluble • weapon system • Weinberg-Salam (in **WS model**) • *international vehicle registration for* Western Samoa • West Saxon • wind speed • *Law* (Scotland) Writer to the Signet

WSA (USA) War Shipping Administration

W. Sam. Western Samoa

WSC World Series Cricket

WSCF World Student Christian Federation

WSF Wake Shield Facility • *Chem.* water-soluble fraction

WSI Writers and Scholars International

WSJ (ital.) (USA) Wall Street Journal

WSL Warren Spring Laboratory (Stevenage, Herts)

WSM Women's Suffrage Movement

WSP water supply point

WSPU Women's Social and Political Union

WSRT Westerbork Synthesis Radio Telescope

WSSA Welsh Secondary Schools Association

WSTN World Service Television News

WSTV World Service Television

WSU Wichita State University

WSW west-southwest

wt warrant • watertight • weight • without

WT *British vehicle registration for* Leeds • *airline flight code for* Nigeria Airways • *Navy* warrant telegraphist • war transport • *Irish fishing port registration for*

Westport • (*or* **W/T**) wireless telegraphy (*or* telephony) • withholding tax

WTA winner takes all • Women's Tennis Association • World Transport Agency

WTAA World Trade Alliance Association

WTAU Women's Total Abstinence Union

WTC Wheat Trade Convention

wtd warranted

Wtf. Waterford

WTG wind turbine generator

WTH whole-tree harvesting

wthr weather

WTIS World Trade Information Service

WTMH watertight manhole

WTN Worldwide Television News

WTO Warsaw Treaty Organization • World Tourism Organization • World Trade Organization

WTP willing(ness) to pay

wtr winter • writer

WTRC Wool Textile Research Council

WTS Women's Transport Service (now amalgamated with FANY)

WTT World Team Tennis

WTTA Wholesale Tobacco Trade Association of Great Britain and Northern Ireland

WTTC World Travel and Tourism Council

WTUC World Trade Union Conference

WTUL Women's Trade Union League

WU *British vehicle registration for* Leeds • Western Union

WUCT World Union of Catholic Teachers

WUF World Underwater Federation

WUJS World Union of Jewish Students

WUPJ World Union for Progressive Judaism

WUR World University Round Table

WUS World University Service

WUSL Women's United Service League

w/v *Chem.* weight in volume (of solution)

WV *airline flight code for* Air South • *British vehicle registration for* Brighton • *international vehicle registration for* (Windward Islands) St Vincent • water valve • *US postcode for* West Virginia • *postcode for* Wolverhampton

W. Va. West Virginia

WVA World Veterinary Association

wvd waived

WVD Wereldverbond van Diamant

Bewerkers (Dutch: Universal Alliance of Diamond Workers)

WVF World Veterans' Federation

WVS Women's Voluntary Service (*see* WRVS)

WVT water vapour transfer (*or* transmission)

WVU West Virginia University

w/w weight for weight • *Chem.* weight in weight

WW *British vehicle registration for* Leeds • (*or* **W/W**) wall-to-wall (in estate agency) • (*or* **W/W**) warehouse warrant • warrant writer • (ital.) Who's Who • World War (*see* WW1, WW2) • worldwide

WW1 (*or* **WWI**) World War One (1914–18)

WW2 (*or* **WWII**) World War Two (1939–45)

WWDC World War Debt Commission

WWDSHEX *Shipping* weather working days, Sundays, and holidays excluded

Wwe Witwe (German: widow)

WWF Worldwide Fund for Nature (formerly World Wildlife Fund) • World Wrestling Federation

WWMCCS World Wide Military Command and Control System

WWO Wing Warrant Officer

WWSSN worldwide standard seismograph network

WWSU World Water Ski Union

WWW (ital.) Who Was Who • World Weather Watch (of the WMO) • World Wide Web

WWY Queen's Own Warwickshire and Worcestershire Yeomanry

WX *British vehicle registration for* Leeds • *Meteorol.* weather • women's extra-large (clothing size)

Wy. Wycliffe (*see* Wycl.) • Wyoming

WY *British vehicle registration for* Leeds • *airline flight code for* Oman Aviation • *British fishing port registration for* Whitby • *US postcode for* Wyoming

Wycl. John Wycliffe (*c.* 1330–84, English religious reformer)

Wyo. Wyoming

WYR West Yorkshire Regiment

wysiwyg (*or* **WYSIWYG**) ('wɪzɪˌwɪg) *Computing* what you see (on the screen) is what you get (from the printer)

Wz. Warenzeichen (German: trademark)

WZ *British vehicle registration for* Belfast

WZO World Zionist Organization

X

x *Genetics, symbol for* the basic number of chromosomes in a genome • (ital.) *Maths., symbol for* a Cartesian coordinate (usually horizontal, as in **x-axis**) • *Bridge, symbol for* any card other than an honour • *symbol for* cross (as in **x-cut, x'd out**) • *Commerce, Finance, etc.* ex • extra • *Meteorol., symbol for* hoar-frost • (ital.) *Chem., symbol for* mole fraction • (ital.) *Maths., symbol for* an algebraic variable

X *Films, symbol for* adults only (former certification, still used in **X-rated, X-rating**) • *symbol for* beer strength (in **XX, XXX**, etc.) • *symbol for* choice (on a ballot paper) • Christ (from Greek letter X (chi), representing *ch*) • *symbol for* Cross (as in **King's X**) • *symbol for* error • experiment(al) • explosive • (ital.) *Physics,*

symbol for exposure dose • extension • extra • extraordinary • *Chem., symbol for* a halogen (as in **MgX**) • *symbol for* a kiss • *symbol for* the location of a place or point on a map, diagram, etc • *symbol for* his (*or* her) mark • (ital.) *Electricity, symbol for* reactance • *Roman numeral for* ten • *symbol for* any unknown, unspecified, or variable thing, factor, number, or person. *See also* X-ray • *Biochem., symbol for* xanthosine • X-ray • *symbol for* xylonite (a plastic) • *Genetics, indicating* a sex chromosome in humans and most animals (as in **X chromosome, X-linked, XYY syndrome**)

X2 *airline flight code for* China Xinhua Airlines

X4 *airline flight code for* Haiti Trans Air

X5 *airline flight code for* Sunworld International Airlines

3X *international civil aircraft marking for* Guinea

4X *international civil aircraft marking for* Israel

5X *international civil aircraft marking for* Uganda

xa *Finance* ex all (without any benefits)

XA *international civil aircraft marking for* Mexico

Xaa *Biochem., symbol for* an unknown amino acid (in a sequence)

xan. *Chem.* xanthene • *Chem.* xanthic

XAS *Chem.* X-ray absorption spectroscopy

xb *Finance* ex bonus (i.e. without bonus shares)

XB *international civil aircraft marking for* Mexico

Xber December

X^{bre} (*or* **xbre**) décembre (French: December)

xbt exhibit

XBT expendable bathythermograph

xc *Finance* ex capitalization (without capitalization) • *Finance* ex coupon (without the interest on the coupon)

XC (*or* **X-C**) (USA, Canada) cross-country (in **XC skiing**) • *international civil aircraft marking for* Mexico

x.c.l. *Insurance* excess current liabilities

xcp (*or* **x.cp.**) *Finance* ex coupon (*see under* xc)

xcpt except

XCT *Med.* X-ray computed tomography

xd *Finance* ex dividend (without dividend)

x'd executed

XD *Physics* X-ray diffraction

xdiv *Finance* ex dividend (without dividend)

XDP *Biochem.* xanthosine 5'-diphosphate

Xdr crusader

XDR extended dynamic range (of cassettes)

Xe *Chem., symbol for* xenon

Xen. Xenophon (*c.* 430–*c.* 356 BC, Greek historian)

Xer. Xerox

XES *Chem.* X-ray emission spectroscopy (*or* spectrum)

XF *airline flight code for* Vladivostok Air

XFA *Chem.* X-ray fluorescence analysis

xfer transfer

xfmr (*or* **xformer**) transformer

xg crossing

XGA extended graphics array

XG Midi extended general musical instrument digital interface

x-height *Printing, indicating* typesize of lower-case letters excluding ascenders and descenders (from the height of the lower-case x)

xhst exhaust

xi (*or* **x in**) *Finance* ex interest (without interest)

XI *British vehicle registration for* Belfast • *Med.* X-ray imaging

XL extra large (clothing size)

xlnt excellent

XLP *Med.* X-linked lymphoproliferative (in **XLP syndrome**)

xlwb extra-long wheelbase

XM experimental missile

Xmas ('ɛksməs) Christmas

xmit transmit

XMP *Biochem.* xanthosine monophosphate

XMS *Computing* extended memory specification

xmsn transmission

xmtr transmitter

xn *Finance* ex new (without right to new shares)

Xn Christian

x/nt excellent

Xnty Christianity

XO executive officer • *Chem.* xylenol orange (indicator) • *indicating* a cognac of superior quality

XOR (*or* **xor**) ('ɛksɔ:) *Electronics, Computing* exclusive-OR (as in **XOR gate**)

XP *symbol or monogram for* Christ *or* Christianity (from X (chi) and P (rho), the first two letters of the Greek word for Christ) • (*or* **x.p.**) express paid • *Med.* xeroderma pigmentosum

XPES *Chem.* X-ray photoelectron (*or* photoemission) spectroscopy (*or* spectrum)

XPG *Computing* X/Open's portability guide

xpl. explosive

xplt exploit

xpn expansion

XPS X-ray photoelectron spectroscopy

xq (*or* **XQ**) cross-question

xr *Finance* ex rights (without rights)

Xr examiner

XR X-ray(s)

X-ray *indicating* electromagnetic radiation

of very short wavelength (*X* refers to its unknown nature at the time of its discovery)

XRD X-ray diffraction

Xrds crossroads

XRE X-ray emission

x ref. (*or* **X ref.**) cross reference

XRF X-ray fluorescence

XRFA *Chem.* X-ray fluorescence analysis

XRFS *Chem.* X-ray fluorescence spectrometry

XRM *Med.* X-ray mammography

XRMA X-ray microprobe analysis

XRT *Med.* X-ray topography

x.rts *Finance* ex rights (without rights)

xs expenses

XS3 code *Computing* excess-3 code

Xt Christ

XT *airline flight code for* Air Exel Netherlands • *international civil aircraft marking for* Burkina Faso • extended technology

xtal crystal

XTE *Astronomy* X-ray Timing Explorer

X-tgd *Building trades* cross-tongued

Xtian Christian

XTP *Biochem.* xanthosine 5'-triphosphate

xtra extra

xtry extraordinary

Xty Christianity

XU *international civil aircraft marking for* Cambodia

XUI *Computing* X user interface

XUV *Astronomy* extreme ultraviolet

XV *airline flight code for* Air Express I Norrköping

xw *Finance* ex warrants (without warrants)

x/wb extra-long wheelbase

X-Windows *Computing, trademark indicating* a precisely defined form of windowing mechanism developed by MIT

XX *Meteorol.* heavy

XY *international civil aircraft marking for* Myanmar (Burma)

xyl. xylophone

Xyl *Biochem.* xylose

XZ *British vehicle registration for* Armagh • *international civil aircraft marking for* Myanmar (Burma)

Y

y (ital.) *symbol for* altitude • (ital.) *Maths., symbol for* a Cartesian coordinate (usually vertical, as in **y-axis**) • *Meteorol., symbol for* dry air • (ital.) *Aeronautics, symbol for* lateral axis • (ital.) *Maths., symbol for* an algebraic variable

y. yacht • yard • year • yellow • young • youngest

Y (ital.) *Electricity, symbol for* admittance • (ital.) *Physics, symbol for* hypercharge • *Biochem., symbol for* tyrosine • *symbol for* yen (Japanese monetary unit) • (*or* **Y.**) Yeomanry • *Colloquial* YMCA, YWCA, YMHA, *or* YWHA • *Irish fishing port registration for* Youghal • *Chem., symbol for* yttrium • *symbol for* yuan (Chinese monetary unit; *see also* RMB) • (*or* **Y.**) Yugoslavia • *Genetics, indicating* a sex chromosome of humans and most animals (as in **Y chromosome**, **Y-linked**, **XYY syndrome**)

Y2 *airline flight code for* Alliance Airlines

Y4 *airline flight code for* Eagle Aviation

Y6 *airline flight code for* Europe Elite

Y9 *airline flight code for* Trans Air Congo

4Y *airline flight code for* Yute Air Alaska

5Y *international civil aircraft marking for* Kenya

6Y *international civil aircraft marking for* Jamaica • *airline flight code for* Nicaraguense de Aviacion

8Y *airline flight code for* Ecuato Guineana de Aviacion

9Y *international civil aircraft marking for* Trinidad and Tobago

YA *international civil aircraft marking for* Afghanistan • *British vehicle registration for* Taunton • (*or* **Y/A**) *Marine insurance* York–Antwerp (Rules) • (USA) young adult

YABA Young American Bowling Alliance

YAC *Genetics* yeast artificial chromosome(s)

YACC (jæk) *Computing* yet another compiler-compiler

YAG (*or* **yag**) (jæg) *Electronics* yttrium–aluminium garnet

YAL Young Australia League

Y & D *US Navy* yards and docks

Y & LR York and Lancaster Regiment

YAR *international vehicle registration for* Yemen • *Marine insurance* York–Antwerp Rules

Y-ARD Yarrow-Admiralty research department

YAS Yorkshire Agricultural Society

YASSR (formerly) Yakut Autonomous Soviet Socialist Republic

YAVIS ('jɑːvɪs) *Colloquial* (USA) young, attractive, verbal, intelligent, and successful

Yb *Chem., symbol for* ytterbium

YB *British vehicle registration for* Taunton • yearbook

YC *airline flight code for* Flight West Airlines • *British vehicle registration for* Taunton • yacht club • Yale College (USA) • Young Conservative • youth club

YCA Youth Camping Association

YC&UO Young Conservative and Unionist Organization

YCL (USA) Young Communist League

YCNAC Young Conservative National Advisory Committee

YCS (International) Young Catholic Students

yct yacht

YCW (*or* **y.c.w.**) *Colloquial* you can't win • (International) Young Christian Workers

yd yard

YD *British vehicle registration for* Taunton • *symbol for* Yemeni dinar (monetary unit)

y'day (*or* **yday**) yesterday

ydg yarding

yds yards

YE *British vehicle registration for* London (central) • Your Excellency

YEA (USA) Yale Engineering Association

yearb. yearbook

yel. yellow

Yel. NP Yellowstone National Park (USA)

Yem. Yemen(i)

yeo. (*or* **Yeo.**) yeoman • yeomanry

YEO Youth Employment Officer

YER (*or* **y.e.r.**) *Finance* yearly effective rate (of interest)

YES Youth Employment Service • Youth Enterprise Scheme

yesty yesterday

YF *British vehicle registration for* London (central)

YFC Young Farmers' Club • Youth for Christ

YFCU Young Farmers' Clubs of Ulster

YFG *Electronics* yttrium–ferrite garnet

YG *British vehicle registration for* Leeds

YH *airline flight code for* Air Baffin • *British vehicle registration for* London (central) • *British fishing port registration for* Yarmouth • *Numismatics* young head (of Queen Victoria) • youth hostel

YHA Youth Hostels Association

YHANI Youth Hostels Association of Northern Ireland

YHWH (*or* **YHVH**) *indicating* the Hebrew name for God (Yahweh *or* Jehovah; as revealed to Moses)

Yi. (*or* **Yid.**) Yiddish

YI *Irish vehicle registration for* Dublin • *international civil aircraft marking for* Iraq

YIG (jɪg) *Electronics* yttrium–iron garnet

YIP (*or* **Yippie**) (jɪp *or* 'jɪpɪ) *Colloquial* (USA) (a member of the) Youth International Party

YJ *British vehicle registration for* Brighton • *airline flight code for* National Airlines • *international civil aircraft marking for* Vanuatu

YK *airline flight code for* Cyprus Turkish Airlines • *British vehicle registration for* London (central) • *international civil aircraft marking for* Syria

YL *international civil aircraft marking for* Latvia • *British vehicle registration for* London (central) • yield limit

YLI Yorkshire Light Infantry

YM *British vehicle registration for* London (central) • (*or* **Y-M**) *Physics* Yang–Mills (as in **YM theory**) • *Colloquial* YMCA (*or* a YMCA hostel)

YMBA Yacht and Motor Boat Association

YMCA Young Men's Christian Association

YMCath.A (USA) Young Men's Catholic Association

YMCU Young Men's Christian Union

YMFS Young Men's Friendly Society

YMHA Young Men's Hebrew Association

YMV *Microbiol.* yellow mosaic virus

YN *British vehicle registration for* London (central) • *international civil aircraft marking for* Nicaragua

YNP Yellowstone National Park (USA)

y.o. *Knitting* yarn over • year(s) old (following a number)

YO *British vehicle registration for* London (central) • *postcode for* York

YOB (*or* **y.o.b.**) year of birth

YOC Young Ornithologists' Club (of the RSPB)

YOD (*or* **y.o.d.**) year of death

YOM (*or* **y.o.m.**) year of marriage

YOP (jɒp) Youth Opportunities Programme

Yorks Yorkshire

y.p. *Finance* year's purchase

YP *British vehicle registration for* London (central) • *Mechanics* yield point • young person (*or* people) • young prisoner

YPA Young Pioneers of America

YPFB (Bolivia) Yacimientos Petrolíferos Fiscales Bolivianos (state petroleum organization)

YPSCE (USA) Young People's Society of Christian Endeavour

YPSL (USA) Young People's Socialist League

YPTES Young People's Trust for Endangered Species

yr year • younger • your

YR *British vehicle registration for* London (central) • *international civil aircraft marking for* Romania

yrbk yearbook

YRls *symbol for* Yemen riyal (monetary unit)

yrly yearly

yrs years • (*or* **Yrs**) yours (in correspondence)

Ys. Yugoslavia(n)

YS *international civil aircraft marking for* El Salvador • *British vehicle registration for* Glasgow • *airline flight code for* Proteus • *Mechanics* yield strength • Young Socialists

YSA (USA) Young Socialist Alliance

YSAG *Electronics* yttrium–scandium–aluminium garnet

YSO *Astronomy* young stellar object

yst youngest

yt yacht

YT *British vehicle registration for* London (central) • *airline flight code for* Skywest Airlines • Yukon Territory

y.t.b. *Knitting* yarn to back

YTD *Accounting* year to date

y.t.f. *Knitting* yarn to front

YTS Youth Training Scheme

YU *British vehicle registration for* London (central) • Yale University (USA) • *international vehicle registration or international civil aircraft marking for* Yugoslavia

Yugo. Yugoslavia

Yuk. Yukon

YUP Yale University Press

yuppie (*or* **yuppy**) ('jʌpɪ) *Colloquial* young urban (*or* upwardly mobile) professional

YV *British vehicle registration for* London (central) • *international vehicle registration or international civil aircraft marking for* Venezuela

YVF Young Volunteer Force

YVFF Young Volunteer Force Foundation

YW *airline flight code for* Air Nostrum • *British vehicle registration for* London (central) • *Colloquial* YWCA (*or* a YWCA hostel)

YWCA Young Women's Christian Association

YWCTU Young Women's Christian Temperance Union

YWF Young World Federalists

YWHA Young Women's Hebrew Association

YWS Young Wales Society • Young Workers' Scheme

YX *British vehicle registration for* London (central) • *airline flight code for* Midwest Express

YY *British vehicle registration for* London (central)

YZ *British vehicle registration for* Londonderry

Z

z (*ital.*) *Maths.*, *symbol for* a Cartesian coordinate (as in **z-axis**) • (*ital.*) *Chem.*, *symbol for* charge number • *Meteorol.*, *symbol for* haze • (*ital.*) *Maths.*, *symbol for* an algebraic variable

z. zero • zone

Z *Irish vehicle registration for* Dublin • (*ital.*) *Electricity*, *symbol for* impedance • (*ital.*) *Physics*, *symbol for* proton number • *international vehicle registration for* Zambia • *international civil aircraft marking for* Zimbabwe • *Genetics*, *indicating* a sex chromosome in birds and some insects • *Physics*, *indicating* a type of fundamental particle

Z. (*or* **Z**) zero (as in **Z-day**) • Zion(ist) • Zoll (German: customs) • zone

Z3 *airline flight code for* Star Airways

Z7 *airline flight code for* Zimbabwe Express Airlines

Z9 *airline flight code for* Aero Zambia

2Z *airline flight code for* Chang-An Airlines

3Z *airline flight code for* Necon Air

4Z *airline flight code for* S. A. Airlink

8Z *airline flight code for* Alaska Island Air

ZA *Irish vehicle registration for* Dublin • *international vehicle registration for* South Africa

Zag. Zagreb

Zam. Zambia

ZAMS *Astronomy* zero-age main sequence

Zan. (*or* **Zanz.**) Zanzibar

ZANU (*or* **Zanu**) ('zænuː) Zimbabwe African National Union

ZANU (PF) (*or* **Zanu (PF)**) ('zænuː 'piː 'ɛf) Zimbabwe African National Union (Patriotic Front)

ZAP *Military* zero anti-aircraft potential

ZAPU (*or* **Zapu**) ('zæpuː) Zimbabwe African People's Union

z.B. zum Beispiel (German: for example; e.g.)

ZB *Irish vehicle registration for* Cork • *airline flight code for* Monarch Airlines • Zen Buddhist

ZBB *Finance* (USA) zero-base(d) budgeting

ZC *Irish vehicle registration for* Dublin • *airline flight code for* Royal Swazi National Airways • Zionist Congress

Z-car *indicating* a police patrol car (from *zulu*, radio call sign)

ZCY *Electronics* zirconia–ceria–yttria

ZD *Irish vehicle registration for* Dublin • *Electronics* Zener diode • *Astronomy* zenith distance

Z-day *Military* zero day

Z-DNA *Biochem.* zigzagged DNA

ZE *Irish vehicle registration for* Dublin • *postcode for* Lerwick

ZEBRA ('ziːbrə) *Nuclear engineering* zero-energy breeder-reactor assembly

Zech. *Bible* Zechariah

Zeep (ziːp) *Nuclear engineering* zero-energy experimental pile

ZEG zero economic growth

zen. zenith

ZENITH ('zɛnɪθ) *Nuclear engineering* zero-energy nitrogen-heated thermal reactor

Zeph. *Bible* Zephaniah

zero-g zero gravity (weightlessness)

ZETA ('ziːtə) *Nuclear engineering* zero-energy thermonuclear apparatus (*or* assembly)

ZETR ('ziːtə) *Nuclear engineering* zero-energy thermal reactor

ZF *airline flight code for* Airborne of Sweden • *Irish vehicle registration for* Cork • (*or* **z.f.**) zero frequency

ZFGBI Zionist Federation of Great Britain and Ireland

ZG *Irish vehicle registration for* Dublin • *airline flight code for* Sabair Airlines • Zoological Gardens

z.H. (*or* **z.Hd.**) zu Händen (German: for the attention of, f.a.o.; *or* care of, c/o)

ZH *Irish vehicle registration for* Dublin • zero hour

ZHR *Astronomy* zenithal hourly rate

ZI *Irish vehicle registration for* Dublin •
Military zone of interior

ZIF (zɪf) *Electronics* zero insertion force (in
ZIF socket)

ZIFT (zɪft) *Med.* zygote intrafallopian
transfer (treatment for infertility)

ZIL *Computing* zigzag in-line

zip (*or* **Zip**) (zɪp) *Computing* zigzag in-line
package • (USA) zone improvement plan
(in **zip code**, US postcode)

ZJ *Irish vehicle registration for* Dublin

ZK *Irish vehicle registration for* Cork • *airline
flight code for* Great Lakes Aviation • *international civil aircraft marking for* New
Zealand

Zl *symbol for* zloty (Polish monetary unit)

ZL *Irish vehicle registration for* Dublin •
international civil aircraft marking for New
Zealand

ZM *Irish vehicle registration for* Galway

Zn *Chem., symbol for* zinc

ZN *airline flight code for* Eagle Airlines •
Irish vehicle registration for Meath • *inter-
national civil aircraft marking for* New
Zealand

ZO *Irish vehicle registration for* Dublin •
Zionist Organization

ZOA Zionist Organization of America

zod. zodiac

zoogeog. zoogeography

zool. zoological • zoologist • zoology

ZP *Irish vehicle registration for* Donegal •
international civil aircraft marking for
Paraguay

ZPG zero population growth

ZQ *airline flight code for* Ansett New
Zealand

Zr *Chem., symbol for* zirconium

ZR *Irish vehicle registration for* Wexford

Zs. Zeitschrift (German: periodical *or*
journal)

ZS *Irish vehicle registration for* Dublin •
international civil aircraft marking for
South Africa • Zoological Society

ZSI Zoological Society of Ireland

ZST zone standard time

z.T. zum Teil (German: partly)

ZT *Irish vehicle registration for* Cork • *airline
flight code for* Satena • *international civil
aircraft marking for* South Africa • zone
time

Ztg Zeitung (German: newspaper)

ZU *Irish vehicle registration for* Dublin •
international civil aircraft marking for
South Africa

Zulu. Zululand

ZUM Zimbabwe Unity Movement

Zur. Zürich

ZV *Irish vehicle registration for* Dublin

zw. zwischen (German: between *or*
among)

ZW *Irish vehicle registration for* Kildare •
international vehicle registration for
Zimbabwe

ZX *Irish vehicle registration for* Kerry

ZY *Irish vehicle registration for* Louth

zz zigzag

Zz. *Med., Pharmacol.* ginger (Latin
zingiber)

ZZ *British vehicle registration for* vehicles
temporarily imported by visitors from
abroad

Alphabets

Arabic

letter forms	translit	
ا ا ا أ	'alif	'
ب ‍ب‍ ‍ب بـ	bā'	b
ت ‍ت‍ ‍ت تـ	tā'	t
ث ‍ث‍ ‍ث ثـ	thā'	th
ج ‍ج‍ ‍ج جـ	jīm	j
ح ‍ح‍ ‍ح حـ	ḥā'	ḥ
خ ‍خ‍ ‍خ خـ	khā'	kh
د ‍د‍ ‍د دـ	dāl	d
ذ ‍ذ‍ ‍ذ ذـ	dhāl	dh
ر ‍ر‍ ‍ر رـ	rā'	r
ز ‍ز‍ ‍ز زـ	zay	z
س ‍س‍ ‍س سـ	sīn	s
ش ‍ش‍ ‍ش شـ	shīn	sh
ص ‍ص‍ ‍ص صـ	ṣād	ṣ
ض ‍ض‍ ‍ض ضـ	ḍād	ḍ
ط ‍ط‍ ‍ط طـ	ṭā'	ṭ
ظ ‍ظ‍ ‍ظ ظـ	ẓā'	ẓ
ع ‍ع‍ ‍ع عـ	'ayn	'
غ ‍غ‍ ‍غ غـ	ghayn	gh
ف ‍ف‍ ‍ف فـ	fā'	f
ق ‍ق‍ ‍ق قـ	qāf	q
ك ‍ك‍ ‍ك كـ	kāf	k
ل ‍ل‍ ‍ل لـ	lām	l
م ‍م‍ ‍م مـ	mīm	m
ن ‍ن‍ ‍ن نـ	nūn	n
ه ‍ه‍ ‍ه هـ	hā'	h
و ‍و‍ ‍و وـ	wāw	w
ي ‍ي‍ ‍ي يـ	yā'	y

Hebrew

א	aleph	'
ב	beth	b,bh
ג	gimel	g,gh
ד	daleth	d,dh
ה	he	h
ו	waw	w
ז	zayin	z
ח	ḥeth	ḥ
ט	ṭeth	ṭ
י	yodh	y
כ ך	kaph	k,kh
ל	lamedh	l
מ ם	mem	m
נ ן	nun	n
ס	samekh	s
ע	'ayin	'
פ ף	pe	p,ph
צ ץ	ṣadhe	ṣ
ק	qoph	q
ר	resh	r
שׂ	śin	ś
שׁ	shin	sh
ת	taw	t,th

Greek

A α	alpha	a
B β	beta	b
Γ γ	gamma	g
Δ δ	delta	d
E ε	epsilon	e
Z ζ	zeta	z
H η	eta	ē
Θ θ	theta	th
I ι	iota	i
K κ	kappa	k
Λ λ	lambda	l
M μ	mu	m
N ν	nu	n
Ξ ξ	xi	x
O o	omicron	o
Π π	pi	p
P ϱ	rho	r,rh
Σ σς	sigma	s
T τ	tau	t
Y υ	upsilon	u
Φ φ	phi	ph
X χ	chi	kh
Ψ ψ	psi	ps
Ω ω	omega	ō

Russian

А а	a	
Б б	b	
В в	v	
Г г	g	
Д д	d	
Е е	e	
Ё ё	ë	
Ж ж	zh	
З з	z	
И и	i	
Й й	ĭ	
К к	k	
Л л	l	
М м	m	
Н н	n	
О о	o	
П п	p	
Р р	r	
С с	s	
Т т	t	
У у	u	
Ф ф	f	
Х х	kh	
Ц ц	ts	
Ч ч	ch	
Ш ш	sh	
Щ щ	shch	
Ъ ъ	'' ('hard sign')	
Ы ы	y	
Ь ь	' ('soft sign')	
Э э	é	
Ю ю	yu	
Я я	ya	

Musical Notation

Values of notes and rests

	notes	rests
1 semibreve		
equals		
2 minims		
or		
4 crotchets		
or		
8 quavers		
or		
16 semiquavers		
or		
32 demisemiquavers		

Some common symbols

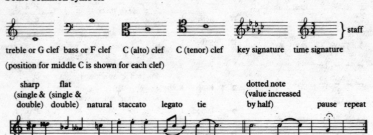

treble or G clef   bass or F clef   C (alto) clef   C (tenor) clef   key signature   time signature   } staff

(position for middle C is shown for each clef)

sharp   flat
(single &   (single &   dotted note
double)   double)   natural   staccato   legato   tie   (value increased
                                                                  by half)   pause   repeat

The circle of fifths

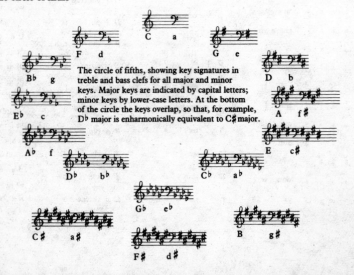

The circle of fifths, showing key signatures in treble and bass clefs for all major and minor keys. Major keys are indicated by capital letters; minor keys by lower-case letters. At the bottom of the circle the keys overlap, so that, for example, D♭ major is enharmonically equivalent to C♯ major.

Monetary symbols

Note: Roman letters that are used as monetary symbols will be found in their alphabetic place in the text.

Symbol	Name	Country
A$	Australian dollar	Australia; Kiribati
B$	Bahamian dollar	Bahamas
	Brunei dollar	Brunei
Bd$	Bermuda dollar	Bermuda
BD$	Barbados dollar	Barbados
BZ$	Belize dollar	Belize
₵	cedi	Ghana
	colón	El Salvador
	Costa Rican colón	Costa Rica
£C	Cyprus pound	Cyprus
C$	Canadian dollar	Canada
	Nicaraguan new córdoba	Nicaragua
Ch$	peso	Chile
CI$	Cayman Islands dollar	Cayman Islands
Col$	peso	Colombia
EC$	East Caribbean dollar	Antigua & Barbuda; Dominica; Grenada; St Kitts and Nevis; St Lucia; St Vincent and the Grenadines
F$	Fiji dollar	Fiji
₲	Paraguayan guaraní	Paraguay
G$	Guyana dollar	Guyana
HK$	Hong Kong dollar	Hong Kong
I£	Irish pound	Ireland
J$	Jamaica dollar	Jamaica
£	pound sterling	UK
L$	Liberian dollar	Liberia
M$	Malaysian dollar (ringgit)	Malaysia
Mex$	peso	Mexico
₦	Nigerian naira	Nigeria
NZ$	New Zealand dollar	New Zealand
NT$	new Taiwan dollar	Taiwan
₱	Philippine peso	Philippines
R$	real	Brazil
RD$	Dominican peso	Dominican Republic
S/.	sucre	Ecuador
S$	Singapore dollar	Singapore
SI$	Solomon Islands dollar	Solomon Islands
T$	pa'anga	Tonga
	Taiwanese dollar	Taiwan
TT$	Trinidad and Tobago dollar	Trinidad and Tobago
Ur$	Uruguayan peso	Uruguay
US$	United States dollar	USA; Guam; Federated States of Micronesia; Puerto Rico; Virgin Islands
WS$	tala	Western Samoa
¥	yen	Japan
Z$	Zimbabwe dollar	Zimbabwe

Proof-correction symbols

(Where appropriate, the marks should also be used by copy-editors in marking up copy)

Instruction to printer	Textual mark	Marginal mark
Correction made in error. Leave unchanged.	- - - - - under character(s) to remain	⊘
Remove extraneous marks or replace damaged character(s)	Encircle marks to be removed or character(s) to be changed	✕
(Wrong fount) Replace by character(s) of correct fount	Encircle character(s) to be changed	⊗
Insert in text the matter indicated in the margin	⋏	New matter followed by ⋏
Delete	Stroke through character(s) to be deleted	♂
Substitute character or substitute part of one or more word(s)	/ through character or ⊢——⊣ through word(s)	New character or New word(s)
Set in or change to italic type	—— under character(s) to be set or changed	⎵⎵⎦
Change italic to roman type	Encircle character(s) to be changed	⎿⎦
Set in or change to capital letter(s)	≡≡≡ under character(s) to be set or changed	≡
Change capital letter(s) to lower-case letter(s)	Encircle character(s) to be changed	≢
Set in or change to small capital letter(s)	══ under character(s) to be set or changed	═
Change small capital letter(s) to lower-case letter(s)	Encircle character(s) to be changed	═̸
Set in or change to bold type	∿∿ under character(s) to be changed	∿
Set in or change to italic type	∿∿∿ under character(s) to be changed	⎵⎵⎦
Invert type	Encircle character to be inverted	↶
Substitute or insert character in 'superior' position	/ through character or ⋏ where required	⌐ under character (e.g. ⅜)
Substitute or insert character in 'inferior' position	/ through character or ⋏ where required	⌐ over character (e.g. ⅙)
Insert full or decimal point	⋏ where required	⊙
Insert colon, semicolon, comma, etc.	⋏ where required	⊙ /;/,/(/)/[/]/ ; , ; ,
Rearrange to make a new paragraph here	⌐ before first word of new paragraph	⌐
Run on (no new paragraph)	⌒ between paragraphs	⌒
Transpose characters or words	⎡⎦ between characters or words to be transposed, numbered where necessary	⎍
Insert hyphen	⋏ where required	⊢⊣
Insert rule	⋏ where required	1 em ‖ 4 mm (i.e. give the size of the rule in the marginal mark)
Insert oblique	⋏ where required	⊘
Insert space between characters	∣ between characters	Y
Insert space between words	Y between words	Y
Reduce space between characters	∣ between characters	⊤
Reduce space between words	⊤ between words	⊤
Equalize space between characters or words	∣ between characters or words	Ⲭ

Hallmarks

All marks shown relate to silver
except where otherwise indicated

A hallmark

| maker's mark | standard mark | Assay Office mark | date letter |

Maker's mark (from 1363)
originally symbols, now initials

  symbol

  symbol and initials

  initials

Assay Office mark (from 1300)
now only London, Birmingham, Sheffield, and Edinburgh

London

gold and silver (leopard's head uncrowned from 1821; mark includes platinum from 1975)

Britannia silver (prior to 1975)

gold and silver (also platinum from 1975)

Edinburgh

gold and silver (also platinum from 1975)

Birmingham

gold (also platinum from 1975)

silver

Sheffield

silver (prior to 1975)

gold (also silver and platinum from 1975)

Some earlier Assay Office marks
(with dates of closure)

Norwich (1702)   York (1856)   Exeter (1883)   Newcastle (1884)   Chester (1962)   Glasgow (1964)

Standard mark (from 1544)
Marks guaranteeing pure metal content of the percentage shown

sterling silver 92.5%

  marked in England

  marked in Scotland (from 1975)

  marked in Scotland (prior to 1975)

Britannia standard silver (1697-1720, also occasional use since) 95.8%

gold
(crown followed by millesimal figure of the standard)

   i.e. 18 carat 75%

  22 carat 91.6%

585 14 carat 58.5%

375 9 carat 37.5%

(prior to 1975 marks incorporated the carat figure, and Scottish 18 and 22 carat gold bore a thistle mark instead of the crown)

Date letter (from 1478)
one letter per year before changing to next style of letter and/or shield
cycles vary between Assay Offices
London date letters (A-U used, excluding J) showing style of first letter and years of cycle

1498-1518		1598-1618		1697-1716		1796-1816		1896-1916	
1518-1538		1618-1638		1716-1736		1816-1836		1916-1936	
1538-1558		1638-1658		1736-1756		1836-1856		1936-1956	
1558-1578		1658-1678		1756-1776		1856-1876		1956-1974	
1578-1598		1678-1697		1776-1796		1876-1896		1975[1]-	

Note 1. Year letter changed with each calendar year; from 1975 all UK Offices use the same date letters and shield shape

Meteorological symbols

Weather conditions

●	rain	✳	snow	⌒	dew
❥	drizzle	▲	hail	≡	fog
▽	showers	⟁	ice pellets	℟	thunderstorm
∞	haze	⌐⌐	smoke	⦔	lightning

Cloud cover

Wind The arrow points in the direction from which the wind is blowing.

knots

Cloud cover		Wind	
◯	clear	◎	calm
◑ (1 okta)	1 okta	◯—	1-2
◕	2	◯—⌐	3-7
◑	3	◯—╱	8-12
◐	4	◯—╱╱	13-17
◑	5	◯—//	18-22
◕	6	◯—///	23-27
◑	7	◯—////	28-32
●	8	◯—⌐///	33-37
⊗	sky obscured (1 okta = ⅛ of the sky)	◯—⌐////	38-42
		◯—⌐/////	43-47
		◯—▲	48-52

Fronts

▼▼▼ **cold front** boundary between overtaking cold air mass and warm air mass.

⌒⌒⌒ **warm front** boundary between overtaking warm air mass and cold air mass.

▼⌒▼⌒ **stationary** boundary between air masses of similar temperature.

▲⌒▲⌒ **occluded** line where a cold front overtakes a warm front.

Mathematical symbols

General symbols

ratio of circumference of circle to its diameter	π
base of natural logarithms	e
imaginary unit: $i^2 = -1$	i, j
infinity	∞
equal to	$=$
not equal to	$\neq$
identically equal to	$\equiv$
corresponds to	$\hat{=}$
approximately equal to	$\approx$
asymptotically equal to	$\simeq$
proportional to	$\propto$
approaches	$\rightarrow$
greater than	$>$
smaller than	$<$
much greater than	$\gg$
much less than	$\ll$
greater than or equal to	$\geq$
less than or equal to	$\leq$
plus	$+$
minus	$-$
plus or minus	$\pm$
a multiplied by b	$ab, a \cdot b, a \times b$
a divided by b	$a/b, \dfrac{a}{b}, ab^{-1}$
a raised to the power n	a^n
magnitude of a	$\lvert a \rvert$
square root of a	$\sqrt{a}, \sqrt[2]{a}, a^{1/2}$
mean value of a	$\bar{a}, \langle a \rangle$
factorial p	$p!$
binomial coefficient: $n!/[p!(n-p)!]$	$\binom{n}{p}$

Symbols for functions

exponential of x	$\exp x, e^x$
logarithm to base a of x	$\log_a x$
natural logarithm of x	$\ln x, \log_e x$
common logarithm of x	$\lg x, \log_{10} x$
binary logarithm of x	$\operatorname{lb} x, \log_2 x$
sine of x	$\sin x$
cosine of x	$\cos x$
tangent of x	$\tan x, \operatorname{tg} x$
contangent of x	$\cot x, \operatorname{ctg} x$
secant of x	$\sec x$
cosecant of x	$\operatorname{cosec} x, \csc x$
inverse sine x	$\sin^{-1} x, \arcsin x$
inverse cosine x	$\cos^{-1} x, \arccos x$
inverse tangent x	$\tan^{-1} x, \arctan x$
integral	$\int$
summation	Σ
product	Π
finite increase of x	Δx
variation of x	δx
total differential of x	$\mathrm{d}x$
function of x	$f(x)$
composite function of f and g	$f \cdot g$
convolution of f and g	$f \star g$
limit of $f(x)$	$\lim_{x \to a} f(x), \lim\limits_{x \to a} f(x)$
derivative of f	$\dfrac{\mathrm{d}f}{\mathrm{d}x}, \mathrm{d}f/\mathrm{d}x, f'$
time derivative of f	$\dot{f}$
partial derivative of f	$\dfrac{\partial f}{\partial x}, \partial f/\partial x, \partial_x f, f_x$
total differential of f	$\mathrm{d}f$
variation of f	δf

Symbols in set theory

is an element of: $x \in A$	$\in$
is not an element of: $x \notin A$	$\notin$
contains as element: $A \ni x$	$\ni$
set of elements	$\{a_1, a_2, \cdots\}$
empty set	$\varnothing$
the set of positive integers and zero	$\mathbb{N}, \mathbf{N}$
the set of all integers, $\{\ldots, -2, -1, 0, 1, 2, \ldots\}$	$\mathbb{Z}, \mathbf{Z}$
the set of rational numbers	$\mathbb{Q}, \mathbf{Q}$
the set of real numbers	$\mathbb{R}, \mathbf{R}$
the set of complex numbers	$\mathbb{C}, \mathbf{C}$

set of elements of A for which $p(x)$ is true	$\{x \in A \mid p(x)\}$
is included in, subset of: $B \subseteq A$	$\subseteq, (\subset)$
contains: $A \supseteq B$	$\supseteq, (\supset)$
is properly contained in	$\subset$
contains properly	$\supset$
union: $A \cup B$ $= \{x \mid (x \in A) \lor (x \in B)\}$	$\cup$
intersection: $A \cap B$ $= \{x \mid (x \in A) \land (x \in B)\}$	$\cap$
difference: $A \backslash B$ $= \{x \mid (x \in A) \land (x \notin B)\}$	$\backslash$
complement of: $\complement A$ $= \{x \mid x \notin A\}$	$\complement$

Logic symbols

conjunction	$\land$
disjunction	$\lor$
negation	$\neg$
implication	$\Rightarrow$
equivalence	$\Leftrightarrow$
universal quantifier	$\forall$
existential quantifier	$\exists$

AND	$\land$	
OR	$\lor$	
NOT	$\neg$	
NAND	Δ	$\mid$
NOR	∇	$\downarrow$
XOR	xor	

Graphical symbols used in electronics

Qualifying graphical symbols

∿ alternating current

↗ variability (noninherent)

⌐ variability in steps

thermal effect

electromagnetic effect

radiation, electromagnetic nonionizing

coherent radiation

ionizing radiation

positive-going pulse

negative-going pulse

pulse of a.c.

positive-going step function

negative-going step function

fault

Graphical symbols

● connection of conductors

○ terminal (circle may be filled in)

junction of conductors

plug & socket (male & female)

earth

primary cell or accumulator (longer line represents +ve pole)

battery of accumulators or primary cells

switch, general symbol; make contact

resistor, general symbol (first form preferred)

variable resistor

resistor with sliding contact

capacitor, general symbol (first form preferred)

IGFET, enhancement type, single gate, p-type channel without substrate connection

amplifier, general symbol

AND gate, general symbol

OR gate, general symbol

inverter (NOT gate)

NAND gate (negated AND)

inductor, coil, winding, choke, general symbol

inductor with magnetic core

transformer, 2 windings

piezoelectric crystal, 2 electrodes

semiconductor diode, general symbol

light-emitting diode, general symbol

photodiode

pnp transistor

npn transistor

JFET, n-type channel

JFET, p-type channel

NOR gate (negated OR)

exclusive-OR gate

indicating instrument (first form) & recording instrument; asterisk is replaced by symbol of unit of quantity being measured (e.g. V for voltmeter, A for ammeter, or by some other appropriate symbol)

antenna, general symbol

Greek letters used as symbols for physical quantities

Note: Roman letters that are used as symbols for physical quantities will be found in their alphabetic place in the text.

Symbol	Name	Physical quantity
α	alpha	absorptance; linear expansion coefficient
γ	gamma	surface tension; cubic expansion coefficient
δ	delta	inclination
ε	epsilon	emissivity; permittivity
η	eta	efficiency; viscosity
θ	theta	angular displacement; Bragg angle; characteristic temperature
θ_N	theta	Néel temperature
λ	lambda	thermal conductivity; decay constant; mean free path; wavelength
μ	mu	coefficient of friction; permeability
ν	nu	kinematic viscosity; frequency
Π	pi (cap.)	osmotic pressure
ϱ	rho	charge density; density; reflectance; resistivity
σ	sigma	electrical conductivity; wave number
τ	tau	mean life
ϕ	phi	luminous flux
Φ	phi (cap.)	magnetic flux; work function
Φ_e	phi (cap.)	radiant flux
χ	chi	magnetic susceptibility
ψ	psi	electric flux
ω	omega	angular frequency; angular velocity
Ω	omega (cap.)	solid angle

Miscellaneous symbols

@ at (in commerce)
& and (ampersand)
© copyright
° degree (angle, temperature, proof of alcohol)
® registered trademark
♂ male
♀ female

Accents, diacritical marks, and special letters

ç	cedilla		ñ	tilde
ü	umlaut		ą	bar-under
ā	macron		à	dot
ǎ	hacek		ạ	dot-under
Å	Swedish boll		ł Ł	Polish barred 'l'
é	acute		ø Ø	Scandinavian barred 'O'
è	grave		æ Æ	a, e ligature
ô	circumflex		œ Œ	o, e ligature
ă	breve		ß	German double 'S'

International vehicle registration marks

Country	Mark	Country	Mark
Afghanistan	AFG	Gambia	WAG
Albania	AL	Georgia	GE
Alderney	GBA	Germany	D
Algeria	DZ	Ghana	GH
Andorra	AND	Gibraltar	GBZ
Argentina	RA	Great Britain	GB
Australia	AUS	Greece	GR
Austria	A	Grenada	WG
Azerbaijan	AZ	Guatemala	GCA
Bahamas	BS	Guernsey	GBG
Bahrain	BRN	Guinea	RG
Bangladesh	BD	Guyana	GUY
Barbados	BDS	Haiti	RH
Belarus	BY	Hong Kong	HK
Belgium	B	Hungary	H
Belize	BZ	Iceland	IS
Benin	DY	India	IND
Bolivia	BOL	Indonesia	RI
Bosnia-Herzegovina	BIH	Iran	IR
Botswana	RB	Iraq	IRQ
Brazil	BR	Ireland	IRL
British Virgin Islands	BV	Isle of Man	GBM
Brunei	BRU	Israel	IL
Bulgaria	BG	Italy	I
Burkina-Faso	BF	Jamaica	JA
Burundi	RU	Japan	J
Cambodia	K	Jersey	GBJ
Cameroon	CAM	Jordan	HKJ
Canada	CDN	Kazakhstan	KZ
Central African Republic	RCA	Kenya	EAK
Chad	TCH	Korea (Republic of)	ROK
Chile	RCH	Kuwait	KWT
Colombia	CO	Kyrgyzstan (Kirghizia)	KS
Congo	RCB	Laos	LAO
Costa Rica	CR	Latvia	LV
Côte d'Ivoire (Ivory Coast)	CI	Lebanon	RL
Croatia	HR	Lesotho	LS
Cuba	C	Liberia	LB
Cyprus	CY	Libya	LAR
Czech Republic	CZ	Liechtenstein	FL
Denmark and Greenland	DK	Lithuania	LT
Dominica	WD	Luxembourg	L
Dominican Republic	DOM	Macedonia	MK
Ecuador	EC	Madagascar	RM
Egypt	ET	Malawi	MW
El Salvador	ES	Malaysia	MAL
Estonia	EST	Mali	RMM
Ethiopia	ETH	Malta	M
Faeroe Islands	FR	Mauritania	RIM
Fiji	FJI	Mauritius	MS
Finland	FIN	Mexico	MEX
France	F	Moldova	MD
Gabon	G	Monaco	MC

Country	Mark	Country	Mark
Morocco	MA	South Africa	ZA
Mozambique	MOC	Spain	E
Myanmar (Burma)	BUR	Sri Lanka	CL
Namibia	NAM	St Lucia	WL
Nauru	NAU	St Vincent	WV
Nepal	NEP	Sudan	SUD
Netherlands	NL	Suriname	SME
Netherlands Antilles	NA	Swaziland	SD
New Zealand	NZ	Sweden	S
Nicaragua	NIC	Switzerland	CH
Niger	RN	Syria	SYR
Nigeria	NGR	Taiwan	RC
Norway	N	Tajikistan	TJ
Pakistan	PK	Tanzania	EAT
Panama	PA	Tanzania (Zanzibar)	EAZ
Papua New Guinea	PNG	Thailand	T
Paraguay	PY	Togo	TG
Peru	PE	Trinidad and Tobago	TT
Philippines	RP	Tunisia	TN
Poland	PL	Turkey	TR
Portugal	P	Turkmenistan	TM
Qatar	Q	Uganda	EAU
Romania	RO	Ukraine	UA
Russian Federation	RUS	United States of America	USA
Rwanda	RWA	Uruguay	ROU
San Marino	RSM	Vatican City	V
Saudi Arabia	SA	Venezuela	YV
Senegal	SN	Vietnam	VN
Seychelles	SY	Western Samoa	WS
Sierra Leone	WAL	Yemen	YAR
Singapore	SGP	Yugoslavia	YU
Slovakia	SK	Zambia	Z
Slovenia	SLO	Zimbabwe	ZW
Somalia	SO		

British vehicle registration marks

Registration office	Marks	Registration office	Marks
Aberdeen	BS, PS, RS, SA, SE, SO, SS	Haverfordwest	BX, DE, EJ
		Huddersfield	HD, CP, CX, JX, VH
Antrim	DZ, IA, KZ, RZ	Hull	AG, AT, KH, RH
Armagh	IB, LZ, XZ	Inverness	AS, BS, JS, SK, ST
Bangor	CC, EY, FF, JC	Ipswich	BJ, DX, GV, PV, RT
Belfast	AZ, CZ, EZ, FZ, GZ, MZ, OI, OZ, PZ, TZ, UZ, WZ, XI	Leeds	BT, DN, NW, UA, UB, UG, UM, VY, WR, WT, WU, WW, WX, WY, YG
Birmingham	DA, JW, OA, OB, OC, OE, OF, OG, OH, OJ, OK, OL, OM, ON, OP, OV, OX, UK, VP	Leicester	AY, BC, FP, JF, JU, NR, RY, UT, BE
Bournemouth	AA, CG, EL, FX, HO, JT, LJ, PR, RU	Lincoln	BE, CT, DO, EE, FE, FU, FW, JL, JV, TL, VL
Brighton	AP, CD, DY, FG, HC, JK, NJ, PN, UF, WV, YJ	Liverpool	BG, CM, DJ, ED, EK, EM, FY, HF, JP, K, KA, KB, KC, KD, KF, LV, TB, TJ, WM
Bristol	AE, EU, FB, HT, HU, HW, HY, OU, TC, WS	London (central)	JD, HM, HV, HX, UC, UL, UU, UV, UW, YE, YF, YH, YK, YL, YM, YN, YO, YP, YR, YT, YU, YV, YW, YX, YY
Cardiff	AX, BO, DW, HB, KG, NY, TG, TX, UH, WO		
Carlisle	AO, HH, RM, SM, SW	London (NE)	MC, MD, ME, MF, MG, MH, MK, ML, MM, MP, MT, MU
Chelmsford	AR, EV, HJ, HK, JN, NO, OO, PU, TW, VW, VX, WC	London (NW)	BY, LA, LB, LC, LD, LE, LF, LH, LK, LL, LM, LN, LO, LP, LR, LT, LU, LW, LX, LY, OY, RK
Chester	CA, DM, FM, LG, MA, MB, TU		
Coventry	AC, DU, HP, KV, RW, VC, WK		
Down	IJ, JZ, SZ, BZ	London (SE)	GU, GW, GX, GY, MV, MX, MY
Dudley	DH, EA, FD, FK, HA, NX, UE, WD	London (SW)	GC, GF, GH, GJ, GK, GN, GO, GP, GT
Dundee	ES, SL, SN, SP, SR, TS	Londonderry	IW, NZ, UI, YZ
Durham	J	Luton	BH, BM, GS, KX, MJ, NK, NM, PP, RO, TM, UR, VS
Edinburgh	FS, RRC, KS, LS, MS, SC, SF, SG, SH, SX		
		Maidstone	FN, JG, JJ, KE, KJ, KK, KL, KM, KN, KO, KP, KR, KT, VB
Exeter	CO, DR, DV, FJ, JY, OD, TA, TK, TT, UN, UO	Manchester	BA, BN, BU, CB, DB, DK, EN, JA, NA, NB, NC, ND, NE, NF, RJ, TD, TE, VM, VR, VU, WH
Fermanagh	IL		
Glasgow	CS, DS, GA, GB, GD, GE, GG, HS, NS, OS, SB, SD, SJ, SM, SU, SW, US, YS		
		Middlesbrough	AJ, DC, EF, HN, PY, VN
Gloucester	AD, CJ, DD, DF, DG, FH, FO, VJ	Newcastle upon Tyne	BB, BR, CN, CU, FT, GR, JR, RG, TN, TY, UP, VK, NL, PT
Guildford	PA, PB, PC, PD, PE, PF, PG, PH, PJ, PK, PL, PM	Northampton	BD, NH, NV, RP, VV

Registration office	Marks	Registration office	Marks
Norwich	AH, CL, EX, NG, PW, VF, VG		RX, TF
		Sheffield	AK, DT, ET, HE, HL,
Nottingham	AL, AU, CH, NN, NU, RA, RB, RC, RR, TO, TV, VO		KU, KW, KY, WA, WB, WE, WF, WG, WJ
Oxford	BW, FC, JO, UD, WL	Shrewsbury	AW, NT, UJ, UX
		Stoke-on-Trent	BF, EH, FA, RE, RF, VT
Peterborough	AV, CE, EB, EG, ER, EW, FL, JE, VA, VE	Swansea	CY, EP, TH, WN
Portsmouth	PO, BK, BP, CR, DL, OR, OT, OW, PX, RV, TP, TR	Swindon	AM, HR, MR, MW
		Taunton	YA, YB, YC, YD
		Truro	AF, CV, GL, RL
Preston	BV, CK, CW, EC, EO, FR, FV, HG, RN	Truro (Isles of Scilly)	SCY
Reading	AN, BL, CF, DP, GM, JB, JH, JM, MO, RD,	Tyrone	HZ, JI, VZ
		Worcester	AB, NP, UY, WP

British postcodes

Area	Postcode	Area	Postcode
Aberdeen	AB	Kirkwall	KW
Bath	BA	Lancaster	LA
Belfast	BT	Leeds	LS
Birmingham	B	Leicester	LE
Blackburn	BB	Lerwick	ZE
Blackpool	FY	Lincoln	LN
Bolton	BL	Liverpool	L
Bournemouth	BH	Llandrindod Wells	LD
Bradford	BD	Llandudno	LL
Brighton	BN	London (east)	E
Bristol	BS	London (east central)	EC
Bromley	BR	London (north)	N
Canterbury	CT	London (northwest)	NW
Cardiff	CF	London (southeast)	SE
Carlisle	CA	London (southwest)	SW
Cambridge	CB	London (west)	W
Chelmsford	CM	London (west central)	WC
Chester	CH	Luton	LU
Cleveland	TS	Manchester	M
Colchester	CO	Medway	ME
Coventry	CV	Milton Keynes	MK
Crewe	CW	Motherwell	ML
Croydon	CR	Newcastle upon Tyne	NE
Darlington	DL	Newport (Gwent)	NP
Dartford	DA	Northampton	NN
Derby	DE	Nottingham	NG
Doncaster	DN	Norwich	NR
Dorchester	DT	Oldham	OL
Dudley	DY	Oxford	OX
Dumfries	DG	Paisley	PA
Dundee	DD	Perth	PH
Durham	DH	Peterborough	PE
Edinburgh	EH	Plymouth	PL
Enfield	EN	Portsmouth	PO
Exeter	EX	Preston	PR
Falkirk	FK	Reading	RG
Galashiels	TD	Redhill	RH
Glasgow	G	Romford	RM
Gloucester	GL	Sheffield	S
Guildford	GU	Salisbury	SP
Halifax	HX	Shrewsbury	SY
Harrogate	HG	Slough	SL
Harrow	HA	Southall	UB
Hemel Hempstead	HP	Southampton	SO
Hereford	HR	Southend-on-Sea	SS
Huddersfield	HD	St Albans	AL
Hull	HU	Stevenage	SG
Ilford	IG	Stoke-on-Trent	ST
Inverness	IV	Stockport	SK
Ipswich	IP	Sunderland	SR
Kilmarnock	KA	Sutton	SM
Kingston-upon-Thames	KT	Swansea	SA
Kirkcaldy	KY	Swindon	SN

Area	Postcode	Area	Postcode
Taunton	TA	Walsall	WS
Telford	TF	Warrington	WA
Tonbridge	TN	Watford	WD
Torquay	TQ	Wigan	WN
Truro	TR	Wolverhampton	WV
Twickenham	TW	Worcester	WR
Wakefield	WF	York	YO

OXFORD

MORE OXFORD PAPERBACKS

This book is just one of nearly 1000 Oxford Paperbacks currently in print. If you would like details of other Oxford Paperbacks, including titles in the World's Classics, Oxford Reference, Oxford Books, OPUS, Past Masters, Oxford Authors, and Oxford Shakespeare series, please write to:

UK and Europe: Oxford Paperbacks Publicity Manager, Arts and Reference Publicity Department, Oxford University Press, Walton Street, Oxford OX2 6DP.

Customers in UK and Europe will find Oxford Paperbacks available in all good bookshops. But in case of difficulty please send orders to the Cash-with-Order Department, Oxford University Press Distribution Services, Saxon Way West, Corby, Northants NN18 9ES. Tel: 01536 741519; Fax: 01536 746337. Please send a cheque for the total cost of the books, plus £1.75 postage and packing for orders under £20; £2.75 for orders over £20. Customers outside the UK should add 10% of the cost of the books for postage and packing.

USA: Oxford Paperbacks Marketing Manager, Oxford University Press, Inc., 200 Madison Avenue, New York, N.Y. 10016.

Canada: Trade Department, Oxford University Press, 70 Wynford Drive, Don Mills, Ontario M3C 1J9.

Australia: Trade Marketing Manager, Oxford University Press, G.P.O. Box 2784Y, Melbourne 3001, Victoria.

South Africa: Oxford University Press, P.O. Box 1141, Cape Town 8000.

Oxford Paperback Reference

OXFORD PAPERBACK REFERENCE

From *Art and Artists* to *Zoology*, the Oxford Paperback Reference series offers the very best subject reference books at the most affordable prices.

Authoritative, accessible, and up to date, the series features dictionaries in key student areas, as well as a range of fascinating books for a general readership. Included are such well-established titles as Fowler's *Modern English Usage*, Margaret Drabble's *Concise Companion to English Literature*, and the bestselling science and medical dictionaries.

The series has now been relaunched in handsome new covers. Highlights include new editions of some of the most popular titles, as well as brand new paperback reference books on *Politics*, *Philosophy*, and *Twentieth-Century Poetry*.

With new titles being constantly added, and existing titles regularly updated, Oxford Paperback Reference is unrivalled in its breadth of coverage and expansive publishing programme. New dictionaries of *Film*, *Economics*, *Linguistics*, *Architecture*, *Archaeology*, *Astronomy*, and *The Bible* are just a few of those coming in the future.

Oxford
Paperback
Reference

THE OXFORD DICTIONARY OF PHILOSOPHY

Edited by Simon Blackburn

* **2,500 entries covering the entire span of the subject including the most recent terms and concepts**

* **Biographical entries for nearly 500 philosophers**

* **Chronology of philosophical events**

From Aristotle to Zen, this is the most comprehensive, authoritative, and up to date dictionary of philosophy available. Ideal for students or a general readership, it provides lively and accessible coverage of not only the Western philosophical tradition but also important themes from Chinese, Indian, Islamic, and Jewish philosophy. The paperback includes a new Chronology.

'an excellent source book and can be strongly recommended . . . there are generous and informative entries on the great philosophers . . . Overall the entries are written in an informed and judicious manner.'
Times Higher Education Supplement

Oxford
Paperback
Reference

THE CONCISE OXFORD DICTIONARY
OF POLITICS

Edited by Iain McLean

Written by an expert team of political scientists from Warwick University, this is the most authoritative and up-to-date dictionary of politics available.

* Over 1,500 entries provide truly international coverage of major political institutions, thinkers and concepts

* From Western to Chinese and Muslim political thought

* Covers new and thriving branches of the subject, including international political economy, voting theory, and feminism

* Appendix of political leaders

* Clear, no-nonsense definitions of terms such as veto and subsidiarity

Oxford
Paperback
Reference

THE CONCISE OXFORD COMPANION TO ENGLISH LITERATURE

Edited by Margaret Drabble and Jenny Stringer

Derived from the acclaimed *Oxford Companion to English Literature*, the concise maintains the wide coverage of its parent volume. It is an indispensable, compact guide to all aspects of English literature. For this revised edition, existing entries have been fully updated and revised with 60 new entries added on contemporary writers.

* Over 5,000 entries on the lives and works of authors, poets and playwrights

* The most comprehensive and authoritative paperback guide to English literature

* New entries include Peter Ackroyd, Martin Amis, Toni Morrison, and Jeanette Winterson

* New appendices list major literary prize-winners

From the reviews of its parent volume:

'It earns its place at the head of the best sellers: every home should have one'
Sunday Times

Oxford Paperback Reference

CONCISE SCIENCE DICTIONARY

New edition

Authoritative and up to date, this bestselling dictionary is ideal reference for both students and non-scientists. Fully revised for this third edition, with over 1,000 new entries, it provides coverage of biology (including human biology), chemistry, physics, the earth sciences, astronomy, maths and computing.

* 8,500 clear and concise entries

* Up-to-date coverage of areas such as molecular biology, genetics, particle physics, cosmology, and fullerene chemistry

* Appendices include the periodic table, tables of SI units, and classifications of the plant and animal kingdoms

'handy and readable . . . for scientists aged nine to ninety'
Nature

'The book will appeal not just to scientists and science students but also to the interested layperson. And it passes the most difficult test of any dictionary—it is well worth browsing through.'
New Scientist

Oxford
Paperback
Reference

THE CONCISE OXFORD DICTIONARY
OF OPERA

New Edition

Edited by Ewan West and John Warrack

Derived from the full *Oxford Dictionary of Opera*, this is the most authoritative and up-to-date dictionary of opera available in paperback. Fully revised for this new edition, it is designed to be accessible to all those who enjoy opera, whether at the opera-house or at home.

* **Over 3,500 entries on operas, composers, and performers**

* **Plot summaries and separate entries for well-known roles, arias, and choruses**

* **Leading conductors, producers and designers**

From the reviews of its parent volume:

'the most authoritative single-volume work of its kind'
Independent on Sunday

'an invaluable reference work'
Gramophone